Springer Proceedings in Physics 311

Indexed by Scopus

The series Springer Proceedings in Physics, founded in 1984, is devoted to timely reports of state-of-the-art developments in physics and related sciences. Typically based on material presented at conferences, workshops and similar scientific meetings, volumes published in this series will constitute a comprehensive up to date source of reference on a field or subfield of relevance in contemporary physics. Proposals must include the following:

- Name, place and date of the scientific meeting
- A link to the committees (local organization, international advisors etc.)
- Scientific description of the meeting
- List of invited/plenary speakers
- An estimate of the planned proceedings book parameters (number of pages/articles, requested number of bulk copies, submission deadline).

Please contact:

For Americas and Europe: Dr. Zachary Evenson; zachary.evenson@springer.com
For Asia, Australia and New Zealand: Dr. Loyola DSilva; loyola.dsilva@springer.com

Sarita Kumari · Amanpal Singh ·
Balram Tripathi · Mahesh Baboo
Editors

Proceedings of the 1st International Conference on Materials and Thermophysical Properties

ICMTP-2024, 21–23 November, Jaipur, India

Springer

Editors
Sarita Kumari
Department of Physics
University of Rajasthan
Jaipur, Rajasthan, India

Amanpal Singh
Department of Physics
University of Rajasthan
Jaipur, Rajasthan, India

Balram Tripathi
Department of Physics
S. S. Jain Subodh PG College
Jaipur, Rajasthan, India

Mahesh Baboo
Department of Physics
Lal Bahadur Shastri PG College
Jaipur, Rajasthan, India

ISSN 0930-8989 ISSN 1867-4941 (electronic)
Springer Proceedings in Physics
ISBN 978-981-96-6794-9 ISBN 978-981-96-6795-6 (eBook)
https://doi.org/10.1007/978-981-96-6795-6

This Springer imprint is published by the registered company Springer Nature Singapore Pte Ltd.
The registered company address is: 152 Beach Road, #21-01/04 Gateway East, Singapore 189721, Singapore

Preface

The **Department of Physics, University of Rajasthan, Jaipur**, successfully organized the International Conference on Materials and Thermophysical Properties (**ICMTP-2024**) in association with the **Thermophysical Society of India** (TPSI), India, under the auspices of the **RUSA 2.0 program** from **November 21–23, 2024**.

The conference attracted **365** participants from across the globe, highlighting the international significance of advancements in materials science and thermophysical research. Over the course of **15** technical sessions, attendees engaged in **66** oral presentations and explored **141** poster presentations, reflecting the cutting-edge developments in this dynamic field.

A total of **311 abstracts** were contributed to the conference, showcasing a broad spectrum of research contributions. Following the event, approximately **70 full-length manuscripts** were submitted for publication under five major thematic areas:

1. **Materials for Energy Applications**
2. **Multifunctional Materials**
3. **Polymeric Materials**
4. **Properties: Dielectric & Electrical Properties**
5. **Properties: Thermophysical Properties**

After a rigorous **peer-review process**, approximately **49 research papers** were accepted for publication in this **Springer Nature** proceedings volume, ensuring the high academic standards of the compilation.

The conference was privileged to host a distinguished panel of **keynote and invited speakers**, including **Prof. J. C. Bernede**, Nantes University, France; **Prof. R. K. Verma**, Magadh University, India; **Prof. P. Predeep**, Mahatma Gandhi University, Kottayam, India; **Prof. Kedar Singh**, JNU, New Delhi, India; **Dr. D. K. Shukla**, UGC-DAE-CSR, Indore, India; **Prof. Ivan Castillo**, University of Puerto Rico, USA; **Prof. Amita Chandra**, University of Delhi, India; **Prof. Arun Pratap**, M.S. University of Baroda, India; **Prof. Cattin Linda**, Nantes University, France; **Prof. Suman K. Pal**, IIT Mandi, India; **Prof. Ambesh Dixit**, IIT Jodhpur, India; **Prof. Kanupriya Sachdev**, MNIT, Jaipur, India; **Prof. Ram J. Choudhary**, UGC-DAE-CSR, Indore, India; **Prof. Alexander Krivchikov**, National Academy of Sciences of Ukraine, Ukraine; **Prof. N.G. Shimpi**, University of Mumbai, India; **Dr. Naveen K. Acharya**, M.S. University of Baroda, India; **Dr. Vaibhav Kulshrestha**, CSIR-CSMRI, Bhavnagar, India; **Dr. Dheeraj K. Singh**, Central University of Gujarat, Gandhinagar, India; **Prof. N. K. Pandey**, Lucknow University, India; **Prof. Ankur Jain**, Suresh Gyan Vihar University, Jaipur, India; **Prof. Ramesh L. Gardas**, IIT Madras, Chennai, India; **Prof. Franz Faupel**, Kiel University, Germany; **Dr. Kamlendra Awasthi**, MNIT, Jaipur, India; and **Prof. Johannes Geirschner**, Madrid Institute for Advanced Studies in Nanoscience, Spain. Their **insightful talks and discussions** significantly enriched the scientific discourse and facilitated valuable collaborations.

We extend our sincere appreciation to the **reviewers** for their **meticulous evaluation of the manuscripts**, ensuring the **scholarly quality** of the proceedings. Our heartfelt gratitude also goes to the **Advisory Committee** of ICMTP-2024 for their **support and guidance**, as well as to the **organizing team** for their **dedicated efforts** in making this conference a resounding success. We also thank **Springer Nature** for publishing this **proceedings volume**, providing a global platform for disseminating these research contributions.

We hope that this compilation serves as a **valuable reference** for researchers, fostering **future collaborations** and driving **continued advancements in materials science and thermophysical research**.

Acknowledgement

We extend our sincere gratitude to the **Rashtriya Uchchatar Shiksha Abhiyan (RUSA)**, the **Anusandhan National Research Foundation (ANRF)**, the **Board of Research in Nuclear Sciences (BRNS)**, and the **Council of Scientific & Industrial Research (CSIR)** for their generous financial support, which played a crucial role in the success of ICMTP-2024.

We also express our deep appreciation to the academic and administrative leadership of the **University of Rajasthan**, especially **Prof. Alpana Kateja, Hon'ble Vice-Chancellor**, whose vision, encouragement, and unwavering support were instrumental in the seamless execution of this conference.

We extend our heartfelt thanks to **Prof. Mangej Singh, Chairman of the Conference and Head, Department of Physics, University of Rajasthan, Jaipur**, for his dedicated leadership and continuous guidance in organizing this event.

We are deeply grateful to **Prof. Shailendra K. Gupta, Nodal Officer, RUSA 2.0**, for his unwavering support in facilitating this conference under the RUSA 2.0 program.

Special recognition goes to **Prof. Kananbala Sharma, Chairperson of ICMTP-2024 and President of TPSI, India**, for her invaluable contributions in shaping the scientific discourse and structure of the conference.

We sincerely appreciate the dedicated contributions of all committee members and volunteers, whose tireless efforts and commitment were pivotal in making ICMTP-2024 a success.

We are also deeply grateful to the presenters, authors, and participants for sharing their valuable research and insights. Your contributions have significantly enriched this conference, and it is our privilege to present your work in these proceedings.

A special note of appreciation goes to the reviewers for their meticulous evaluation and constructive feedback, ensuring the academic integrity and quality of the research presented in this volume.

A heartfelt acknowledgment is due to **Mr. Jitin Kapuria, Mr. Shubham Gupta, and Mr. Mukesh Kilak** for their dedicated efforts in compiling and organizing these proceedings. Their hard work and meticulous attention to detail have played a crucial role in ensuring a high-quality publication.

Finally, we extend our sincere thanks to **Springer Nature** for providing a prestigious platform to publish this scholarly work.

This conference and its proceedings stand as a testament to the collective effort, collaboration, and unwavering commitment of numerous individuals and institutions. We truly appreciate your continued support and look forward to future endeavors in advancing scientific knowledge.

Contents

Contributors

Aakansha Department of Physical Sciences, Banasthali Vidyapith, Banasthali, Rajasthan, India

Sadhana Agrawal Department of Physics, National Institute of Technology Raipur, Raipur, Chhattisgarh, India

Parvez Ahmed Alvi Department of Physical Sciences, Banasthali Vidyapith, Banasthali, Rajasthan, India

Anchal Department of Physics, Banasthali Vidyapith, Banasthali, Rajasthan, India

Anita Department of Physical Sciences, Banasthali Vidyapith, Banasthali, Rajasthan, India

Anuradha Department of Physics, University College of Science, Mohanlal Sukhadia University, Udaipur, Rajasthan, India

Gunjan Arora Department of Physics, University College of Science, Mohanlal Sukhadia University, Udaipur, Rajasthan, India

V. Balaprakash Hindusthan College of Arts & Science, Coimbatore, Tamilnadu, India

Shuvabrata Bandophyay Department of Physical Sciences, Banasthali Vidyapith, Banasthali, Rajasthan, India

R. K. Beniwal Department of Physics, Banasthali Vidyapith, Banasthali, Rajasthan, India

Shivraj Bhardwaj Department of Physics, B. S. A. College (Dr. Bhimrao Ambedkar University, Agra), Mathura, India

Shovit Bhattacharya Technical Physics Division, Bhabha Atomic Research Centre, Mumbai, India;
Homi Bhabha National Institute, Mumbai, India

N. K. Bhatt Department of Physics, Maharaja Krishnakumarsinhji Bhavnagar University, Bhavnagar, Gujarat, India

Ranu Bhatt Technical Physics Division, Bhabha Atomic Research Centre, Mumbai, India;
Homi Bhabha National Institute, Mumbai, India

Priya Chandra Department of Physics, National Institute of Technology Raipur, Raipur, Chhattisgarh, India

Anurag Chandrakar Department of Physics, National Institute of Technology Raipur, Raipur, India

Shraddha Chaudhari Faculty of Science and Humanities, Sankalchand Patel University, Visnagar, Gujarat, India;
M. N. College, Visnagar, Gujarat, India

B. L. Choudhary Department of Physics, Banasthali Vidyapith, Banasthali, Rajasthan, India

Nidhi Choudhary Department of Electrical Engineering, Manipal University Jaipur, Jaipur, Rajasthan, India

Vipin Choudhary Department of Physics, University of Rajasthan, Jaipur, Rajasthan, India

Abhay Dasadia A. D. Patel Institute of Technology, The Charutar Vidya Mandal (CVM) University, New Vallabh Vidyanagar, Gujarat, India

Niyanta Datta Technical Physics Division, Bhabha Atomic Research Center, Mumbai, India

M. S. Dhaka Department of Physics, Mohanlal Sukhadia University, Udaipur, India

Khushbu Dhaked Department of Physics, IIS (Deemed to be University), Jaipur, Rajasthan, India

A. Dhina Infant Maria Hindusthan College of Arts and Science, Coimbatore, Tamilnadu, India

D. K. Dhruv Natubhai V. Patel College of Pure and Applied Sciences, The Charutar Vidya Mandal (CVM) University, Vallabh Vidyanagar, Anand, India

Anju Dixit Department of Physics, School of Basic Sciences, CSJM University, Kanpur, India

Pramod S. Dobal Department of Physics, VSSD College, Kanpur, India

S. N. Dolia Department of Physics, University of Rajasthan Jaipur, JLN Marg, Jaipur, Rajasthan, India;
Center for Non-conventional Energy Resources, University of Rajasthan Jaipur, JLN Marg, Jaipur, Rajasthan, India

Anusha Dubey Department of Physics, Noida Institute of Engineering and Technology, Greater Noida, India

Preksha Dubey Department of Physics, B. S. A. College (Dr. Bhimrao Ambedkar University, Agra), Mathura, India

Kailasa Ganapathi Technical Physics Division, Bhabha Atomic Research Center, Mumbai, India

Yogita Gendre Department of Physics, National Institute of Technology Raipur, Chandigarh, India

Kiran Githala Department of Physics, University of Rajasthan, Jaipur, Rajasthan, India

Sneh Lata Goyal Department of Physics, Guru jambheshwar University of Science and Technology, Hisar, Haryana, India

Manisha Gupta Department of Physics, University of Rajasthan, Jaipur, Rajasthan, India

Ritu Gupta Department of Physics, National Institute of Technology Raipur, Raipur, Chhattisgarh, India

Saral Kumar Gupta Department of Physical Sciences, Banasthali Vidyapith, Banasthali, Rajasthan, India

Vivek Gupta Department of Physics, Guru jambheshwar University of Science and Technology, Hisar, Haryana, India

Himanshu University Centre for Research & Development, Chandigarh University, Mohali, India

Ankur Jain Centre for Renewable Energy, Suresh Gyan Vihar University, Jaipur, India; Natural Science Centre for Basic Research and Development, Hiroshima University, Hiroshima, Japan

Karishma Jain Department of Physics, School of Physical Sciences, Manipal University, Jaipur, India

M. K. Jangid Department of Physics, Vivekananda Global University, Jaipur, India

Nairaj Jat Department of Physics, University of Rajasthan, Jaipur, Rajasthan, India

R. K. Jhakal Department of Physics, Govt Dungar College, Bikaner, Rajasthan, India; Department of Physics, SBD Govt College, Sardarshahar, Rajasthan, India

Komal Kaliraman Department of Physics, Panjab University, Chandigarh, India

Shradha Kalkhundiya Department of Physics, School of Applied and Life Science, Uttaranchal University, Dehradun, India

Hansraj Karwasara Department of Physics, Manipal University Jaipur, Jaipur, Rajasthan, India

Manmeet Kaur Technical Physics Division, Bhabha Atomic Research Center, Mumbai, India

Shaminder Kaur Department of Electronics and Communication Engineering, Chitkara University Institute of Engineering and Technology, Chitkara University, Rajpura, Punjab, India

Simranjeet Kaur Department of Physics, Center for Advanced Study in Physics, Panjab University, Chandigarh, India

Ayush Khare Department of Physics, National Institute of Technology Raipur, Raipur, India

Akhilesh Kumar Department of Physics, School of Basic Sciences, CSJM University, Kanpur, India;
Narain College Shikohabad, Firozabad, India;
DBRA University Agra, Agra, India

Avinash Kumar Department of Physics, Centre for Advanced Study in Physics, Panjab University, Chandigarh, India

Devender Kumar Department of Physics, Himachal Pradesh University, Shimla, Himachal Pradesh, India

Kishor Kumar Department of Physics, Manipal University Jaipur, Jaipur, Rajasthan, India

Manoj Kumar Department of Physics, Amity School of Applied Sciences, Amity University, Gurugram, Haryana, India

Rajiv Kumar Department of Physics, University of Rajasthan, Jaipur, India

Rishikesh Kumar Technical Physics Division, Bhabha Atomic Research Centre, Mumbai, India;
Department of Mechanical Engineering, Indian Institute of Technology, Patna, India

Suresh Kumar Ceramics and Ceramic Matrix Composites Division, DMSRDE, Kanpur, India

Arjun Kumawat Department of Physics, Vivekananda Global University, Jaipur, India

Sunil Kumawat Department of Physics, University College of Science, Mohanlal Sukhadia University, Udaipur, Rajasthan, India

Chhagan Lal Department of Physics, University of Rajasthan, Jaipur, Rajasthan, India;
Center for Non-Conventional Energy Resources, University of Rajasthan, Jaipur, Rajasthan, India

Chinmaya Mahapatra Department of Physics, National Institute of Technology Raipur, Chandigarh, India

Harish Mahla Department of Physics, Vivekananda Global University, Jaipur, India

Mamta Department of Physical Sciences, Banasthali Vidyapith, Banasthali, Rajasthan, India

Nirmal Manyani Department of Physics, Center for Advanced Study in Physics, Panjab University, Chandigarh, India

G. Mariya Gnana Merlin Hindusthan College of Arts and Science, Coimbatore, Tamilnadu, India

Arti Meena Department of Physics, University of Rajasthan, Jaipur, India

Deepak K. Meena Department of Physics, University College of Science, Mohanlal Sukhadia University, Udaipur, Rajasthan, India

Dhara Singh Meena Department of Physics, Vivekananda Global University, Jaipur, India

Harish Kumar Meena Department of Physics, University of Rajasthan, Jaipur, India

Pavan Kumar Meena Department of Physics, Vivekananda Global University, Jaipur, India

Prashant Meena Department of Physics, Vivekananda Global University, Jaipur, India

Ved Prakash Meena Department of Physics, Vivekananda Global University, Jaipur, India

Vinita Mirdha Department of Physics, University of Rajasthan, Jaipur, India

Mahitha Mohan Hindusthan College of Arts & Science, Coimbatore, Tamilnadu, India

Bhavya Monga Centre for Renewable Energy, Suresh Gyan Vihar University, Jaipur, India;
Department of Physics, S. S. Jain Subodh P.G. (Autonomous) College, Jaipur, India

Monika Department of Physics, University of Rajasthan, Jaipur, India

C. M. S. Negi Department of Physical Sciences, Banasthali Vidyapith, Banasthali, Rajasthan, India

K. S. Ojha Department of Physics, National Institute of Technology Raipur, Raipur, Chhattisgarh, India

Vishwa Padia Dr. K C Patel R and D Centre, Charotar University of Science and Technology (CHARUSAT), Changa, Anand, India

Rishi Pal Department of Applied Science, Kalpana Chawla Government Polytechnic for Women, Ambala, Haryana, India

Naincy Pandit Department of Physics, School of Allied Sciences, Dev Bhoomi Uttarakhand University, Dehradun, India

Amit Patel Government Science College, Gujarat University, Ahmedabad, Gujarat, India

Anita Patel Dr. K C Patel R and D Centre, Charotar University of Science and Technology (CHARUSAT), Changa, Anand, India

Chirag Patel Faculty of Science and Humanities, Sankalchand Patel University, Visnagar, Gujarat, India

Margi Patel Department of Basic Science and Humanities, Mukesh Patel School of Technology Management and Engineering, SVKM's NMIMS Deemed University, Mumbai, Maharashtra, India

Mitesh H. Patel Dr. K C Patel R and D Centre, Charotar University of Science and Technology (CHARUSAT), Changa, Anand, India

Pulkesh Prajapati A. D. Patel Institute of Technology, The C.V.M University, New Vallabh Vidyanagar, Gujarat, India

R. Prema Hindusthan College of Arts and Science, Coimbatore, Tamilnadu, India

Ashutosh Raj National Institute of Technology Karnataka, Surathkal, India

Rajiv Department of Physics, Hindu College, Sonepat, Haryana, India

Abhimanyu Singh Rana Centre for Advanced Materials and Devices, SoET, BML Munjal University, Gurugram, Haryana, India

Monika Rani Department of Physics, University College of Science, Mohanlal Sukhadia University, Udaipur, Rajasthan, India

R. Ridhi Mehr Chand Mahajan DAV College for Women, Chandigarh, India

Ritu Department of Physics, Banasthali Vidyapith, Banasthali, Rajasthan, India

Ruchi Department of Physics, Guru jambheshwar University of Science and Technology, Hisar, Haryana, India

Bhogendra Kumar Sah St. Xavier's College, Kathmandu, Nepal

Jagrati Sahariya Department of Physics, National Institute of Technology, Srinagar, Uttarakhand, India

Yogita Sahu Department of Physics, National Institute of Technology Raipur, Raipur, Chhattisgarh, India

Summaiyya Saleem Department of Physics, University of Rajasthan, Jaipur, India

Soumen Samanta Technical Physics Division, Bhabha Atomic Research Center, Mumbai, India

Vibhav Kumar Saraswat Department of Physics, Army Cadet College, Indian Military Academy, Dehradun, India

Yogendra Kumar Saraswat Department of Chemistry, B S A College, Mathura, India

Sarita Department of Physics, Banasthali Vidyapith, Banasthali, Rajasthan, India

K. Sathiyaseelan Hindusthan College of Arts and Science, Coimbatore, Tamilnadu, India

Mitesh Shakya Department of Physics, VSSD College, Kanpur, India

Anshu Sharma Chitkara University School of Engineering and Technology, Chitkara University, Baddi, Himachal Pradesh, India

Arushi Sharma Department of Physics, Mohanlal Sukhadia University, Udaipur, India

Dheeraj Sharma Department of Physics, Himachal Pradesh University, Shimla, Himachal Pradesh, India

Karishma Sharma Department of Physics, Manipal University Jaipur, Jaipur, Rajasthan, India

Krishna S. Sharma Department of Physics, IIS (Deemed to be University), Jaipur, Rajasthan, India

Lalit Kumar Sharma Department of Mechanical Engineering, SoET, BML Munjal University, Gurugram, Haryana, India

M. D. Sharma Department of Physics, Govt Dungar College, Bikaner, Rajasthan, India

Neeraj Kumar Sharma Department of Mechanical Engineering, SoET, BML Munjal University, Gurugram, Haryana, India

Ramanand Sharma Department of Physics, University of Rajasthan, Jaipur, Rajasthan, India;
Govt. Kamla Modi Girls College, Neem Ka Thana, Rajasthan, India

Ramphal Sharma Department of Physics, IIS (Deemed to be University), Jaipur, Rajasthan, India;
Center of NanoScience and Technology, IIS (Deemed to be University), Jaipur, Rajasthan, India

S. S. Sharma Department of Physics (H & S), Government Mahila Engineering College, Ajmer, India

Smriti Sharma Department of Physics, Government College for Women, Gurugram, Haryana, India

Heena Rohit Shevde Department of Basic Science and Humanities, Mukesh Patel School of Technology Management and Engineering, SVKM's NMIMS Deemed University, Mumbai, Maharashtra, India

Rimjhim Shrimal Department of Physics, University of Rajasthan, Jaipur, Rajasthan, India

Rimpy Shukla Department of Physics, IIS (Deemed to be University), Jaipur, Rajasthan, India

Mohd. Gulman Siddiqui Department of Physical Sciences, Banasthali Vidyapith, Banasthali, Rajasthan, India

Simran Department of Physical Sciences, Banasthali Vidyapith, Banasthali, Rajasthan, India

Ajay Singh Technical Physics Division, Bhabha Atomic Research Centre, Mumbai, India;
Homi Bhabha National Institute, Mumbai, India

Amarjeet Singh Department of Physics, Himachal Pradesh University, Shimla, Himachal Pradesh, India

Arvind Kumar Singh Department of Mathematics, Institute of Science, Banaras Hindu University, Varanasi, Uttar Pradesh, India

Bhanu Pratap Singh Department of Physics, School of Applied and Life Science, Uttaranchal University, Dehradun, India

Jasveer Singh Department of Physics, Himachal Pradesh University, Shimla, Himachal Pradesh, India

Mamraj Singh Department of Physics, University of Rajasthan, Jaipur, India

Neha Singh Department of Physical Sciences, Banasthali Vidyapith, Banasthali, Rajasthan, India

Monu Singhal Department of Physics, University of Rajasthan, Jaipur, Rajasthan, India

Vanaraj Solanki Dr. K C Patel R and D Centre, Charotar University of Science and Technology (CHARUSAT), Changa, Anand, India

Amit Soni Department of Electrical Engineering, Manipal University Jaipur, Jaipur, Rajasthan, India

Yashaswi Soni Department of Physics, Rajeev Gandhi Govt PG College, Mandsaur, Madhya Pradesh, India

Pooja Srivastava Department of Physical Sciences, Banasthali Vidyapith, Banasthali, Rajasthan, India

Supreet Department of Physics, Amity School of Applied Sciences, Amity University, Gurugram, Haryana, India

K. Thangavel Hindusthan College of Arts and Science, Coimbatore, Tamilnadu, India

B. Tripathi Department of Physics, S. S. Jain Subodh P.G. (Autonomous) College, Jaipur, India

S. K. Tripathi Department of Physics, Center for Advanced Study in Physics, Panjab University, Chandigarh, India

Prateek Uniyal Department of Physical Sciences, Banasthali Vidyapith, Banasthali, Rajasthan, India

Aditi Upadhyaya Department of Physical Sciences, Banasthali Vidyapith, Banasthali, Rajasthan, India

Alok Upadhyay Department of Physics, B. S. A. College (Dr. Bhimrao Ambedkar University, Agra), Mathura, India

Khyati Upadhyay A. D. Patel Institute of Technology, The C.V.M University, New Vallabh Vidyanagar, Gujarat, India

Shikha Varma Institute of Physics, Sachivalaya Marg, Bhubaneswar, Odisha, India

Ajay Singh Verma Division of Research & Innovation, School of Applied and Life Sciences, Uttaranchal University, Dehradun, India;
University Centre for Research & Development, Department of Physics, Chandigarh University, Mohali, Punjab, India

Jitendra Kumar Verma Department of Physics, University of Rajasthan, Jaipur, India

Seema Verma Department of Physical Sciences, Banasthali Vidyapith, Banasthali, Rajasthan, India;
Department of Computer Science Engineering, Vivekanand Global University, Jaipur, Rajasthan, India

Madhu Yadav Department of Physics, University of Rajasthan Jaipur, Jaipur, Rajasthan, India;
Center for Non-conventional Energy Resources, University of Rajasthan Jaipur, Jaipur, Rajasthan, India

Pooja Yadav Department of Physics, Banasthali Vidyapith, Banasthali, Rajasthan, India

Shekhar Yadav Department of Physical Sciences, Banasthali Vidyapith, Banasthali, Rajasthan, India

List of Abbreviations

AC	Activated Charcoal
ACS	Auto-Combustion Synthesis
AFM	Atomic Force Microscopy
ALD	Atomic Layer Deposition
AM 1.5G	Standard Solar Spectrum
ANSYS	ANSYS Software (Engineering Simulation Software)
AOPs	Advanced Oxidation Processes
APS	Ammonium Persulphate
CAL	Cesium-Based Perovskite Absorber Layer
CBO	Conduction Band Offset
CCD	Central Composite Design
CE	Cohesive Energy
CFSE	Crystal Field Stabilization Energy
CH	Cluster Head
CMOS	Complementary Metal-Oxide-Semiconductor
CNT	Carbon Nanotube
COMSOL	COMSOL Multiphysics (Simulation Software)
CST	Computer Simulation Technology (Software)
CV	Cyclic Voltammetry
CZO	Cobalt-doped Zinc Oxide
DETF	Double-Ended Tuning Fork
DFT	Density Functional Theory
DG	Diversity Gain
DGS	Defected Ground Structure
DIBL	Drain-Induced Barrier Lowering
DOS	Density of States
DSC	Differential Scanning Calorimetry
EDLCs	Electric Double-Layer Capacitors
EDS	Energy Dispersive X-ray Spectroscopy
EDTA	Ethylenediaminetetraacetic Acid
EDX	Energy Dispersive X-ray Spectroscopy
EMEECP-IoT	Enhanced Multitier Energy-Efficient Clustering Protocol for IoT
EMI	Electromagnetic Interference
EOS	Equation of State
ETC	Effective Thermal Conductivity
ETL	Electron Transport Layer
Eg	Band Gap Energy
FEA	Finite Element Analysis
FESEM	Field Emission Scanning Electron Microscopy
FF	Fixed-Fixed Resonator
FFR	Fixed-Free Resonator

FTIR	Fourier Transform Infrared Spectroscopy
FTO	Fluorine-doped Tin Oxide
FWHM	Full Width at Half Maximum
GA	Genetic Algorithm
GGA	Generalized Gradient Approximation
GGA-PBE	Generalized Gradient Approximation-Perdew Burke Ernzerhof
HER	Hydrogen Evolution Reaction
HTL	Hole Transport Layer
J-V	Current-Voltage
JCPDS	Joint Committee on Powder Diffraction Standards
JLFET	Junctionless Field-Effect Transistor
JSC	Short-Circuit Current Density
Kmax	Maximum Plane Wave Cutoff
L(ω)	Energy Loss Function
LDA	Local Density Approximation
LEACH	Low-Energy Adaptive Clustering Hierarchy
LEL	Lower Explosive Limit
LH	Light Hole
LO	Linear Optical
LSDA	Local Spin Density Approximation
M.W.	Molecular Weight
MAWCA	Mobility-Aware Weighted Clustering Algorithm
MAX	A-phase Layered Structure of Transition Metal Carbides/Nitrides
MB	Methylene Blue
MEA	Monoethanolamine
MEMS	Micro-Electro-Mechanical Systems
MOFGSA	Multi-Objective Fractional Gravitational Search Algorithm
MOFs	Metal-Organic Frameworks
MOSFET	Metal-Oxide-Semiconductor Field-Effect Transistor
MPCVD	Microwave Plasma Enhanced Chemical Vapor Deposition
MXene	Two-Dimensional Transition Metal Carbides and Nitrides
NAM	Network Animator
NCs	Nanocrystals
NLO	Nonlinear Optical
NPs	Nanoparticles
ORR	Oxygen Reduction Reaction
PANI	Polyaniline
PBE	Perdew-Burke-Ernzerhof Approximation
PDOS	Partial Density of States
PEG	Polyethylene Glycol
PEM	Proton Exchange Membrane
PEO	Polyethylene Oxide
PSC	Perovskite Solar Cell
PSO	Particle Swarm Optimization
PVA	Polyvinyl Alcohol
QCM	Quartz Crystal Microbalance

QD	Quantum Dot
QE	Quantum Efficiency
QoS	Quality of Service
RMT	Muffin-Tin Radius
RhB	Rhodamine B
S	Seebeck Coefficient
SBD	Schottky Barrier Diode
SCF	Self-Consistent Field
SEM	Scanning Electron Microscopy
SRH	Shockley-Read-Hall Recombination
SS301	Stainless Steel 301
TCO	Transparent Conducting Oxide
TE	Total Energy
TEEN	Threshold Sensitive Energy Efficient Sensor Network Protocol
TENG	Triboelectric Nanogenerator
TGA	Thermogravimetric Analysis
TMDs	Transition Metal Dichalcogenides
TX	Surface Terminations (-O, -OH, -F, Cl)
UV	Ultraviolet
VBO	Valence Band Offset
VLSI	Very Large-Scale Integration
WBAN	Wireless Body Area Network
WF	Work Function
XPS	X-ray Photoelectron Spectroscopy
XRD	X-ray Diffraction
ZT	Figure of Merit
$L(\omega)$	Energy Loss Function
$k(\omega)$	Extinction Coefficient
$n(\omega)$	Refractive Index
Ω	Ohm (Resistance Unit)
α	Optical Absorption Coefficient
ε	Dielectric Constant
$\varepsilon(\omega)$	Dielectric Function
$\sigma(\omega)$	Optical Conductivity
ΔG	Free Energy of Change of Decomposition
ΔS	Entropy of Activation

Revisiting Optical Properties of $BaLiX_3$(X = Cl, Br, I) Using First-Principles Calculations with Full-Potential DFT Code WIEN2k

Jasveer Singh[1], Amarjeet Singh[1](✉), Naincy Pandit[2], Anusha Dubey[3], and Ajay Singh Verma[4]

[1] Department of Physics, Himachal Pradesh University, Shimla 171005, Himachal Pradesh, India
amarjeet.sirohi@gmail.com
[2] Department of Physics, School of Allied Science Dev Bhoomi Uttarakhand University, Dehradun 248007, India
[3] Noida Institute of Engineering and Technology, Greater Noida 201306, India
[4] Division of Research and Innovation, School of Applied and Life Sciences, Uttaranchal University, Dehradun 248007, India

Abstract. This research utilizes density functional theory (DFT) to explore the electronic and optical characteristics of $BaLiX_3$ (X = Br, Cl, I) using the WIEN2K simulation package which uses full potential linear augmented plane wave method (FP-LAPW). Investigation of electronic properties is done by calculating band structure of $BaLiX_3$ using Local Spin Density Approximation (LSDA) and Generalized Gradient Approximation (GGA-PBE). Density of States (DOS) of $BaLiX_3$ have been thoroughly investigated in this research work. This study found that BaLiCl3, $BaLiBr_3$ and $BaLiI_3$ have indirect energy band gaps at R → Γ symmetry points. To understand optical properties, our calculation provides detailed insight into the material's dielectric function, absorption coefficient, and refractive index. This study is also establishing the comparison between pseudo-potential and full potential approaches in the Density Functional Theory (DFT).

Keywords: DFT · WIEN2k · FP-LAPW · LSDA · GGA · Band Structure

1 Introduction

The 1950s marked a significant increase in studies focused on optoelectronics devices, spurred by the identification and progress in silicon technology. Researchers have invested in numerous optoelectronic devices concepts, driven by the advancements in solar photovoltaic materials. Perovskites materials are extensively investigated since their potential for the photovoltaic and optoelectronic applications have been discovered [1–4]. The first perovskites solar cell (PSC) using $CH_3NH_3PbI_3$ was demonstrated by Kojima [5] in 2009, with an initial efficiency of 3.81% [6], marking a significant milestone in the development of perovskites-based solar cells. In 2014, it reached 13.2% and now has a power conversion efficiency of 25.1% [7]. Al-Ashouri et al. [8] have achieved

S. Kumari et al. (Eds.): ICMTP 2024, SPPHY 311, pp. 1–9, 2025.
https://doi.org/10.1007/978-981-96-6795-6_1

efficiency of over 29% in perovskites/silicon tandem solar cells through enhanced hole extraction. Fehr et al. [9] have explored the integration of halide perovskites into photochemical cells, paving the way for new applications beyond traditional solar energy harvesting. To ensure long-term operational stability, Zhu et al. [10] conducted an in-depth review highlighting the significance of stability in perovskite photovoltaics. Their study explores the key challenges and potential strategies for enhancing the durability of perovskite solar cells (PSCs) over extended periods. Considerable research efforts have been dedicated to the discovery of lead-free perovskites, [11, 12] driven by the need for more environmentally friendly and sustainable alternatives. Computational modelling of materials gives us the advantage of testing various materials for different physical properties before their actual synthesis in the laboratory. In the present research work, we focused on halide perovskites $BaLiX_3$ (X = Cl, Br, I), we invested the variation of electronic and optical properties as an anion atom Cl replaced by Br and I. The study on this system has already been done using a pseudo-potential approach [13]. We revisited the study using Full-potential code to enrich data. The experimental study on this perovskite is yet to be achieved. The enrichment of computational data is very useful to train machine learning [14] models for more accurate prediction. In this research work, we investigated structure, Density of states (DOS), Partial Density of States (PDOS) using Full Potential approach. For optical properties we have investigated the real and the imaginary part of Dielectric Function, refractive index, and absorption coefficient. This study is also helpful to establish the comparison between the results from full-potential and pseudo-potential [15] approaches in DFT [16]. These DFT calculations generate high-quality data that can be used to train machine learning models [17], enabling more accurate predictions of material properties. This integration of DFT with artificial intelligence/machine learning (AI/ML) will accelerate the discovery and design of novel materials.

2 Computational Details

All of the first-principles DFT calculations were performed with WIEN2k [18] simulation code. The WIEN2k package supports the Full Potential Linear Augmented Plane Wave Method (FP-LAPW). In the FP-LAPW method, wave functions for charge density and potential are represented using spherical harmonic functions inside the non-overlapping muffin-tin spheres around atomic sites, while plane waves are utilized in the interstitial region [13]. Structure optimization carried by the Wu Cohen-Generalized Gradient approximation (WC-GGA) potential. Once structure is optimized, Electronic and Optical properties calculated by Local Spin Density Approximation (LSDA) [19] and GGA-PBE [20] exchange correlation potentials. The software's fundamentals parameters are fine-tuned as follows: - $RMT \times K_{max} = 8$, Here RMT [21] represents muffin-tin sphere radius and K_{max} corresponds to the wave vector. Electronic properties are accessed using 3000 K points in the irreducible Brillouin zone. Optical properties are derived by using 10,000 K points with the OPTIC program supported by the WIEN2k simulation package using LSDA and GGA exchange correlation potential.

3 Structure

The perovskites materials $BaLiX_3$ (X = Cl, Br, I) adopt a simple cubic structure and belong to the space group pm-3 m (#221). Structures are optimized using WC-GGA potential. Ba atoms are positioned at the Wyckoff's coordinates of (0, 0, 0) and Li is located at (1/2, 1/2, 1/2) and X (Cl, Br and I) are at (1/2, 0, 1/2) as shown in the Fig. 1.

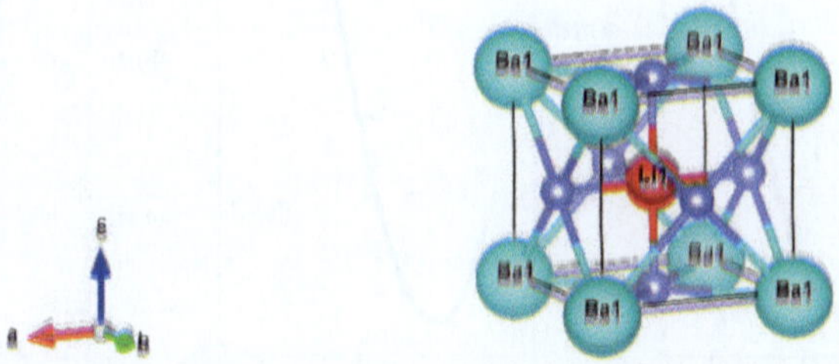

Fig. 1. Cubic structure of the halide perovskites $BaLiX_3$. Where blue colour atoms represent anion atom X = Cl, Br, I.

4 Results and Discussion

4.1 Structural Properties

A DFT code first required to optimize the coordinates of atoms to obtain equilibrium structure. Structure optimization was done using WC-GGA potential. Structural parameters of $BaLiCl_3$, $BaLiBr_3$ and $BaLiI_3$ have been calculated by using Birch- Murnaghan's [22] equation of the states which is expressed as,

$$E(V) = E_0 + \frac{9V_0}{16}\left\{\left[\left(\frac{V_0}{V}\right)^{\frac{2}{3}} - 1\right](B_0') + \left[\left(\frac{V_0}{V}\right)^{\frac{2}{3}} - 1\right]^2 \times \left[6 - 4\left(\frac{V_0}{V}\right)^{\frac{2}{3}}\right]\right\} \quad (1)$$

The geometry optimization of these compounds is given in Fig. 2. Our findings in this research are consistent [13] with previous studies, albeit with some variations. We used a full potential approach in this research work. We find that, the lattice constant of $BaLiCl_3$, $BaLiBr_3$, $BaLiI_3$ are 4.79, 5.06 and 5.54 Å respectively, and results shows that there is increase in the lattice constant as we replace anion atom Cl with Br and I. This happens because of an increase in ionic radius when the halide ions are replaced as Cl → Br → I. The ionic radii of Cl, Br and I are 1.82, 1.97, 2.06 Å respectively. Bulk modulus of $BaLiCl_3$, $BaLiBr_3$, and BaLiI3 was found at 40.05, 31.08 and 24.04 GPa respectively. The results have been compared with previous study on $BaLiX_3$ (X = Cl, Br, I) using pseudo potential method [13]. Pressure derivative bulk modulus B′ of $BaLiCl_3$, $BaLiBr_3$ and $BaLiI_3$ are 6.53, 4.44 and 4.72 respectively. The optimized structural parameters from this study are summarized in Table 1.

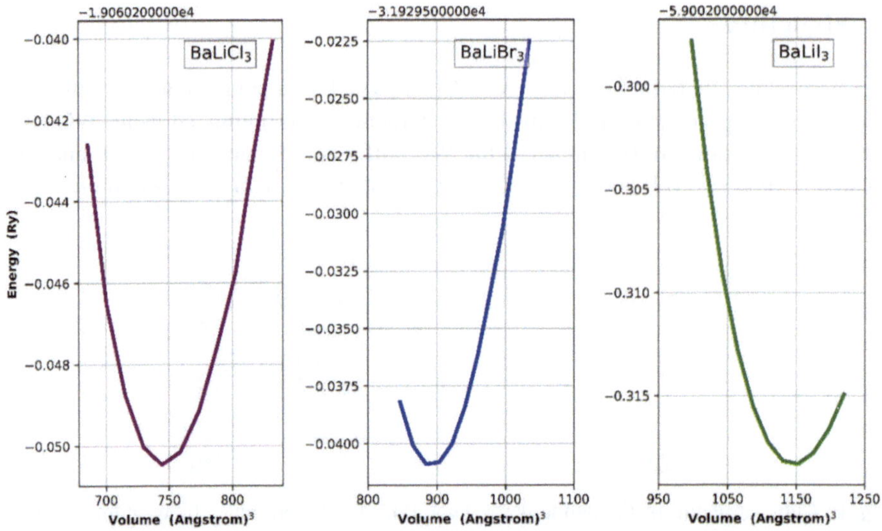

Fig. 2. Geometry optimization (E vs V) of $BaLiCl_3$, $BaLiBr_3$, and $BaLiI_3$. Note: - x-axis represents the volume, and y-axis represents the value of energy that is added with the value given at the top of each plot.

Table 1. Lattice parameters of the optimized structure of BaLiX3(X = Cl, Br, I) using FP LAPW, where a:- Lattice constant, B:- Bulk Modulus, B': - Pressure derivatives of Bulk modulus, FP-LAPW: -Full Potential Linear Augmented Plane Wave (Present Work), and PP: - Pseudo Potential [13].

Compound	a (Å)	B(Gpa)	B'	E_0 (Ry)	V $(Å)^3$
$BaLiCl_3$					
FP-LAPW	4.79	40.05	6.53	−19060.25	745.05
PP	4.90	33.86	4.54	−185.434	795
$BaLiBr_3$					
FP-LAPW	5.06	31.08	4.437	−31929.54	892.15
PP	5.21	27.14	4.53	−225.98	954.67
$BaLiI_3$					
FP-LAPW	5.54	24.037	4.72	−59002.32	1146.55
PP	5.66	20.67	4.25	−292.230	1227.41

4.2 Electronic Properties

After optimization, electronic properties have been studied using LSDA and GGA potentials. Electronic band structure of $BaLiX_3$ is given in the Fig. 3. Electronic band gap of $BaLiCl_3$, $BaLiBr_3$, and $BaLiI_3$ are 4.05, 3.26 and 2.40 eV respectively using LSDA, and GGA-PBE gives the energy band gap of 4.27, 3.46 and 2.56 eV respectively as shown in Table 2. We have seen that the electronic energy band decreases as we move $BaLiCl_3 \rightarrow BaLiBr_3 \rightarrow BaLiI_3$, this is because as the bond length increases, its lattice parameter also increases, which causes its valence electron to be loosely bound and these valence electrons require less energy to be free, which makes the energy band gap decrease [8]. This study found that the band gap decreases as the lattice constant increases, aligning with previous research findings [23]. The further band structure of these compounds showed relatively flat band gaps. This different property might be useful for thermoelectric devices.

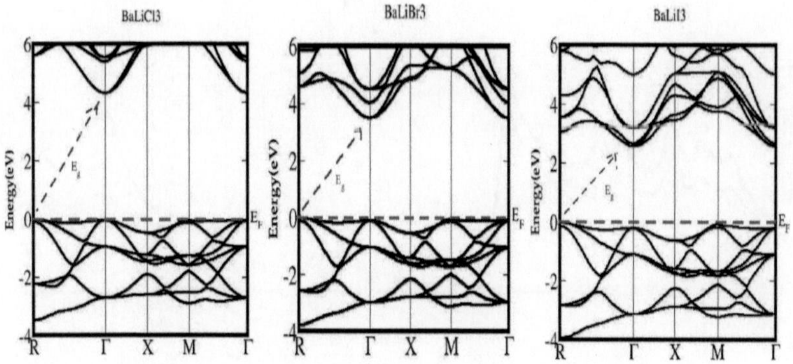

Fig. 3. Band structure diagram of $BaLiX_3$ (X = Cl, Br, I).

Table 2. Computed Band gap of the $BaLiX_3$ (X = Cl, Br, I). FP: –Full Potential, PP: – Pseudo Potential.

Compounds	Band Gap (eV)		
	FP-LAPW LSDA (This Work) **(R → Γ)**	FP-LAPW GGA (This Work) **(R → Γ)**	PP [8] GGA **(R → Γ)**
$BaLiCl_3$	4.05	4.24	4.29
$BaLiBr_3$	3.26	3.46	3.50
$BaLiI_3$	2.40	2.56	2.58

A high power factor is achieved when the band structure exhibits both flat band gaps and high dispersion, as this combination optimizes electronic transport properties [13, 24]. In these compounds, the top of the valence band lies within the symmetry point R, while the bottom edge of the conduction band is at symmetry point Γ, as shown in the Fig. 3. This system suggests an indirect transition, which improves thermoelectric properties but may slightly adversely affect its optical performance [19]. Total density of states (TDOS) and partial density of states (PDOS) diagram of these compounds implies that Barium (Ba) exhibits a dominant contribution to the density of states in the conduction band, and valence band has density of states mainly due to anion atom (X = Cl, Br, and I) (Fig. 4).

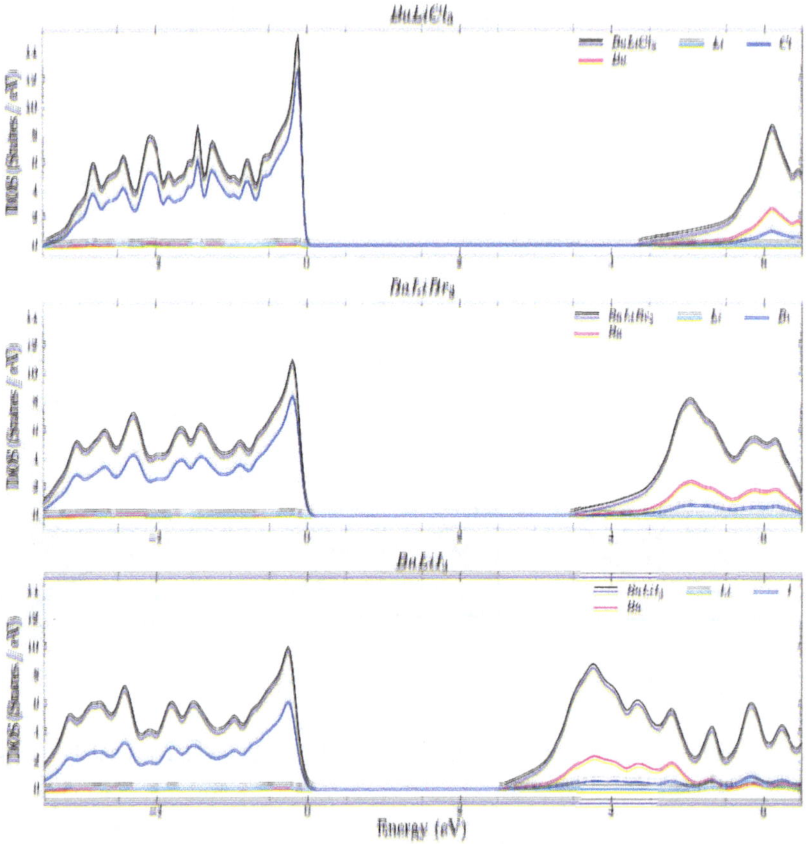

Fig. 4. Total Density of States (TDOS) and Partial Density of States (PDOS) diagram of $BaLiX_3(X = Cl, Br, I)$.

4.3 Optical Properties

Optical Properties of the $BaLiX_3$ (X = Cl, Br and I) are investigated by the OPTIC program supported by the WIEN2k simulation package. The real part of the dielectric function (ε_1) for $BaLiCl_3$, $BaLiBr_3$, and $BaLiI_3$ exhibits a noticeable displacement of the primary peak to the low value of the energy, along with an increase in the static dielectric function (Fig. 5).

The primary peaks for these perovskites are observed at 5.59 eV, 4.63 eV, and 3.84 eV respectively. The static dielectric function ϵ_0 is inversely proportional to band gap [24]. Imaginary part ϵ_2 represents the absorption behaviour of the compounds [25]. It contains all the information on the inter-band and intra-band transitions. The threshold value of the dielectric function ϵ_2 shifts to lower energies as Cl is replaced by Br and I. The threshold value of ϵ_2 determines the energy at which the material begins to absorb light energy. The imaginary part of the dielectric function corresponds to the absorption characteristics of the materials. It has been observed that static refractive index n(0) for $BaLiCl_3$, $BaLiBr_3$ and $BaLiI_3$ are 1.79, 1.97 and 2.21 respectively. It has observed that the refractive index

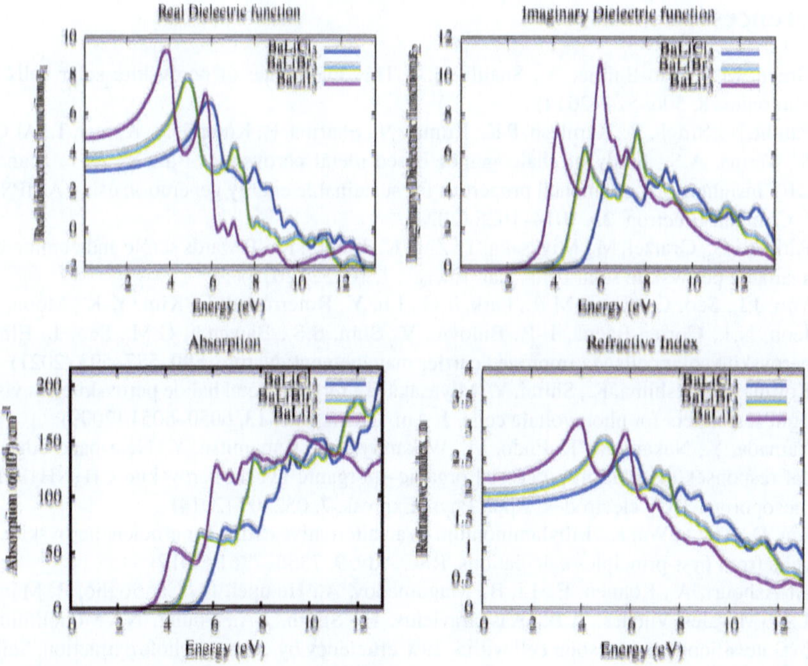

Fig. 5. Real dielectric function, imaginary dielectric function, absorption coefficient and refractive index diagram of $BaLiX_3$ (X = Cl, Br, I).

has reached a peak value of 2.6 at 6.03 eV, for $BaLiCl_3$, 2.9 at 4.94 eV, for $BaLiI_3$ and 3.2 at 3.99 eV for $BaLiI_3$. The absorption coefficient was found to be in the order of 10^4 cm^{-1}, which is suitable for the photovoltaic applications and all of these compounds exhibit a broad absorption range, indicating their potential for numerous applications in the optoelectronic systems.

5 Conclusion

Theoretical DFT prediction of the structural, electronic and optical properties is done by the full potential method. It was seen the pseudopotential codes slightly over predict the band gap, or it may happen that full potential code under predicts which could only be verified by experimental findings. The optimized lattice constant increased when Cl was replaced by Br and I, while the opposite trend was observed for their band gaps. The excellent optical properties of these materials indicate their potential for a wide range of applications. This study also compares the pseudo potential and full potential approaches in DFT, highlighting the trade-offs between computational efficiency and accuracy. The data generated can be used to train AI/ML models to predict material properties, optimizing the choice of DFT method for different materials. This enables active learning and more efficient high-throughput screening in materials discovery. Overall, it aids in developing smarter AI-driven tools for material design and simulation automation.

References

1. Green, M.A., Ho-Baillie, A., Snaith, H.J.: The emergence of perovskite solar cells. Nat. Photonics. **8**, 506–514 (2014)
2. Pandit, N., Singh, R., Kamlesh, P.K., Kumar, N., Sharma, P., Kumari, S., Kumar, T., Al-Qaisi, S., Verma, A.S.: Study of chalcogenide-based metal perovskites $BaZrX_3$ (X = S and se): DFT insight into fundamental properties for sustainable energy generation using AMPS 1-D. J. Comput. Electron. **23**, 1014–1028 (2024)
3. Park, N.-G., Gratzel, M., Miyasaka, T., Zhu, K., Emery, K.: Towards stable and commercially available perovskite solar cells. Nat. Energy. **1**, 16152 (2016)
4. Yoo, J.J., Seo, G., Chua, M.R., Park, T.G., Lu, Y., Rotermund, F., Kim, Y.-K., Moon, C.S., Jeon, N.J., Correa-Baena, J.-P., Bulovic, V., Shin, S.S., Bawendi, G.M., Seo, J., Efficient perovskite solar cells via improved carrier management: Nature. **590**, 587–593 (2021)
5. Kojima, A., Teshima, K., Shirai, Y., Miyasaka, T.: Organometal halide perovskites as visible-light sensitizers for photovoltaic cells. J. Am. Chem. Soc. **13**, 6050–6051 (2009)
6. Yamada, Y., Nakamura, T., Endo, M., Wakamiya, A., Kanemitsu, Y.: Near-band-edge optical responses of solution-processed organic–inorganic hybrid perovskite $CH_3NH_3PbI_3$ on mesoporous TiO_2 electrodes. Appl. Phys. Express. **7**, 032302 (2014)
7. Liu, D., Li, Q., Wu, K.: Ethylammonium as an alternative cation for efficient perovskite solar cells from first-principles calculations. RSC Adv. **9**, 7356–7361 (2019)
8. Al-Ashouri, A., Köhnen, E., Li, B., Magomedov, A., Hempel, H., Caprioglio, P., Márquez, J.A., Morales Vilches, A.B., Kasparavicius, E., Smith, J.A., Phung, N.: Monolithic perovskite/silicon tandem solar cell with> 29% efficiency by enhanced hole extraction. Science. **370**, 1300–1309 (2020)
9. Fehr, A.M., Agrawal, A., Mandani, F., Conrad, C.L., Jiang, Q., Park, S.Y., Alley, O., Li, B., Sidhik, S., Metcalf, I., Botello, C.: Integrated halide perovskite photoelectrochemical cells with solar-driven water-splitting efficiency of 20.8%. Nat. Commun. **14**, 3797 (2023)
10. Zhu, H., Teale, S., Lintangpradipto, M.N., Mahesh, S., Chen, B., McGehee, M.D., Sargent, E.H., Bakar, O.M.: Long-term operating stability in perovskite photovoltaics. Nat. Rev. Mater. **8**, 569–586 (2023)
11. Dubey, A., Mishra, K., Srivastava, R., Kumari, A., Jangra, P., Joshi, T., Choudhary, B., Verma, A.: Structural, optoelectronic, thermal and transport properties of hybrid perovskite ($EAGeCl_3$) material. J. Ovonic Res. **20**, 381–394 (2024)
12. Kumari, S., Kamlesh, P.K., Kumari, L., Kumar, S., Kumari, S., Singh, R., Gupta, R., Chauhan, M.S., Rani, U., Verma, A.S.: Progress in theoretical study of lead-free halide double perovskite Na_2AgSbX_6 (X= F, Cl, Br, and I) thermoelectric materials. J Mol. Model. **29**, 195 (2023)
13. Pingak, R.K., Bouhmaidi, S., Setti, L., Pasangka, B., Bernandus, B., Sutaji, H.I., Nitti, F., Ndii, M.Z.: Structural, electronic, elastic, and optical properties of cubic $BaLiX_3$ (X= F, cl, Br, or I) perovskites: an ab-initio DFT study. Indonesian J. Chem. **23**, 843–862 (2023)
14. Schleder, G.R., Padilha, A.C., Acosta, C.M., Costa, M., Fazzio, A.: From DFT to machine learning: recent approaches to materials science–a review. J Physics: Mater. **2**, 032001 (2019)
15. Garrity, K.F., Bennett, J.W., Rabe, K.M., Vanderbilt, D.: Pseudopotentials for high-throughput DFT calculations. Comput. Mater. Sci. **81**, 446–452 (2014)
16. Burke, K., Wagner, L.O.: DFT in a nutshell. Int. J. Quantum Chem. **113**, 96–101 (2013)
17. Fiedler, L., Shah, K., Bussmann, M., Cangi, A.: Deep dive into machine learning DFT for materials science and chemistry. Phys. Rev. Mater. **6**, 040301 (2022)
18. Blaha, P., Schwarz, K., Tran, F., Laskowski, R., Madsen, G.K., Marks, L.D.: WIEN2k: an APW+lo program for calculating the properties of solids. J. Chem. Phys. **152**, 074101 (2020)
19. Karasiev, V.V., Sjostrom, T., Dufty, J., Trickey, S.: Accurate homogeneous electron gas exchange-correlation free energy for local spin-density calculations. Phys. Rev. Lett. **112**, 076403 (2014)

20. Madsen, G.K.: Functional form of the generalized gradient approximation for exchange: the PBE α functional. Physical review B—condensed matter and materials. Physics. **75**, 195108 (2007)
21. Schwarz, K.: DFT calculations of solids with LAPW and WIEN2k. J. Solid State Chem. **176**, 319–328 (2003)
22. Katsura, T., Tange, Y.: A simple derivation of the Birch–Murnaghan equations of state (EOSs) and comparison with EOSs derived from other definitions of finite strain. Fortschr. Mineral. **9**, 745 (2019)
23. Mousa, A., Abu-Jafar, M., Dahliah, D., Shaltaf, R., Khalifeh, J.: Investigation of the perovskite $KSrX_3$ (X= Cl and F) compounds, examining the optical, elastic, electronic and structural properties: FP-LAPW study. J. Electron. Mater. **47**, 641–650 (2018)
24. Hariharan, M., Eithiraj, R.D.: Comprehensive study on structural, electronic, optical, elastic, and transport properties of natural mercury sulphohalides via DFT computation. Sci. Rep. **14**, 18593 (2024)
25. Bertsch, G.F., Iwata, J.-I., Rubio, A., Yabana, K.: Real-space, real-time method for the dielectric function. Phys. Rev. B. **62**, 7998 (2000)

Synthesis of V_2C MXene: Effects of Etchant Concentration on Material Applications

Heena Rohit Shevde and Margi Patel(✉)

Department of Basic Science and Humanities, Mukesh Patel School of Technology Management and Engineering, SVKM's NMIMS Deemed University, Mumbai 400 056, Maharashtra, India
margi23102000@gmail.com

Abstract. MXenes, a rapidly growing family of two-dimensional (2D) transition metal carbides and nitrides, have garnered significant attention due to their unique properties and potential applications across various fields. Vanadium-based MXenes (V-MXenes) are particularly noteworthy due to their excellent electrochemical performance. Their special qualities, which include high surface area and capacity to promote quick ion transport, help explain why they work so well in electro-catalysis, supercapacitors, and batteries, among other applications. This study reports the synthesis of V_2C MXene through controlled etching using different HF concentrations. X-ray diffraction (XRD) confirmed the removal of aluminum layers and successful V_2C formation, with characteristic peaks of the (002) plane. UV-visible spectroscopy indicated promising optical properties, suggesting suitability for supercapacitors. These findings contribute to optimizing MXene synthesis methods for applications in catalysis, energy storage, and advanced electronics. Improved synthesis methods reduce the reliance on hazardous hydrofluoric acid (HF) and energy intensive processes in turn making fabrication of MXene more environmental friendly and cost effective.

Keywords: MXene · 2D Material · V_2C · Energy Storage · Supercapacitors

1 Introduction

MXenes are a novel class of 2D materials derived from MAX phases by selectively removing the A layer [1]. Since their discovery in 2011 by Yury Gogotsi and Michel Barsoum, MXenes have attracted significant attention due to their exceptional properties, including high mechanical strength and hydrophilicity [2]. MXenes are represented by the general formula $M_{n+1}X_nT_x$, where M represents transition metals, X is carbide/nitride, n is an integer between 1–3 and T_x is the surface terminations, such as −O, −OH, −F, and/or −Cl. Several different MXene compositions having either titanium or vanadium as transition metals are reported [1, 2]. The one with vanadium as a transition metal, Vanadium carbide (V_2C) is well known for its unique combination of high electrical conductivity and surface tunable chemistry [3]. These materials are found to have the potential for advanced applications like energy storage specially in supercapacitors, optoelectronics, and catalysis [4].

S. Kumari et al. (Eds.): ICMTP 2024, SPPHY 311, pp. 10–20, 2025.
https://doi.org/10.1007/978-981-96-6795-6_2

Vanadium MXenes exhibit exceptional conductivity because of their metallic nature. The layered structure of V_2C allows them for extensive surface area while their surface termination (−OH, −O, or −F) can be well balanced to enhance their interactions with other elements hence enhancing their performance in catalysis and adsorption-based applications [5]. Vanadium MXene also indicates high mechanical properties and flexibility, making them suitable for wearable electronics, and flexible energy storage systems [6]. Their unique conductivity and 2D structure make them best for electromagnetic interference (EMI) shielding, which is essential for modern communications devices [7]. The potential of this material to move ions rapidly and keep several ions such as Li+, Na+, and K+ have made them essential for hybrid energy devices [8]. V_2C exhibits remarkable catalytic activity, making it ideal for hydrogen evolution reaction (HER) and CO_2 reduction [9]. Their flexible band gap and high conductivity are promising for optoelectronics devices [10]. Theoretical study shows V_2C can have high purity and can demonstrate exceptional performance as compared to other MXene.

Hydrofluoric acid (HF) has proven to be the most efficient etchant for synthesizing V_2C MXene because of its strong ability to selectivity remove the 'Al' from Vanadium aluminum carbide (V_2AlC) and resulting MXene 2D layers. Literature lists a number ways to synthesize V_2C but the most common synthesis method reported is etching through a HF solution at higher temperature and long duration [11–14]. Though these methods successfully etch Al but harshness of HF cannot be neglected. In 2017 HF free etching technique was reported which used sodium fluoride/hydrochloric acid (NaF + HCl) as etchants at 90 °C.This method was successful in synthesizing pure V_2C, but required high temperature and long duration with a specialized setup [11]. Similarly, various other fluoride salts like LiF and KF are used along with acid for producing V_2C, however each of these salts have drawbacks, such as high temperature and low yield [12]. Hence, the choice of etchant with correct concentration of and temperature are the critical factors that impact synthesis of V_2C MXenes [15]. Literature proves that the best results are obtained by using HF as the etchant, which not only simplifies the synthesis process but also results in MXene with good properties suitable for variety of applications. However, these processes are unsafe for the environment because of high concentrations of HF and long duration for etching. To address this gap, this study systematically examines the influence of low HF concentration and reduced duration for etching on V_2C synthesis and its material properties.

2 Experimental

2.1 Materials

In this etching method, vanadium aluminum carbide V_2AlC (99%) was used as the MAX phase, procured from Amratlal Bhurabhai & Co., Mumbai, India. Etching solution (40% M.W. concentrated HF) was procured from high purity Lab Chemicals Pvt. Ltd. Mumbai. 30% HF M.W. was made with distilled water.

2.2 Synthesis

Synthesis was done by selective etching of "Al" of V_2AlC through lower HF concentrations at normal and elevated temperatures for a shorter duration.

2.2.1 Synthesis of Vanadium Carbide at Room Temperature (V_2C)

In this method, V_2CT_X MXene was prepared by using different concentrations of Hydrofluoric Acid (HF) as etchant [18]. Solutions was prepared for 30% HF M.W. and 40% HF M.W. concentration. 2 g V_2AlC powder was immersed in both solutions and kept both reactions on simultaneously for 24 hours with magnetic stirring in fume-hood.

Following that, 30 mL of distilled water was added to the solution to neutralize residual HF until the pH of the washing solution was close to neutral. After that, the solution was transferred into a centrifuge tube. The etched solution was washed multiple times with distilled water via centrifugation. This process was repeated 3 times at 25 °C at 10,000 rpm for 10 min to remove possible absorbed ions and remaining precursors and until a neutral pH was achieve. After the final centrifugation, MXene was collected by vacuum filtration with whatman filter paper.

Later, MXene was transferred into heat-resistant pellet and was placed at rest at room temperature for 24 h. The semi-dried powder was further put in the oven for 2 h at 60 °C, which influence the oxidation level of stability of V_2C. The final black powder was collected as the resulting MXene [18, 19].

2.2.2 Synthesis of Vanadium Carbide at Higher Temperature (V_2C)

V_2C T_X was synthesized at elevated temperatures and different concentrations. Again solution was prepared for 30% HF M.W. and 40% HF M.W. concentration and 1 gm of V_2AlC powder was slowly added into both solutions. The prepared mixtures were stirred at 80 °C for 8 h for etching. After 8 h V_2C was separated by centrifugation and washed with distilled water. The washing cycle was repeated 4 times until the pH of the mixture reached neutral approximately pH 6–7 to ensure the removal of HF. The semi-dried material was kept in the oven and dried at 70 °C for 3 h. Obtained black powder is V_2CT_X MXene [9, 11].

Table 1. Performed Synthesis.

MAX + HF	Condition	Time (Hour)
2 gms **(V_2AlC)** + 30 mL **(40% M.W. HF) +** 10 ml **(distilled water)**	Room temperature	24
2 gms **(V_2AlC)** + 40 ml **(40% M.W. HF)**	Room temperature	24
1 gm **(V_2AlC)** + 15 ml **(40% M.W. HF)** + 5 mL **(distilled water)**	80° temperature	8
1 gm **V_2AlC** + 20 mL **40% M.W. HF**	80° temperature	8

The following Fig. 1 shows (A) Synthesis V_2C at room temperature (B) Synthesis V_2C at elevated temperature and Table 1 shows the details of performed synthesis

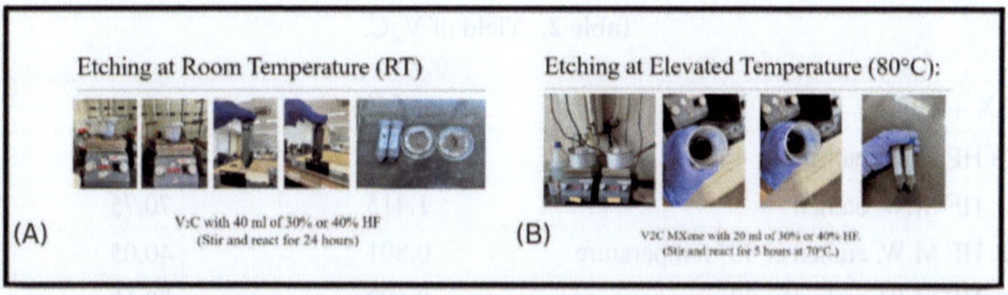

Fig. 1. (**A**) Synthesis V_2C at room temperature (**B**) Synthesis V_2C at elevated temperature.

2.3 Characterization

X-Ray Diffraction patterns of samples were obtained with PANalytical X-PERT PRO X-Ray diffractometer with Cu-Kα (radiation 50 kV/100 m A°) for structural studies, phase identification and, lattice parameters for conforming V_2C. The optical study was carried out using a UV-vis Spectrophotometer (Shimadzu, UV-1900).

3 Result and Discussions

3.1 Measurement of the Yield (Y.)

The yield was calculated by utilizing the following Eq. (1).

$$\text{Yield}\ (\%) = \left(\mathbf{mMAX} \big/ \mathbf{mMXene} \right) \times 100 \tag{1}$$

For the experiment conducted at room temperature, the yield calculated for 40% HF M.W. is 70.75% whereas for 30% HF M.W. yield is 80.05%. The computed yield values are comparatively less for both concentrations of HF for the experiments done at higher temperatures. This indicates variability in the efficiency of synthesis process [18]. Reduced yield at higher temperature is due to over etching and unusual side reactions. Etching at elevated temperature leads to structural defects while secondary reactions like oxidation and fluorination, further lowering the yield. Controlled etching at moderate or room temperatures are essential for maintaining MXene integrity and maximizing output.

Calculated yields for each concentration are shown in Table 2 [18].

3.2 X-Ray Diffraction (XRD)

The XRD patterns of MAX phase V_2AlC sample before and after exfoliation of V_2C with different concertation of HF at room temperature and higher temperature are shown in Fig. 2.

V_2AlC, pattern (A), show distinct sharp peaks that demonstrate a stable crystalline structure, characterized by strong peaks attributed to its typical layered arrangements. The presence of several sharp peaks within the 2θ range, specially near ~13°, ~39°,

Table 2. Yield of V_2C.

MAX + HF	m MXene	Yield %
30% HF M.W. etched	1.601	80.05
40% HF M.W. etched	1.415	70.75
30% HF M.W. etched at 80° temperature	0.801	40.05
40% HF M.W. etched at 80 temperature	0.602	30.10

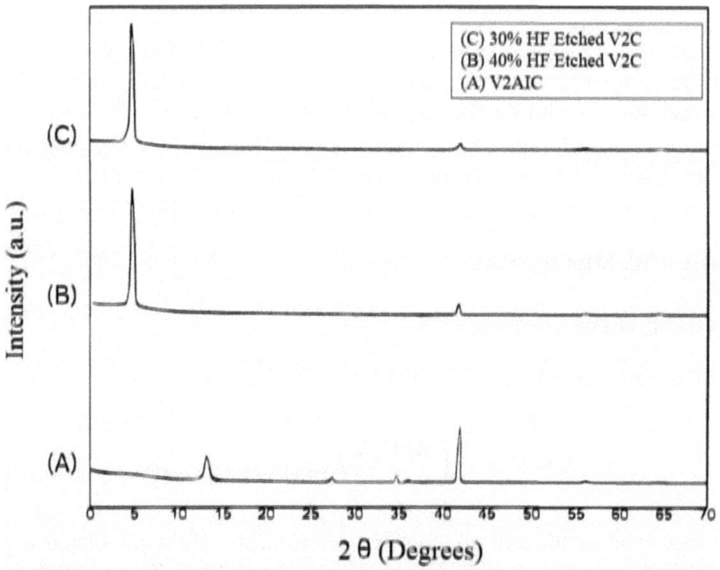

Fig. 2. XRD Patterns (**a**) V_2AlC MAX phase (**b**) 40% HF M.W. etched (**c**) 30% HF M.W. etched.

and ~ 45° corresponds to the well-ordered crystalline MAX phase. The peaks derive from the (002), (101), (103), and (104) planes of V_2AlC. Absence of strong peak at low 2θ (~6–8°) confirms that this is a non-layered structure before etching. 30% HF M.W. etched V_2C, pattern (C), shows a prominent peak at a lower 2θ (~6–8°), which corresponds to the characteristic (002) plane of V_2C MXene. This shift to a lower angle indicates interlayer expansion, indicating that the functional groups (−OH, −F, −O) caused by the removal of Al increased the distance between V_2C layers. The absence of several V_2AlC phase peaks at ~13° and ~ 39° confirms successful etching of Al. 40% HF M.W. etched V_2C shows (002) peak at ~6–8° shifts slightly more to the left compared to the 30% HF M.W. sample. This suggests even greater interlayer expansion due to stronger etching, leading to more functionalization. The intensity of the other peaks further reduces, indicating an increased exfoliation effect. Additional small peaks might correspond to unetched residual MAX phase or other secondary phases, which can subtly influence conductivity, electrochemical performance limiting material in advanced applications.

Following Fig. 3 shows for 30% HF M.W. etched at 80° temperatures peak significantly reduce and shift to lower angle, this suggests incorporation of additional flicks. Peaks are weak due to excessive etching, leading V_2C to loss of long-range order. An increase in background noise in 40% HF M.W. etched at 80° temperature indicates the presence of amorphous material [20].

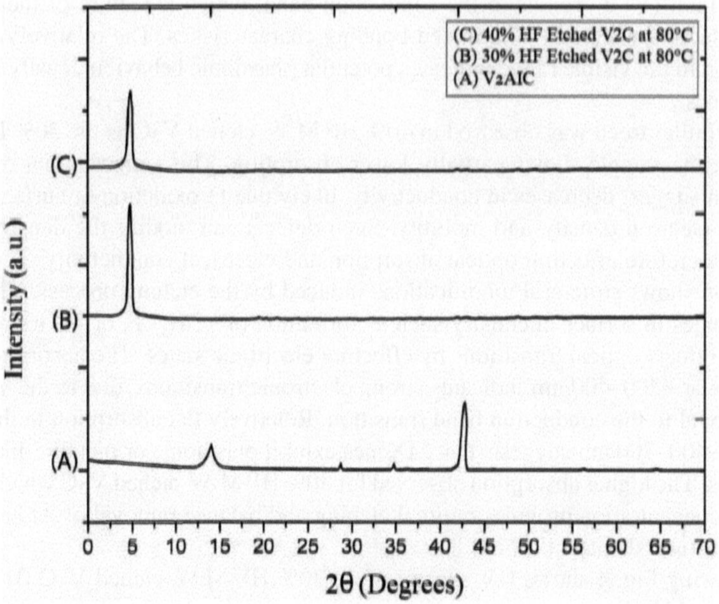

Fig. 3. XRD Patterns (**a**) V_2AlC (**b**) 30% HF M.W. etched at 80° C (**c**) 40% HF M.W. etched at 80° C.

Following Table 3 shows key observations of XRD of V_2C.

Table 3. Key observations of XRD.

Sample	Key Observation
30% HF M.W. etched at room temp	Good crystallinity with well-defined peaks
40% HF M.W. etched at room temp	Good crystallinity and precisely stated peaks
30% HF M.W. etched at 80° temp	Reduce crystallinity; lower yield
40% HF M.W. etched at 80° temp	Loss of 2-D structure; lower yield

3.3 UV Visible Spectroscopy

The peaks in the UV-Vis spectra shows the electronic transitions and optical properties of the materials. V_2AlC MAX Phase provide the lower absorption across the spectrum, which is indicating minimum interaction with incident light because of more metallic nature and limited available electronic transitions.

After etching with 30% HF M.W. V_2C indicates significantly higher absorption, showing stronger light-matter interaction. A gradual increase in absorption in the UV region (~200–400 nm) identify inter band transitions which shows to the metallic vanadium carbide layers. These transitions are initially associated with excitations from the vanadium (V) d-orbitals to the conduction band, which is feature of metallic and semiconducting materials with mixed bonding characteristics. The relatively stable of absorption in the visible range indicates potential plasmonic behaviour, easily shown in 2D MXenes.

The similar trend was observed in 40% HF M.W. etched V_2C as the 30% HF M.W. However, the sample shows partially lower absorption. This suggests that over etching which suggest decreases in conductivity, likely due to oxidation or surface defects affecting electron density and mobility. Such defects can modify the density of free carriers, therefore affecting optical absorption and electrical conductivity. The shift in absorption shows structural modifications induced by the etching process. This structural changes in surface chemistry such as formation of –OH, –F, or –O terminations, which is effects optical transitions by effecting electronic states. The absorption in the UV range at ~200–400 nm indicates strong electronic transitions, due to the vanadium (V) d-orbital to the conduction band transition. Relatively flat absorption in the visible range at ~400–700 nm suggests that MXenes exhibit plasmonic or metallic-like optical properties. The higher absorption observed for 30% HF M.W. etched V_2C which suggest that this concentration provides optimal etching and balance removal of Al layers with lower structural damage [22, 23].

Following Fig. 4 shows UV spectra of A) 30% HF M.W. etched V_2C B) 40% HF M.W. etched V_2C.

4 Future Direction

This study provides a strong foundation for exploring the application of energy storage and supercapacitors. Results indicate that concentrations of etchants and reaction conditions particularly enhance the properties of V_2C. Synthesis conditions especially with the use of 30% HF M.W. at room temperature created V_2C MXene with enhanced structural integrity, conductivity, and stability [15]. These properties directly influence charge capacity, energy density, and cycle life, which are crucial for high-performance supercapacitors. The following parameters of the results demonstrate the potential of V_2C for supercapacitors and energy storage development which can help focus on future studies [24].

V_2C MXene exhibits a wide range of applications; which are impacted due to its layered structure and optical stability. A brief outline of how the characterization results prove these properties of MXene is discussed below.

- **Layered Structured**

  XRD shows the successful synthesis of well-defined 2D layered V_2C MXene. This suggests their suitability for high surface area electrodes, a key requirement for many applications such as supercapacitors and battery anodes [25, 26]. The high

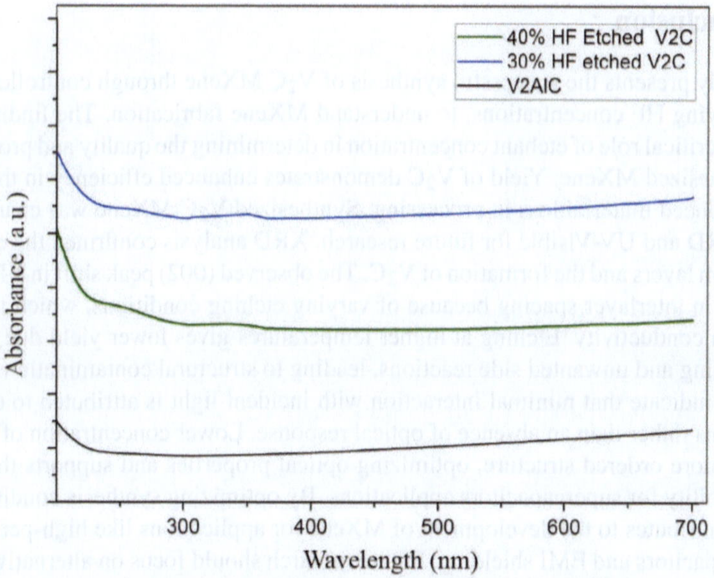

Fig. 4. UV Visible Spectra.

conductivity and surface area influence to explore V_2C as anode in lithium-ion batteries, furthermore, assessing the productivity of hydrogen evolution reaction (HER) or Oxygen reduction reaction (ORR) [27].

- **Optical Stability**

  The UV-Visible spectra demonstrate well-preserved optical properties under optimal conditions further supporting the durability of material, which is required for long term stability. These results suggest strong plasmonic behavior in V_2C, leveraging their tunable optical properties for light detection and integrating V_2C into solar cell designs to enhance light absorption and charge transport [28].

Following Table 4 shows specific applications outlining actionable next steps for future direction.

Table 4. Future Directions for Applications.

Sample	Key Properties	Potential Application	Future Proposed Work
30% HF M.W. etched	High conductivity	Batteries, supercapacitors, Energy storage	Investigate electrochemical performance in energy storage devices [29, 30]
40% HF M.W. etched	Layered structured, over-etching	EMI shielding, catalysis	Evaluation of EMI shielding, enhance (HER) or (ORR) [27, 31]

5 Conclusion

This study presents the successful synthesis of V_2C MXene through controlled etching with varying HF concentrations, to understand MXene fabrication. The findings highlight the critical role of etchant concentration in determining the quality and properties of the synthesized MXene. Yield of V_2C demonstrates enhanced efficiency in the synthesis or reduced material loss in processing. Synthesized V_2C MXene was characterized using XRD and UV-Visible for future research. XRD analysis confirmed the etching of aluminum layers and the formation of V_2C. The observed (002) peak shift in XRD shows changes in interlayer spacing because of varying etching conditions, which is directly effect on conductivity. Etching at higher temperatures gives lower yield due to excessive etching and unwanted side reactions, leading to structural contaminations. UV-Vis analysis indicate that minimal interaction with incident light is attributed to electronic transitions rather than an absence of optical response. Lower concentration of HF gives yield a more ordered structure, optimizing optical properties and supports the materials suitability for supercapacitors applications. By optimizing synthesis conditions, this study contributes to the development of MXene for applications like high-performance super capacitors and EMI shielding. Future research should focus on alternative etching strategies to minimize HF usage, ensuring a more sustainable and scalable synthesis process. These insights cover the way for next-generation MXene with enhanced functional properties tailored for real-world applications.

References

1. Naguib, M., et al.: Two-dimensional nanocrystals produced by exfoliation of Ti_3 AlC_2. Adv. Mater. **23**(37), 4248–4253 (2011). https://doi.org/10.1002/adma.201102306
2. Naguib, M., et al.: Two-dimensional nanocrystals produced by exfoliation of Ti_3 AlC_2. Adv. Mater. **23**(37), 4248–4253 (2011). https://doi.org/10.1002/adma.201102306
3. Naguib, M., Mochalin, V.N., Barsoum, M.W., Gogotsi, Y.: 25th anniversary article: MXenes: a new family of two-dimensional materials. Adv. Mater. **26**(7), 992–1005 (2014). https://doi.org/10.1002/adma.201304138
4. Naguib, M., et al.: New two-dimensional niobium and vanadium carbides as promising materials for Li-ion batteries. J. Am. Chem. Soc. **135**(43), 15966–15969 (2013). https://doi.org/10.1021/ja405735d
5. Naguib, M., et al.: Two-dimensional transition metal carbides. ACS Nano. **6**(2), 1322–1331 (2012). https://doi.org/10.1021/nn204153h
6. Rahman, U.U., et al.: MXenes as emerging materials: synthesis, properties, and applications. Molecules. **27**(15), 4909 (2022). https://doi.org/10.3390/molecules27154909
7. Xu, B., Gogotsi, Y.: MXenes: from discovery to applications. Adv. Funct. Mater. **30**(47), 1 (2020). https://doi.org/10.1002/adfm.202007011
8. Ampong, D.N., et al.: MXene: Fundamentals to Applications in Electrochemical Energy Storage. Springer, Cham (2023). https://doi.org/10.1186/s11671-023-03786-9
9. Zhou, W., Sun, L., Li, K., Tian, S.: Enhanced photocatalytic activity of V2C MXene-coupled ZnO porous Nanosheets with increased surface area and effective charge transfer. Materials. **17**(11), 2529 (2024). https://doi.org/10.3390/ma17112529
10. Otgonbayar, Z., Yang, S., Kim, I.-J., Oh, W.-C.: Recent advances in two-dimensional MXene for supercapacitor applications: progress challenges, and perspectives. Nanomaterials. **13**(5), 919 (2023). https://doi.org/10.3390/nano13050919

11. Wu, M., Wang, B., Hu, Q., Wang, L., Zhou, A.: The synthesis process and thermal stability of V2C MXene. Materials. **11**(11), 2112 (2018). https://doi.org/10.3390/ma11112112
12. Borah, A.J., Natu, V., Biswas, A., Srivastava, A.: A review on recent progress in synthesis, properties, and applications of MXenes. Oxford Open Materials Science. **5**, itae017 (2024). https://doi.org/10.1093/oxfmat/itae017
13. Li, X., Ran, F., Yang, F., Long, J., Shao, L.: Advances in MXene films: synthesis, assembly, and applications. Transactions of Tianjin University. **27**(3), 217–247 (2021). https://doi.org/10.1007/s12209-021-00282-y
14. Wu, Y., Sun, M.: Recent progress of MXene as an energy storage material. Nanoscale Horiz. **9**(2), 215–232 (2024). https://doi.org/10.1039/D3NH00402C
15. Ronchi, R.M., Arantes, J.T., Santos, S.F.: Synthesis, structure, properties and applications of MXenes: current status and perspectives. Elsevier (2019). https://doi.org/10.1016/j.ceramint.2019.06.114
16. Gogotsi, Y., Anasori, B.: The rise of MXenes. ACS Nano. **13**(8), 8491–8494 (2019). https://doi.org/10.1021/acsnano.9b06394
17. Wei, Y., Zhang, P., Soomro, R.A., Zhu, Q., Xu, B.: Advances in the synthesis of 2D MXenes. John Wiley and Sons Inc. (2021). https://doi.org/10.1002/adma.202103148
18. Mirkhani, S., Sundararaj, U., Zeraati, A. S., Sun, P., and Naguib, M.: 'Improved synthesis of Ti3C2Tx MXene resulting in exceptional electrical conductivity, high synthesis yield, and enhanced capacitance'
19. Shan, Q., et al.: Two-dimensional vanadium carbide (V2C) MXene as electrode for supercapacitors with aqueous electrolytes. Electrochem. Commun. **96**, 103–107 (2018). https://doi.org/10.1016/j.elecom.2018.10.012
20. Mohseni-Salehi, M.S., Taheri-Nassaj, E., Babaei, A., Ghazvini, A.S., Soleimanzade, M.: Effect of temperature and atmosphere on V2AlC etching for V2CTx MXenes synthesis used as anode for Li-ion storage systems. J. Energy Storage. **66**, 107462 (2023). https://doi.org/10.1016/j.est.2023.107462
21. Luo, K., et al.: First-principles study on the electrical and thermal properties of the semiconducting Sc3(CN)F2 MXene. RSC Adv. **8**(40), 22452–22459 (2018). https://doi.org/10.1039/c8ra03424a
22. Liu, C., et al.: Vanadium carbide MXene: as a reductant for the synthesis of gold nanoparticles and its biosensing application. Amino Acids. **54**(8), 1173–1181 (2022). https://doi.org/10.1007/s00726-022-03173-1
23. Gogotsi, Y.: MXenes, 1st edn. Jenny Stanford Publishing, New York (2023). https://doi.org/10.1201/9781003306511
24. Jeon, J., Yang, Y., Choi, H., Park, J.-H., Lee, B.H., Lee, S.: MXenes for future nanophotonic device applications. Nano. **9**(7), 1831–1853 (2020). https://doi.org/10.1515/nanoph-2020-0060
25. Pang, J., et al.: Applications of 2D MXenes in energy conversion and storage systems. Royal Society of Chemistry. **48**, 72 (2019). https://doi.org/10.1039/c8cs00324f
26. Nahirniak, S., Ray, A., Saruhan, B.: Challenges and future prospects of the MXene-based materials for energy storage applications. Batteries. **9**, 126 (2023). https://doi.org/10.3390/batteries9020126
27. He, L., et al.: Advances and challenges in MXene-based electrocatalysts: unlocking the potential for sustainable energy conversion. Mater. Horiz. **11**(18), 4239–4255 (2024). https://doi.org/10.1039/D4MH00845F
28. Lee, J., Kwon, S.-Y., Lee, J.H.: Investigation on the nonlinear optical properties of V_2C MXene at 1.9 μm. J. Mater. Chem. C Mater. **9**(42), 15346–15353 (2021). https://doi.org/10.1039/D1TC03610F

29. Awan, H.T.A., Kumar, L., Wong, W.P., Walvekar, R., Khalid, M.: Recent Progress and challenges in MXene-based phase change material for solar and thermal energy applications. Energies. **16**, 1977 (2023). https://doi.org/10.3390/en16041977
30. Pang, S.Y., et al.: Universal strategy for HF-free facile and rapid synthesis of two-dimensional MXenes as multifunctional energy materials. J. Am. Chem. Soc. **141**(24), 9610–9616 (2019). https://doi.org/10.1021/jacs.9b02578
31. Han, M., et al.: Efficient microwave absorption with Vn+1CnT MXenes. Cell Rep. Phys. Sci. **3**(10), 101073 (2022). https://doi.org/10.1016/j.xcrp.2022.101073

Synthesis and Characterization of ZnO Nanoparticles by Green Synthesis Method

Ramanand Sharma[1,2], Madhu Yadav[1,3], Rimjhim Shrimal[1], and Chhagan Lal[1,3](✉)

[1] Department of Physics, University of Rajasthan, Jaipur 302004, Rajasthan, India
clsaini52@gmail.com

[2] Govt. Kamla Modi Girls College, Neem Ka Thana 332713, Rajasthan, India

[3] Center for Non-Conventional Energy Resources, University of Rajasthan, Jaipur 302004, Rajasthan, India

Abstract. With industrial development mainly paper, textile, medicine, rubber and plastic industry, as a contaminant, the number of organic dyes enhanced. The consistent discharge of toxic dyes into water bodies has escalated into a significant issue. Due to the carcinogenic and mutagenic properties of the majority of dyes, they pose a variety of health risks. Hence its required to find out solution of this problem and efficient methods to the remediation of sewage that contains toxic dyes. One of the effective solutions to this issue is the use of nanoparticles as a photocatalyst. Various physical, chemical, and biological methods can be employed to generate the nanoparticles. Among various environmental applications of nanoparticles, they can be used as for decontamination of dyes from polluted water bodies. The practical application of metal and metal oxide nanoparticles is on the rise alongside their nanocomposites as one of the powerful solutions of photocatalytic degradation of various hazardous dyes. In this paper, we fabricate ZnO nanoparticles through a green route, in which Aloe vera leaf (Aloe Barbadensis) extract is employed as a reducing and capping agent. The synthesized nanoparticles were subsequently characterized using a variety of analytical techniques, including XRD and UV-Visible. The band gap is 2.63 eV.

Keywords: Nanoparticles · dyes · natural · organic · metal

1 Introduction

Dyes are a major source of water pollution, originating from various industries such as textiles, plastics, cosmetics, food processing, pharmaceuticals, and printing [1–3]. These synthetic dyes, due to their complex chemical structure and high solubility, are challenging to degrade and pose significant environmental and health hazards [4–6]. Even at low concentrations, dyes can be toxic, affecting aquatic ecosystems by reducing sunlight penetration, which disrupts photosynthesis and aquatic life [7–9]. Additionally, exposure to synthetic dyes has been linked to severe health issues, including respiratory problems, skin disorders, and even carcinogenic effects [10–12]. Various treatment methods, such as adsorption, coagulation, oxidation, and microbial degradation, have

S. Kumari et al. (Eds.): ICMTP 2024, SPPHY 311, pp. 21–27, 2025.
https://doi.org/10.1007/978-981-96-6795-6_3

been explored; however, they often suffer from limitations such as high costs, secondary pollution, and inefficiency in degrading persistent dyes [13–15]. Therefore, the effective degradation and removal of dyes from wastewater is a crucial area of environmental research, requiring innovative and sustainable solutions [16–18].

Metal and metal oxide nanoparticles have gained significant attention for their potential in dye degradation due to their high surface area, catalytic efficiency, and unique optical properties [19–22]. Among these, nanoparticles of gold, silver, titanium, tin, copper, manganese, iron, zinc, cerium, and zirconium have been widely studied for their photocatalytic and adsorption capabilities in wastewater treatment [23, 24]. The ability of these nanoparticles to generate reactive oxygen species (ROS) under light irradiation makes them highly effective in breaking down toxic organic pollutants into less harmful byproducts [25, 26]. Specifically, metal oxide nanoparticles, such as ZnO, TiO_2, and Fe_2O_3, have been extensively investigated due to their stability, reusability, and cost-effectiveness [27, 28]. Compared to conventional dye removal techniques, such as activated carbon adsorption and advanced oxidation processes, metal oxide nanoparticles offer superior efficiency and environmental sustainability [29, 30]. These materials provide a promising alternative for industrial wastewater treatment, minimizing environmental impact while ensuring high degradation efficiency [31].

In this study, ZnO nanoparticles were synthesized using a green synthesis method, providing an environmentally friendly and cost-effective alternative to conventional chemical and physical synthesis techniques. Green synthesis involves the use of plant extracts, microorganisms, or biopolymers as reducing and stabilizing agents, eliminating the need for hazardous chemicals. This approach not only supports sustainable practices but also improves the biocompatibility and stability of the synthesized nanoparticles, making them suitable for various applications.

The study focuses on the green synthesis of ZnO nanoparticles, emphasizing their structural and optical characterization. Phytogenic synthesis offers several advantages, including cost-effectiveness, environmental compatibility, scalability, and reduced chemical waste. However, the formation of nanoparticles through biological routes is often limited to specific metals, oxides, and sulfides, highlighting the need for innovative methods to synthesize metal oxide-based nanomaterials with controlled properties.

To address these challenges, ZnO nanoparticles were synthesized using a plant-mediated approach, ensuring controlled size and morphology without the use of hazardous chemicals. The synthesized nanoparticles were systematically characterized using X-ray diffraction (XRD) to determine their crystalline structure and phase purity, while UV-Vis spectroscopy was employed to analyze their optical properties, including band gap estimation. Although this study does not investigate dye degradation or photocatalytic activity, it provides a fundamental understanding of the structural and optical characteristics of green-synthesized ZnO nanoparticles. These findings contribute to the development of eco-friendly synthesis techniques for future applications in various fields, including environmental and biomedical research.

2 Synthesis Process of ZnO Nanoparticles

The most prevalent method of making ZnO NPs from plant extracts necessitates the thorough cleaning of the plants with clean or distilled water. Plant extracts are subsequently obtained through direct soaking, dissolving in solvent, grinding into powder, or drying. Zinc salt solution is added to the extracts to serve as the metal precursor. Subsequently, a precipitate is produced through the reaction. In the final step, the precipitate is calcined to produce ZnO NPs.

3 Experimental

3.1 XRD Analysis

The structural characterization of the synthesized ZnO nanoparticles was carried out using X-ray diffraction (XRD) with a CuKα radiation source (λ = 1.5406 Å). The diffraction pattern, presented in Fig. 1, exhibits prominent peaks at 2θ values of 31.72°, 34.40°, and 36.20°, which correspond to the (100), (002), and (101) crystal planes of hexagonal wurtzite ZnO (JCPDS card no. 36-1451). These observed peaks are in close agreement with reported literature values, confirming the successful formation of ZnO nanoparticles. In addition to the major diffraction peaks, several weaker peaks were detected at 37.78°, 45.07°, 47.49°, 56.67°, 62.94°, 66.27°, 67.87°, 69.12°, 72.62°, and 76.96°, suggesting the presence of minor impurities or secondary phases. The presence of these peaks could be attributed to residual precursor materials or structural defects introduced during the synthesis process.

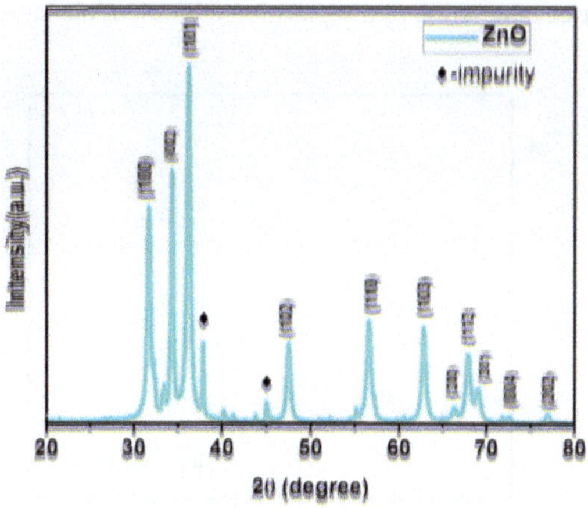

Fig. 1. XRD of ZnO Nanoparticles.

Debye-Scherrer's formula was employed to determine the average particle size [22]:

$$D = \frac{0.9\,\lambda}{\beta\, cos\theta}$$

where λ is the X-ray wavelength, β is the full-width at half-maximum (FWHM) of the diffraction peak, and θ is the Bragg angle. The calculated crystallite size was found to be approximately 40 nm, indicating the nanoscale nature of the synthesized ZnO particles. The formation of agglomerates and the stability of nanoparticles are significantly influenced by environmental conditions and surface interactions.

3.2 UV-Visible Spectroscopy

Optical characterization of the synthesized ZnO nanoparticles was performed using UV-Visible absorption spectroscopy. The absorption spectrum and Tauc plot of ZnO nanoparticles are shown in Fig. 2. UV-Visible spectroscopy was used to analyze the electronic transitions and band gap energy of the nanoparticles. The absorption spectrum of ZnO nanoparticles exhibited a strong absorption edge in the UV region, which is characteristic of ZnO due to its wide band gap and excitonic transitions. The optical band gap (Eg) was determined using Tauc's relation:

$$(\alpha h\vartheta)^n = A\left(h\vartheta - E_g\right) \quad (1)$$

where α is the absorption coefficient, hv is the photon energy, A is a proportionality constant, and n depends on the nature of the electronic transition (for ZnO, n = 2 for a direct band gap semiconductor). From the Tauc plot, the extrapolated band gap energy of the synthesized ZnO nanoparticles was found to be 2.63 eV, which is slightly lower than the bulk ZnO band gap (~3.37 eV) [32]. This reduction in band gap energy is attributed to the quantum confinement effect, structural defects, and possible impurity incorporation during the green synthesis process. The optical properties of ZnO nanoparticles make them suitable for photocatalytic applications, particularly in dye degradation, where efficient charge carrier generation and separation are essential.

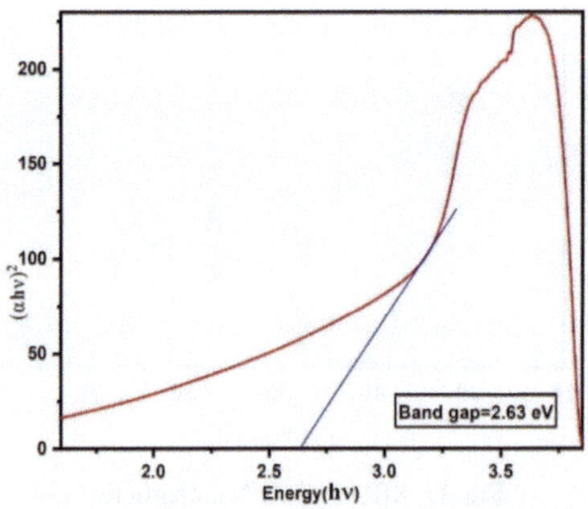

Fig. 2. UV-Visible plot of ZnO nanoparticles.

4 Conclusions

Environmental pollution remains a major concern due to the continuous release of industrial pollutants into ecosystems, particularly water sources. Effective remediation strategies should be environmentally sustainable, cost-effective, and free from hazardous byproducts. Green synthesis of nanoparticles presents a promising alternative, as it utilizes plant extracts or biological agents to produce stable and eco-friendly nanomaterials. This study focused on the green synthesis of ZnO nanoparticles and their structural and optical characterization. X-ray diffraction (XRD) analysis confirmed the successful synthesis of ZnO nanoparticles, with characteristic diffraction peaks observed at 2θ values of 31.72°, 34.40°, and 36.20°, which are in good agreement with standard ZnO patterns. Additionally, UV-Vis spectroscopy analysis revealed an estimated band gap of 2.63 eV, attributed to quantum confinement effects. These findings establish the structural purity and optical properties of the synthesized ZnO nanoparticles. Although this study did not explore the photocatalytic degradation of dyes, the results provide a fundamental understanding of the crystallinity and optical behavior of green-synthesized ZnO nanoparticles. Future research could extend these findings by investigating their potential applications in environmental remediation, including wastewater treatment. Developing cost-effective, stable, and recyclable metal oxide-based materials for large-scale environmental applications remains a key focus for sustainable nanotechnology.

Acknowledgments. I am grateful to SS Jain Jaipur for XRD. And Department of Physics for UV-Visible.

References

1. Khan, S., Mansoor, S., Rafi, Z., Kumari, B.: A review on nanotechnology: properties, applications, and mechanistic insights of cellular uptake mechanisms. J. Mol. Liq. **348**, 0167–7322 (2022)
2. Jeevanandam, J., Barhoum, A., Chan, Y.S., Dufresne, A., Danquah, M.K.: Review on nanoparticles and nanostructured materials: history, sources, toxicity and regulations. Beilstein J. Nanotechnol. **9**, 1050–1074 (2018)
3. Mobasser, S., Firoozi, A.A.: Review of nanotechnology applications in science and engineering. J. Civil Eng. Urbanism. **6**, 84–93 (2016)
4. Deepika, B., Neeta, R.S., Joginder, S., Rameshwar, S.K.: Biological methods for textile dye removal from wastewater: a review. Crit. Rev. Environ. Sci. Technol. **47**, 1836–1876 (2017)
5. Sharma, J., Sharma, S., Soni, V.: Classification and impact of synthetic textile dyes on aquatic Flora: a review. Reg. Stud. Mar. Sci. **45**, 101802 (2021)
6. Kishor, R., Purchase, D., Saratale, G.D., Saratale, R.G., Ferreira, L.F.R., et al.: Ecotoxicological and health concerns of persistent coloring pollutants of textile industry wastewater and treatment approaches for environmental safety. J. Environ. Chem. Eng. **9**(2), 105012 (2021)
7. Hu, H., Li, X., Wu, S., Yang, C.: Sustainable livestock wastewater treatment via phytoremediation: current status and future perspectives. Bioresour. Technol. **315**, 123809 (2020)
8. Tkaczyk, A., Mitrowska, K., Posyniak, A.: Synthetic organic dyes as contaminants of the aquatic environment and their implications for ecosystems: a review. Sci. Total Environ. **717**, 137222 (2020)

9. Alireza, S., Farid, M.: A novel and simple micro emulsion method for synthesis of biocompatible functionalized gold nanoparticles. J. Mol. Liq. **268**, 849–853 (2018)
10. Jishma, P., Thomas, R., Snigdha, S., Radhakrishnan, E.K.: Kinetic study of gold nanoparticle mediated photocatalytic degradation of Victoria blue. 3 Biotech. **8**(2), 97 (2018)
11. Antonio, F., Alessandro, M., Rita, M., Elisa, P., Enrico, B., Luca, M.: Green synthesis of Ag nanoparticles using plant metabolites. AIP Conf. Proc. **1873**, 020012 (2017)
12. Ashok, D., Raju, S., Rama, K.: Phytomediated synthesis of silver nanoparticles using Dicrostachys cinerea leaf extract and evaluation of its antibacterial and photocatalytic activity of textile dye. Int. J. Chem Tech Res. **10**, 302–314 (2017)
13. Goutam, S.P., Saxena, G., Singh, V., Yadav, A.K., Bharagava, R.N., Thapa, K.B.: Green synthesis of TiO_2 NPs using leaf extract of Jatropha curcas L. for photocatalytic degradation of tannery wastewater. Chem. Eng. J. **336**, 386–396 (2018)
14. Mobeen, A.A., Sundaram, R.: Green synthesis of TiO2 nanoparticles using orange peel extract for anti-bacterial, cytotoxicity and humidity sensor applications. Mater. Today Proc. **8**, 323–331 (2019)
15. Barkha, R., Sathyanarayanan, P., Niroj, K.S.: Polyol asserted hydrothermal synthesis of SnO_2 nanoparticles for the fast adsorption and photocatalytic degradation of methylene blue cationic dye. New J. Chem. **42**, 943–954 (2018)
16. Bhavana, S., Gubbiveeranna, V., Kusuma, C.G., Ravikumar, H., Sumachirayu, C.K., Nagabhushana, H., Nagaraju, S.: Facile green synthesis of SnO_2 NPs using Vitex altissima (L.) leaves extracts: characterization and evaluation of antibacterial and anticancer properties. J. Clust. Sci. **30**(2), 431–437 (2019)
17. Amrutha, S., Sridhar, S.: Green synthesis of magnetic iron oxide nanoparticle using leaves of Glycosmis Mauritiana and their antibacterial activity against human pathogens. J. Innov. Pharm. Biol. Sci. **5**, 22–26 (2018)
18. Bishnoi, S., Kumar, A., Selvaraj, R.: Facile synthesis of magnetic iron oxide nanoparticles using inedible Cynometra ramflora fruit extract waste and their photocatalytic degradation of methylene blue dye. Mater. Res. Bull. **97**, 121–127 (2018)
19. Mahich, S., Saini, Y.K., Devra, V., Aggarwal, K., Kumar, A., Kumar, D., Singh, A., Arya, Y.: Metal-free adsorption and photodegradation methods for methylene blue dye removal using different reduction grades of graphene oxide. Heliyon. **10**(11), e31702 (2024). https://doi.org/10.1016/j.heliyon.2024.e31702
20. Fuku, X., Thovhogi, N., Maaza, M.: Photocatalytic effect of green synthesized CuO nanoparticles on selected environmental pollutants and pathogens. AIP Conf. Proc. **1962**, 040006 (2018)
21. Yadav, M., Yadav, D.K., Dolia, S.N., Lal, C.: Improved hydrogen de/absorption kinetics of magnesium (Mg) via addition of catalyst Y_2O_3. Orient. J. Chem. **40**(5), 1470 (2024). https://doi.org/10.13005/ojc/400530
22. Singh, A., Kumar, D., Khanna, P.K., Kumar, M.: Reduction in point defects of sol–gel derived ZnO thin films with oxygen ambient. Mater. Lett. **183**, 365–368 (2016). https://doi.org/10.1016/j.matlet.2016.07.055
23. Ahmar, R.M., Mohammad, O., Ravikant, R., Faraz, A., Nazoora, K., Swaleha, Z.: Bio mimetically synthesized ZnO nanoparticles attain potent antibacterial activity against less susceptible S. aurous skin infection in experimental animals. RSC Adv. **7**(58), 36361–36373 (2017)
24. Aminuzzaman, M., Ying, L.P., Goh, W.S., Watanabe, A.: Green synthesis of zinc oxide nanoparticles using aqueous extract of Garcinia mangostana fruit pericarp and their photocatalytic activity. Bull. Mater. Sci. **41**, 50 (2018)
25. Balaji, S., Kumar, M.B.: Facile green synthesis of zinc oxide nanoparticles by Eucalyptus globulus and their photocatalytic and antioxidant activity. Adv. Powder Technol. **28**, 785–797 (2017)

26. Mengstu, E.A., Gebrekidan, M.T., Gebretinase, Y.N., Tesfakiros, W.G.: Structural, optical, and photocatalytic activities of Ag-doped and Mn-doped ZnO nanoparticles. J. Nanomater. **1220**, 9425938 (2018)
27. Adinaveen, T., Karnan, T., Selvakumar, S.A.S.: Photocatalytic and optical properties of NiO added Nephelium lappaceum L. peel extract: an attempt to convert waste to a valuable product. Heliyon. **5**(5), e01751 (2019)
28. Kumar, A., Singh, A., Kumar, D., Garg, A., Dutta, V., Swami, S.K., Chaturvedi, N., Kumar, M.: Spray deposited carbon nanotube embedded ZnO as an electron transport layer in inverted organic solar cells. Hybrid Adv. **4**, 100088 (2023). https://doi.org/10.1016/j.hybadv.2023.100088
29. Behbahani, A., Rowshanzamir, S., Esmaeilifar, A.: Hydrothermal synthesis of zirconia nanoparticles from commercial zirconia. Proc. Eng. **42**, 908–917 (2012)
30. Anitha, R., Ramesh, K.V., Sudheerkumar, K.H.: Photocatalytic activity of CeO2 nanoparticles, synthesis using Artocarpus Gomezianus fruit mediated facile green combustion method. Int J Pharm. Bio. Sci. **8**, 933–936 (2017)
31. Fatemeh, J., Mohammad, E.T.Y., Mohsen, B., Ali, E.H.: Biosynthesis, characterization of cerium oxide nanoparticles using Ceratonia siliqua and evaluation of antioxidant and cytotoxicity activities. Mater. Res. Express. **6**, 0654 (2019)
32. Singh, A., Joshi, B.C., Kumar, D., Khanna, P.K., Kumar, M.: Effect of band gap offset on ZnO (Cd, Mg)/ZnO:P multi-quantum well light-emitting diodes: a simulation study. Energy Environ. Focus. **3**, 1–4 (2014). https://doi.org/10.1166/eef.2014.1095

The Structural and Optical Properties of ZnS_XTe_{1-X}

R. K. Jhakal[1,2(✉)] and M. D. Sharma[1]

[1] Department of Physics, Govt Dungar College, Bikaner, Rajasthan, India
jhakalraj@gmail.com

[2] Department of Physics, SBD Govt College, Sardarshahar, Rajasthan, India

Abstract. The Structural and optical properties of ternary II-VI ternary alloys are calculated. The properties are calculated by merging the Epm with Harrison bond orbital model. The approximates of elastic constants are calculated following the empirical pseudopotential method. Elastic constants are calculated by relating with the Harrison bond orbital model. The refrective index, static and high frequency dielectric constants are reported to reveal optical behavior of alloys. The refractive indices (n) are determined from moss formula, which is related to ε_∞ to ε_0. The properties of the material based on the Zn: Zn S Te were calculated. The Concept of pseudopotentials was introduced by Fermi to study high lying atomic states. Pseudopotential be used for calculating the energy levels of the alkali metals. Alloying effects are considered through modified virtual crystal approximation. Calculations are performed for semiconducting alloys ZnS_XTe_{1-X} ($0 < x < 1$).

Keywords: Structural Properties · Optical Properties · Semiconducting alloys

1 Introduction

The empirical pseudopotential method is generally used in computing the structure and optical properties of semiconductor materials [1]. The II–VI mixed-anion semiconductor alloys have been studied relatively less than the mixed-cation alloys [2]. II–VI semiconductors have significant scientific applications in manufacturing photonic devices, detecting systems for environmental pollution, and color arrangement modules [3]. The II–VI wide-bandgap semiconducting ternary alloys exhibit better operational stability and increased lifetime in optoelectronic devices used in the blue-green to UV region [4]. To investigate the suitability of materials exhibiting minimal lattice mismatch and other parameters, it is essential to study the electronic structure and bonding behavior. The bulk modulus and elastic constants are crucial structural parameters for semiconductor materials [5]. Semiconductor alloys provide an effective way of adjusting the bandgap to improve and enhance the applications of semiconductor devices. The II–VI semiconductor alloys have enabled new generations of device applications over the past few years. III–V and II–VI semiconductors typically crystallize in one of two structures: hexagonal wurtzite (WZ) or cubic zinc blende (ZB). The ZB and WZ structures share the same local tetrahedral coordination but begin to differ only in their third-nearest-neighbor atomic arrangement. Despite the structural similarities, there are significant differences in their electronic and optical properties [6, 7].

S. Kumari et al. (Eds.): ICMTP 2024, SPPHY 311, pp. 28–32, 2025.
https://doi.org/10.1007/978-981-96-6795-6_4

II–VI alloys are currently being extensively investigated through various theoretical approaches. In particular, the kinetic properties of CdTe have been recently studied using first-principles calculations within the projector-augmented wave (PAW) formalism [8]. The electronic, phonon, optical, and thermodynamic properties of the CdTe crystal were analyzed based on density functional theory (DFT) using the pseudopotential method [9]. The same theoretical approach has been successfully applied to study the electronic structure of CdSeTe solid solutions [10].

Ternary semiconductor alloys ZnS_XTe_{1-X} have attracted considerable attention due to their efficient light emission and widely tunable direct bandgap. At room temperature, ZnS_XTe_{1-X} ($0 < x < 1$) exhibits direct bandgaps in the range of 2.26 eV–3.60 eV, making the alloy a promising candidate for green and blue light-emitting materials. Additionally, the highly mismatched ZnS_XTe_{1-X} alloy with a sulfur composition of $0 < x \leq 0.2$ is particularly noteworthy because its electronic and optical properties are primarily influenced by an anticrossing interaction between the energy states of the minority sulfur species and the conduction band of the host ZnTe semiconductor. However, studies on ZnS_XTe_{1-X} are limited, as high-sulfur-content alloys are challenging to synthesize.

According to the literature, ZnTe and ZnS are compatible in forming ZnS_XTe_{1-X} bulk single crystals over only a limited composition range [11]. II-VI alloys have various applications, particularly in the development of sensing devices for imaging systems and multi-energy X-ray radiography. The narrow solubility of elements poses a challenge in producing ZnS_XTe_{1-X} solid solutions. Among different II-VI alloys, ZnS_XTe_{1-X} solid solutions are especially interesting because their components exhibit a broad solubility range [12, 13]. The synthesis of ZnS_XTe_{1-X} mixed crystals is challenging due to the limited miscibility of their constituent elements and can only be achieved at 8–10% concentrations, as dictated by the ZnS_XTe_{1-X} phase diagram [14]. Ternary semiconductor alloys ZnS_XTe_{1-X} have garnered significant interest due to their potential for efficient light emission and widely tuneable direct bandgap. At room temperature, ZnS_XTe_{1-X} ($0 < x < 1$) exhibits direct bandgaps in the range of 2.26 eV–3.60 eV [15].

2 Methods and Calculations

The Elastic constant C_{11}, C_{12} are calculated using the same procedure as that used by Bouarissa and Kassali [16] based essentially on Baranowski [17]

$$C_{11} = \frac{(\sqrt{3})}{4d^5}\left[4.37\,\frac{\hbar^2}{m}\,(5+\lambda)\left(1-\alpha_p^2\right)^{3/2} - 0.6075\frac{\hbar^2}{m}\left(1-\alpha_p^2\right)^{1/2}\right] \quad (1)$$

$$C_{12} = \frac{(\sqrt{3})}{4d^5}\left[4.37\,\frac{\hbar^2}{m}\,(3-\lambda)\left(1-\alpha_p^2\right)^{\frac{3}{2}} + 0.6075\frac{\hbar^2}{m}\left(1-\alpha_p^2\right)^{1/2}\right] \quad (2)$$

Where $d = \frac{\sqrt{3}}{4}a$ is the nearest-neighbour distance (a is the lattice constant) and is dimensionless parameter with constant value of 0.738, m is the mass of electron ħ is the plank constant and α_p is the polarity that is obtained according to the vogal definition [18] *h* elastic constant C_{44} is calculated by valence force field model [18, 19] as

$$C_{44} = \frac{3(C_{11} + 2C_{12}\ )(C_{11} - C_{12})}{(7C_{11} + 2C_{12})} \quad (3)$$

The bulk modules (B_s) and the shear wave modulus (C_s) are given by [19]

$$B_s = \frac{(C_{11} + 2C_{12})}{3} \quad (4)$$

$$C_s = \frac{(C_{11} - C_{12})}{2} \quad (5)$$

The bandgap is calculated [20], the optical properties such as refractive index and dielectric constants are computed. The wavelength independent refractive index (n) is calculated by band gap (E_g) using the mass's formula [21] given as.

$$n^4 = \frac{E_g}{95}\ (Ev) \quad (6)$$

The high frequency dielectric constant ε_∞ can be obtained using the following relation $\varepsilon_\infty = n^2$ according to the Harrison bond orbital model the static dielectric constant ε_0 is directly related to ε_∞ as [22].

$$n^4 = \frac{E_g}{95}\ (Ev) \quad (7)$$

α_p is the polarity and αc is the covalency of the material denied as

$$\alpha_c{}^2 = 1 - \alpha_p{}^2 \quad (8)$$

The calculated elastic constant C_{11}, C_{12}, C_{44}, Bulk Modulus Bs, Shear modulus Cs, refractive index(n) (Table 1).

Table 1. Various elastic constants for different alloy composition.

X	Alloys	C_{11}	C_{12}	C_{44}	B_S	C_S
0.0	ZnTe	95.9223	41.2302	38.8217	59.4609	27.3460
0.1	$ZnS_{0.1}Te_{0.9}$	102.0868	43.86756	41.32134	63.27399	29.10964
0.2	$ZnS_{0.2}Te_{0.8}$	108.6504	46.67656	43.98238	67.33451	30.98693
0.3	$ZnS_{0.3}Te_{0.7}$	115.6518	49.67373	46.8206	71.66642	32.98904
0.4	$ZnS_{0.4}Te_{0.6}$	123.1328	52.87692	49.85295	76.29554	35.12792
0.5	$ZnS_{0.5}Te_{0.5}$	131.1385	56.30550	53.09778	81.24984	37.41651
0.6	$ZnS_{0.6}Te_{0.4}$	139.7183	59.98052	56.57504	86.55978	39.8689
0.7	$ZnS_{0.7}Te_{0.3}$	148.9260	63.92501	60.30654	92.25866	42.50047
0.8	$ZnS_{0.8}Te_{0.2}$	158.8203	68.16421	64.31617	98.38292	45.32806
0.9	$ZnS_{0.9}Te_{0.1}$	169.4662	72.72584	68.63015	104.9726	48.37018
1.0	ZnS	180.9348	77.64044	73.27737	112.0719	51.64719

The calculated refractive index(n), The high frequency dielectric constant ε_∞ the static dielectric constant ε_0 as shown in Table 2, (Fig. 1).

Table 2. Various refractive index values for different composition.

X	Alloys	n	ε_∞	ε_0
0.0	ZnTe	2.598697655	6.753229504	6.926405
0.1	$ZnS_{0.1}Te_{0.9}$	2.499478853	6.247394538	6.386987
0.2	$ZnS_{0.2}Te_{0.8}$	2.385652818	5.691339368	5.801882
0.3	$ZnS_{0.3}Te_{0.7}$	2.26389448	5.125218215	5.211481
0.4	$ZnS_{0.4}Te_{0.6}$	2.140052345	4.579824037	4.646365
0.5	$ZnS_{0.5}Te_{0.5}$	2.0187841	4.075489244	4.126372
0.6	$ZnS_{0.6}Te_{0.4}$	1.903631652	3.623813468	3.662496
0.7	$ZnS_{0.7}Te_{0.3}$	1.796435014	3.22717876	3.256463
0.8	$ZnS_{0.8}Te_{0.2}$	1.698602248	2.885249599	2.905869
0.9	$ZnS_{0.9}Te_{0.1}$	1.610586319	2.593988292	2.610689
1.0	ZnS	1.532295702	2.347930118	2.360543

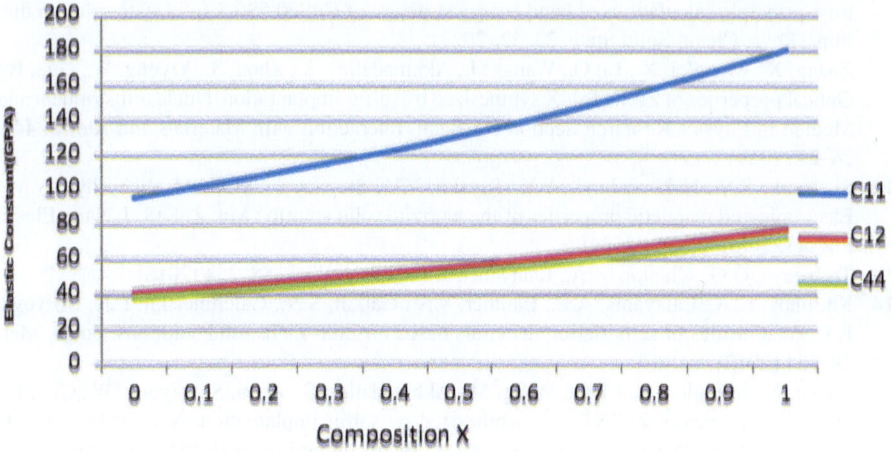

Fig. 1. Various elastic constants for different alloy composition.

3 Conclusion

The structural and optical property are performed for alloy ZnS_XTe_{1-X} with a combination of x with value from 0 to 1 using the empirical pseudopotential under the virtual crystal approximation combined with the Harrison bond orbital model. Our result are predictions and may serve as a reference.

References

1. Chelikowsky, J.R., Cohen, M.L.: Phys. Rev. B. **14**, 556 (1976)

2. Swarnkar, C.B., Pandya, R.K., Paliwal, U., Patel, N.N., Joshi, K.B.: Study of charge density, density of state and electron momentum density of ZnSxSe1-x semiconductor alloys. Chalcogenide Lett. **6**, 137–148 (2009)
3. Swarnkar, C.B., Paliwal, U., Patel, N.N., Joshi, K.B.: Electronic properties and charge density of $Be_xZn_{1-x}Te$ alloys. Bull. Mater. Sci. **34**(3), 499–506 (2011)
4. Adachi, S.: Properties of Semiconductor Alloys Group IV, III-V and II-VI Semiconductors. Wiley, West Sussex (2009)
5. Sharma, M.D.: Electron momentum dencity and elastic properties of Be $_x$ Zn$_{1-x}$ Se alloys. In: Proceedings of the 54 th DAE Solid State Physics Symposium (2009)
6. Yeh, C.-Y., Wei, S.-H., Zunger, A.: Relationship between the band gaps of the zinc-blende and wurtzite modification of semiconductors. Phys. Rev. B. **50**, 2715 (1994)
7. van Schlfgaarde, M., Sher, A., Chen, A.-B.: Theory of AIN, GaN, InN and their alloys. J. Crystal Growth. **178**(1–2), 8–31 (1997)
8. Malyk, O.P., Syrotyuk, S.V.: Heavy hole scattering on intrinsic acceptor defects in cadmium telluride, calculation from the first principles. Phys. Chem. Solid State. **23**(1), 89 (2022)
9. Ilchuk, H.A., Nykyruy, L.I., Kashuba, A.I., Semkiv, I.V., Solovyov, M.V., Naidych, B.P., Kordan, V.M., Deva, L.R., Karkulovska, M.S., Petrus, R.Y.: Electron, phonon, optical and thermodynamic properties of CdTe crystal calculated by DFT. Phys. Chem. Solid State. **23**(2), 261 (2022)
10. Kashuba, A.I., Andriyevsky, B., Semkiv, I.V., Ilchuk, H.A., Petrus, R.Y., Storozhuk, Y.M.: First-principle calculations of band energy structure of CdSe0.5S0.5 solid state solution thin films. Phys. Chem. Solid State. **23**, 52 (2022)
11. Zhang, X., Menglei, X., Li, Q., Wang, M., Akhmadaliev, S., Zhou, S., Yiyong, W., Guo, B.: Optical properties of ZnSXTe1-X synthesized by sulfur implantation. Nuclear Instrument and Method in Physics Research Section B : Beam Interaction with Materials and Atoms. **442**, 24–27 (2019)
12. Syrotyuk, S.V., Nakonechnyi, A.Y., Klysko, Y.V., Stepanyak, M.V., Myshchyshyn, V.M.: Electronic and magnetic properties of the wurtzite solid solution Mn: ZnSeS. J. Nano Electr. Physics. **16**(5), 05–33 (2024)
13. Trubaieva, O.G., Chaika, M.A., Lalayants, A.I.: Lith. J. Phys. **58**, 254 (2018)
14. Katrunov, K.A., Lalayants, A.L., Baumer, V.N., Galkin, S.N., Galchinetskii, L.P., Brilyova, E.J.: Peculiarities of scintillation materials based on ZnS–ZnTe solid solutions. Funct. Mat. **20**, 384 (2013)
15. Zhang, X., Menglei, X., Li, Q., Wang, M., Akhmadaliev, S., Zhou, S., Yiyong, W., Guo, B.: Optical properties of ZnSXTe1-X synthesized by sulfur implantation. Nuclear Instruments and Methods in Physics Research Section B: Beam Intractions with Materials and Atoms. **442**(1), 24–28 (2019)
16. Bouarissa, N., Kassali, K.: Mechnical properties and elastic constants of zinc-blende $Ga_{1-x}In_xN$ alloys Phys. Stat. Sol. B. **228**, 663–670 (2001)
17. Baranowski, J.M.: Bon lengths, force constants, and local impurity distortions in semiconductors. J. Physics C: Solid State Physics. **17**, 6287–6302 (1984)
18. Vogal: J. Physics C: Solid State Phys. **11**, 251 (1978)
19. Levinshtein, M., Rumyantsev, S., Shur, M.: Handbook Series on Semiconductor Parameters, vol. 2. World Scientific (1999)
20. Adachi, S.: Properties of Semiconductor Alloys, vol. 166. Wiley (2009)
21. Moss, T.S.: Proc. Phys. Soc. London B. **63**, 167–174 (1950)
22. Oussaifi, Y., Fredj, A.B., Debbichi, M., Bouarissa, N., Said, M.: Semicond. Sci. Technol. **22**, 641–646 (2007)

Synthesis and Characterization of Carbides for Fuel Cells

Akhilesh Kumar[1,2,3] and Anju Dixit[1](✉)

[1] Department of Physics, School of Basic Sciences, CSJM University, Kanpur 208024, India
anjudixit@csjmu.ac.in
[2] Narain College Shikohabad, Firozabad, India
[3] DBRA University Agra, Agra 283135, India

Abstract. Titanium carbide exhibits exceptional electrical conductivity (1.5 × 10^4 S/cm) and thermal stability, along with impressive mechanical properties, including hardness (28–35 GPa). This makes it a promising candidate for structural materials in next-generation fuel cell electrodes. Utilizing TiC as a support to reduce platinum loading presents a significant opportunity for cost savings in fuel cell technology. Nanosized titanium carbide powders were produced through a sol-gel process. This process facilitates the production of extremely uniform materials with precise control over composition and structure at the molecular scale, leading to more reliable properties. Also, the sol-gel process may be more economically viable than conventional high-temperature synthesis techniques, especially for large-scale production. The precursor gels (Ti-O-C) were synthesized using titanium tetra isopropoxide, sucrose ($C_{12}H_{22}O_{11}$), and acetic acid (CH_3COOH). The produced xerogels were then subjected to calcination in a consistent argon environment at temperatures range from 800 °C to 1350 °C, maintaining this condition for 1 h. A range of analytical methods, such as TGA-DSC, XRD, SEM, and cyclic voltammetry, were utilized to characterize this during the entire process. The cyclic voltammetry result refers to the material's ability to maintain its electrochemical performance over multiple charge-discharge cycles without significant degradation. The TiC nanoparticles produced were approximately 45 nm in size. The formation of TiC structure was confirmed by XRD, alongside trace amounts of TiO_2 and free carbon, with fluctuations in oxygen and carbon content influenced by the synthesis parameters.

Keywords: Nano Carbides · TGA · DSC · XRD · SEM · cyclic voltametry

1 Introduction

Transition metal (TM) carbides display unique bonding properties when there is a simultaneous presence of covalent, ionic, and metallic interactions. The metallic linkage enhances conductivity, whereas the ionic interactions contribute to improved thermal stability. Additionally, the covalent interactions at the surface enhance adsorption characteristics, rendering it appropriate for catalytic applications. Consequently, TM-carbides demonstrate exceptional mechanical properties, including hardness, in conjunction with

S. Kumari et al. (Eds.): ICMTP 2024, SPPHY 311, pp. 33–44, 2025.
https://doi.org/10.1007/978-981-96-6795-6_5

elevated electrical conductivity and thermal stability [1, 2]. They can also act as the building material for fuel cell electrodes in the future. Reducing Pt loading with TM-Carbides as supports presents a great opportunity to decrease fuel cell costs. Among these carbides Titanium carbide (TiC) is highly stable under fuel cell operating conditions (high temperature and corrosive environments). In contrast, platinum can degrade over time due to sintering or poisoning by impurities in the fuel or air supply. Platinum, a precious metal, is a vital element in the catalysts of proton exchange membrane (PEM) fuel cells, but it's not readily available. By reducing the amount of platinum needed in the fuel cell design, TiC can help minimize the environmental impact of mining, refining, and processing platinum. Platinum mining is energy-intensive and often associated with significant environmental degradation. By using TiC as a support material, we decrease the need for platinum extraction, which helps conserve this resource and reduces the environmental footprint associated with mining. So, TiC is more resistant to degradation, making it an attractive long-term alternative. TiC can be used as a promising material for fuel cell electrodes, solid oxide fuel cells (SOFCs), and other types of electrochemical devices. TiC also exhibits a high conductivity, ranging from 300 to 800 S/m [3]. Characteristics such as excellent chemical stability, resistance to oxidation, resistance to deformation and cavitation at interfaces, medium volatility, sufficient toughness at room temperature, and thermal shock resistance are present in common transition metal carbides like TiC [4]. TiC provides superior electrical conductivity compared to many carbon-based materials and is an ideal support material for enhancing the performance of catalysts.

Due to these attributes, TiC is a valuable man-made engineering substance that can be used as a strengthening component in super alloys and composites [5, 6]. Titanium carbide (TiC) is remarkable carbide known for its existing properties, including a high melting point, remarkable hardness, able to resist wear and corrosion better and having outstanding electrical and thermal conductivity [4–8]. These attributes classify it as an ultra-high-temperature ceramic. The distinctive properties of TiC render it exceptionally valuable for various applications. Furthermore, when employed as a retentive element in composite materials, TiC may greatly enhance the characteristics of the matrices [9, 10]. Titanium carbide has been produced conventionally through sintering processes [11–19], either alone or in addition with other techniques like mechanical alloying [11], thermal plasma processing [12], liquid magnesium reduction of titanium tetrachloride ($TiCl_4$) and carbon tetrachloride (CCl_4) [13–15], gas phase reaction of $TiCl_4$ and gaseous hydrocarbons [16], direct chemical reaction of Titanium (Ti) element and carbon (C), [15, 18] Carbon and titanium oxide (TiO_2) undergo carbothermal reduction in a controlled environment, such as a vacuum or inert atmosphere [19, 20].

The sol-gel method, when integrated with carbothermal reduction, offers an enhanced approach for the fabrication of carbide ceramics that exhibit high purity and a fine grain structure at comparatively low temperatures, differing from the methods previously examined. Additionally, the sol-gel approach serves an essential function in determining the composition of the final product, and the overall process is practical. The ability to mix the initial components at the molecular or atomic level allows the sol-gel process to attain a significant level of consistency in the chemical composition of the final product. The heating of the precursor during the process leads to the creation of the final products. The extensive surface area of the reactants present in the precursor allows the reaction to take place at considerably reduced temperatures, leading to a final product that is typically characterized by a fine particle size and high purity [21, 22]. In our research, we are focusing on studying the chemical processing methods used to create nanoparticles of refractory carbides that have a uniform particle size distribution.

2 Synthesis and Experimental Details

2.1 Starting Materials

Titanium tetra isopropoxide TTIP ($Ti[OCH(CH_3)_2]_4$, Thermo Scientific, 95%), acetic acid (CH_3COOH, Sisco Research Lab, 99.5%), and sucrose ($C_{12}H_{22}O_{11}$, AnalaR, BDH, 98%) were employed as the initial materials. In this study, acetic acid served as both a chelating agent and a chemical modifier. Nanosized titanium carbide was synthesized through a straightforward sol-gel method, utilizing TTIP as the precursor, acetic acid as the chemical modifier, and sucrose as the carbon source.

2.2 Readiness of Ti-O-C Precursor

In a typical method, the molar proportions of Titanium (IV) isopropoxide (Ti $[OCH(CH_3)_2]_4$), acetic acid (CH_3COOH), and sucrose ($C_{12}H_{22}O_{11}$) were established at 1:1:4.8. The molar ratio of TTIP is typically optimized based on prior studies and experiments to ensure complete conversion of titanium to TiC while avoiding excess titanium or unreacted precursor. The acetic acid concentration is often chosen to balance the rate of hydrolysis and condensation of TTIP, which has been optimized in many studies for the sol-gel synthesis of titanium-based materials. The ratio of sucrose is optimized based on prior studies, where the amount of carbon required to react with titanium precursors and form TiC is determined. This also ensures that the desired stoichiometry for TiC formation is maintained.

The preparation of the Ti-O-C precursor was conducted in the following manner. First, to achieve the desired molar ratio of carbon to titanium, a specific quantity of sucrose was dissolved in acetic acid while being magnetically stirred at a temperature of 80 °C with the aid of a Thermostat Magnetic stirrer. This stirring process continued until the solution developed a light-yellow color, ensuring homogeneity. Initially, the yellow color is due to titanium in the form of titanium oxide (TiO_2) or titanium hydroxide complexes formed during hydrolysis and condensation.

While stirring vigorously, TTIP was added drop by drop, maintaining a constant temperature of 80 °C. Throughout this active process, the solution changed color from clear to a yellow tint upon the first addition of TTIP, eventually becoming brown after a few minutes of mixing. This change in color signifies the hydrolysis and condensation of TTIP.As the reaction proceeds, sucrose decomposes, introducing carbon into the system. The interaction between titanium and carbon leads to the formation of carbon-containing species or titanium oxycarbide, turning the material dark brown. When the precursor undergoes further heating, carbonization occurs, and titanium carbide (TiC) is formed, which is typically associated with a darker color, marking the complete transformation to TiC. So, after several hours of uninterrupted stirring, a consistent dark brown sol was achieved. This color change is a visible indicator of the complex chemical transformations happening during the precursor preparation, including the formation of the titanium-carbon bond and the evolution of TiC from titanium oxide.

The precursor sol was dried in an oven at a temperature of 110 °C for about 24 h. During this period, the evaporation of water and various organic solvents transformed the solution into a gel, ultimately leading to the formation of xerogels. These xerogels were subsequently ground into powders, referred to as the Ti-O-C precursor, in a mortar to facilitate the next stages of the process. Subsequently, carbothermal reduction was carried out in the temperature ranging from 800 °C to 1200 °C. The heating rate was kept at 5 °C/min, and the samples were kept at the maximum temperature for 1 h. The synthesis procedure is depicted in Fig. 1.

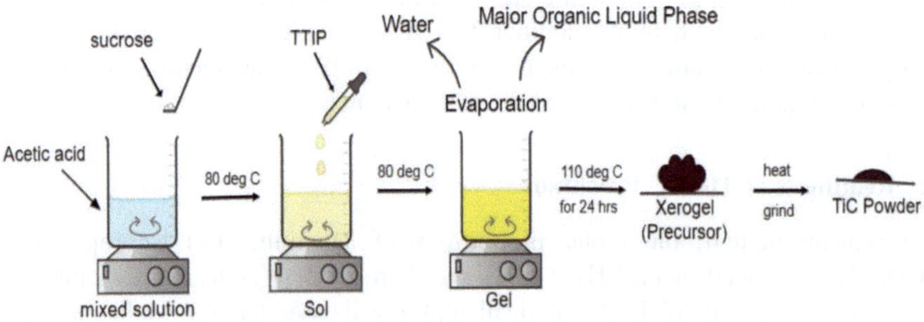

Fig. 1. Synthesis of Titanium Carbide by Sol-gel Method.

2.3 Experimental Details

To investigate the reactions occurring in the precursor during the heating process, thermos-gravimetric analysis (TGA) as well as the differential scanning calorimetry (DSC) measurements were done in the temperature range 25–1100 °C in a pure nitrogen atmosphere, employing a heating rate 10 °C/min via a simultaneous thermal analyzer (Tolendo-Mettler, USA). The phase composition and crystallinity were evaluated through powder X-ray diffraction (X'pert Pro PANalytical, USA), which operated at 40.0 kV and 40.0 mA (Cu Kα radiation, $\lambda = 1.540598$ Å). XRD data were gathered over a 2θ range of 10.006 to 100.001 degrees (°) with a step size of 0.0131303°. The crystallite size was determined from the full width at half maximum (FWHM) using the Debye-Scherrer equation. Additionally, the microstructure and morphology were examined using a scanning electron microscope (FESEM), specifically the FEI Nova NanoSEM230. The electrochemical properties of titanium carbide were assessed using a cyclic voltameter.

3 Result and Discussion

3.1 Thermal Analysis

To comprehend the prepared material's behavior and chemical reactions during heat treatment, the simultaneous thermal analysis curves TGA-DSC are presented in Fig. 2. The weight loss processes and their estimations are illustrated in Fig. 3. The entire weight loss process can be categorized into four distinct stages according to the TGA curve. The initial stage occurs at temperatures below 200 °C, where a weight loss of 9.84% is recorded, alongside an endothermic peak at 111 °C. During this phase, only residual water and volatile organic compounds are released. The next range, spanning from 200 °C to 550 °C, exhibits a significant weight loss of 42.28%, which is primarily due to the decomposition of sucrose into amorphous carbon and water.

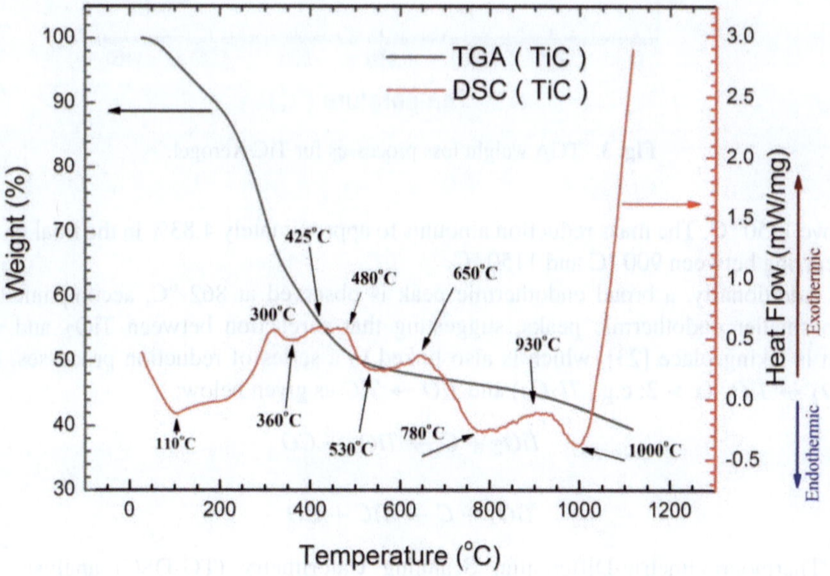

Fig. 2. TGA-DSC of Xerogel powders of TiC under nitrogen atmosphere.

The DSC curve observed in this interval exhibits five prominent exothermic peaks alongside five smaller endothermic peaks. The exothermic peaks are primarily attributed to the thermolysis of sucrose, residual organic substances, and the crystallization of TiO2. In contrast, the endothermic peaks are associated with the release of structural water. The third stage, which spans from 550 °C to 800 °C, indicates a minor weight loss of 3.52%. The observed curve in this range displays two exothermic peaks and two endothermic peaks. Relatively broader exothermic peak at about 640 °C is attributed to the carbonization of glucose, as indicated in the corresponding equation: $C_6H_{12}O_6 \rightarrow 6C + 6H_2O$. It can be noted that during this phase, there is a transformation in the crystallinity and crystal structure of TiO_2, with minimal involvement of other reactions [23]. The weight keeps decreasing once the temperature exceeds 800 °C, eventually stabilizing

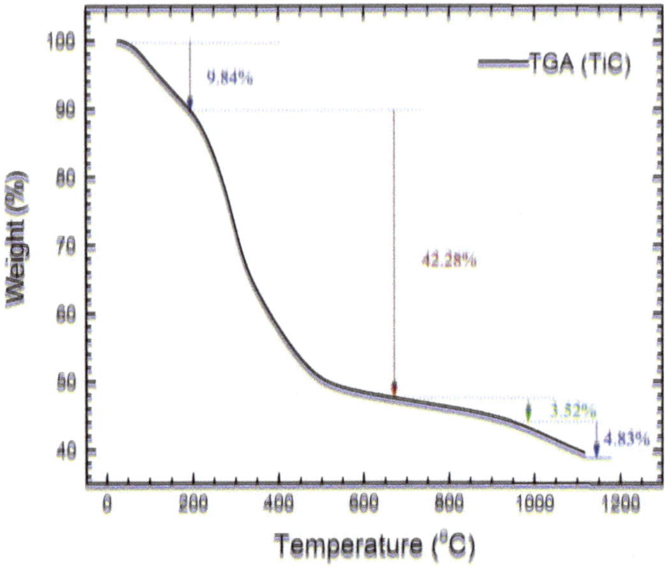

Fig. 3. TGA weight loss processes for TiC Xerogel.

above 1150 °C. The mass reduction amounts to approximately 4.83% in the final phase, occurring between 900 °C and 1150 °C.

Additionally, a broad endothermic peak is observed at 862 °C, accompanied by two smaller endothermic peaks, suggesting that a reaction between TiO_2 and carbon is taking place [23], which is also linked to a series of reduction processes, like $TiO_2 \rightarrow TiO_x$ ($x > 2$; e.g., Ti_2O_3) and $TiO \rightarrow TiC$ as given below:

$$TiO_2 + C \rightarrow TiO_x + CO \quad (1)$$

$$TiO_x + C \rightarrow TiC + CO \quad (2)$$

In Thermogravimetric-Differential Scanning Calorimetry (TG-DSC) analysis, the observed weight loss stages provide valuable insights into the thermal behavior and chemical transformations occurring during the heating of the material, such as titanium carbide (TiC). Each weight loss stage corresponds to a particular physical or chemical process that can impact the final TiC product. Here's a breakdown of these stages and their significance:

Initial Weight Loss (Low-Temperature Stage). This stage typically involves the loss of adsorbed water, residual solvents, or low-molecular-weight organic compounds. For TiC, this could include the removal of surface moisture or volatile impurities. The presence of such impurities can affect the purity of the final TiC product. If not removed during this stage, these volatile substances could lead to contamination or undesired phases in the final carbide.

Mid-Temperature Weight Loss. In some cases, this weight loss is related to the decomposition of organic binders or precursors used in the synthesis of TiC (e.g., during the synthesis of TiC from titanium-based precursors). It might also correspond to the loss of carbon or oxygen-containing species. This stage is crucial for ensuring that the carbon content required to form TiC is not lost prematurely or insufficiently. Incomplete removal of organic precursors could lead to the formation of undesirable byproducts or alter the stoichiometry of TiC, potentially impacting its properties.

High-Temperature Weight Loss or Stability. At higher temperatures, weight loss may be minimal, indicating the stability of TiC formation. However, if the weight continues to decrease, it could suggest the loss of excess carbon or the reduction of titanium oxide (TiO_2) to form TiC. This may also indicate decomposition reactions that could be linked to the final carbide formation.

This stage is critical for controlling the precise stoichiometry of the TiC product. If there's too much weight loss due to excess carbon, the carbide might be substoichiometric (e.g., TiC_x where $x \ll 1$), leading to a product with lower titanium content or altered crystalline structure. On the other hand, insufficient weight loss or an incorrect balance could lead to the formation of other titanium phases, such as TiO_2 or Ti_2C, which may affect the overall material properties.

3.2 Structural Analysis

Figure 4 illustrates the diffraction patterns of titanium carbide-based powder samples that were reduced at temperatures of 800 °C, 1000 °C, and 1200 °C, without the presence of an Ar/N_2 atmosphere, and with a holding time of 1 h. The diffractograms for all samples display sharp and intense peaks, primarily indicating the presence of TiO_2 and titanium sub-oxides, including TiO and $Ti_{.992}O_2$, with variations in phase content. Due to the intense nature of the diffraction peaks, key rutile TiO_2 peaks are labeled R(hkl) and the same was not possible to for each of the phases observed. Notably, there is no evidence of the formation of the Magneli phases Ti_2O_3, Ti_3O_5, and Ti_4O_7 after thermal treatment at the specified temperatures. The TiC sample reduced at 800 °C predominantly exhibits the rutile phase (TiO_2) [JCPDS card no. 01-078-1510], with a crystallite size of 35.81 nm. Other phases present include hongquiite (TiO) [JCPDS card no. 00-029-1361] and graphite (C) [JCPDS card no. 00-032-0567]. The peak at $2\theta =$ 90.79° (*400*) corresponds to TiC—khamrabaevite TiC [JCPDS card no. 00-032-1383], while TiO is associated with peaks at 36.18° (*111*) and 90.79° (*400*). The TiC sample reduced at 1000 °C also shows rutile as the dominant phase (crystallite size = 20.73 nm), along with khamrabaevite TiC, hongquiite (TiO), and brookite (TiO_2) [JCPDS card no. 00-002-0514].

Notably, graphite (C) was not detected in the TiC-1000 °C sample, indicating that the carbon element contributed to the reduction of rutile, resulting in additional TiC peaks at $2\theta = 72.51°$ (*311*) and 90.79° (*400*). The lattice constants for the TiC phase in the TiC-800 °C and TiC-1000 °C samples were measured at 0.4327 nm and 0.4323 nm, respectively, showing a decrease with increasing temperature. Using the XRD data, lattice constant (*a*) and crystallite sizes (*D*) were calculated using Debye-Scherrer equation [24] and the values are given in Table 1. At higher temperatures, the nucleation rate may surpass the growth rate of the crystallites. This is because due to elevated temperatures

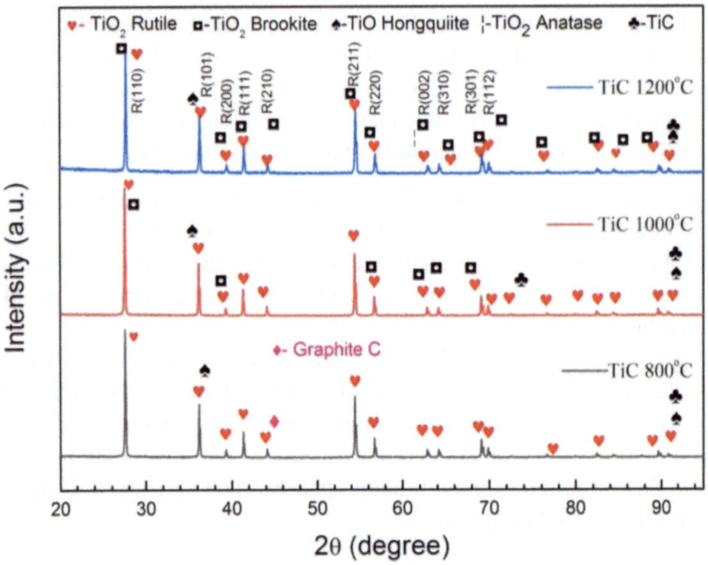

Fig. 4. XRD patterns of TiC nanopowders reduced at temperatures 800, 1000, and 1200 °C.

can increase the energy available for nucleation and rapid nucleation leads to the formation of many small crystallites rather than a few large ones. The diffraction peaks corresponding to the brookite phase were found.

Table 1. Peak positions, indices, lattice constants and crystallite sizes of TiC nanopowders reduced at 800, 1000, and 1200 °C.

Sample	Peak Position 2θ (degree)	(*h k l*)	Lattice Constant (*nm*)		Crystallite Size (*nm*)	
			a	Average	*D*	average
TiC-800 °C	90.7964	*(4 0 0)*	0.4327	0.4327	97.80	97.80
TiC-1000 °C	72.5144	*(3 1 1)*	0.4319	0.4323	66.7	81.80
	90.7900	*(4 0 0)*	0.4327		96.9	
TiC-1200 °C	90.8867	*(4 0 0)*	0.4325	0.4325	42.45	42.45

There are some potential implications of the observed changes in size of crystallites and phase composition with temperature variations. Such as a reduction in crystallite size as temperature increases often leads to larger surface area, which can enhance the material's reactivity and improve electrochemical performance. In batteries, for instance, smaller crystallites can provide more active sites for ion insertion or extraction, potentially improving the rate capability and capacity. On the other hand, larger crystallites at the lower temperature i.e. 800 °C can reduce surface area, which may limit the number of active sites for electrochemical reactions, potentially reducing the capacity or performance of the cell. Additionally, larger crystals could lead to slower ion diffusion rates due to reduced surface-to-volume ratios, decreasing the overall efficiency of ion transport within the material.

3.3 SEM Analysis

It is noteworthy that during the initial phase of the carbothermal reduction, the partially reduced TiO_2 particles aggregated to form large spherical particles, as illustrated in Fig. 5. This indicates that the vapour reaction was present in the creation of TiC_xO_y at 800 °C. It is clear that a significant number of micropores develop on the surface of the particles, a consequence of the CO released during the process of carbothermal reduction. At this stage, several cavities within the particles resembling a non-pen hollow structure were noted. SEM of sample reduced at 1000 °C demonstrates that the surfaces of the uncrushed coarse particles were relatively flat and smooth (b). However, the powders produced at 1200 °C lead to significant expansion of the grains and can be seen in Fig. 5(c). Smaller particles appear to consolidate into larger particles in order to form the bulk particles.

Fig. 5. SEM images of TiC particles reduced at (**a**) 800 °C, (**b**) 1000 °C, and (**c**) 1200 °C.

3.4 Electrochemical Behaviour of Titanium Carbide

The cyclic voltammetry (CV) curves of 1200 °C TiC recorded in 1 M hydrochloric acid electrolyte solutions are illustrated in Fig. 6 at various scan rates. The experimental data present that the shape of V-*I* curves is independent of the number of a cycle [25, 26]. It is observed that there is a small difference in peak height and width for different electrodes in the same potential region indicating that the amount of carbon impurities and oxidized TiC products can be responsible for this behaviour. Removal of oxygen from the solution did not influence the basic electrochemistry of the electrodes.

At low scan rates, specifically below 10 mVs^{-1}, the voltage-current (V-I) curves exhibit nearly symmetrical current responses resembling mirror images. The nanoporous carbon electrodes presented almost smooth curves in the potential range of −0.7 to +0.7 V, indicating their ideal polarizability in this region [27, 28]. As the scan rate increases, the V-I curves begin to deviate from this mirror image symmetry. This distortion is attributed to the high resistivity of the electrolyte within the porous material. Concurrently, the diffusion layer at the catalyst's surface becomes thinner, which enhances

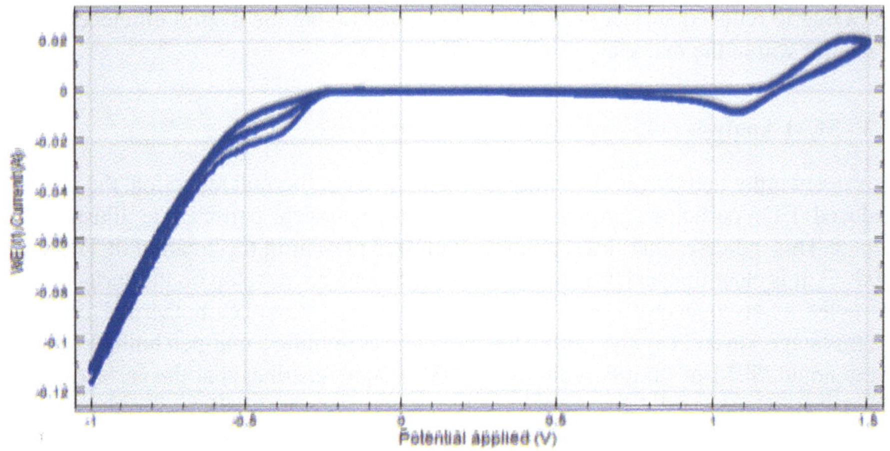

Fig. 6. Cyclic voltamograms of TiC at different scan rates.

the diffusion rate, resulting in a higher peak current density and increased electrochemical polarization. This phenomenon leads to elevated overpotential and diminished reversibility.

The separation between the oxidation and reduction peaks narrows, suggesting improved reversibility and a heightened electron-transfer rate. Consequently, this enhances the electrochemical catalytic activity. It is important to highlight that a less viscous electrolyte correlates with a higher "effective" diffusion coefficient for ions within the nanopores. Notably, the current densities observed in the negative potential region were unexpectedly low. In contrast, there was minimal variation in current density values among different carbon materials in the positive potential region. Previous studies have indicated that cyclic voltammetry experiments should be conducted at low scan rates to accurately reveal the underlying electrochemical mechanisms. The present work is underway to quantify the CV results and assess the electrochemical performance of the material, so that specific values for current densities, peak-to-peak separation, I_a/I_k ratio, cycling stability, scan rate dependence, and capacitance/energy density can be reported. These quantitative metrics not only provide insights into the material's electrochemical behavior but also help benchmark its performance for specific applications, such as energy storage devices. A material with high current densities, low peak separation, and excellent reversibility will generally demonstrate superior performance in terms of efficiency, stability, and capacity retention over multiple cycles.

4 Conclusion

In summary, TiC nanocomposites were prepared from in situ synthesized titania through sol-gel processing of mineral precursors and titanium tetraisopropoxide (TTIP). The lowering of the synthesized TiC at different temperatures resulted in the creation of khamrabaevite TiC polytypes. The observed weight loss stages in TG-DSC provide a means of monitoring the key steps in TiC formation. Ensuring the proper removal of volatile components and maintaining the correct stoichiometric balance of carbon and titanium during heating are critical for achieving high-purity TiC with the desired structural and functional properties. The lattice constants for the TiC phase in samples subjected to treatments at 800 °C, 1000 °C, and 1200 °C were about 0.4327 nm, 0.4323 nm, and 0.4325 nm, respectively. The crystallite dimensions, determined by the Debye-Scherrer technique at these temperatures were 97.8 nm, 81.8 nm, and 54.10 nm, in that order. Elevated temperatures could additionally trigger the diffusion of impurities or defects within the substance. These can anchor grain boundaries or hinder the expansion of individual crystallites, thereby constraining their size. This shows that nanosized TiC particles can be designed with enhance surface area and outstanding electrical conductivity, which is essential for materials utilized in fuel cell electrodes. TiC offers a promising alternative to commercially available electrode materials due to its high electrical conductivity, exceptional thermal and chemical stability, and excellent cycling performance. Compared to traditional electrode materials like graphite, TiC has superior performance in high-temperature and high-power applications.

References

1. Hu, C., Huang, J., Sumpter, B.J., Meletis, E., Dumitrica, T.: Ab initio predictions of strong interfaces in transition-metal carbides and nitrides for Superhard nanocomposite coating applications. ACS Appl. Nano Mater. **1**, 2029–2035 (2018). https://doi.org/10.1021/acsanm.8b00227
2. Liu, Y., Kelly, T.G., Chen, J.G., Mustain, W.E.: Metal carbides as alternative Electrocatalyst supports. ACS Catal. **3**, 1184–1194 (2013). https://doi.org/10.1021/cs4001249
3. Chandra, N., Sharma, M., Singh, D.K., Amritphale, S.S.: Synthesis of nano-TiC powder using titanium gel precursor and carbon particles. Mater. Lett. **63**, 1051–1053 (2009). https://doi.org/10.1016/j.matlet.2009.02.004
4. Rahaei, M.B., Rad, R.Y., Kazemzadeh, A., Ebadzadeh, T.: Mechanochemical synthesis of nano TiC powder by mechanical milling of titanium and graphite powders. Powder Technol. **217**, 369–376 (2012). https://doi.org/10.1016/j.powtec.2011.10.050
5. Koc, R., Meng, C., Swift, G.A.: Sintering properties of submicron TiC powders from carbon coated titania precursor. J. Mater. Sci. **35**, 3131–3141 (2000). https://doi.org/10.1023/A:1004876121000
6. Vallauri, D., Adrián, I.C.A., Chrysanthou, A.: TiC–TiB_2 composites: a review of phase relationships, processing and properties. J. Eur. Ceram. Soc. **28**, 1697–1713 (2008). https://doi.org/10.1016/j.jeurceramsoc.2007.11.011
7. Liu, N., Zhou, J., Zhang, X.B., Rong, C.L.: Effect of powder size on microstructure and mechanical properties of Ti(C,N) based cermets. Adv. Appl. Ceram. **106**, 247–254 (2007). https://doi.org/10.1179/174367607X202636
8. Yuan, X., Cheng, L., Kong, L., Yin, X., Zhang, L.: Preparation of titanium carbide nanowires for application in electromagnetic wave absorption. J. Alloys Comp. **596**, 132–139 (2014). https://doi.org/10.1016/j.jallcom.2014.01.022
9. Durlu, N.: Titanium carbide based composites for high temperature applications. J. Eur. Ceram. Soc. **19**, 2415–2419 (1999). https://doi.org/10.1016/S0955-2219(99)00101-6
10. Alvaredo, P., Mari, D., Gordo, E.: High temperature transformations in a steel-TiCN cermet. Int. J. Refract. Met. Hard Mater. **41**, 115–120 (2013). https://doi.org/10.1016/j.ijrmhm.2013.02.012

11. Mahmoodian, R., Hamdi, M., Hassan, M.A., Akbari, A.: Mechanical and chemical characterization of a TiC/C system synthesized using a focus plasma arc. PLoS One. **10**, e0130836 (2015). https://doi.org/10.1371/journal.pone.0130836
12. Tong, L., Reddy, R.G.: Synthesis of titanium carbide nano-powders by thermal plasma. Scripta Mater. **52**, 1253–1258 (2005)
13. Lee, D.W., Kim, B.K.: Synthesis of nano-structured titanium carbide by mg-thermal reduction. Scripta Mater. **48**, 1513–1518 (2003). https://doi.org/10.1016/S1359-6462(03)00130-1
14. Lee, D.W., Alexandrovskii, S.V., Kim, B.K.: Novel synthesis of substoichiometric ultrafine titanium carbide. Mater. Lett. **58**, 1471–1474 (2004). https://doi.org/10.1016/j.matlet.2003.10.011
15. Benoit, C., Ellen, H., Nikhil, K., Dominique, V., Sylvain, D.: TiC nucleation/growth processes during SHS reactions. Powder Technol. **157**, 92–99 (2005). https://doi.org/10.1016/j.powtec.2005.05.035
16. Niyomwas, S.: Synthesis and characterization of TiC and TiC-Al_2O_3Composite from wood dust by self-propagating high temperature synthesis. Energy Procedia. **9**, 522–531 (2011). https://doi.org/10.1016/j.egypro.2011.09.060
17. Woo, Y.C., Kang, H.J., Kim, D.J.: Formation of TiC particle during carbothermal reduction of TiO_2. J. Eur. Ceram. Soc. **27**, 719–722 (2007). https://doi.org/10.1016/j.jeurceramsoc.2006.04.090
18. Razavi, M., Rahimipour, M.R., Kaboli, R.: Synthesis of TiC nanocomposite powder from impure TiO_2 and carbon black by mechanically activated sintering. J. Alloys Compd. **460**, 694–698 (2008). https://doi.org/10.1016/j.jallcom.2007.06.080
19. Zhang, H., Li, F., Jia, Q., Ye, G.: Preparation of titanium carbide powders by sol-gel and microwave carbothermal reduction methods at low temperature. J. Sol-Gel Sci. Technol. **46**, 217–222 (2008). https://doi.org/10.1007/s10971-008-1697-0
20. Wang, K., Cai, R., Yuan, T., Yu, X., Ran, R., Shao, Z.P.: Process investigation, electrochemical characterization and optimization of $LiFePO_4$/C composite from mechanical activation using sucrose as carbon source. Electrochim. Acta. **54**, 2861–2868 (2009). https://doi.org/10.1016/j.electacta.2008.11.012
21. Wang, C.L., Hwang, W.S., Chu, H.L., Lin, H.J., Ko, H.H., Wang, M.C.: Kinetics of anatase transition to rutile TiO_2 from titanium dioxide precursor powders synthesized by a sol-gel process. Ceram. Int. **42**, 13136–13143 (2016). https://doi.org/10.1016/j.ceramint.2016.05.101
22. Lin, H., Tao, G., Xiong, J., Li, Q.: Using a cobalt activator to synthesize titanium carbide (TiC) nanopowders. Int. J. Refract. Met. Hard Mater. **41**, 363–365 (2013). https://doi.org/10.1016/j.ijrmhm.2013.05.010
23. Permann, L., Latt, M., Leis, J., Arulepp, M.: Electrical double layer characteristics of nanoporous carbon derived from titanium carbide. Electrochim. Acta. **51**, 1274–1281 (2006). https://doi.org/10.1016/j.electacta.2005.06.024
24. Holzwarth, U., Gibson, N.: The Scherrer equation versus the Debye-Scherrer equation. Nat. Nanotech. **6**, 534 (2011). https://doi.org/10.1038/nnano.2011.145
25. Lupu, A., Compagnone, D., Orlanducci, S., Terranova, M., Magearu, V., Palleschi, G.: Titanium carbide thin-film electrodes: characterization and evaluation as working electrodes. Electroanalysis. **16**, 1704–1710 (2004). https://doi.org/10.1002/elan.200303017
26. Wang, Y., Wu, M., Lin, X., Hagfeldt, A., Ma, T.: Optimization of the performance of dye-sensitized solar cells based on Pt-like TiC counter electrodes. Eur. J. Inorg. Chem. **2012**, 3557–3561 (2012). https://doi.org/10.1002/ejic.201200329
27. Ramasamy, E., Lee, J.: Ferrocene-derivatized ordered mesoporous carbon as high performance counter electrodes for dye-sensitized solar cells. Carbon. **48**, 3715–3720 (2010)
28. Nicholson, R.S.: Theory and application of cyclic voltammetry for measurement of electrode reaction kinetics. Anal. Chem. **37**, 1351–1355 (1965). https://doi.org/10.1021/ac60230a016

Numerical Simulation of Pb Free $MGeI_3$ (M = Cs and Rb) Based Perovskite Solar Cells: Theoretical Insights into Performance Optimization by SCAPS-1D

Arushi Sharma[1](✉), Himanshu[2], and M. S. Dhaka[1]

[1] Department of Physics, Mohanlal Sukhadia University, Udaipur 313001, India
anshu26667@gmail.com

[2] University Centre for Research & Development, Chandigarh University, Mohali 140413, India

Abstract. The efficiency of perovskite solar cells (PSCs) has reached an astounding mark of 26.7% thanks to recent developments, particularly in those which utilize organic-inorganic Lead halides based perovskite light harvesting absorbers. However, Lead toxicity is still a foremost barrier towards their wider commercialization owing to the associated environmental issues. The present research delves at novel, environmental friendly PSCs those are based upon Lead free $CsGeI_3$ and $RbGeI_3$ perovskite absorbers. The proposed PSCs comprise device architecture as $FTO/TiO_2/CsGeI_3/GO/Au$ and $FTO/TiO_2/RbGeI_3/GO/Au$ those are numerically simulated by SCAPS-1D software. The parameters pertaining to the absorber layer, including its thickness, defect density and acceptor density, are optimized and the $CsGeI_3$ based device achieves outstanding performance with V_{oc} of 1.3960 V, FF of 79.13%, J_{sc} of 24.04 mA/cm^2, PCE of 26.56%, at absorber layer's optimized condition of thickness 1000 nm, acceptor density 2×10^{21} cm^{-3}, defect density 1×10^{12} cm^{-3} and temperature 300 K. Whereas the $RbGeI_3$ based device accomplishes V_{oc} of 640.5 mV, FF of 78.38%, J_{sc} of 28.81 mA/cm^2 and PCE of 14.46% where, the optimized conditions bear thickness 400 nm, acceptor density 1×10^{16} cm^{-3}, defect density 1×10^{12} cm^{-3} and temperature 300 K. The present work intended to optimize Lead free Germanium based halide PSCs by using SCAPS-1D in order to lay the foundation for developing advanced Pb free, Ge based PSCs those could offer substantial potential for achievement of high conversion efficiencies and subsequent commercialization.

Keywords: Perovskite solar cells · Numerical simulation · SCAPS-1D · Power conversion efficiency · Device designing

1 Introduction

The expanding worldwide need for energy, in addition to the pressing need to limit the release of greenhouse gases and decrease dependency on fossil fuels, have led to a renewed emphasis on renewable and sustainable energy alternatives. Among these, solar energy stands out as a highly promising alternative, offering a clean, abundant, and

S. Kumari et al. (Eds.): ICMTP 2024, SPPHY 311, pp. 45–61, 2025.
https://doi.org/10.1007/978-981-96-6795-6_6

renewable power source. Solar energy offers environmental benefits and apt adaptability, which make it a strong contender in the renewable energy sector. Solar cell technology's versatility allows for seamless integration into diverse energy systems, making it suitable for various applications ranging from residential to industrial needs. Continuous advancements in solar technology are enhancing its efficiency and cost-effectiveness, bringing it closer to widespread adoption. As innovations make solar energy increasingly accessible and scalable, it is poised to become a cornerstone of sustainable energy solutions, driving global efforts toward a cleaner future [1–5]. Perovskite solar cells (PSCs), heralded as the forthcoming era of solar technology, created a significant breakthrough in this field. PSCs have sparked attention because to their outstanding optical and electrical attributes, including higher light absorption, efficient charge carrier generation, tunable optical energy band gap, optimal diffusion lengths, and excellent charge carrier mobility. Additionally, these offer lower density of trap states and lesser energy loss during operation. Since their debut in 2009, PSCs have demonstrated remarkable growth in power conversion efficiency (PCE), rising from 3.8% to an exemplary 26.7% today [6]. This rapid development underscores their potential to transform the renewable energy landscape, offering new possibilities for efficient and cost-effective solar power generation. Building on these innovations, the integration of inorganic materials, such as Cesium (Cs), into PSC absorbers has further enhanced their performance and durability.

Replacing organic components like methylammonium (MA) and formamidinium (FA) with Cesium in prevalent perovskite absorber has significantly improved the stability of PSCs, reducing degradation rates and enabling them to endure high temperatures of up to 150 °C while maintaining efficiency and structural integrity over extended periods. Furthermore, the incorporation of Cesium narrows the perovskite material's bandgap, substantially increasing light absorption. This improvement supports the development of advanced Cesium-based PSC architectures, achieving higher power conversion efficiencies and greater longevity in solar applications. By optimizing these designs, researchers can harness the unique benefits of Cesium to drive the evolution of perovskite technology and expand its role in the future of renewable energy [7, 8].

Recent research has increasingly focused on exploring alternative ions, especially Germanium (Ge^{+2}) and Tin (Sn^{+2}), to replace hazardous Lead (Pb^{+2}) in traditional perovskite materials. These substitutes are gaining considerable attention due to their potential to reduce environmental impact while maintaining several desirable attributes of Pb-based perovskites, Since they share comparable chemical traits alongside their proximity on the periodic table. Despite these facts, Tin faces a significant challenge: its propensity to oxidize from Sn^{2+} to Sn^{4+}, which undermines both the structural stability and performance of Tin-based PSCs. As a result, Germanium has emerged as a more stable and promising alternative. By addressing the limitations associated with Tin, Germanium offers a pathway for developing high-performing and environmentally friendly PSCs, further advancing the potential of perovskite technology in sustainable energy solutions [9]. Germanium's smaller ionic radius (0.73 Å) compared to Tin (1.35 Å) and Lead (1.49 Å) plays a crucial role in enhancing conductivity within perovskite materials. Research indicates that perovskites incorporating Germanium exhibit better conductivity and stability than those based on Lead or Tin elements. This is largely because the replacement of Sn^{+2} or Ge^{+2} for Pb^{+2} minimally disrupts the perovskite crystal structure, given their comparable ionic sizes. Furthermore, the perovskite crystal structure is not much altered when Sn^{+2} and Ge^{+2} replaced for Pb^{+2}. However, the Ge^{+2} offers a distinct advantage due to its smaller ionic radius, which makes it less susceptible to oxidation, thereby maintaining stability more effectively. As a result, non toxic Ge based perovskites ($CsGeI_3$, $RbGeI_3$) show enhanced conductivity compared to both the Pb and Sn based variants [8, 10, 11].

The $CsGeI_3$ perovskite absorber layer demonstrates not only improved stability and efficiency of associated solar cells but also possess a bandgap that optimally matches light absorption and energy conversion requirements. Their enhanced thermal and environmental resilience makes $CsGeI_3$ a promising candidate for next generation solar technologies, potentially overcoming many challenges associated with organic and Pb based PSCs [12]. The notion of inorganic PSCs using $CsGeI_3$ as the absorber layer has been proposed and experimental synthesis of $CsGeI_3$ crystals with stable rhombohedral structures has been achieved, showing no phase transitions within the operational temperature range. Additionally, Sherburne and group synthesized $MAGeI_3$ and $FAGeI_3$ crystals and analyzed their crystal structures, band gap, and thermal stabilities. Solar cells made from $CsGeI_3$ and $MAGeI_3$ demonstrated photocurrent values of 5.7 mA/cm^2 and 4 mA/cm^2, respectively [7]. Tara et al. [13] executed SCAPS-1D simulations to examine the influence of different charge transport layers on functionality of $CsGeI_3$-based PSCs, where by utilizing ZnOS, CuI as ETL and HTL, respectively, much better performance have been observed. They also optimized absorber layer's thickness, defect density, and ZnOS electron affinity parameters in their device architecture viz. FTO/ZnOS/$CsGeI_3$/CuI/Au, provided outstanding results with an PCE of 23.10%.

The $RbGeI_3$ perovskite material also has benefits over other alkali metals like Cs, including decreased reactivity, which leads to increased stability and simpler processing. It has promising features such as a 1.31 eV band gap, 3.9 eV electron affinity, and 23.1 dielectric permittivity. These features make the $RbGeI_3$ a promising absorber material for PSCs, potentially increasing performance and stability. The PSC having device architecture Al/ITO/TiO_2/$RbGeI_3$/Sb_2S_3/Pt was investigated by varying various physical parameters employing SCAPS-1D code, where the optimized device demonstrated an efficiency of 30.35% [14]. The influence of several HTMs on current-voltage (J-V) characteristics and external quantum efficiency (QE) was analyzed by utilizing SCAPS-1D code, where CuI was found out to be the superior HTM. The device viz. FTO/TiO_2/$RbGeI_3$/Spiro-OMeTAD/Ag was optimized and a significant improvement in PCE from 10.11% to 18.10% was observed by considering $RbGeI_3$ thickness, light transmission/reflection, metal contact work function, defect states, and working temperature [15].

Considering the utmost significance of Cs and Rb based perovskite absorbers, in this study, we model inorganic PSCs employing $CsGeI_3$ and $RbGeI_3$ material as light harvesting layer, TiO_2 as the electron transport layer (ETL) and graphene oxide (GO) as the hole transport layer (HTL) within the SCAPS-1D simulation framework. The $CsGeI_3$ is chosen for its non-hazardous, inorganic character and its bandgap of 1.65 eV, being nearer to $MAPbI_3$ (1.55 eV) for apt power conversion efficiency. The smaller bandgap of $CsGeI_3$ enhances light absorption, potentially improving overall device performance.

To ensure the stability and efficiency of $MGeI_3$ (M = Cs and Rb) based PSCs, a detailed simulation is conducted, where an organized optimization strategy is employed for optimising cells structure variables, including layer thickness, doping concentration, defect density and operating temperature. These factors are thoroughly examined to enhance the performance of different PSC structures. Additionally, a comparative analysis of $MGeI_3$ based PSCs is performed to evaluate their performance characteristics, providing insights into their potential advantages and limitations in solar cell applications.

2 Device Modeling and Numerical Simulation

Numerical simulations play a vital role in exploring systems characterized by complex mathematical models that cannot be solved analytically. These simulations provide critical insights into an extensive variety of parameters with no need for actual physical experimentation, making them indispensable for device design and manufacturing processes to save the valuable time and materials costs. In the context of solar cell development, researchers can leverage these simulations to enhance performance by systematically varying essential parameters, including defect density, temperature, donor and acceptor concentrations, and various transport layer thicknesses. Each of these factors has a substantial effect on electricity generation, enabling researchers to identify optimal configurations that maximize efficiency. Furthermore, numerical simulations allow for rapid iteration and testing of theoretical designs, opening the path for revolutionary improvements in PV technology and more effective energy solutions [16, 17].

Variety of numerical simulation software packages, such as SETFOS, AMPS, SCAPS, PC1D, SILVACO, ATLAS and, COMSOL, are available for computationaly analyzing solar cell performance. Amongst these, SCAPS-1D is particularly advantageous for designing heterostructure systems with up to seven layers, making it a vesatile tool for researchers. For this research, we use the Solar Cell Capacitance Simulator-One Dimensional (SCAPS-1D) variant 3.3.0.11, crafted by Burgelman's team at the University of Gent using LabWindows/CVI, mainly because of its simple to use interface, robust control features, and ability to simulate equally in illuminated and dark conditions. The software predicts a variety of key metrics, including power conversion efficiency (PCE), fill factor (FF), open circuit voltage (V_{oc}), short circuit current density (J_{sc}), quantum efficiency (QE), capacitance-voltage (C-V) profiles, capacitance-frequency (C-f) data, electric field distributions, band diagrams and structures, carrier densities, recombination profiles, carrier transport phenomena, and spectral response ($Q(\lambda)$). These capabilities enable detailed investigation and performance optimization of device. Herein, the electron and hole continuity equations (Eqs. 2 and 3) and Poisson's equation (Eq. 1) form the basis of the associated algorithm used for modelling the solar cells numerically [18], which is presented below.

Poisson's equation

$$\frac{\partial}{\partial x}\left(-\varepsilon(x)\frac{\partial V}{\partial x}\right) = q\left[p(x) + N_D^+ + p_t(x) - n(x) - N_A^- - n_t(x)\right] \tag{1}$$

Electron continuity equation

$$\frac{\partial n}{\partial x} = \frac{1}{q}\frac{\partial J_n}{\partial x} + \text{Gn} - \text{Rn} \tag{2}$$

Hole continuity equation

$$\frac{\partial p}{\partial x} = -\frac{1}{q}\frac{\partial J_p}{\partial x} + \mathrm{Gp} - \mathrm{Rp} \tag{3}$$

Here ε, q and V denote the respective dielectric permittivity, charge and potential. The $p(x)$, N_D^+, $p_t(x)$ corresponds to free hole concentration, ionized donor concentration and hole trap density, respectively whereas the symbols $n(x)$, N_A^-, $n_t(x)$ represent the free electron concentration, ionized acceptor concentration and electron trap density. The current density of electrons and holes is shown by J_n and J_p respectively. While the recombination rates of holes and electrons are designated by R_p and R_n, respectively, and the generation rates of electrons and holes are symbolized by G_n and G_p.

The SCAPS-1D programme is employed to simulate Pb-free all inorganic PSCs utilising design as glass/FTO/TiO_2/$MGeI_3$/GO/Ag. Figure 1(a), (b) depicts the PSCs architecture employed in this work. In this study, several factors including band gap, electron affinity, donor and acceptor densities, and defect density have been acquired from prior research, as cited in Table 1. Additionally, some values have been adjusted within acceptable ranges based on theoretical considerations. To simulate the desired device, neutral defects with a density of 10^{15} cm^{-3} have been incorporated into the various layers. The following table lists the fundamental physical parameters of every layer required for numerical simulation of $MGeI_3$ based PSCs, the optical reflectance at each interface is taken to be zero, and a single distribution is used to represent the nature of the imperfections at both the ETL/absorber and absorber/HTL interfaces. Whereas, Table 2 gives a brief description of defects introduced at different interfaces. The PSCs are simulated with conventional parameters viz. 300 K temperature, AM (air mass) 1.5G spectra, and irradiance of 100 mWcm^{-2} where light enters from FTO coated glass side.

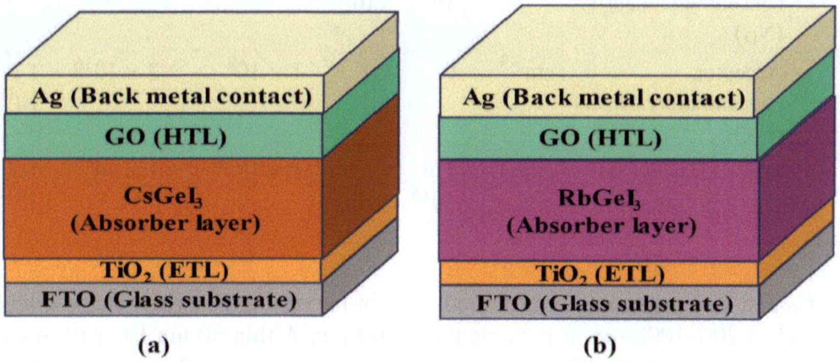

Fig. 1. Device architecture employed in the simulation of (a) $CsSnI_3$ (b) $RbGeI_3$ absorber based PSCs.

In the proposed device, a 500 nm thick Fluorine doped Tin oxide (FTO) layer, known for its excellent transparency, serves as the transparent conducting oxide (TCO). A 50 nm thin TiO_2 layer, serving as ETL, is incorporated to form an asymmetric heterojunction with the inorganic perovskite absorber film. The Lead free all inorganic $MGeI_3$ (where

Table 1. Various physical parameters associated with different constituent layers employed during the simulation.

S. No.	Parameters	Unit	FTO [12]	TiO_2 [12]	$RbGeI_3$ [15]	$CsGeI_3$ [12]	GO [19]
1	Layer thickness (W)	nm	500	50	200–1000	200–1000	50
2	Bandgap energy (E_g)	eV	3.5	3.2	1.310	1.6	2.48
3	Electron affinity (χ)	eV	4	4	3.9	3.523	2.3
4	Relative permittivity (ε_r)		9	10	23	18	10
5	Effective conduction band density of states (N_c)	cm^3	2.2×10^{18}	5×10^{19}	2.8×10^{19}	1×10^{18}	2.2×10^{18}
6	Effective valence band density of states (N_V)	cm^3	1.8×10^{19}	5×10^{19}	1.4×10^{19}	1×10^{19}	1.8×10^{18}
7	Electron thermal velocity	cm/s	1×10^7	1×10^7	1×10^7	1×10^7	5.2×10^7
8	Hole thermal velocity	cm/s	1×10^7	1×10^7	1×10^7	1×10^7	5×10^7
9	Electron mobility (n)	cm^2/Vs	2×10^1	6 × 10–3	2.86×10^1	2.0 × 10	2.6×10^1
10	Hole mobility (p)	cm^2/vs	1×10^1	6 × 10–3	2.73×10^1	2.0 × 10	1.23×10^2
11	Donor concentration/density (N_D)	cm^{-3}	1×10^{19}	1×10^{18}	1×10^9	0	0
12	Acceptor concentration/ density (N_A)	cm^{-3}	0	0	1×10^9	2×10^{16}	1×10^{19}
13	Defect density (N_t)	cm^{-3}	1×10^{15}	1×10^{18}	1×10^{15}	1×10^{15}	1×10^{14}

M = Cs and Rb) perovskite is exploited as light harvesting absorber whose thickness is altered in 200–1000 nm in increments of 200 nm. A thin 50 nm layer of graphene oxide (GO) serves as HTL. The device structure is finalized with a Silver (Ag) metal back contact, featuring a work function of 4.73 eV. Influence of $MGeI_3$ thickness, temperature, acceptor and defect density on device performance are explored in depth in the following section.

Table 2. Description of defects introduced at different interfaces.

Interface	$TiO_2/CsGeI_3$	$TiO_2/RbGeI_3$	$CsGeI_3$/GO	$RbGeI_3$/GO
Energy distribution	Single	Single	Single	Single
Defect type	Neutral	Neutral	Neutral	Neutral
Total density/concentration (cm^{-2})	1×10^{10}	1×10^{10}	1×10^{10}	1×10^{10}
Energy distribution	Single E_t = 0.60 eV	Single E_t = 0.60 eV	Single E_t = 0.60 eV	Single E_t = 0.60 eV
Reference for defect energy level	Above the VB maximum	Above the VB maximum	Above the VB maximum	Above the VB maximum

3 Results and Discussion

3.1 Thickness Optimization in Absorber Layer

The performance of PSCs is highly dependent on the thickness of the absorber layer. Using the parameters outlined in Table 1, the simulation investigates absorber layer thicknesses between 200 nm and 1000 nm. Figure 2(a, b) portrays the J-V traits, while Fig. 2 (c, d) showcases the impact of $CsGeI_3$ thickness on J_{sc}, V_{oc}, FF, and PCE. The V_{oc} and J_{sc} of the devices show a notable increase with the rise in absorber layer thickness, spanning from 1.1120 V to 1.1272 V and 15.82 mA/cm^2 to 23.68 mA/cm^2, respectively. This improvement pertained to a higher carrier concentration [20]. In contrast, the FF decreases from 78.80% to 75.92% as the thickness of the perovskite absorber layer (PAL) surges. The reduction in FF is possibly related to a boost in series resistance (R_s) and enhanced charge recombination within the PAL. Although the rise in both V_{oc} and J_{sc} contributes to an increase in P_{max}, the rate of increase in the product of V_{oc} and J_{sc} exceeds the growth in P_{max}, resulting in a decline in FF with increasing absorber layer thickness [21, 22]. The enhancement in V_{oc} and J_{sc} led to boost of the PCE from 13.86% to 20.27% of corresponding PSCs with the enhancement in absorber layer thickness from 200 nm to 1000 nm which suggests that one micron is an optimal thickness to achieve higher performance of modeled devices.

An analysis of $RbGeI_3$ based PSCs revealed optimal performance at an absorber layer thickness of 400 nm. The open-circuit voltage (V_{oc}) improved marginally from 0.6684 V to 0.6733 V, whereas the J_{sc} exhibited a significant increase from 22.45 mA/cm^2 to 27.44 mA/cm^2. The observed enhancement in J_{sc} is a result of improved light absorption and greater carrier generation at thicker layers, leading to higher carrier concentration. However, the FF decreased from 64.28% to 59.06% as the thickness increased. This decline is primarily due to the rise in series resistance (R_s) caused by the longer carrier transport paths, as well as increased charge recombination within the absorber layer. The overall maximum power output (P_{max}) improved as a result of increment in both the J_{sc} and V_{oc}. However, J_{sc} rose more substantially than P_{max}, leading to a reduction in FF due to reduced charge extraction efficiency at higher thicknesses.

Consequently, the effieciency of the $RbGeI_3$-based PSCs improved from 9.65% to 10.91% as the thickness increased reaching its peak at 400 nm. However, further increases in thickness beyond this point led to a decline in PCE, driven by increased recombination losses and excessive resistance, which outweighed the benefits of additional light absorption. These trends are illustrated in Fig. 3, highlighting the critical balance between optical absorption and carrier transport in determining the optimal absorber thickness.

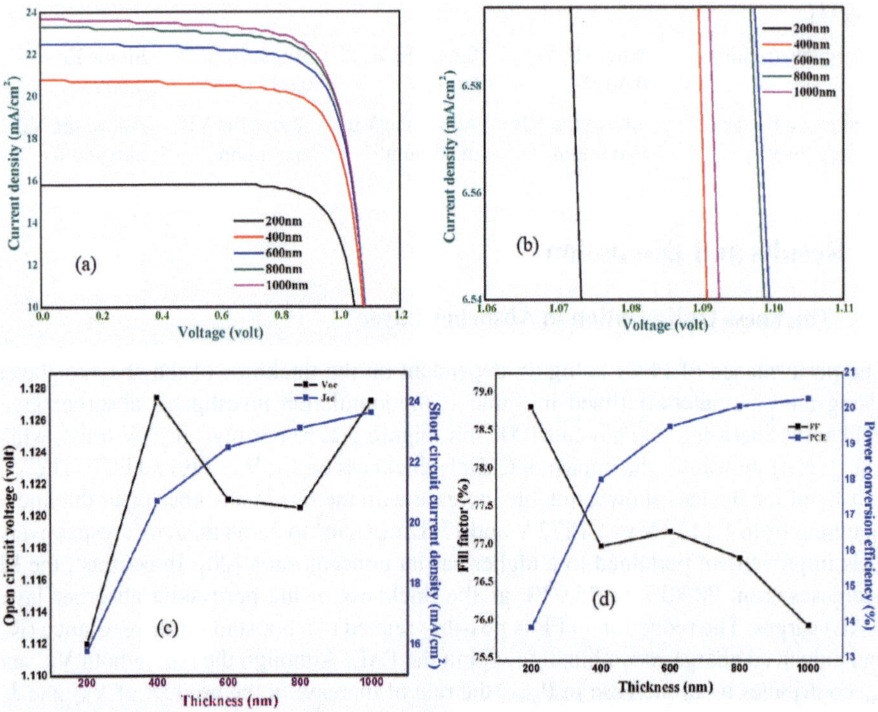

Fig. 2. (a) J-V characteristics of simulated device with varying thickness, (b) enlarged view of J-V curves, behavior of (c) open circuit voltage and short circuit current density, (d) fill factor and power conversion efficiency of PSCs with increased $CsGeI_3$ absorber layer thickness.

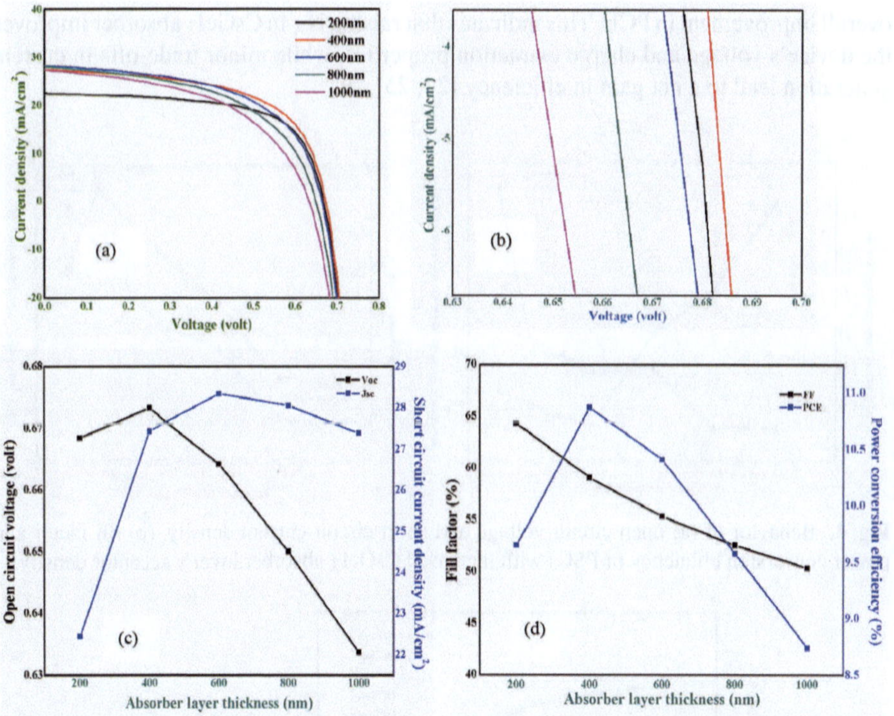

Fig. 3. (a) J-V characteristics of simulated device at different thickness, (b) enlarged view of J-V curves, behavior of (c) open circuit voltage and short circuit current density, (d) fill factor and power conversion efficiency of PSCs with increased $RbGeI_3$ absorber layer thickness.

3.2 Acceptor Density Optimization in the Absorber Layer

In PSCs, the acceptor density (N_A) of the absorber layer is a key factor influencing PV performance. A higher charge carrier density improves the built-in electric field, aiding in charge separation and suppressing recombination under open circuit conditions. This, in turn, enhances the V_{oc}, FF, and PCE. Additionally, increasing acceptor density boosts quasi-Fermi level splitting, further contributing to a rise in V_{oc}. However, this benefit comes with a trade-off, as elevated acceptor densities in certain materials may increase non-radiative recombination losses, thereby decreasing the J_{sc} [23]. The impact of varying N_A in the absorber film from 2×10^{12} cm^{-3} to 2×10^{21} cm^{-3} is analyzed through simulations and revealed a rise in V_{oc} accompanied by a decline in J_{sc}, as apparent in Fig. 4(a). This change also resulted in an increase FF, as shown in Fig. 4(b). Despite the decline in J_{sc}, PCE improved overall due to the dominant contributions of V_{oc} and FF. The acceptor density is optimized at 2×10^{21} cm^{-3}, providing the best balance between enhanced PV performance and cost-effectiveness, while mitigating the risks of large conduction band offset (CBO) and valence band offset (VBO). The observed trend in performance parameters is a result of optimizing charge transport and suppressing recombination losses at an expenditure of some loss in photocurrent generation. The improved V_{oc} and FF outweigh the minor decline in J_{sc}, resulting in an overall improvement in PCE. This indicates that raising N_A in $CsGeI_3$ absorber improves the device's voltage and charge extraction properties, while minor trade-offs in current generation lead to a net gain in efficiency [24, 25].

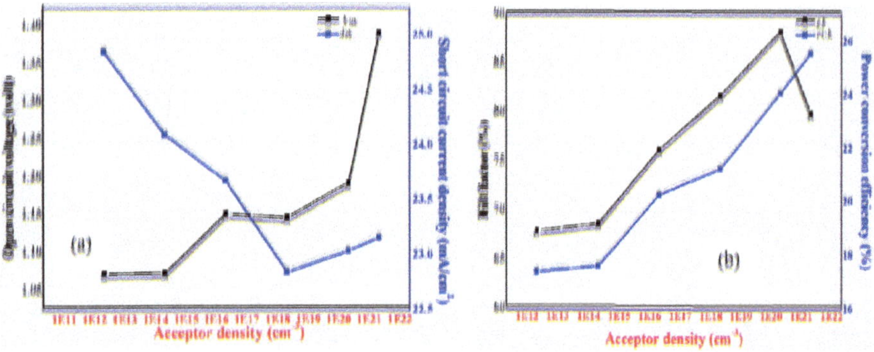

Fig. 4. Behavior of (a) open circuit voltage and short circuit current density, (b) fill factor and power conversion efficiency of PSCs with increased $CsGeI_3$ absorber layer's acceptor density.

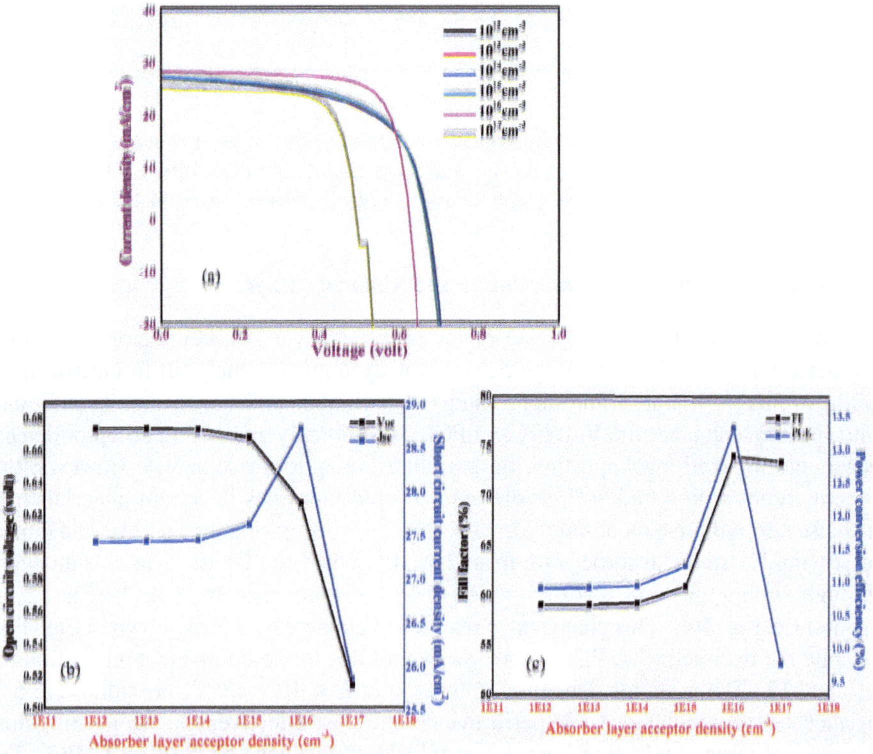

Fig. 5. (a) J-V characteristics of simulated device with varying acceptor density, behavior of (b) open circuit voltage and short circuit current density, (c) fill factor and power conversion efficiency of PSCs with increased $RbGeI_3$ absorber layer's acceptor density.

The impact of altering the acceptor density in the $RbGeI_3$ film on PSC performance is examined, with the corresponding J-V curves displayed in Fig. 5(a). The simulations reveal that N_A rises from 1×10^{12} cm^{-3} to 1×10^{17} cm^{-3}, the P_{max} progressively improves. A decrease in V_{oc} is observed (Fig. 5(b)) which is likely due to bandgap narrowing or shifting of the Fermi level in $RbGeI_3$, leading to a reduction in the built-in potential and open-circuit voltage. An increment in J_{sc} is also observed in Fig. 5(b), this could be due to improved light absorption, better charge generation, and enhanced hole transport resulting from increased acceptor density. The increased hole density leads to improve photocurrent generation. Figure 5(c) depicts an increment in both FF and PCE which results from better charge extraction, reduced recombination, and lower series resistance due to higher acceptor density in $RbGeI_3$. Despite the decrease in V_{oc} the upsurge in J_{sc} and FF contributes to a net increase in PCE, since the resultant efficiency depends more on current and fill factor than on voltage alone. The opposite trend observed for $RbGeI_3$ compared to $CsGeI_3$ can be explained by the unique material properties and doping characteristics of each absorber. In $RbGeI_3$, higher acceptor density seems to enhance charge generation and transport, producing better photocurrent (J_{sc}) and improved charge extraction (FF), even though the V_{oc} decreases slightly. These enhancements compensate for the V_{oc} reduction, resulting in a net gain in PCE.

3.3 Defect Density Optimization in the Absorber Layer

Defects within the absorber layer significantly affect their PV performance by influencing e–h recombination, migration, and trapping. These defects comprise point defects (vacancies, interstitials, antisites), extended defects (grain boundaries, dislocations). Aditionally, impurities introduced during material synthesis or processing are also significant contributors to defect formation [26]. Collectively referred to as the total defect density (N_t), these factors are crucial in influencing both the efficiency and stability of the device. High defect densities lead to pinholes and increased non-radiative recombination, reducing J_{sc}, V_{oc}, and PCE. These also accelerate degradation, compromising device longevity [27]. Defects in the absorber layer, such as Schottky and Frenkel defects, tend to be inevitable because of the constraints of deposition and material preparation methods [28]. Modeled devices with realistic defect densities, both within the bulk and at interfaces, helps to provide accurate insights into their impact. When the defect is high, it creates pinholes, reducing stability. It causes high recombination, reducing overall efficiency [28, 29]. Such studies are key to optimizing the output and robustness of PSCs. In this study, N_t of $CsGeI_3$ is adjusted from 1×10^{12} cm^{-3} to 1×10^{18} cm^{-3} while maintaining acceptor density fixed to 2×10^{21} cm^{-3} and thickness fixed to 1000 nm. Figure 6(a) displays that both J_{sc} and V_{oc} decrease as defect density increases. Moreover, Fig. 6(b) highlights the trends in FF and PCE, where FF improves and PCE decline with total defect density of perovskite layer. Figure 6(b) further illustrates FF increases, while the PCE declines with rising N_t in the perovskite film.

Defect states in PSCs can be categorized as either interruptions or impurities within the pristine crystal lattice. These flaws frequently occur during deposition, material prepping, or growth procedures, rendering to their inevitable incorporation into the system [30, 31]. To achieve more accurate results, device modeling should consider defect density both in the bulk and at the interface. The effect of altering absorber $RbGeI_3$

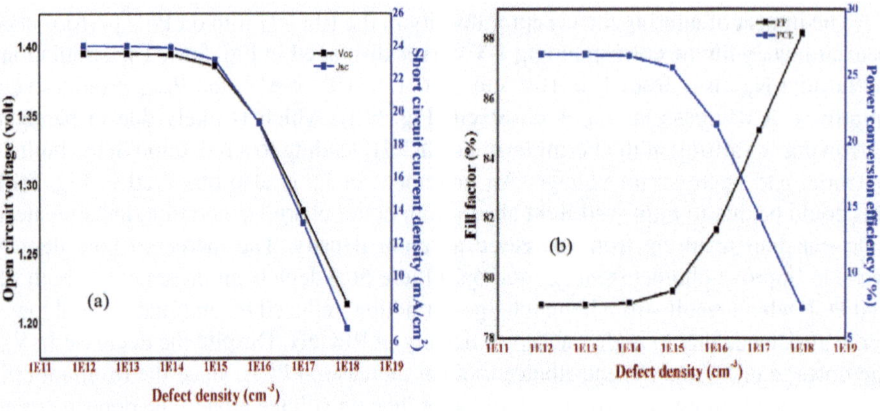

Fig. 6. Behavior of (a) open circuit voltage and short circuit current density, (b) fill factor and power conversion efficiency of PSCs with increased $CsGeI_3$ absorber layer's defect density.

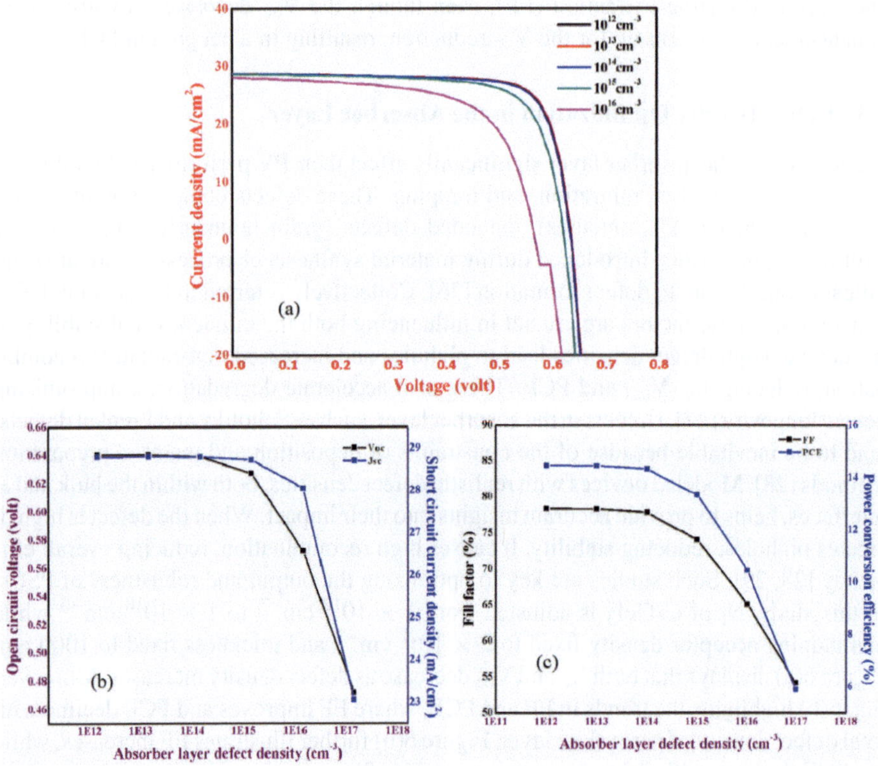

Fig. 7. J-V characteristics of simulated device with varying defect density, behavior of (b) open circuit voltage and short circuit current density, (c) fill factor and power conversion efficiency of PSCs with increased $RbGeI_3$ absorber layer's defect density.

defect density from 10^{12} cm^{-3} to 10^{18} cm^{-3} whilst maintaining acceptor density set at 10^{16} cm^{-3} and thickness at 400 nm was investigated in the present study. In Fig. 7(a), the J-V curves corresponding to varying defect densities in absorber layer are presented. Figure 7(b, c) demonstrate the performance degradation with increasing defect density in the light absorbing layer. The decline in device's working is primarily caused by the presence of Shockley-Read-Hall (SRH) non-radiative recombination, a phenomenon that significantly reduces the device's longevity by promoting carrier recombination and leading to overall performance losses [32]. While lower defect densities typically result in higher efficiency, it is practically unfeasible to achieve these in actual devices [33]. Considering these factors, the optimal N_t for the absorber is set at 10^{12} cm^{-3} for both devices.

3.4 Effect of Temperature on the Performance of $MGeI_3$ Based PSCs

Temperature has a vital impact on the performance and efficiency of PV devices. To assess the real-world performance of PSCs, we examine their behavior within a temperature range of 300 K to 500 K. Figure 8(a, b) depicts how temperature affects device's performance. V_{oc} declines from 1.396 V to 1.278 V as temperature rises, primarily due to bandgap narrowing and increased recombination. In contrast, J_{sc} rises marginally from 24.04 mA/cm^2 to 24.16 mA/cm^2 with temperature, probably attributed to enhanced charge carrier mobility at moderate temperatures. The FF increases from 79.13% to 81.32% with temperature because of reduced series resistance and improved charge transport efficiency. In addition the PCE declines from 26.56% to 25.11% with temperature due to the combined effect of decreasing V_{oc}, along with thermal degradation and increased recombination losses. Consequently, the overall performance of the $CsGeI_3$ based PSC diminishes as temperature rises, largely driven by the decrease in V_{oc} and possible thermal instability or degradation of the perovskite material.

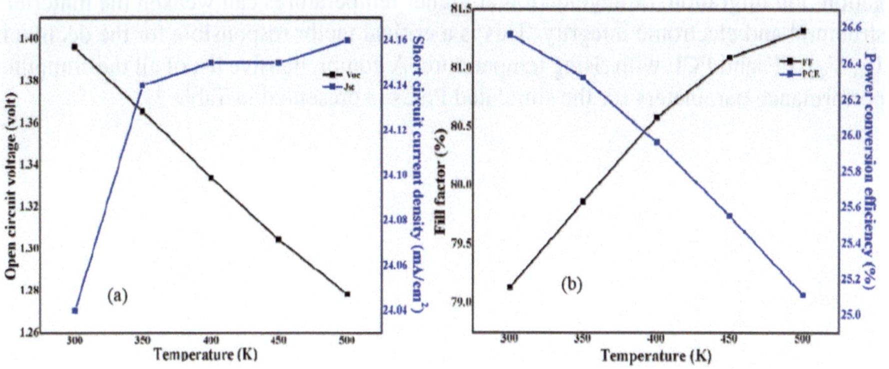

Fig. 8. Behavior of (a) open circuit voltage and short circuit current density, (b) fill factor and power conversion efficiency with temperature for $CsGeI_3$ absorber based PSCs.

Fig. 9(a, b) depicts how temperature affects the PV characteristics of $RbGeI_3$ based PSCs across a range of 300 to 500 K. With higher temperatures, the reverse saturation

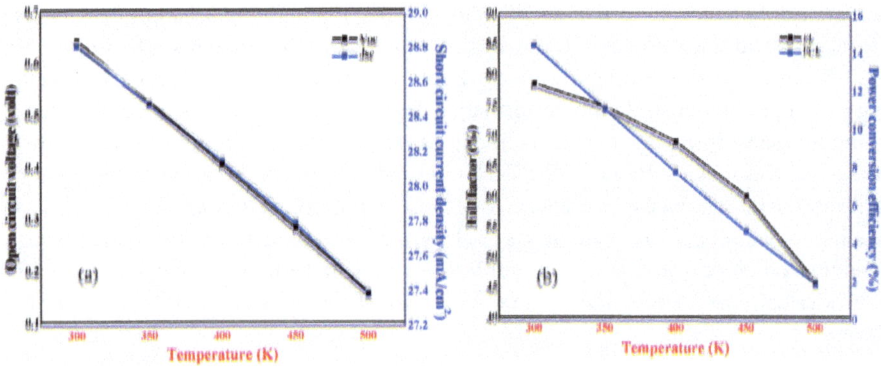

Fig. 9. Behavior of (a) open circuit voltage and short circuit current density, (b) fill factor and power conversion efficiency with temperature for $RbGeI_3$ based PSC.

current density escalates, leading to a decline in V_{OC} and FF. A reduction in V_{OC} from 0.641 V to 0.157 V is observed, likely due to bandgap narrowing, increased recombination, and possible thermal degradation of the material. The J_{sc} also declines from 28.49 mA/cm^2 to 27.49 mA/cm^2, probably caused by enhanced recombination and possibly reduced light absorption, FF decreases from 78.38% to 45.88% because of increased series resistance and reduced charge transport efficiency, additionally, the PCE experiences a drastic drop from 14.46% to 1.98% with rising temperatures, owing to increased series resistance and reduced charge transport efficiency. The most important reason for decreasing the efficiency is that the velocity of charge particles increases when the temperature increases which leads to more recombination of electrons and holes [18, 34]. The $RbGeI_3$ is a relatively less studied perovskite material compared to other halide perovskites, and it may have greater thermal instability. The occurrence of phase segregation, ion migration, or degradation at higher temperatures can weaken the material's structural and electronic integrity. This is a critical factor responsible for the decline in J_{sc}, V_{OC}, FF, and PCE with rising temperature. A comprehensive list of all the computed performance parameters for the simulated PSCs is presented in Table 3.

Table 3. Performance overview of optimized simulated device.

Optimized Parameters ($CsGeI_3$)	V_{oc} (V)	J_{sc} (mA/cm^2)	FF (%)	PCE (%)	Optimized Parameters ($RbGeI_3$)	V_{oc} (V)	J_{sc} (mA/cm^2)	FF (%)	PCE (%)
Absorber thickness (1000 nm)	1.127	23.68	75.92	20.27	Absorber thickness (400 nm)	0.673	27.45	59.06	10.91
Absorber acceptor density (2×10^{21} cm^{-3})	1.387	23.16	79.59	25.56	Absorber acceptor density (10^{16} cm^{-3})	0.627	28.73	74.01	13.33
Absorber defect density (10^{12} cm^{-3})	1.396	24.04	79.13	26.56	Absorber defect density (10^{12} cm^{-3})	0.641	28.81	78.38	14.46
Temperature (300 K)	1.396	24.04	79.13	26.56	Temperature (300 K)	0.641	28.81	78.38	14.46

4 Conclusion

The present study uses SCAPS-1D program to examine the behavior of $MGeI_3$ (M = Cs, Rb) based PSCs, emphasising on investigating the influence of important parameters including absorber layer's thickness, acceptor density, defect density and temperature. The simulation results have been optimized to determine the maximum achievable efficiency, and the findings are well described where maximum power conversion efficiency of 26.56% and 14.46% is achieved for respective $CsGeI_3$ (at thickness of 1000 nm) and $RbGeI_3$ (at thickness of 400 nm) based PSCs. In conclusion, $CsGeI_3$ outperforms the $RbGeI_3$ for perovskite solar cells due to its superior crystallographic stability, more favorable bandgap, higher charge carrier mobility, and better environmental resilience, leading to higher efficiency and longer device lifetimes. The results confirm that $MGeI_3$ based perovskite solar cells show promising theoretical performance. This work is valuable for the future design of a free from Lead, non hazardous Germanium based PSCs, contributing to the development of sustainable solar technologies.

Acknowledgement. Our research team appreciates the funding support from the Ministry of Education, Government of India, and the Ministry of Higher Education, Government of Rajasthan, via the RUSA 2.0 Research and Innovation Project. Special thanks are extended to Prof. M. Burgelman of the University of Gent for granting access to the SCAPS-1D program, which was instrumental in conducting the research on $FTO/TiO_2/MGeI_3/GO/Ag$ solar cells.

References

1. Obeidat, F.: A comprehensive review of future photovoltaic systems. Sol. Energy. **163**, 545–551 (2018)
2. Patel, A.K., Mishra, R., Soni, S.K.: Performance analysis of a cu $(In_{1-x}Ga_x)Se_2$ solar cell with nontoxic WS_2 and WSSe buffer layers. J. Electron. Mater. **51**(11), 6168–6179 (2022)

3. Yadav, S., Lohia, P., Sahu, A.: Enhanced performance of double perovskite solar cell using WO_3 as an electron transport material. J. Opt. **52**(2), 776–782 (2022)
4. Patel, A.K., Rao, P.K., Mishra, R., Soni, S.K.: Numerical study of a high-performance thin film CIGS solar cell with a-Si and $MoTe_2$ hole transport layer. Optik. **243**, 167498 (2021)
5. Kumar, S., Chauhan, R.K.: Performance up-gradation of CIGS solar cell using Ag_2S quantum dot as buffer layer. J. Mater. Res. **38**(10), 2689–2700 (2023)
6. Home page. https://www.nrel.gov/pv/interactive-cell-efficiency.html. Accessed 19 December 2024
7. Sherburne, M., Li, S., Asta, M., Mathews, N., Mhaisalkar, S.G.: Lead-free germanium iodide perovskite materials for photovoltaic application. J. Mater. Chem. A. **3**(47), 23829–23832 (2015)
8. Raj, A., Kumar, M., Bherwani, H., Gupta, A., Anshul, A.: Evidence of improved power conversion efficiency in lead-free CsGeI3 based perovskite solar cell heterostructure via scaps simulation. J. Vac. Sci. Technol. B. **39**(1), 012401 (2021)
9. Stoumpos, C.C., Frazer, L., Clark, D.J., Kim, Y.S., Rhim, S.H., Freeman, A.J., Kanatzidis, M.G.: Hybrid germanium iodide perovskite semiconductors: active lone pairs, structural distortions, direct and indirect energy gaps, and strong nonlinear optical properties. J. Am. Chem. Soc. **137**(21), 6804–6819 (2015)
10. Noel, N.K., Stranks, S.D., Abate, A., Wehrenfennig, C., Guarnera, S., Haghighirad, A.A., Snaith, H.J.: Lead-free organic–inorganic tin halide perovskites for photovoltaic applications. Energy Environ. Sci. **7**(9), 3061–3068 (2014)
11. Sarkar, D.K., Mottakin, M., Hasan, A.M., Selvanathan, V., Sobayel, K., Khan, M.N.I., Akhtaruzzaman, M.: A comprehensive study on RbGeI3 based inorganic perovskite solar cell using green synthesized $CuCrO_2$ as hole conductor. J. Photochem. Photobiol. A Chem. **439**, 114623 (2023)
12. Uddin, M.S., Al Mashud, M.A., Toki, G.I., Pandey, R., Zulfiqar, M., Saidani, O., Hossain, M.K.: Lead-free Ge-based perovskite solar cell incorporating TiO_2 and Cu_2O charge transport layers harnessing over 25% efficiency. J. Opt. **53**(4), 3726–3742 (2024)
13. Tara, A., Bharti, V., Sharma, S., Gupta, R.: Computational approach to explore suitable charge transport layers for all inorganic $CsGeI_3$ perovskite solar cells. Opt. Mater. **128**, 112403 (2022)
14. Rahman, M.M., Ali, M.H., Haque, M.D., Islam, A.Z.M.T.: Numerical modeling and extensive analysis of an extremely efficient $RbGeI_3$-based perovskite solar cell by incorporating a variety of ETL and HTL materials to enhance PV performance. Energy Adv. **3**(9), 2377–2398 (2024)
15. Samaki, S., Tchangnwa Nya, F., Dzifack Kenfack, G.M., Laref, A.: Materials and interfaces properties optimization for high-efficient and more stable $RbGeI_3$ perovskite solar cells: optoelectrical modelling. Sci. Rep. **13**(1), 15517 (2023)
16. Burgelman, M., Decock, K., Khelifi, S., Abass, A.: Advanced electrical simulation of thin film solar cells. Thin Solid Films. **535**, 296–301 (2013)
17. Verschraegen, J., Burgelman, M.: Numerical modeling of intra-band tunneling for heterojunction solar cells in SCAPS. Thin Solid Films. **515**(15), 6276–6279 (2007)
18. Pindolia, G., Shinde, S.M., Jha, P.K.: Optimization of an inorganic lead free $RbGeI_3$ based perovskite solar cell by SCAPS-1D simulation. Sol. Energy. **236**, 802–821 (2022)
19. Chaudhary, A.K., Verma, S., Chauhan, R.K.: Design of a low-cost, environment friendly perovskite solar cell with synergic effect of graphene oxide-based HTL and $CH_3NH_3GeI_3$ as ETL. Eng. Res. Express. **5**(3), 035039 (2023)
20. Anwar, F., Mahbub, R., Satter, S.S., Ullah, S.M.: Effect of different HTM layers and electrical parameters on ZnO nanorod-based lead-free perovskite solar cell for high-efficiency performance. Int. J. Photoenergy. **2017**(1), 9846310 (2017)
21. Du, H.J., Wang, W.C., Zhu, J.Z.: Device simulation of lead-free $CH_3NH_3SnI_3$ perovskite solar cells with high efficiency. Chin. Physics B. **25**(10), 108802 (2016)

22. Chakraborty, K., Choudhury, M.G., Paul, S.: Numerical study of Cs_2TiX_6 ($X= Br^-, I^-, F^-$ and cl^-) based perovskite solar cell using SCAPS-1D device simulation. Sol. Energy. **194**, 886–892 (2019)
23. Yang, D., Wang, Y., Sano, T., Gao, F., Sasabe, H., Kido, J.: A minimal non-radiative recombination loss for efficient non-fullerene all-small-molecule organic solar cells with a low energy loss of 0.54 eV and high open-circuit voltage of 1.15 V. J. Mater. Chem. A. **6**(28), 13918–13924 (2018)
24. Minemoto, T., Murata, M.: Theoretical analysis on effect of band offsets in perovskite solar cells. Sol. Energy Mater. Sol. Cells. **133**, 8–14 (2015)
25. Singh, N., Agarwal, A., Agarwal, M.: Study the effect of band offsets on the performance of lead-free double perovskite solar cell. Opt. Mater. **125**, 112112 (2022)
26. Kumar, A., Rana, A., Vashistha, N., Garg, K.K., Singh, R.K.: Defect states influencing hysteresis and performance of perovskite solar cells. Sol. Energy. **211**, 345–353 (2020)
27. Johnston, M.B., Herz, L.M.: Hybrid perovskites for photovoltaics: charge-carrier recombination, diffusion, and radiative efficiencies. Acc. Chem. Res. **49**(1), 146–154 (2016)
28. Rai, S., Pandey, B.K., Dwivedi, D.K.: Modeling of highly efficient and low cost $CH_3NH_3Pb(I_{1-x}Cl_x)_3$ based perovskite solar cell by numerical simulation. Opt. Mater. **100**, 109631 (2020)
29. Zhao, X., Yao, C., Gu, K., Liu, T., Xia, Y., Loo, Y.L.: A hole-transport material that also passivates perovskite surface defects for solar cells with improved efficiency and stability. Energy Environ. Sci. **13**(11), 4334–4343 (2020)
30. Dzifack Kenfack, G.M., Tchangnwa Nya, F., Laref, A.: Towards high tin-based halide organic-inorganic perovskite photovoltaic cells efficiency improvement: SCAPS 1D modeling. Int. J. Energy Res. **46**(7), 8962–8976 (2022)
31. Ochoa-Martinez, E., Ochoa, M., Ortuso, R.D., Ferdowsi, P., Carron, R., Tiwari, A.N., Saliba, M.: Physical passivation of grain boundaries and defects in perovskite solar cells by an isolating thin polymer. ACS Energy Lett. **6**(7), 2626–2634 (2021)
32. Alam, I., Mollick, R., Ashraf, M.A.: Numerical simulation of $Cs_2AgBiBr_6$-based perovskite solar cell with ZnO nanorod and P3HT as the charge transport layers. Phys. B Condens. Matter. **618**, 413187 (2021)
33. Patel, P.K.: Device simulation of highly efficient eco-friendly $CH_3NH_3SnI_3$ perovskite solar cell. Sci. Rep. **11**(1), 3082 (2021)
34. Sarker, K., Sumon, M.S., Orthe, M.F., Biswas, S.K., Ahmed, M.M.: Numerical simulation of high efficiency environment friendly $CuBi_2O_4$-based thin-film solar cell using SCAPS-1D. Int. J. Photoenergy. **2023**(1), 7208502 (2023)

Electronic, Structural and Optical Properties of Magnesium Based Perovskite Chalcogenides [$MgZrS_3$]

Alok Upadhyay[1(✉)], Preksha Dubey[1], Naincy Pandit[2], Anusha Dubey[3], Shivraj Bhardwaj[1], and Ajay Singh Verma[4,5]

[1] Department of Physics, B. S. A. College (Dr. Bhimrao Ambedkar University, Agra), Mathura, India
alokupadhyay1020@gmail.com
[2] Department of Physics, School of Allied Sciences, Dev Bhoomi Uttarakhand University, Dehradun, India
[3] Department of Physics, Noida Institute of Engineering and Technology, Greater Noida, India
[4] Division of Research & Innovation, School of Applied and Life Sciences, Uttaranchal University, Dehradun, India
[5] University Centre for Research & Development, Department of Physics, Chandigarh University, Mohali, Punjab, India

Abstract. Chalcogenide perovskites are anticipated to be a significant topic of study in the upcoming years. This is due to a variety of factors involving the possibility of making photovoltaic cells with high efficacy, which have demonstrated great potential in a variety of applications for energy conversion devices. In the present work, we used magnesium-based chalcogenide perovskite compounds and we have computed different properties, namely structural, electronic, and optoelectronic, of $MgZrS_3$ by using the theory called density functional theory, implemented in the WIEN2k code. Additionally, this compound shows an adequate energy band gap and excellent absorbance coefficients for photovoltaic transformation. From our research, we found that $MgZrS_3$ can be used in the photovoltaic applications.

Keywords: Density Functional Theory · WIEN2k · Chalcogenide Perovskites · Optoelectronic · Optical Properties

1 Introduction

Solar energy contains a large amount of potential for PV and thermal utilization, which includes cooking, crop drying, and water heating [1, 2]. ABX_3 is the generic form of the perovskite structure, where A and B are known as metal cations and X is known as an anion (generally halogens and oxides). Inside an ideal unit cell, atom A is placed at the cube corners, atom B is positioned at the center, and atom X is positioned at the faces [3, 4]. Chalcogenide perovskites exist in both cubic and orthorhombic phases [5] and have an ABX_3 structure. These chalcogenide perovskites are the best materials for

S. Kumari et al. (Eds.): ICMTP 2024, SPPHY 311, pp. 62–66, 2025.
https://doi.org/10.1007/978-981-96-6795-6_7

photovoltaic applications and optoelectronic properties [6, 7]. Han et al. investigated that magnesium-based chalcogenides are a perfect choice for optoelectronics because they overcome the stability and toxicity problems that lead-based chalcogenides face. These materials show high carrier mobility and low defect density, which makes magnesium-based chalcogenides very suitable for photovoltaic applications [8, 9]. In an attempt to eliminate the toxicity of lead and in light of the previously mentioned literature, we conducted our study on $MgZrS_3$ utilizing DFT theory by applying the WIEN2K software.

2 Computational Details

The material's stability and electrical and optical properties may all be predicted using DFT. In order to swiftly screen materials for required properties and concentrate the search for materials with the best potential for a given application, predictive models are utilized in this process. The great precision and adaptability of WIEN2k are well recognized, making it an important tool for materials science researchers [10, 11]. Researchers use WIEN2k to examine the electrical characteristics and related properties of solid-state materials, such as metals, semiconductors, and insulators [12]. Cut-off parameters for convergence $R_{mt}K_{max} = 7$ are the parameter for the convergence of energy and charge. Here, R_{mt} is the lowest muffin tin radius, and K_{max} is the wave function cut-off. To prevent load leakage into the core states, the cut-off energy for the separation of valence and core states was established at −6.0 Ry. The energy difference between successive iterations is smaller than 10-5 Ry throughout the SCF process convergence. The charge convergence is up to 0.001 e; here, in this case, e denotes electron charge. We have simulated optical characteristics using a revised k-mesh of 104 k-points in the Irreducible Brillouin Zone (IBZ).

3 Results and Discussion

3.1 Structural Properties

The symmetry of the chalcogenide perovskite $MgZrS_3$, characterized by a deformed orthorhombic arrangement and associated with the space group (62_Pnma), is examined. The optimization of the structural characteristics has been carried out using the Wu-Cohen exchange-correlation potential (WC) [13]. It becomes crucial to optimize the structure in DFT [14, 15]. The entire energy variation was established at 0.001 eV/atom, with the charge adjusted to 0.0001 C/formula unit [15]. Volume optimization of the magnesium-based chalcogenide $MgZrS_3$ shows a parabolic curve (Fig. 1(a)). The optimized volume that minimizes the total energy of $MgZrS_3$ is found to be 3177.1928 ($bohr^3$). By using Murnaghan's equation of state, we find the stable structure file of $MgZrS_3$, which is very necessary for SCF calculations and also important for further properties like the optical and electronic properties of $MgZrS_3$ [16]. After the volume optimization, the lattice parameters of the stable crystal structure of $MgZrS_3$ are found to be (a = 12.792 bohr, b = 13.029 bohr, c = 18.215 bohr), and the value of the bulk modulus is found to be 68.776 GPa, with its first derivative being 3.9876.

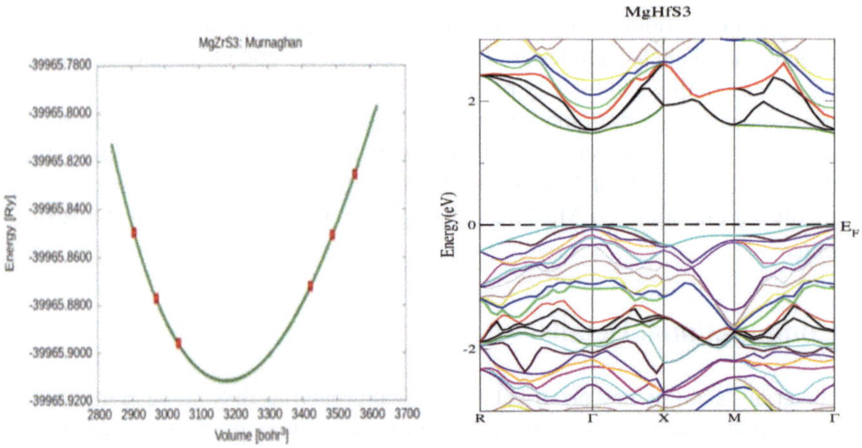

Fig. 1. Volume optimization curve and, (b) Band structure of $MgZrS_3$.

3.2 Electronic Properties

With the help of the modified Becke-Johnson potentials (TB-mBJ), the bandgap value for $MgZrS_3$ is 1.2 eV. The spectral gap of MgZrS3 is illustrated in Fig. 1(b). As we can see in the figure, the valence band maxima and conduction band minima are located in the same Brillouin zone (Γ), indicating a direct transition between electrons, which is advantageous for optoelectronics and photovoltaic applications [17]. Thus, these chalcogenide perovskites based on magnesium turn out to be the finest option [18, 19]. According to the band gap investigation, the d-states of Zr and the p-states of S lie at the Γ points, and it shows the direct gap of $MgZrS_3$.

3.3 Optical Properties

Optical properties of materials include optical conductivity σ (ω), optical absorption coefficient α(ω), and dielectric function ε(ω). These properties are strongly related to the interaction of substances with radiation. The dielectric constant ε(ω) can be expressed by the given Eq. (1) [20]:

$$\varepsilon(\omega) = \varepsilon_1(\omega) + i\varepsilon_2(\omega) \quad (1)$$

In Eq. (1), $\varepsilon_1(\omega)$ represents the real part of the dielectric function, and $\varepsilon_2(\omega)$ stands for the imaginary part of the dielectric function; these two parts of the dielectric functions are illustrated in Fig. 2(c, d). The optical conductivity σ(ω) of $MgZrS_3$ perovskite is affected by the energy of incoming photons, as illustrated in Fig. 2(a). In the figure, we can see that the optical conductivity is increasing in the visible region, which is very important for photovoltaic applications because the higher the optical conductivity, the lower the Joule heating. For $MgZrS_3$, the optical conductivity is approximately lying between (2×10^3) and (5×10^3) (1/Ω cm) in the visible range. Electrons are raised from the valence band maxima to the conduction band minima states as a result of radiation absorption within the material whenever α(ω) approaches its highest level [21, 22]. After entering the visible region, the value of the absorption coefficient can be seen increasing for $MgZrS_3$; the absorption coefficient quickly exceeds 1×10^5 cm^{-1} to 7×10^5 cm^{-1}, which is represented in Fig. 2(b). These values indicate that magnesium-based chalcogenides have remarkable parameters for use in photovoltaic cells and optoelectronics.

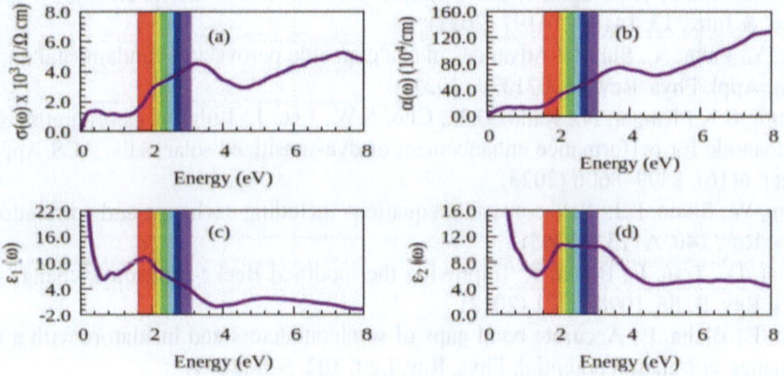

Fig. 2. Optical properties of $MgZrS_3$.

4 Summary and Conclusion

We discussed the structural, electronic, and optical properties of the orthorhombic chalcogenide perovskite structure $MgZrS_3$, with the help of WIEN2k code. Its crystal structure is orthorhombic with lattice parameters (a = 12.792 bohr, b = 13.029 bohr, c = 18.215 bohr). This compound exhibits a direct band gap, having attractive optical properties for the optical region. Elastic parameters of $MgZrS_3$ validated the mechanical stability, and its optical properties demonstrated information about their applicability in optoelectronic devices. Based on their optical, electronic, and structural characteristics, we determine that magnesium-based chalcogenide perovskites are promising for applications in energy conversion devices.

References

1. Rabaia, M.K.H., Abdelkareem, M.A., Sayed, E.T., Elsaid, K., Chae, K.J., Wilberforce, T., Olabi, O.G.: Environmental impacts of solar energy systems: a review. Sci. Total Environ. **754**, 141989 (2021)
2. Swami, S.K., Chaturvedi, N., Kumar, A., Dutta, V.: Effect of zinc precursor on Cu_2ZnSnS_4 nanoparticles synthesized by the solvothermal method and its application in dye-sensitized solar cells as the counter electrode. Mater. Today. **9**, 377–382 (2018)
3. Graef, M.D., McHenry, M.E.: Structure of Materials: an Introduction to Crystallography, Diffraction and Symmetry. Cambridge University Press, Cambridge ISBN.: 978-1-107-00587-7 (2012)
4. Swami, S.K., Barman, B., Kumar, A., Dutta, V.: Multilayer and thin transport conducting oxide fabrication using RF magnetron sputtering on flexible substrates. Organic Mater. **5**(01), 59–65 (2023)

5. Oumertem, M., Maouche, D., Berri, S., Bouarissa, N., Rai, D.P., Khenata, R., Ibrir, M.: Theoretical investigation of the structural, electronic and thermodynamic properties of cubic and orthorombhic $XZrS_3$ (X=Ba, Sr, Ca) compounds. J. Comp. Elec. **18**, 415–427 (2019)
6. Tiwari, D., Hutter, O.S., Longo, G.: Chalcogenide perovskites for photovoltaics: current status and prospects. J. Physics: Energy. **3**, 034010 (2021)
7. Shahmohamadi, H., Naghavi, S.S.: Sulfide perovskites for thermoelectricity. ACS Appl. Mate. & Inter. **13**, 14189–14197 (2021)
8. Han, Y., Fang, X., Shi, Z.: Advances in chalcogenide perovskites fundamental and applications. Appl. Phys. Rev. **11**, 021338 (2024)
9. Swami, S.K., Kumar, N., Radu, D.R., Cho, S.W., Lee, J.: Lithium incorporation into TiO_2 photoanode for performance enhancement of dye-sensitized solar cells. ACS Appl. Energy Mater. **6**(16), 8599–8606 (2023)
10. Kohn, W., Sham, L.J.: Self-consistent equations including exchange and correlation effects. Phys. Rev. **140**, A1133 (1965)
11. Koller, D., Tran, F., Blaha, P.: Improving the modified Becke-Johnson exchange potential. Phys. Rev. B. **85**, 1098–0121 (2012)
12. Tran, F., Blaha, P.: Accurate band gaps of semiconductors and insulators with a semilocal exchange-correlation potential. Phys. Rev. Lett. **102**, 5–8 (2009)
13. Sharma, S., Verma, A.S., Jindal, V.K.: First principles studies of structural, electronic, optical, elastic and thermal properties of Ag-chalcopyrites ($AgInX_2$: X= S, Se). Phys. B Condens. Matter. **438**, 97–108 (2014)
14. Birch, F.: Finite elastic strain of cubic crystals. Phys. Rev. **71**, 809–824 (1947)
15. Pandit, N., Singh, R., Kumar, A., Joshi, T.K., Shukla, A., Rani, U., Kamlesh, P.K., Kumar, T., Priyanka, Verma, A.S.: Physical properties and power conversion efficiency of SrZrX3 (X= S and Se). Modern Physics Lett. B. **38**, 2450345 (2024)
16. Murghan, F.D.: On the theory of the tension of an elastic cylinder. Proc. Natl. Acad. Sci. **30**, 244 (1944)
17. Kumar, A., Singh, A., Kumar, D., Garg, A., Dutta, V., Swami, S.K., Chaturvedi, N., Kumar, M.: Spray deposited carbon nanotube embedded ZnO as an electrons transport layer in inverted organic solar cells. Hybrid Adv. **4**, 100088 (2023)
18. Tauc, R., Girigorovici, R., Vancu, R.: Optical properties and electronic structure of amorphous germanium. Physical Status Solid. **2**, 627–637 (1996)
19. Sharma, S., Verma, A.S., Jindal, V.K.: Ab initio studies of structural, electronic, optical, elastic and thermal properties of silver gallium dichalcogenides ($AgGaX_2$: X= S, Se, Te). Mater. Res. Bull. **53**, 218–233 (2014)
20. Rani, U., Kamlesh, P.K., Agrawal, R., Kumari, J., Verma, A.S.: Electronic and thermophysical properties of double antiperovskites $X6SOA_2$ (X=Na, K and A= Cl, Br, I): a non-toxic and efficient energy storage materials. Int. J. Quantum Chem. **121**, e26759 (2021)
21. Nishigaki, Y., Nagi, T., Nishiwaki, M., Aizawa, T., Kozawa, M., Hanzawa, K., Kato, Y., Sai, H., Hiramatsu, H., Hosono, H., Fujiwara, H.: Extraordinary strong band –edge absorption in distorted chalcogenide perovskites. Sol. RRL. **4**, 1900555 (2020)
22. Tiwari, D., Hutter, O.S., Longo, G.: Chalcogenide perovskites for photovoltaic current status and prospects. J. Phys. Energy. **3**, 034010 (2021)

Structural, Electronic and Optical Properties of Formamidinium $FAGeCl_3$ Hybrid Halide Perovskites

Preksha Dubey[1](✉), Alok Upadhyay[1], Naincy Pandit[2], Anusha Dubey[3], Shivraj Bhardwaj[1], and Ajay Singh Verma[4,5]

[1] Department of Physics, B. S. A. College (Dr. Bhimrao Ambedkar University, Agra), Mathura, India
prekshad102@gmail.com

[2] Department of Physics, School of Allied Sciences, Dev Bhoomi Uttarakhand University, Dehradun, India

[3] Department of Physics, Noida Institute of Engineering and Technology, Greater Noida, India

[4] Division of Research & Innovation, School of Applied and Life Sciences, Uttaranchal University, Dehradun, India

[5] University Centre for Research & Development, Department of Physics, Chandigarh University, Mohali, Punjab, India

Abstract. In the present time the demands of the photovoltaic applications is very high because they are eco-friendly and avoid the danger caused by carbon emission due to fossil fuels. Hybrid halide Perovskite solar cells are viewed as an ecofriendly technique over the other techniques used earlier along with the improved materials for devices and mainly high efficiency of power conversion. In this work for the capacity for betterment in photovoltaic applications we investigate the Structural, electronic and optical properties of formamidinium $FAGeCl_3$ Hybrid Halide Perovskites with the help of density functional theory (DFT) based WIEN2k simulation software. It may be believed that formamidinium $FAGeCl_3$ Hybrid Halide perovskites have good potential for enhancement in photovoltaic applications.

Keywords: Hybrid Halide perovskites · DFT · WIEN2k · Band gap · Optical properties · Photovoltaic device

1 Introduction

Material scientists have intensified their quest for efficient materials in energy conversion devices with low-cost, high-stability, and non-toxic for human life. Herein, emerging materials inspire us to study the Organometallic- based hybrid perovskites [1, 2]. Perovskites solar cells (PSCs) are viewed as an eco-friendly option over the other used earlier along with the improved materials for devices and mainly high efficiency of power conversion up to 23.7% [3, 4]. The molecular formula for hybrid perovskite halide is ABX_3, in which A is a monovalent hybrid cation; B is the metallic cation, while X is an anion

S. Kumari et al. (Eds.): ICMTP 2024, SPPHY 311, pp. 67–72, 2025.
https://doi.org/10.1007/978-981-96-6795-6_8

halide [5]. Formamidinium hybrid halide perovskites are perfect for solar cell with outstanding electronic, optical properties and power conversion efficiency [6]. Even while lead (Pb) is the ideal material for solar cells, it is very poisonous and deadly to the environment [7]. We can replace lead with germanium because they have the same electron configuration and optoelectronic properties like lead [8]. So these lead-free perovskites are unstable in nature but Ge^{2+} have high carrier mobility with moisture resistance[9] and theoretical research predicts that the power efficiency of lead-free Ge halide perovskites up to 10% [10]. Formamidinium ($CH(NH_2)_2$) has recently been exploited as a viable cationic to develop effective perovskites with a good band gap in the visible region and an excellent coefficient of absorption [11, 12]. In order to improve the photovoltaic capabilities of perovskites and eliminate lead toxicity, Formamidinium Pb-free halides might be a perfect choice. Saini et al. [13] have reported Germanium based halide perovskites can be used in the list of lead-free halide perovskite. Ge-based perovskite have bandgap around 1.6 eV and PCE greater than 0.57% with ionic radius (73 pm). In this concern, According to Yin et al. [14], these materials' lengthy electron-hole diffusion length may be caused by their strong light absorption, low carrier efficient masses, and prominent point abnormalities. Using the WIEN2K software, we attempt our investigation on $FAGeCl_3$ with the aid of DFT theory. We investigated the optical, electrical, and structural properties of $FAGeCl_3$.

2 Computational Details

Density functional theory (DFT) can be used to predict the stability of the material as well as its electronic and optical properties [15, 16]. WIEN2k enables simulations under a variety of circumstances, including pressure, strain, and the existence of imperfections or flaws. It enables scientists to compute electrical characteristics, optimize crystal formations, display the results, and contrast them with experimental data [17, 18]. The WIEN2k algorithm has its foundation in the full-potential linearized augmented plane wave (FP-LAPW) approach of addressing the Kohn Sham equations. It was the initial LAPW code created accessible. The LAPW approach divides a single cell with interstitial and distinct atomic spheres that are centered on the atomic sites. The Muffin Tin Approximation (MTA) assumes that a potential is spherically symmetrical inside of atomic spheres and uniform outside for constructing constituent functions [19]. Therefore, $R_{mt}K_{max} = 3$ is the cutoff value for the parameters for the convergence of charge and energy. In this instance, Rmt is the lowest muffin tin radii, and K_{max} is the wave function cut off. In order to separate the valence and core states and stop load leakage into the core state, cut-off energy of −6.0 Ry was used. The energy difference between iterations is smaller than 10^{-5} Ry during the convergence of the SCF process. Even at 0.001 Coulomb/formula units, the charge converges. We have simulated optical properties in the Irreducible Brillouin Zone (IBZ) using a revised k-mesh of 10^4 k-points [20].

3 Results and Discussion

3.1 Structural Properties

$FAGeCl_3$ perovskites are structurally cubic. $FAGeCl_3$ has a space group of Pm_6m and lattice parameter of $FAGeCl_3$ is 10.5231 (bohr). To find the most stable unit cell layout, volume optimization calculations have been performed in Fig. 1(a). It Shows energy optimization curves that illustrate the total energy of a material at different lattice volumes. The equilibrium volume is determined using the Murnaghan equation of state [21, 22]. Its optimized volume is 1152.4975 $bohr^3$ and its optimized energy is −7265.418779 (Ry). The bulk modulus (B) of FAGeCl3 changes by 4.1701 at room pressure (BP), and its bulk modulus (B) is 69.4025 GPa.

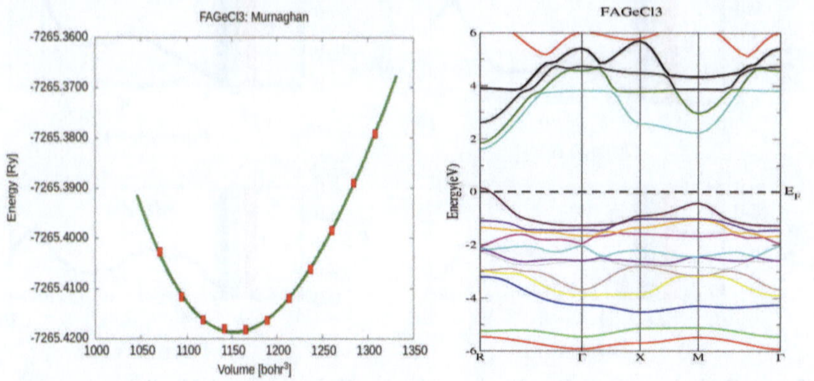

Fig. 1. (a) Volume optimization curve and, (b) The band structure of perovskites $FAGeCl_3$.

3.2 Electronic Properties

It is crucial to understand a material's electrical structure since it indicates which materials are suitable for a given application. Therefore, we looked into the band structure of $FAGeCl_3$, where we displayed the Fermi level using a dotted line. It was found that $FAGeCl_3$ has a direct band gap with a value of 1.731 eV. The band gap looks to be accurate because the Fermi energy has been adjusted to zero. The valence band maximum and the conduction band minimum are located in the R brillouin zone, which shows a direct transition between electrons in Fig. 1(b). The precise band gap of $FAGeCl_3$ is shown by its band structure, which makes it appropriate for solar applications [23, 24].

3.3 Optical Properties

Any material's optical characteristics are mostly controlled by the energy used in the visible light spectrum. This is often accomplished by figuring out the material's absorption coefficient and dielectric constants. The effects of incoming photon energy on the optical conductivity σ(ω) and optical absorption coefficient α (ω) of $FAGeCl_3$ perovskite are shown in Fig. 2(a, b). The visible region's maximum α (ω) value is 10^5 cm^{-1}. It is determined that the corresponding dependency of optical conductivity σ (ω) increases to a maximum value of 0.4 × 10^3 (1/Ω cm) in visible region. The dielectric constant ε (ω) is expressed as eq. (1) [25].

$$\varepsilon(\omega) = \varepsilon_1(\omega) + i\varepsilon_2(\omega) \quad (1)$$

Where ε_1 (ω) represents real part and ε_2 (ω) showing the imaginary part of dielectric constant [26]. Figure 2(c, d) shows the relative values of ε_1 (ω) and ε_2 (ω) are 5.7 and 1.6 in visible spectrum.

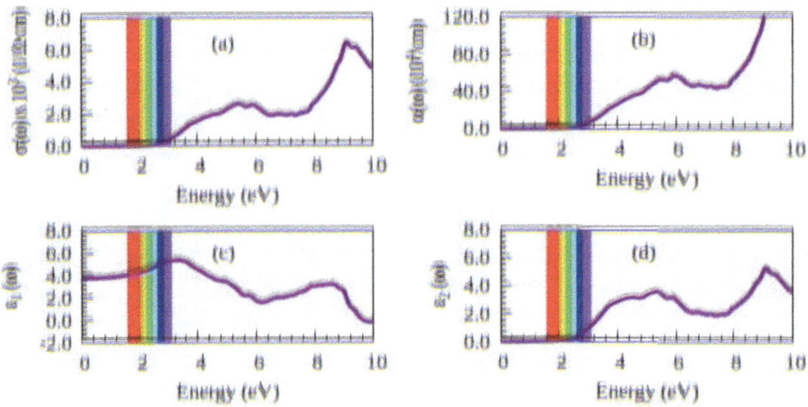

Fig. 2. (a) Optical conductivity, (b) absorption coefficient (c) and (d) dielectric constants of $FAGeCl_3$.

4 Summary and Conclusion

Using the WIEN2K simulation software, the electronic, optical, and structural characteristics of the cubic-shaped $FAGeCl_3$ hybrid halide perovskite with a lattice constant of 10.5231 (bohr) were investigated. The first-order differential of the bulk modulus relative to pressure was 4.1701, its optimized volume was 1152.4975 ($bohr^3$), and its bulk modulus (B) was 69.4025 (GPa), according to our volume optimization for stable crystal structure measurements. A thorough examination of its electrical and optical characteristics shows that it act as a semiconductor with a direct band gap of 1.731 eV and performs well in the optical spectrums. Its excellent optical conductivity and high absorption coefficient further demonstrate its potential for solar applications.

References

1. Pachori, S., Agrawal, R., Choudhary, B.L., Verma, A.S.: An efficient and stable lead-free organic–inorganic tin iodide perovskite for photovoltaic device: Progress and challenges. Energy Rep. **8**, 5753–5763 (2022)

2. Kumar, A., Singh, A., Kumar, D., Garg, A., Dutta, V., Swami, S.K., Chaturvedi, N., Kumar, M.: Spray deposited carbon nanotube embedded ZnO as an electrons transport layer in inverted organic solar cells. Hybrid Adv. **4**, 100088 (2023)
3. Chaturvedi, N., Swami, S.K., Dutta, V.: Plasmonic effect of spray-deposited Au nanoparticles on the performance of inverted organic solar cells. Nanoscale. **6**, 10772–10778 (2014)
4. Rani, U., Kamlesh, P.K., Agrawal, R., Shukla, A., Verma, A.S.: Emerging study on lead-free hybrid double perovskite $(CH_3NH_3)_2AgInBr_6$: potential material for energy conversion between heat and electricity. Energ. Technol. **10**, 2200002 (2022)
5. Pachori, S., Agarwal, R., Prakash, B., Kumari, S., Verma, A.S.: Fundamental physical properties of nontoxic tin-based Formamidinium $FASnX_3$ (X= I, Br, Cl) hybrid halide perovskites: future opportunities in photovoltaic applications. Energ. Technol. **10**, 2100709 (2022)
6. Joshi, T.K., Sharma, G., Agarwal, R., Verma, A.S.: Emerging potential photovoltaic absorber hybrid halide perovskites ($CH_3CH_2NH_3PbX_3$; X= Br, Cl) materials: an ab-initio calculation. Int. J. Energy Res. **45**, 15231–15244 (2021)
7. Chaudhary, J., Gupta, S.K., Verma, A.S., Negi, C.M.S.: Impact of electron transport layer material on the performance of $CH_3NH_3PbBr_3$ perovskite-based photo detectors. J. Mater. Sci. **55**, 4345–4357 (2020)
8. Flora, G., Gupta, D., Tiwari, A.: Toxicity of lead: a review with recent updates. Interdiscip. Toxicol. **5**, 47–58 (2012)
9. Liang, L., Gao, P.: Lead-free hybrid perovskite absorbers for viable application: can we eat the cake and have it too. Adv. Sci. **5**, 1700331 (2018)
10. Ju, M.G., Dai, J., Ma, L., Zeng, X.C.: Lead-free mixed tin and germanium perovskites for photovoltaic application. J. Am. Chem. Soc. **139**, 8038–8043 (2017)
11. Filip, M.R., Giustino, F.: Computational screening of homovalent lead substitution in organic–inorganic halide perovskites. J. Physical Chem. C. **120**, 166–173 (2016)
12. Chaturvedi, N., Swami, S.K., Dutta, V.: Spray deposition of poly (3- hexylthiophene) and [6,6]—phenyl—C61 - butyric acid methyl ester blend under electric field for improved interface and organic solar cell characteristics. Thin Solid Films. **598**, 82–87 (2016)
13. Liu, D., Li, Q., Wu, K.: Ethyl ammonium as an alternative cation for efficient perovskite solar cells from first-principles calculations. RSC Adv. **9**, 7356–7361 (2019)
14. Sani, F., Shafie, S., Lim, H.N., Musa, A.O.: Advancement on lead-free organic-inorganic halide perovskite solar cells: a review. Materials. **11**, 1008 (2018)
15. Yin, W.J., Shi, T., Yan, Y.: Unique properties of halide perovskites as possible origins of the superior solar cell performance. Adv. Mater. **26**, 4653–4658 (2014)
16. Burke, K., Wagner, L.O.: DFT in a nutshell. Int. J. Quantum Chem. **113**(2), 96–101 (2013)
17. Schwarz, K., Blaha, P., Trickey, S.B.: Electronic structure of solids with WIEN2k. Mol. Phys. **108**, 3147–3166 (2010)
18. Schwarz, K., Blaha, P., Madsen, G.K.H.: Electronic structure calculations of solids using the WIEN2k package for material sciences. Comput. Phys. Commun. **147**, 71–76 (2002)
19. Sjostedt, E., Nordstrom, L., Singh, D.J.: An alternative way of linearizing the augmented plane-wave method. Solid State Commun. **114**, 15 (2000)
20. Fu, Y., Hautzinger, M.P., Luo, Z., Wang, F., Pan, D., Aristov, M.M., Guzei, I.A., Pan, A., Zhu, X., Jin, S.: Incorporating large A cations into lead iodide perovskite cages: relaxed Goldschmidt tolerance factor and impact on exciton–phonon interaction. ACS. Cent. Sci. **5**, 1377–1386 (2019)
21. Joshi, T.K., Sharma, G., Verma, A.S., Sharma, Y.R.: A density functional study of electronic and optical properties of perovskite ($CH_3CH_2NH_3PbI_3$) for photovoltaic cell application. Comp. Cond. Matter. **30**, e00635 (2022)
22. Murnaghan, F.D.: On the theory of the tension of an elastic cylinder. Proc. Natl. Acad. Sci. Unit. States Am. **30**, 382–384 (1944)

23. Tran, F., Blaha, P.: Accurate band gaps of semiconductors and insulators with a semi local exchange-correlation potential. Phys. Rev. Lett. **102**, 5–8 (2009)
24. Chaturvedi, N., Alam, F., Swami, S.K., Dutta, V.: Effect of electric field on the spray deposited poly (3, 4- ethylenedioxythiophene): poly (Styrenesulfonate) layer and its use in organic solar cell. J. Appl. Phys. **114**, 184501 (2013)
25. Rani, U., Kamlesh, P.K., Agrawal, R., Kumari, J., Verma, A.S.: Electronic and thermophysical properties of double antiperovskites X_6SOA_2 (X=Na, K and A= Cl, Br, I): a non-toxic and efficient energy storage materials. Int. J. Quantum Chem. **121**, e26759 (2021)
26. Sharma, S., Verma, A.S., Jindal, V.K.: Ab initio studies of structural, electronic, optical, elastic and thermal properties of silver gallium dichalcogenides ($AgGaX_2$: X= S, Se, Te). Mater. Res. Bull. **53**, 218–233 (2014)

Theoretical Investigation of Structural, Electronic and Optical Properties of Rb_2AgBiX_6 (X = Cl, Br, I) Double Perovskite for Multiple Applications

Shradha Kalkhundiya[1], Yashaswi Soni[2], Naincy Pandit[3], Vibhav Kumar Saraswat[4], and Ajay Singh Verma[1,5](✉)

[1] Department of Physics, School of Applied and Life Science, Uttaranchal University, Dehradun 248007, India
ajay_phy@rediffmail.com

[2] Department of Physics, Rajeev Gandhi Govt PG College, Mandsaur 458001, Madhya Pradesh, India

[3] Department of Physics, School of Allied Science Dev Bhoomi Uttarakhand University, Dehradun 247232, India

[4] Department of Physics, Army Cadet College, Indian Military Academy, Dehradun 248007, India

[5] University Centre for Research and Development, Department of Physics, Chandigarh University, Mohali 140143, Punjab, India

Abstract. Rb_2AgBiX_6 (X = Cl, Br, I) double perovskites have been studied theoretically using DFT (density functional theory) as implemented in the WIEN2k algorithm. These materials' tuneable band gaps and intriguing electrical characteristics have drawn interest in light of their possible uses in optoelectronic systems. Both $Rb_2AgBiCl_6$, $Rb_2AgBiBr_6$ and Rb_2AgBiI_6 crystallise in a cubic perovskite structure, retaining stability and displaying minor differences in lattice parameters because of the substitution of halides, according to the structural study. Due to the halides differing electronegativity and atomic sizes, the electronic structure calculations reveal that these double perovskites have indirect band gaps, with the band gap energy dropping as the halide transitions from Cl to I. These materials are appropriate for photovoltaic and photodetector applications due to a notable absorption in the visible region, as revealed by optical property investigation. The potential of Rb_2AgBiX_6 double perovskites in next-generation optoelectronic devices is highlighted by the combination of these features.

Keywords: Density Functional Theory · WIEN2k · Double Perovskites · Optoelectronic

S. Kumari et al. (Eds.): ICMTP 2024, SPPHY 311, pp. 73–80, 2025.
https://doi.org/10.1007/978-981-96-6795-6_9

1 Introduction

DPs (Double perovskites) have attracted significant attention due to structural elasticity, miscellaneous electronic properties and ecologically friendly composition. Contrasting traditional lead-based perovskites, which raise toxicity issues, lead-free double perovskites display a safer option while maintaining splendid material performance. Double perovskites offer a more environmentally friendly composition despite having the ability to retain the desirable optoelectronic characteristics of traditional perovskites, which is one of the most important key advantages of DPs [1]. The toxicity of lead is a major drawback otherwise lead-based perovskites such as $MAPbI_3$ have demonstrated outstanding efficiency in solar cells [2]. As a result, lead-free alternatives like bismuth and silver have arisen as promising replacements due to their similar electronic configurations and capacity to form stable structures [3]. Bi-based perovskites have drawn significant attention, where bismuth replaces lead at the B site with its heavy atomic mass and relatively low toxicity, preserving many of the desirable electronic properties obtained in lead-based materials [4]. Overcoming the inherent toxicity of lead-based materials is one of the key challenges in advancing perovskites-based technologies [5]. Tripathi et al. [6] have used DFT to study the halide-substituted double perovskite Cs_2AgBiX_6 (X = Br, Cl, I) free of lead. Strong resilience was indicated by the greatest elastic constants and modulus values reported in $Cs_2AgBiBr_6$. Rehman et al. [7] have obtained that as halide varies in Rb_2AgBiX_6 (X = Cl, Br, I) DPs the band gap also decreases, which is a vital factor for tuning the efficiency of materials. Ali et al. [8] have used DFT which glorifies the maximum light absorption occurring in UV-IR ranges for Rb_2 $AlAgX_6$ (X = Cl, I). Double perovskites based on these elements like Rb_2AgBiX_6, represent a promising possibility for future applications. The exploration for Substitute materials based on bismuth and silver has directed to investigation of Rb_2AgBiX_6 as an ideal candidate for lead-free perovskites for optoelectronic application. While Rb_2AgBiX_6 materials are more stable than organic-inorganic perovskites, further improvements in environmental stability are desirable for long-term use in devices [9].

2 Computational Methodology

To explore the structural, optical and electronic properties of double perovskites DFT is employed. It is a quantum mechanical modelling method utilizing Hohenberg-Kohn theorems, where the lowest state energy is the function of the electron density for a complex electron system. To perform the DFT WIEN2k code, an implementation of the Full-Potential Linearized Augmented Plane wave method was employed. To perform highly accurate calculations of structure and electronic properties WIEN2k is well suited for complex materials like double perovskites. Structural optimization was done by Wu-Cohen Generalized Gradient Approximation (WC-GGA) potential. Once, the structure is optimized, Generalized Gradient Approximation -Perdew Burke Ernzerhof (GGA-PBE) was preferred as the exchange-correlation function for characterisation studies [10]. The cubic perovskite's structure was assumed for each material in which atoms were positioned fitting to symmetry constraints. Along high-symmetry points in $10 \times 10 \times 10$ in the Brillouin zone using WIEN2k the electronic band structure of Rb_2AgBiX_6

was calculated. The optical properties of Rb_2AgBiX_6 were investigated using an optic program. The optical absorption coefficient, reflectivity, energy loss and refractive index were appraised as functions of photon energy. These properties are critical for evaluating the suitability of the materials for photovoltaic and optoelectronic applications [11–12].

3 Results and Discussions

3.1 Structural Properties

Rb_2AgBiX_6 (X = Br, I, Cl) possess a cubic crystal structure that belongs to space group number 225 (fm_3m). Optimization is performed on the cubic structures for all materials, yielding optimized energy (R_Y) values illustrated against ground state volume (a.u.^{3}) exhibited in Fig. 1(a, b and c). The optimized structure parameters of Rb_2AgBiX_6 (X = Cl, Br, I), derived from the volume minimization method via the Murnaghan equation of state [13] are tabulated in Table 1. With varying halide substitution, both the volume

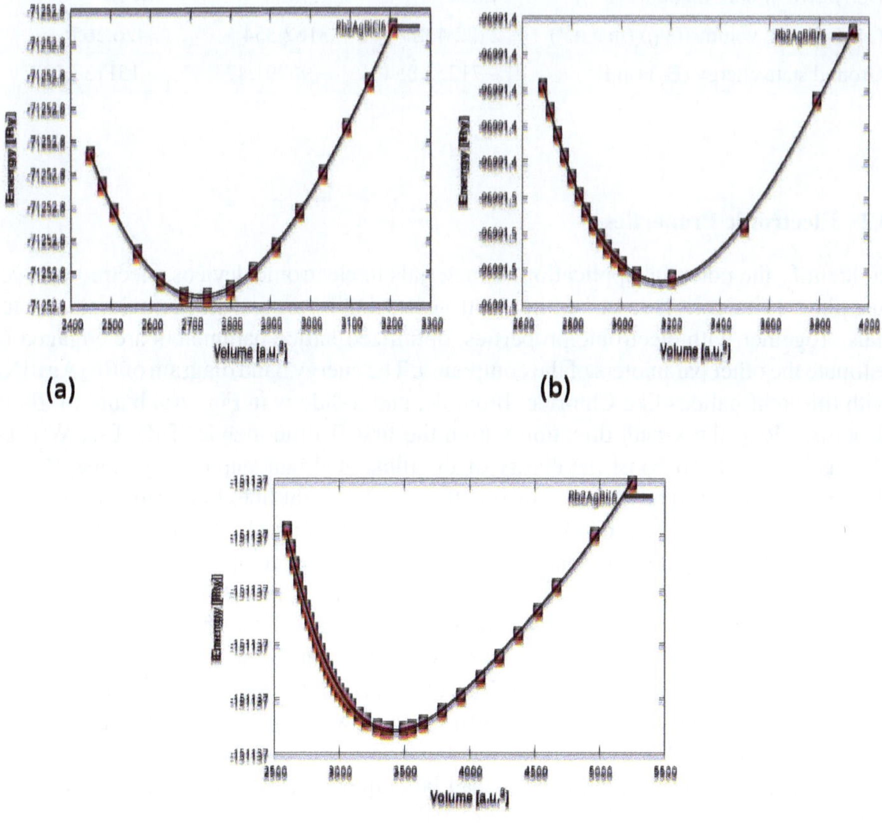

(c)

Fig. 1. Geometry optimization (E vs V) (**a**) $Rb_2AgBiCl_6$, (**b**) $Rb_2AgBiBr_6$, (**c**) Rb_2AgBiI_6.

and lattice constant of Rb_2AgBiX_6 (X = Br, Cl, I) decreased consequently, due to reduced interatomic distance. Figure 1(a–c) illustrates the variation in lattice constant and unit cell volume resulting from halide variation. As the halide changes from Cl to I, the increasing lattice constant suggests a relaxation of the atomic structure. This larger interatomic spacing decreases the overlap of atomic orbitals, which stimuli the material's electronic band structure as well as its bandgap. The larger lattice constant for Rb_2AgBiI_6 suggests a narrower bandgap, which is further confirmed by the electronic structure analysis. Larger lattice constants for Rb_2AgBiI_6 suggest a narrower bandgap, confirmed by electronic structure analysis.

Table 1. The lattice parameter of the optimized structure of Rb_2AgBiX_6 (X = Cl, Br, I).

Compounds	$Rb_2AgBiCl_6$	$Rb_2AgBiBr_6$	Rb_2AgBiI_6
Lattice constant (a_0) (in Å)	12.079	13.262	14.597
Bulk modulus (B_0) (in GPa)	19.246	16.459	15.904
Derivative of bulk modulus (B'_0)	4.634	4.461	4.727
Ground state volume (V_0) (in a.u.^{3})	2722.432	3162.354	3426.265
Ground state energy (E_0) (in R_y)	−71252.854	−96991.473	−151137.127

3.2 Electronic Properties

To identify the potential application of materials in electronic devices, electronic structure plays a vital role. It describes the electron distribution around the Fermi level in materials. Together with electronic properties, optimized lattice parameters are engaged to estimate the other parameters of the compound. The energy band diagram of Rb_2AgBiX_6 with different halides like Chloride, bromide, and iodide is in Fig. 2(a, b and c). Band structure along the k-path direction within the first Brillouin zone (BZ): Γ-X-W-Γ-L-W was bordered. To boost the clarity of the illustrated bandgap energy value, the E_F (Fermi level) is set to be at zero shown with a solid straight line. Bandgap was obtained from the difference between CBM (conduction band minima) from VBM (valence band maxima). The computational band structural study of the compounds Rb_2AgBiX_6 (X = Cl, Br, I) shows the existence of an indirect band gap with VBM positioned between the X-W - points for compounds with halide Chloride and Bromide and for Iodide, the position of VBM is at Γ. CBM is located at the L-point in the first Brillouin zone of the reciprocal lattice. The Bandgap of $Rb_2AgBiCl_6$ and $Rb_2AgBiBr_6$ is calculated to be 1.503 eV and 1.198 eV while for Rb_2AgBiI_6, which was obtained as 0.6909 eV i.e., due to the substitution halide (Cl to I) the band gap decreases simultaneously. Due to these critical characteristics, these materials can be employed as semiconductors, rendering them appropriate for optoelectronics applications [14].

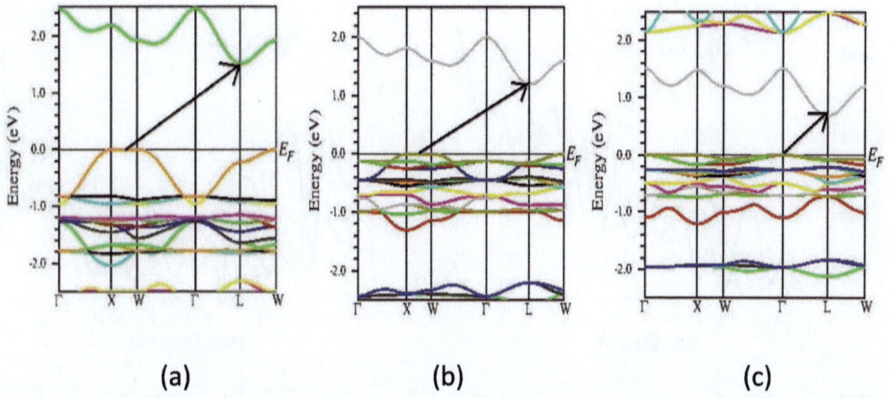

Fig. 2. Energy band diagram of (**a**) $Rb_2AgBiCl_6$, (**b**) $Rb_2AgBiBr_6$, and (**c**) Rb_2AgBiI_6.

3.3 Optical Properties

Some key optical properties that can illustrate how light interacts through a material are absorption, reflection, refraction and energy loss. They disclose critical information about the material structure, energy dissipation and electronic transitions. As these crucial optical parameters govern the rate of charge carrier recombination within a solid, they highlight the feasibility of the material for use in optoelectronic devices [15]. Refractive index, energy loss, absorption and reflectivity are among the optical properties of Rb_2AgBiX_6 (X = Cl, I, Br), which have been thoroughly investigated. These properties arise from the interaction between wave-matter as well as the interaction between the substance and waves, which are influenced by the frequency of electromagnetic waves [16].

In evaluating a material's light absorption capabilities, the optical absorption coefficient (α) is a crucial parameter used for optimizing solar energy conversion efficiency. Figure 3(a) displays each compound exhibits a unique absorption line, suggesting differences in their electronic structures. For all three compounds absorption begins at rather low photon energies around 2 eV, which specifies that these materials have relatively small bandgaps to allow them to absorb visible light. Multiple absorption peaks are shown in the spectra. A vibrant absorption edge is obtained in the spectra, particularly for $Rb_2AgBiCl_6$ and $Rb_2AgBiBr_6$, this edge marks the onset of strong absorption, beyond which the material becomes highly absorbing. Among the three $Rb_2AgBiCl_6$ compound exhibits the highest absorption intensity in higher photon energies. It has a well-defined absorption edge and multiple distinct peaks. $Rb_2AgBiBr_6$ also has a similar outline to $Rb_2AgBiCl_6$ but with low intensities as well as a less pronounced absorption edge whereas Rb_2AgBiI_6 has the lowest absorption intensity out of all three.

Optical reflectivity delivers insights into its surface characteristics, by calculating the amount of light energy reflected by the surface can outline the surface properties of the solids. As, shown in Fig. 3(b) all three compounds illustrated relatively low reflectivity throughout the entire photon energy range, with a value below 0.4. This shows that these materials absorb a substantial portion of incident light. It can arise from electronic transitions within the materials, which indicate incident light is absorbed

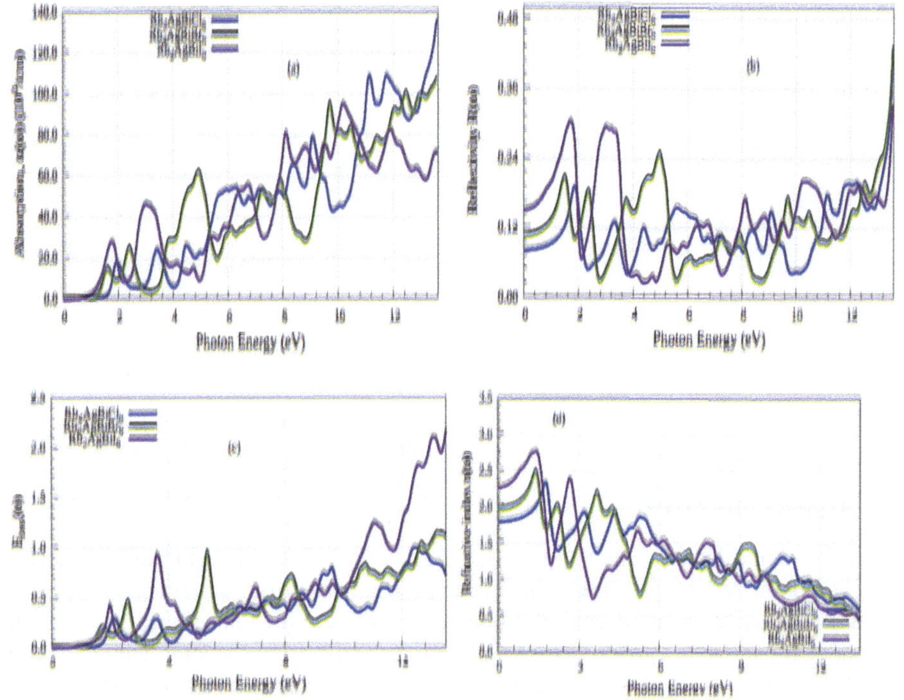

Fig. 3. Optical Properties (a) Absorption Coefficient (b) Reflectivity (c) Energy loss (d) Refractive index of Rb_2AgBiX_6 (X = Cl, Br, I).

and re-emitted from that region. $Rb_2AgBiCl_6$ and $Rb_2AgBiBr_6$ exhibit similar spectral profiles but the $Rb_2AgBiCl_6$ compound shows the lowest reflectivity among the three whereas Rb_2AgBiI_6 has the highest reflectivity in the Visible region and $Rb_2AgBiBr_6$ shows a gradual increase in reflectivity towards higher photon energies.

During the collision of a photon and a fast-moving electron, energy loss is an essential factor that determines the degradation of the material. They cause heating effects and dispersion phenomena [17]. The plot of the energy loss function E_{loss} (ω), is illustrated in Fig. 3(c). Peak with the highest energy loss is observed at 13 eV for Rb_2AgBiI_6, 9.58 eV for $Rb_2AgBiCl_6$ and 5.34 eV for $Rb_2AgBiBr_6$.

Since the refractive index n(ω) determines the velocity with which light propagates through a solid, it supports predictions of the stability of materials for device application. Moreover, it enables the calculation of the phase velocity within the medium for an electromagnetic wave. After analyzing Fig. 3(d) it has been concluded that the refractive index of Rb_2AgBiI_6 originates at 2.3 and gradually increases to a maximum value of 2.8 at 1.4 eV. In contrast, the refractive index of $Rb_2AgBiCl_6$ and $Rb_2AgBiBr_6$ originates at 1.8 and 2, respectively, and reaches 2.36 and 2.54 at 1.75 eV and 1.4 eV. Nevertheless, the value of the refractive index in the high photon energy region for all compounds falls below unity, resulting in a group velocity i.e. directed towards the negative domains [18, 19]. The compounds under examination display distinguished absorption peaks, with the greatest peak obtained at 13.6 eV. Due to its electrical and optical properties, this substance is possibly used in light-emitting semiconductor diodes and solar cells. Rb_2AgBiI_6 plays a vital role in this context, which can facilitate high power efficiency conversion.

4 Conclusion

This paper provides an overview of the structural, electronic, and optical characteristics of cubic double perovskites Rb_2AgBiX_6 (X = Cl, Br, I) using DFT computations. The values of lattice constants 12.079 Å, 13.262 Å, and 14.597 Å and bandgaps 1.503 eV, 1.198 eV, and 0.6909 eV for $Rb_2AgBiCl_6$, $Rb_2AgBiBr_6$ and Rb_2AgBiI_6 respectively. By the halide substitution, the lattice constants and band gaps are influenced. All compounds have an indirect band gap, which decreases from Cl to I. This is an assessment of the electrical properties; this tunability can be used for some specific applications. The optical characteristics, including reflectivity, absorption, refractive index and energy loss, were examined within the energy range of 0–13.5 eV. Significantly, these compounds are suitable for optoelectronic devices due to their large absorption spectra within the UV energy range and minimal reflectance.

References

1. Chu, L., Ahmad, W., Liu, W., Yang, J., Zhang, R., Sun, Y., Yang, J., Li, X.: Lead-free halide double perovskite materials: a new superstar toward green and stable optoelectronic applications. Nano-Micro Lett. **11**, 16 (2019)
2. Lu, R., Liu, Y., Zhang, J., Zhao, D., Guo, X., Li, C.: Highly efficient (200) oriented $MAPbI_3$ perovskite solar cells. Chem. Eng. J. **433**, 133845 (2021)
3. Cucco, B., Pedesseau, L., Katan, C., Even, J., Kepenekian, M., Volonakis, G.: Silver–bismuth halide double salts for Lead-free photovoltaics: insights from symmetry-based modeling. Sol. RRL. **6**, 2200718 (2022)
4. Chen, H., Zhang, C., Liu, Z., Gong, J., Zhang, M., Wu, Y., Chen, Y., Chen, H.: Optoelectronic properties of Rb-doped inorganic double perovskite $Cs_2AgBiBr_6$. Chem. Phys. Lett. **771**, 138501 (2021)
5. Ikram, M., Malik, R., Raees, R., Imran, M., Wang, F., Ali, S., Khan, M., Khan, Q., Maqbool, M.: Recent advancements and future insight of lead-free non-toxic perovskite solar cells for sustainable and clean energy production: a review. Sustain Energy Technol Assess. **53**, 102433 (2022)
6. Tripathi, M.N., Saha, A., Singh, S.: Structural, elastic, electronic and optical properties of lead-free halide double perovskite Cs_2AgBiX_6 (X = Cl, Br, and I). Mater. Res. Express. **6**, 115517 (2019)
7. Rehman, M.A., Rehman, Z.U., Usman, M., Bano, N., Riaz, M., Ali, S.M., Farid, A.: Identification of lead-free Rb2AgBiX6 (X=Cl, Br, I) double halide perovskites for promising photovoltaic applications: first-principles investigations. Phys. B Condensed Matter. **695**, 416499 (2024)
8. Ali, M., Ahmed, F., Shahzad, H.N.: Investigation of structural, electronic, optical and thermoelectric properties of halide double perovskite Rb2AlAgX6 (X = Cl, I): a first-principles DFT study. J. Physics Chem. Solids. **192**, 112097 (2024)

9. Guo, S., Chen, C., Zhang, Z., Cai, Y., Wang, S., Gao, S., Chen, W., Abduryim, E., Dong, C., Guan, X., Liu, Y., Lu, P.: Theoretical investigations of double perovskite Rb_2AgBiX_6 (X = Cl, Br, I) for solar cell and photocatalysis applications. Mater. Sci. Semicond. Process. **177**, 108373 (2024)
10. Perdew, J.P., Burke, K., Ernzerhof, M.: Generalized gradient approximation made simple. Phys. Rev. Lett. **77**, 3865 (1996)
11. Okoye, C.M.I.: Theoretical study of the electronic structure, chemical bonding and optical properties of $KNbO_3$ in the paraelectric cubic phase. J. Physics: Condensed Matter. **15**, 5945 (2003)
12. Sharma, S., Verma, A.S., Jindal, V.K.: Ab initio studies of structural, electronic, optical, elastic and thermal properties of silver gallium dichalcogenides ($AgGaX_2$: X= S, Se, Te). Mater. Res. Bull. **53**, 218–233 (2014)
13. Murnaghan, F.D.: The compressibility of media under extreme pressures. Proc. Natl. Acad. Sci. **30**, 244–247 (1944)
14. Bakeer, A.M., Mahmoud, N.T., Shaltaf, R.: Coupled insights into structural, electronic, and optical properties of $GeTMS_3$ (TM= Sc, Fe, Zn) sulfide perovskite across different phases: Ab initio study. Int. J. Energy Res. **2024**, 2316636 (2024)
15. Liu, X., Xie, B., Duan, C., Wang, Z., Fan, B., Zhang, K., Lin, B., Colberts, F.J.M., Ma, W., Janssen, R.A.J., Huang, F., Cao, Y.: A high dielectric constant non-fullerene acceptor for efficient bulk-heterojunction organic solar cells. J. Mater. Chem. A. **6**, 395–403 (2018)
16. Shahzad, M.K., Zhang, Y., Cui, L., Liu, L., Butt, M.K., Li, H.: Dispersing upconversion nanocrystals in PMMA microfiber: a novel methodology for temperature sensing. RSC Adv. **8**, 19362–19368 (2018)
17. Ilyas, I., Zafar, A.A., Ullah, H.M.N., Rizwan, M., Usman, Z., Gillani, S.: Pressure-induced elastic, mechanical and optoelectronic response of $RbCdF_3$: a comprehensive computational approach. J. Phys. Chem. Solid. **165**, 110642 (2022)
18. Wang, L.J., Kuzmich, A., Dogariu, A.: Gain-assisted superluminal light propagation. Nature. **406**, 277–279 (2000)
19. Mugnai, D., Ranfagni, A., Ruggeri, R.: Observation of superluminal behaviours in wave propagation. Phys. Rev. Lett. **84**, 4830–4833 (2000)

Investigation of Structural, Electronic and Optical Properties of $ASiH_3$ (A = Li, K, Na, Mg) Hydrogen Based Perovskites

Naincy Pandit[1], Anusha Dubey[2], Yogendra Kumar Saraswat[3], and Ajay Singh Verma[4](✉)

[1] Department of Physics, School of Allied Sciences, Dev Bhoomi Uttarakhand University, Dehradun 248007, India

[2] Department of Physics, Noida Institute of Engineering and Technology, Greater Noida 201306, India

[3] Department of Chemistry, B S A College, Mathura 281004, India

[4] Division of Research & Innovation, School of Applied and Life Sciences, Uttaranchal University, Dehradun 248007, India

ajay_phy@rediffmail.com

Abstract. During these years chalcogenide perovskites was proven to be the best choice because of its great potential in a diversity of applications for the energy conversion and ecological stability. This study includes substantial properties of $ASiH_3$ hydrogen based perovskites along with the help of density functional theory (DFT). The theoretical calculation was completed with WIEN2k simulation code. All these compounds are mechanically, dynamically and thermally stable. The lattice constant for $LiSiH_3$, $NaSiH_3$, $MgSiH_3$ and $KSiH_3$, is 3.65, 3.75, 3.86 and 3.65 Å, respectively. All the materials are brittle in nature having zero bandgap with ionic bonding. The optical properties show the absorption coefficient exceeds 10^5 cm^{-1}, which is suitable for the optical absorption of visible light. Some thermodynamic parameters has been calculated namely Debye temperature and melting temperature and has been discussed.

Keywords: Density Functional Theory · WIEN2k · Hydrogen Storage · Optoelectronic

1 Introduction

Different factors are responsible for the increasing energy demands, including population growth and improving living standards. International energy Agency said that utilization of fossil fuels is responsible for the majority of demands [1–3]. This fossil fuels dependency conduct to the emission of human caused CO_2 and cause severe climate occurrence like floods; droughts and deforestation. Scientists already warn that if the global temperature increases 2 degree Celsius more poses a momentous treat livelihood being on planet. Renewable hydrogen or clean hydrogen can be applied in the industrial sectors for the decarbonization of steel and cements. As per the entire situation by 2050;

S. Kumari et al. (Eds.): ICMTP 2024, SPPHY 311, pp. 81–87, 2025.
https://doi.org/10.1007/978-981-96-6795-6_10

the hydrogen globally demand is awaited to increase up to 700% as compared current creation of 70 million tons annually [4, 5]. The requirement for the temperature is low for physically bound hydrogen and high for chemically bound hydrogen for achieving the high storage capacity [6]. The transition metal halides $LiSiH_3$; $KSiH_3$; $NaSiH_3$; and $MgSiH_3$ are having the exceptional physical properties which attracted researcher's interest. The possible uses of these perovskite materials in upcoming hydrogen fuel cells and associated electronic components have been the talking point of much research as a result of this interest In current paper we have used the Density Functional Theory (DFT) for the calculation of different parameters of $ASiH_3$ (A = Li; K; Na and Mg) hydride perovskites. Structural properties includes the lattice constants and ground state energy calculation, Electronic properties shows the Bandgap and density of states calculation, inside the optical properties we will find out the absorption coefficient, reflectivity and conductivity. The results of this study have a great deal of potential to direct further theoretical and experimental investigations into the hydride materials.

2 Computational Details

Wien-2 k simulation packages has been used for the calculation by using density functional theory (DFT) by utilizing the Full-potential linearized augmented plane wave (FP-LAPW) [7, 8]. Exchange correlation potentials which is generalized gradient approximation of Wu- Cohen (WC-GGA) Optimization has been used for the calculation of structure its optimization and SCF which is self-consistent-field has been complete [9–12] while for calculating electronic properties and further calculation LDA potential has been used. The total charge and the energy criteria must be achieved to 0.1 mRy and 0.0001 e until this the repetition process continuously runs. The k-mesh values is 10,000 k-points in the BZ (Brillouin zone) for the optical parameters. Some calculated optical parameters such as refractive index, reflectivity, loss functions and the optical conductivity all are computed using the WIEN2k simulation code.

3 Results and Discussion

3.1 Structure Properties

The hydride perovskites $ASiH_3$ (A = Li, K, Na and; Mg) shows Pm-3 m space group. The hydride crystal structure of cubic $ASiH_3$ compounds has been pictured in Fig. 1. The formation of these compounds unit cells by fractional coordinates, with all the atoms A atoms pinpoint at corner side point (0; 0; 0); and Si is placed at the center points (0.5; 0.5; 0.5); and all the Hydrogen atoms is placed at the face centered (0.5, 0.5, 0); (0.5, 0, 0.5); and (0, 0.5, 0.5) octahedral sites, respectively.

For the better understanding of phase transition and structural stability the energy volume curve has been shown in Fig. 2. Since these compounds have not yet been investigated, their computed outcomes may be contrasted with those of subsequent research. The different parameters studied during optimization have been listed in Table 1. By having a look into lattice constants we can conclude that lattice constant of $KSiH_3$ is largest than that of other compounds.

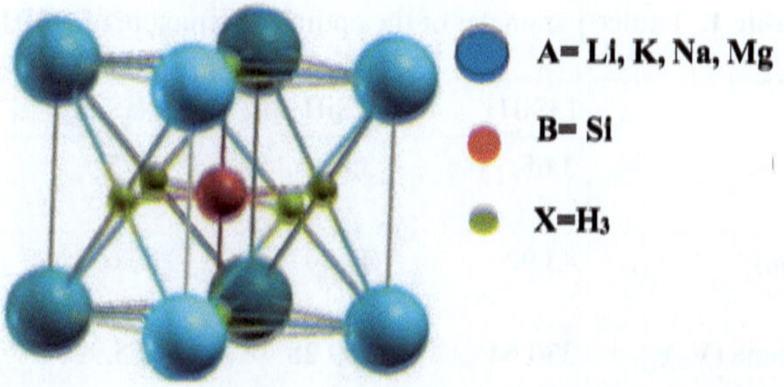

Fig. 1. Cubic structure of hydrides $LiSiH_3$; $KSiH_3$; $NaSiH_3$ and $MgSiH_3$ perovskites.

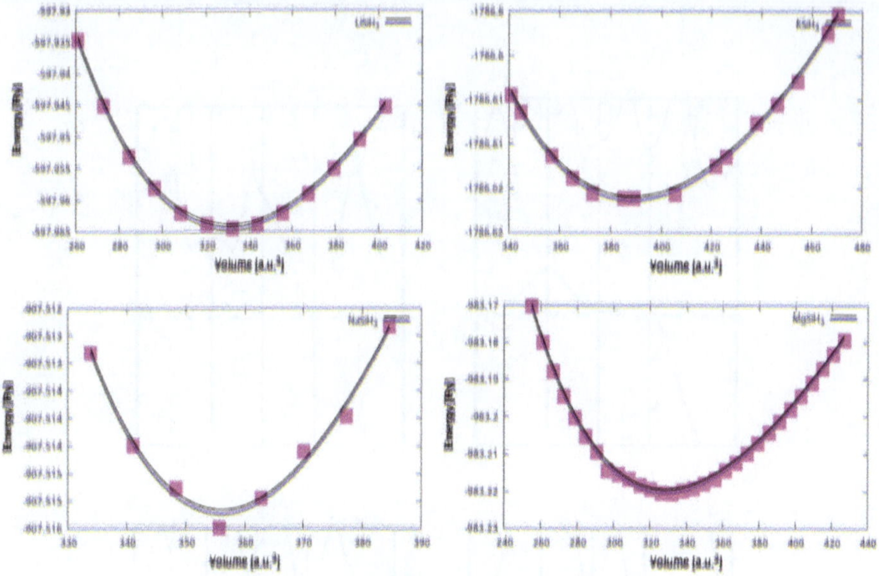

Fig. 2. Optimization curve of $ASiH_3$.

3.2 Electronic Properties

In this section all the $LiSiH_3$; $NaSiH_3$; $KSiH_3$ and $MgSiH_3$ hydrides beside the higher symmetric directions inwards very first or initial Brillouin zones were survived for the determination of band structure. Through the computational method all these computation were performed based on the equilibrium lattice constant shown in Fig. 3. The electronic band structure graph shows the metallic behaviour which means higher electrical conductivity. The electronic conductivity of perovskites affects the kinetics of hydrogen diffusion under the materials. The calculated TDOS and PDOS for $ASiH_3$ were shown in Fig. 4 which offers the extensive inspection of density of states (DOS) behind the hydrogen bonding inside the structure.

Table 1. Lattice parameter of the optimized structure of $ASiH_3$.

Compounds	$LiSiH_3$	$KSiH_3$	$NaSiH_3$	$MgSiH_3$
Lattice constant (a_o) (in Å)	3.65	3.86	3.75	3.65
Bulk modulus (B_o) (in GPa)	45.92	47.80	52.65	61.45
Ground state volume (V_o) (in a.u.^{3})	330.64	389.28	355.99	328.50
Ground State Energy (E_o) (in Ry)	−597.96	−1786.62	−907.51	−983.22

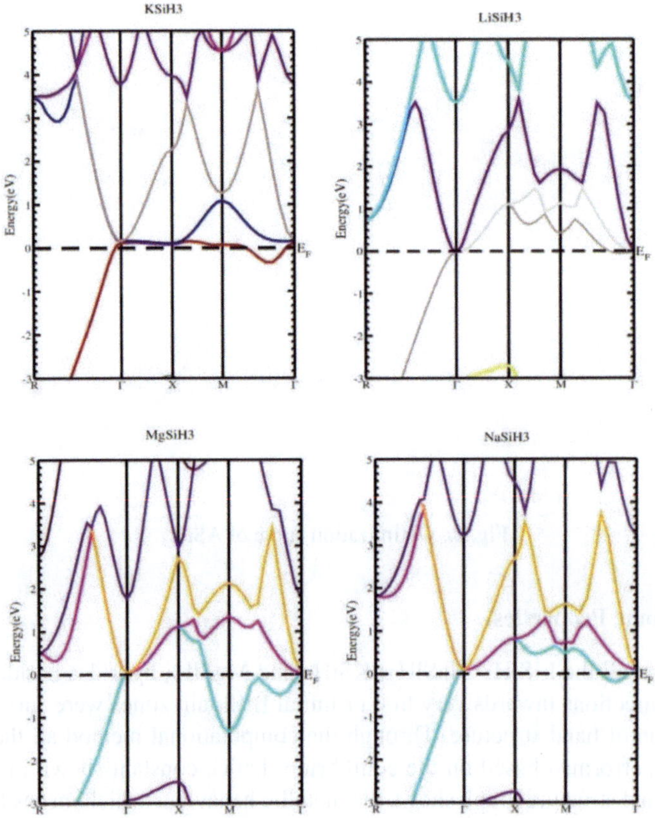

Fig. 3. Band structure of $LiSiH_3$, $NaSiH_3$, $KSiH_3$ and $MgSiH_3$.

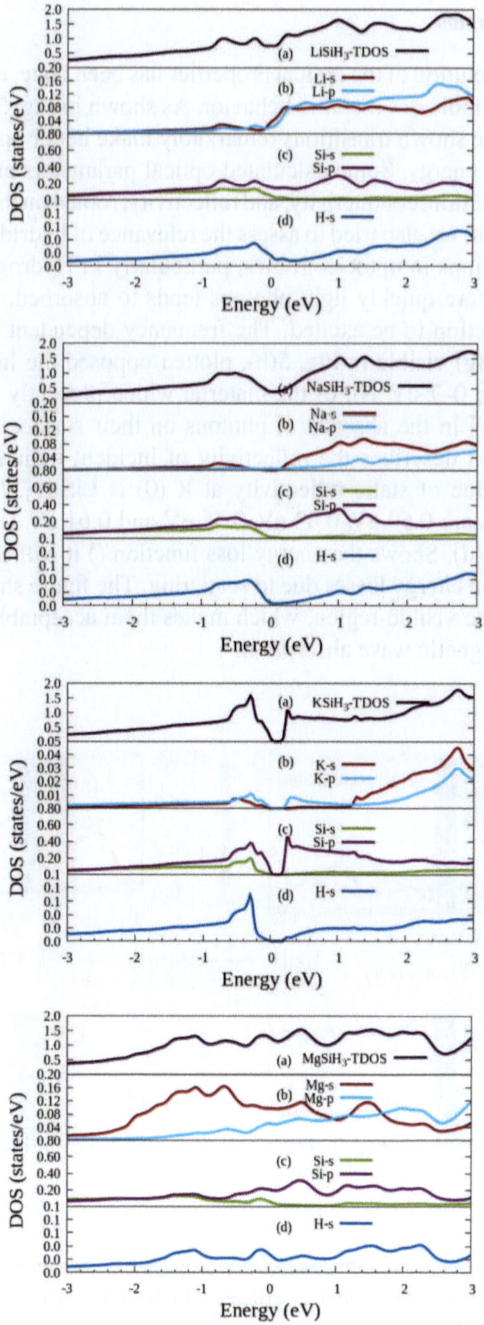

Fig. 4. Total (TDOS) and (PDOS) Partial Density of States of Hydrides.

3.3 Optical Properties

A comprehensive evolution of the optical properties has been done, usually on intra-band transition because it indicates metallic behavior. As shown in Fig. 5(a), the electromagnetic spectrum all the shown transitions remarkably make better optical conductivity in the lowest region of energy. Some calculated optical parameters are absorption coefficient, energy loss function, conductivity, and reflectivity, rooted on the radiated frequency which is incident on it, we also tried to assess the relevance of hydrides perovskites materials for the applications in optoelectronics, particularly in Hydrogen storages. Higher value of () means more quickly light photons tends to absorbed, which causes material electrons conduction to be excited. The frequency dependent () of perovskites (A = Li; K; Na; and Mg) visible in Fig. 5(b), plotted opposed the incident frequency of photon ranging from 0–7 eV. Nil of the material which presently examined displayed a piece of absorption in the absence of photons on their surface. Figure 5(c), Shows the reflectivity which describes the reflectivity of incident radiations from the material surface. The value of static reflectivity at R (0) is $LiSiH_3$, $NaSiH_3$, $KSiH_3$ and $MgSiH_3$ perovskites are 0.59 eV, 0.63 eV, 0.55 eV and 0.61 eV. Lowest value for R is for $KSiH_3$. Figure 5(d), Shows the energy loss function () it will explain the transition dispersion or electron energy losses due to scattering. The figure shows the least energy loss in $MgSiH_3$ in the visible-region, which makes them acceptable candidate for well organized electromagnetic wave absorption.

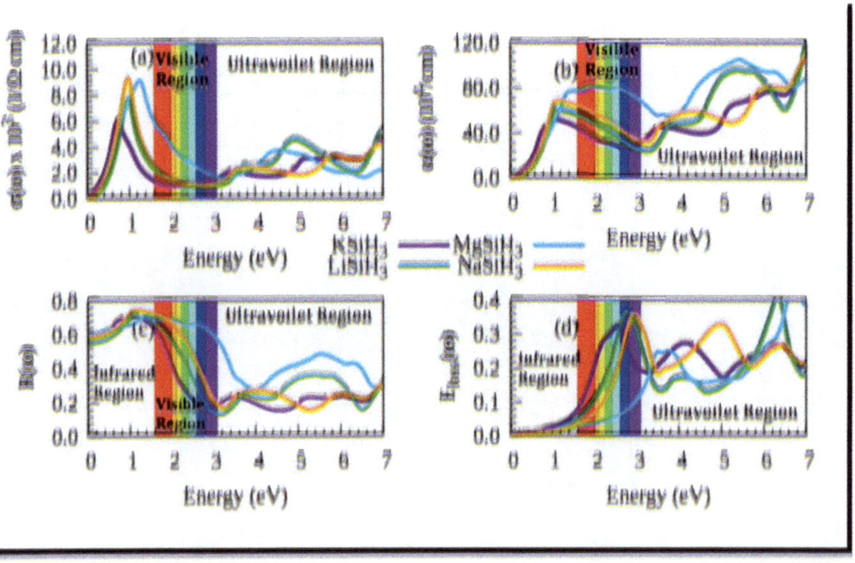

Fig. 5 Optical conductivity; absorption coefficient; Reflectivity; and loss function of LiSiH3, $NaSiH_3$, $KSiH_3$ and $MgSiH_3$.

4 Conclusion

We have investigated the physical effects of $LiSiH_3$, $NaSiH_3$, $KSiH_3$ and $MgSiH_3$ for solid state hydrogen storage for the sake of advancement of hydrogen technology. The results show that the compound is dynamically and mechanically stable. All the compounds possess metallic nature and hence ionic bonding having brittle nature. The optical properties of these compounds shows that, all expose noteworthy electrical conductivity for the photons of suitable energy, and hence making it favourable choice for applications in hydrogen storage.

References

1. Pandit, N., Singh, R., Joshi, T.K., Shukla, A., Kamlesh, P.K., Dubey, A., Kumar, T., Chauhan, M.S., Verma, A.S.: Investigation of a potential photovoltaic absorber based on first principles spectroscopic screening of chalcogenide perovskites: CaZrX3 (X=S, Se). J. Comp. Elect. **24**, 1–19 (2025)
2. Pandit, N., Singh, R., Kamlesh, P.K., Kumar, N., Sharma, P., Kumari, S., Kumar, T., Al-Qaisi, S., Verma, A.S.: Study of chalcogenide-based metal perovskites bazrx3 (x= s and se): Dft insight into fundamental properties for sustainable energy generation using amps-1d. J. Comp. Elect. **23**, 1014–1028 (2024)
3. Dubey, A., Mishra, K., Srivastava, R., Kumari, A., Jangra, P., Joshi, T., Choud-Hary, B., Verma, A.: Structural, optoelectronic, thermal and transport properties of hybrid perovskite (EAGeCl3) material. J. Ovonic Res. **20**, 381–394 (2024)
4. Mao, S.X., Song, J.Y., Zhu, W.S., Li, H.M., Pang, J.Y., Dang, D.B., Bai, Y.: Heterogeneous oxidative desulfurization of fuels using amphiphilic mesoporous phosphomolybdate-based poly (ionic liquid) over a wide temperature range. Fuel. **352**, 128982 (2023)
5. Eicke, L., Blasio, N.D.: Green hydrogen value chains in the industrial sector-geopolitical and market Implications. Energy Res. Soc. Sci. **93**, 102847 (2022)
6. Dou, Y., Wang, A., Zhao, L., Yang, X., Wang, Q., Sudi, M.S., Zhu, W., Shang, D.: Boosted hydrogen evolution reaction for a nitrogen-richazo-bridged metallated porphyrin network. J. Colloid Interface Sci. **650**, 943–950 (2023)
7. Xiangbo, F., Lvwei, S., Wang, W., Zhao, Y., Shi, J.W.: Construction of CdS@ ZnO coreeshell nanorod arrays by atomic layer deposition for efficient photoelectrochemical H_2 evolution. Sep. Purif. Technol. **324**, 124520 (2023)
8. Zhao, S., Rongjun, W.Z., Hongwei, L., Yu, W., Toan, S., Zhiqiang, S.: Boosting hydrogen production via deoxygenation-sorption-enhanced biomass gasification. Bioresour. Technol. **382**, 129197 (2023)
9. Blaha, P., Schwarz, K., Madsen, G.K.H., Laskowsk, R., Tran, F., Marks, L.D.: WIEN2K: an augmented plane wave plus local orbitals program for calculating properties of solid. J. Chem. Phys. **152**, 074101 (2020)
10. Singh, D.J., Nordstrom, L.: Planewaves, Pseudopotentials, and the LAPW Method. Springer, Berlin (2006)
11. Perdew, J.P., Ruzsinszky, A., Csonka, G.I., Vydrov, O.A., Scuseria, G.E., Constantin, L.A., Zhou, X., Burke, K.: Restoring the density gradient expansion for exchange in solids and surfaces. Phys. Rev. Lett. **100**, 136406 (2008)
12. Wu, Z., Cohen, R.E.: More accurate generalized gradient approximation for solids. Phys. Rev. B. **73**, 235116 (2006)

A Low Temperature Chemiresistive Hydrogen Sensor Based on Pd: Iron Pthalocyanine (FePc) Thin Films

Niyanta Datta[1(✉)], Ashutosh Raj[2], Kailasa Ganapathi[1], Soumen Samanta[1], and Manmeet Kaur[1]

[1] Technical Physics Division, Bhabha Atomic Research Center, Mumbai, India
niyantadatta@gmail.com

[2] National Institute of Technology Karnataka, Surathkal, India

Abstract. Energy transition is advancing rapidly, with hydrogen (H_2) gas emerging as a leading clean fuel. Various stakeholders are engaged in building H_2-compatible infrastructure for production, storage, and transport. Due to H_2 gas explosive nature, reliable systems are essential for safe operations. In this context, our group has already developed a Hydrogen Sensor Device (0.5% to LEL working at 150 °C) based on Pd thin film working on Pellister sensing principle. As a step forward, we have investigated Pd modified FePc thin films as low temperature H_2 sensor device based on chemiresistive sensing principle. The films were prepared by thermal evaporation of FePc at 200 °C (80 nm, β phase) substrate temperature followed by sputter deposition of Pd film (10 nm). The Pd: FePc films exhibited room temperature (RT) sensing (lowest detection limit of 0.3% H_2) with best sensing characteristics at 100 °C (sensitivity 1.4 times for 2% H_2 with a response time of 200 s). The sensing behavior may be attributed to the catalytic behavior of Pd in facilitation of oxygen adsorption on low coordinated Fe atoms in FePc. The same is inferred using Kelvin probe studies on pure and Pd modified FePc films.

Keywords: H_2 sensor · Pd film · FePc film · composite structure · low temperature sensing

1 Introduction

H_2 gas is an alternative fuel under the *Energy Policy Act of 1992* [1] due to its high heat of combustion (240 KJ/mol) and zero emission (power fuel cells). H_2 is a small size molecule (120 pm) with highest NFPA rating of four on the flammability scale. It has a very small ignition energy of 0.02 mJ and hence combustible in a wide range of H_2-Air (4–75%) and H_2-O_2 mixture (4–95%) [2]. Table 1 depicts methods for detection of H_2 gas and their respective shortcomings. Most of the methods show good response; however, they suffer from shortcomings like complex system, high cost, time-consuming process, requirement of trained personnel to operate those, maintenance and portability issues etc. The above-mentioned limitations restrict the continuous operation of such instruments. An ideal solution is a combination of sensors catering to specific requirements,

S. Kumari et al. (Eds.): ICMTP 2024, SPPHY 311, pp. 88–92, 2025.
https://doi.org/10.1007/978-981-96-6795-6_11

for ex: - optical sensors for high precision, RT sensors for portable device triboelectric for self-powered device etc. The chemiresistive sensors in particular offer low power, scalable, and room-temperature operation, making them ideal for next-generation hydrogen detection systems.

Table 1. H_2 gas sensing technologies.

Type of Sensor	Working Principle	Limitations
Catalytic combustion	H_2 reacts with a catalyst (e.g., Pt, Pd), generating heat that changes resistance.	High power consumption Risk of explosion at high concentrations Cannot detect very low H_2 levels
Electrochemical	H_2 undergoes redox reactions at electrodes, generating a measurable current.	Limited lifetime (electrode degradation) Requires frequent calibration Sensitive to humidity
Optical fiber	H_2 changes the refractive index or absorbance of a coated fiber, affecting light transmission.	Expensive Complex setup Requires precise alignment
Acoustic sensors	H_2 alters the resonant frequency of an acoustic wave propagating through a sensing layer.	Requires complex fabrication Susceptible to environmental noise
Triboelectric sensors	H_2-induced charge transfer in a triboelectric nanogenerator (TENG) generates a voltage signal.	Low maturity (still under development) May have long-term stability issues
Chemiresistive sensors (Pd:FePc)	H_2 adsorption changes the electrical resistance of nanomaterials and thin films.	Selectivity issues (cross-sensitivity to other gases) Long-term stability needs improvement

Recent advancements in chemiresistive hydrogen sensors have focused on enhancing sensitivity, selectivity, response time, and operational stability. Researchers have explored various nanomaterials and hybrid structures to achieve these improvements. For ex: - Antony et al. reported Pd metal polymer hydbrid sensors operating at room temperature with a sensor response of 30% for 0.5% H_2 conc [3]. Nanostructured thin film sensor based on Cu-$CuFe_2O_4$ was reported by S. De et al. [4] with response of 79% towards 500 ppm of H_2 at 400 °C. In this paper, we are presenting the H_2 gas sensing results for Pd modified Iron Pthalocyanine (FePc) thin films. Pd:FePc films exhibit low temperature chemiresistive sensing characteristics that are measured in a 2-probe geometry. Pd is a well-known catalyst for H_2 sensing and FePc has been reported for RT and low temperature sensing in low power consumption devices [5]. FePc is a p-type material with low co-ordinated metal center that enables oxygen adsorption and makes it suitable for gas sensing applications.

2 Materials and Methods

2.1 Sample Preparation

Pd: FePc samples were prepared in two steps. Firstly, FePc (Fig. 1 (a)) was thermally evaporated onto Si/SiO_2 (100 orientation) substrates mounted on heater at 200 °C to ensure formation of stable FePc ß phase [6]. Secondly, Pd thin film was sputtered on top of the FePc films. The depositions were carried out for various thicknesses but an optimized configuration of Pd: FePc was 10 nm: 80 nm (measured using QCM). Figure 1(b) shows the sensor response at various Pd thickness values for 2% H_2 at 100 °C. The maxima can be attributed to optimized coverage of FePc wherein metallic Pd is not a continuous film but sparsely distributed particles [7].

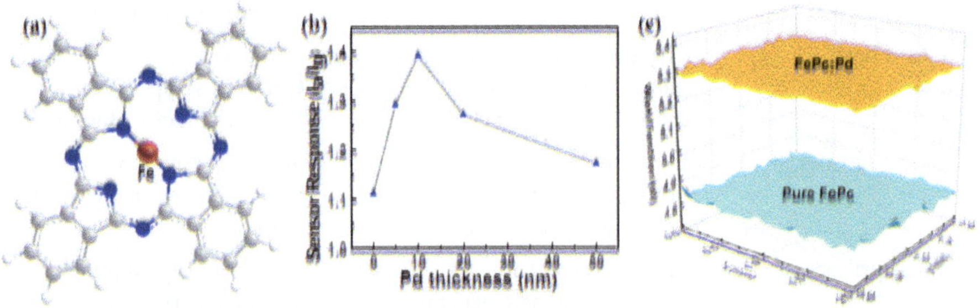

Fig. 1. (a) FePc structure (b) Pd film thickness optimization and (c) Kelvin probe contour plot of pure and Pd: FePc sensor films.

Kelvin Probe Studies. Kelvin probe studies were carried out to characterize the heterostructure. As shown in the Fig. 1(b), addition of Pd lead to increase in work function value (~4.8 [3] to 5.3 eV). This can be attributed to the formation of space charge layer at the Pd-FePc interface due to transfer of e^- from metallic Pd to p-type FePc, hence reducing the majority hole carriers and increasing band bending in FePc.

3 Gas Sensing Studies

3.1 Measurements Methodology

To carry out the sensing measurements, Au electrodes were thermally evaporated onto the sensor films and Ag wires were connected to integrate the sensor device with Keithley meter for current/resistance measurement. The Pt-100 heaters were connected at the back side of the silicon substrates to investigate the gas sensing properties at various temperatures (RT to 150 °C) and various H_2 gas concentrations (0.3–4%) in a dynamic gas sensing set-up [2].

Sensing Mechanism. In this sensor configuration, catalytic role of Pd is two-fold. It facilitates enhanced oxygen adsorption in FePc by transferring e^- to FePc (corroborated by Kelvin probe) followed by catalytic dissociation of H_2 gas, enabling its reaction with adsorbed O_2 (O_2^- < 150 °C).

$$\alpha H^* + \beta O_2^- \rightarrow \alpha/2H_2O + \beta e^- \quad (1)$$

The above reaction reduces the majority holes in p-type FePc resulting in decrease of current values in presence of H_2 gas (Fig. 2(c) and (d)). As shown in Fig. 2(a) and (b), sensing characteristics at 100 °C are more stable and exhibit highest sensor response (ratio of current value in air, I_a, to its value in presence of gas, I_g in p-type materials [8]). The sensor films depict sensitivity of 1.4 times for 2% H_2 with a response time of 200 s at 100 °C. To the best of our knowledge, this is the first-time reporting of RT and low temperature H_2 sensing in FePc thin films.

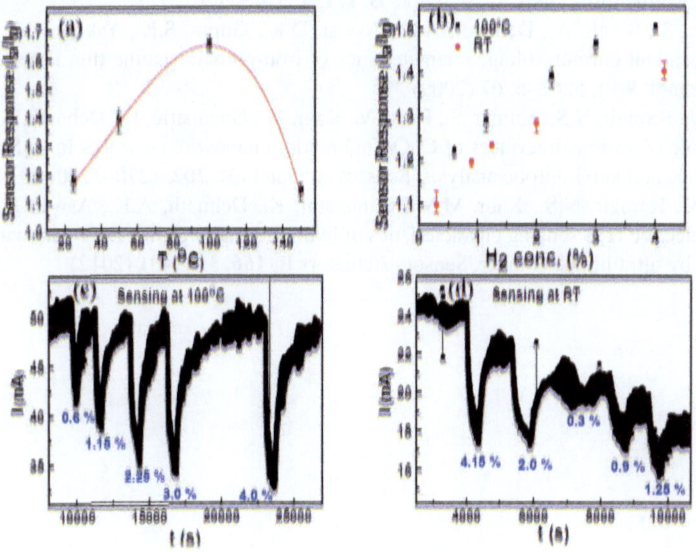

Fig. 2. (a) Sensor response of Pd: FePc sensor film as a function of temperature, (b) sensor response at RT and 100 °C, (c) current variations in presence of H_2 gas as a function of time at 100 °C, and (d) RT.

Future Scope of the Studies. To further improve the gas sensing properties (sensitivity and response kinetics), FePc is being functionalized with Pt and combination of Pd:Pt and the investigations are going on. In addition, integration of the low temperature and RT sensors onto flexible substrates is a logical step forward in the direction of portable sensors. The sensor device discussed in this paper is suitable for ambient H_2 gas monitoring in industry (refineries, aviation, ammonia production etc.) H_2 production, storage and transport facilities as well as nuclear reactors to monitor H_2 build-up.

References

1. Department of Energy Homepage, https://afdc.energy.gov/laws/key_legislation. Accessed 21 November 2022
2. Datta, N., Samanta, S., Bhattacharya, S., Pathak, A., Ramgir, N., Ganpathi, K., Kaur, M., Muthe, K.P., Debnath, A.K., Bhambure, D., Abhichandani, P.: Integrated hydrogen gas sensing device based on sputter-deposited Pd thin films. BARC Newsletter. **375**, 29–33 (2021)
3. Christina, A.E., Gaana, K., Praveen, S.G., Jayakumar, A., Yadav, A., Sivakumar, N.S., Kamath, N., Suma, M.N., Kamble, V.B., Nagar, J.P.: Polyvinylpyrrolidone-stabilized palladium nanocrystals as Chemiresistive sensors for low-concentration hydrogen gas detection. ACS Appl. Nano Mater. **2**(4), 1643–1653 (2021)
4. De, S., Venkatramani, N., Prasad, S., Dusane, R.O., Presmanes, L., et al.: Ethanol and hydrogen gas-sensing properties of $CuO–CuFe_2O_4$ nanostructured thin films. IEEE Sensors J. **18**(17), 6937–6945 (2018)
5. Kumar, A., Debnath, A.K., Samanta, S., Singh, A., Prasad, R., Putta, V., Singh, S., Basu, S., Aswal, D.K., Gupta, S.K.: Enhanced Cl_2 response of ultrathin bi-nuclear (cobalt–iron) phthalocyanine films. Sensors Actuators B. **171**, 423–430 (2012)
6. Samanta, S., Singh, A., Debnath, A.K., Aswal, D.K., Gupta, S.K., Yakhmi, J.V.: Temperature dependent current-voltage characteristics of iron-phthalocyanine thin films. J. Nanosci. Nanotechnol. **9**(9), 5262–5267 (2009)
7. Datta, N., Ramgir, N.S., Kumar, S., Putta, V., Kaur, M., Ganapathi, K., Debnath, A.K., Gupta, S.K.: Role of various interfaces of CuO/ZnO random nanowire networks in H_2S sensing: an impedance and kelvin probe analysis. Sensors Actuators B. **202**, 1270–1280 (2014)
8. Datta, N., Ramgir, N.S., Kaur, M.M., Ganapathi, K., Debnath, A.K., Aswal, D.K., Gupta, S.K.: Selective H_2S sensing characteristics of hydrothermally grown ZnO-nanowires network tailored by ultrathin CuO layer. Sensors Actuators B. **166**, 394–401 (2012)

Hydrothermal Synthesis and Characterization of $MoSe_2$/rGO Nanocomposite for Supercapacitor Applications

Monika, Rajiv Kumar, and Mamraj Singh(✉)

Department of Physics, University of Rajasthan, Jaipur 302004, India
bmamrajsingh@gmail.com

Abstract. In this study, a molybdenum selenide ($MoSe_2$/rGO) nanocomposite was synthesized using a hydrothermal method. The X-ray diffraction (XRD) pattern of the $MoSe_2$/rGO nanocomposite reveals the presence of $MoSe_2$ and rGO, with a broad diffraction peak corresponding to the (100) plane, which confirms that the $MoSe_2$ nanostructure has a two-dimensional sheet-like morphology and a hexagonal crystal structure. X-ray photoelectron spectroscopy (XPS) and UV-visible spectroscopy analysis further confirms the formation of $MoSe_2$, $MoSe_2$/rGO, and Se. This study highlights the potential application of the nanocomposite as an electrode material for supercapacitors.

Keywords: Transition metal dichalcogenides · $MoSe_2$/rGO · Nanocomposite · Hydrothermal synthesis · XPS analysis · Electrode · Energy storage

1 Introduction

The growing global energy demand, driven by a rising world population, has historically been met by fossil fuels [1, 2]. While effective, fossil fuels are major contributors to carbon dioxide emissions, making them unsustainable and environmentally harmful [3, 4]. Global energy consumption reached approximately 27,000 terawatt-hours in 2023, contributing to over 37.2 gigatons of CO_2 emission annually. In recent years, renewable energy sources have become pivotal in addressing energy needs. However, their intermittent nature necessitates efficient energy storage systems to ensure reliability and sustainability [5, 6]. Energy storage devices, including solar cells, batteries, and supercapacitors, play a vital role by storing excess energy during peak production and releasing it during high demand [7, 8]. These systems improve grid stability, support electric vehicle infrastructure, enable effective load management, and enhance overall energy efficiency [9].

Supercapacitors, also known as ultracapacitors, bridge the gap between traditional capacitors and batteries [10]. They offer fast charge and discharge cycles, high power density, and extended operational lifespans, making them ideal for applications such as regenerative braking in electric vehicles and energy storage in renewable energy

S. Kumari et al. (Eds.): ICMTP 2024, SPPHY 311, pp. 93–102, 2025.
https://doi.org/10.1007/978-981-96-6795-6_12

systems [11]. Supercapacitors are classified into two types: electric double-layer capacitors (EDLCs), which rely on electrostatic charge storage in a non-faradaic process, and pseudocapacitors (PCs), which store charge electrochemically through faradaic redox reactions [12]. The electrodes of EDLCs are made from carbon-based materials, such as activated carbon, graphene, and carbon nanosheets, among others [13, 14]. These materials enable long charge-discharge cycles but offer lower energy density [15]. PCs, on the other hand, employ electrode materials such as metal oxides, metal chalcogenides, metal nitrides, and conducting polymers [16, 17]. While PCs provide higher energy density, they often suffer from limited lifetimes and lower power density [18].

Hybrid supercapacitors (HSCs) combine the strengths of EDLCs and PCs [1, 19]. The dual mechanism of these HSCs provides high energy density and high-power density [20]. Hybrid electrodes enhance electrical conductivity, mechanical flexibility, cost-effectiveness, and chemical stability, making them promising for advanced energy storage solutions [21].

Despite these advances, conventional materials often lack the scalability and mechanical flexibility required for next-generation supercapacitors.

The present study focuses on transition metal dichalcogenides (TMDs) due to their two-dimensional sheet-like morphology. The general formula for TMDs is given by MX_2, where M is a transition metal and X = S, Se, or Te. Sulfur-based materials have been more extensively studied compared to selenium (Se)-based materials for supercapacitor applications. Molybdenum diselenide ($MoSe_2$) has gained attention for its unique properties, including high surface area, excellent electrical conductivity, and robust mechanical stability, in comparison to other TMDs [22]. $MoSe_2$, composed of molybdenum atoms sandwiched between selenium atoms via strong covalent bonds, offers high theoretical capacitance, multiple oxidation states, and cost-effective, abundant constituent elements [23, 24]. These attributes make TMDs highly suitable for enhancing the electrochemical performance, cycling stability, and energy storage capacity of supercapacitors [25].

Graphene is a promising material due to its remarkable electrical conductivity, thermal stability, and mechanical strength [27]. However, its challenging synthesis has led to the adoption of reduced graphene oxide (rGO) as a scalable alternative [28]. Graphene oxide (GO) is easier and more economical to produce and serves as a precursor material for making rGO [29]. This makes rGO a promising material for various applications, including energy storage devices such as supercapacitors. rGO offers high electrical conductivity and large mechanical strength, which can improve the electrochemical and structural stability of $MoSe_2$ [30]. The $MoSe_2$/rGO nanocomposite enhances electron transport, ion diffusion, and improves structural stability [31]. This study emphasizes the hydrothermal synthesis and characterization of the $MoSe_2$/rGO nanocomposite as electrode materials for supercapacitors, aiming to deliver superior performance in energy storage applications.

2 Materials and Method

2.1 Hydrothermal Synthesis of $MoSe_2$ Nanostructure

A total of 1.90 g of sodium molybdate dihydrate ($Na_2MoO_4{\cdot}2H_2O$), 1.243 g of selenium powder, and 0.595 g of sodium borohydride ($NaBH_4$) were dissolved in 80 mL of DI water at room temperature while stirring continuously for 30 min [32]. The homogeneous solution was then transferred into a 100 mL Teflon-lined stainless-steel autoclave, sealed, and heated in an oven at 200 °C for 24 h. After the reaction, final product was washed and dried.

For the synthesis of $MoSe_2$/rGO nanocomposite, GO was first prepared using the modified Hummer's method. Above process repeated for it [33, 34].

3 Results and Discussion

3.1 Structural Characterization

X-ray diffraction analysis of pristine $MoSe_2$ is a key technique for evaluating their crystalline properties, phase purity, and structural integrity. Figure 1 shows the XRDpattern of $MoSe_2$. The diffraction peaks at 2θ values 12.83, 31.70, 34.70, 55.71, 56.22^0 corresponding to (002), (100), (102), (110) and (008) planes (JCPDS No. 00-029-0914), indicates hexagonal or layered structure of $MoSe_2$, with lattice constants of a and b = 3.28 Å and c = 13.11 Å [35–37]. Moreover, XRD pattern shows some other diffraction peaks which are attribute to presence of Se emerging as a result of precipitation of Se atom during hydrothermal synthesis [38, 39]. Further annealing is required to remove surfactants. Furthermore, a peak near 25° is present, corresponding to the (002) plane of rGO [40].

3.2 Optical Characterization

Figure 2 illustrates the UV-Vis absorption spectra of $MoSe_2$ and $MoSe_2$/rGO nanocomposite. The UV- Vis spectra of $MoSe_2$ exhibits a broad peak in 400-700 nm range, which is attributed to excitonic transitions from the valence band to the conduction band [41, 42].

The absorption spectra of the $MoSe_2$/rGO nanocomposite reveal an absorption peak at 257.29 nm in the UV region, which confirms the incorporation of rGO [43].

The composite exhibits a less distinct shoulder around 400 nm, along with enhanced absorption in the visible and near-infrared regions [44].

This synergistic combination of $MoSe_2$ and rGO leads to improved material properties, establishing the composite as a potential candidate for high-performance supercapacitor electrodes [45].

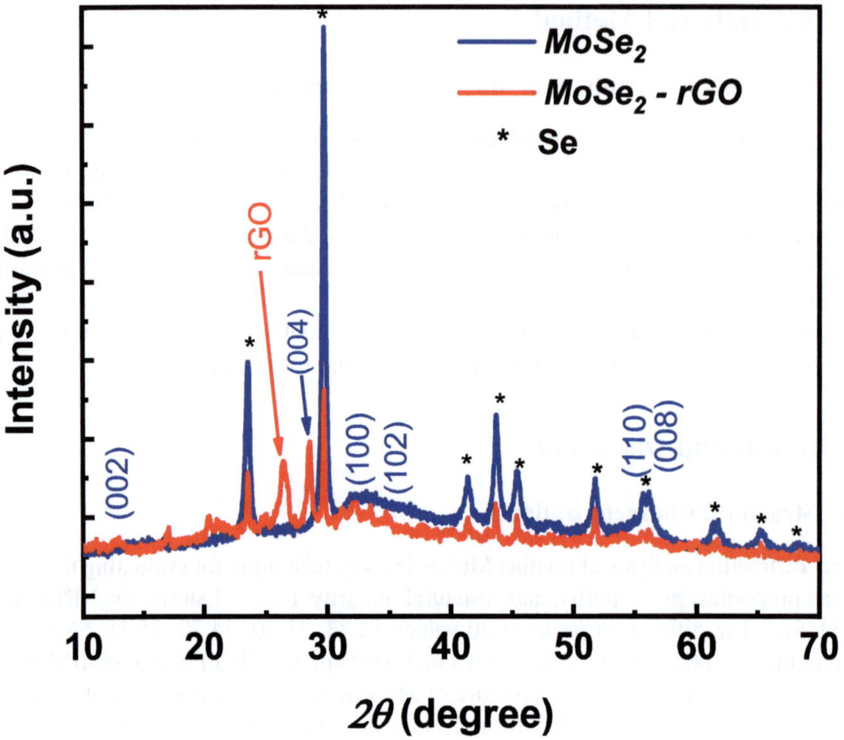

Fig. 1. Powder X-ray Diffraction pattern of Pristine $MoSe_2$, $MoSe_2$-rGO.

3.3 Electronic Structure of $MoSe_2$

High-resolution X-ray photoelectron spectroscopy (XPS) analysis offers detailed information about the electronic structure and oxidation states of molybdenum and selenium in $MoSe_2$ [46]. XPS survey scan of $MoSe_2$ is shown in Fig. 3, which indicates the presence of Mo, O, Se at their corresponding binding energies value [47]. Binding energies of the elements were collected by using reference peak of C 1 s at 284.8 eV. The relative atomic percentages of the element are presented in Table 1.

Spectrum of the Mo 3d region (see Fig. 4(a)) reveals four prominent peaks, indicating the presence of molybdenum in multiple oxidation states: Peaks at binding energies of 228.4 eV and 231.5 eV are assigned to Mo^{4+} $3d_{5/2}$ and Mo^{4+} $3d_{3/2}$, respectively [48]. Peaks at 232.0 eV and 234.8 eV correspond to Mo^{6+} $3d_{5/2}$ and Mo^{6+} $3d_{3/2}$, respectively [32].

The presence of Mo^{6+} (Fig. 4(a)) suggests formation of MoO_3 phase [49]. An additional feature at 229.4 eV is attributed to the overlap of the Se^{2-} 3s state with Mo^{4+} 3d states, adding complexity to the electronic structure [50].

Figure 4(b) presents the XPS spectrum of the Se 3d region for $MoSe_2$, highlighting two distinct peaks at: 54.3 eV for Se^{2-} $3d_{5/2}$, and 55.2 eV for Se^{2-} $3d_{3/2}$ [48]. These peaks represent the spin-orbit components of selenium. The intensity ratio ($I_{3/2}/I_{5/2}$) of these peaks is approximately 0.69, and the measured spin-orbit splitting is 0.86 eV,

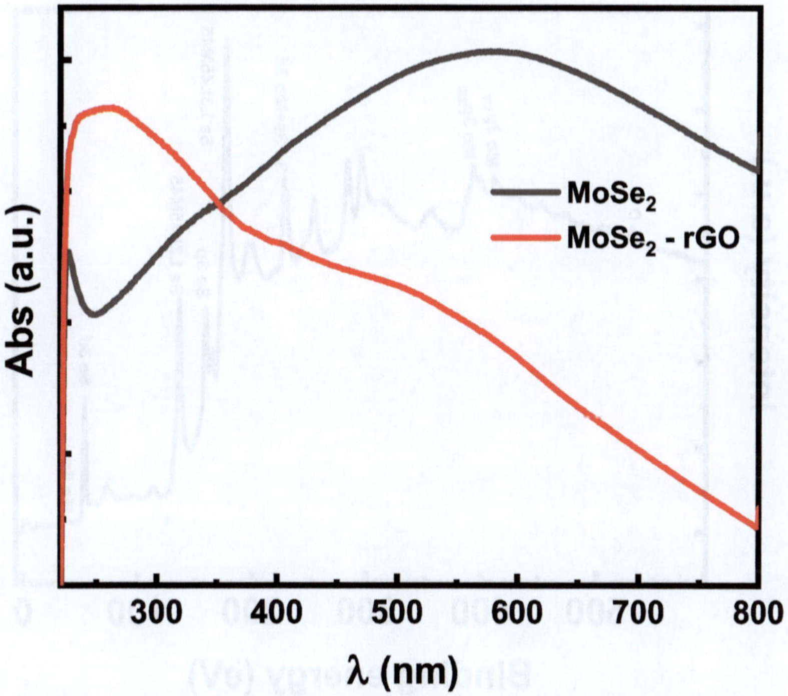

Fig. 2. UV-Vis absorption spectra of MoSe$_2$ and MoSe$_2$/rGO nanocomposite.

consistent with the characteristic binding energies of Mo^{4+} and Se^{2-} in the material [51]. An additional peak at 56.0 eV is linked to Se–Se metallic bonding, indicating the presence of unreacted selenium [47].

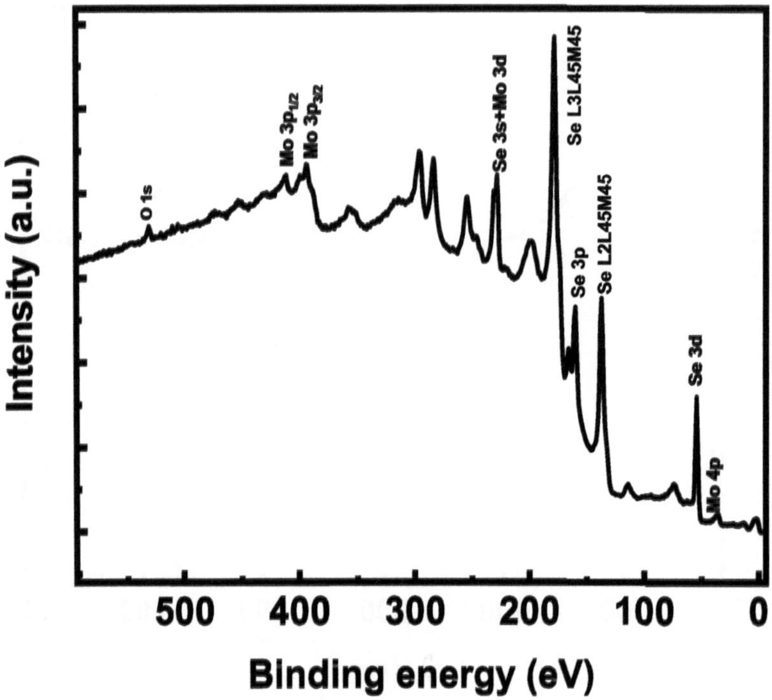

Fig. 3. XPS survey scan of $MoSe_2$ nanostructure.

Table 1. Atomic Percentage of the elements from XPS data

Sr. No.	Element	Atomic (%)
1.	Mo	6.45
2.	O	4.69
3.	Se	36.57

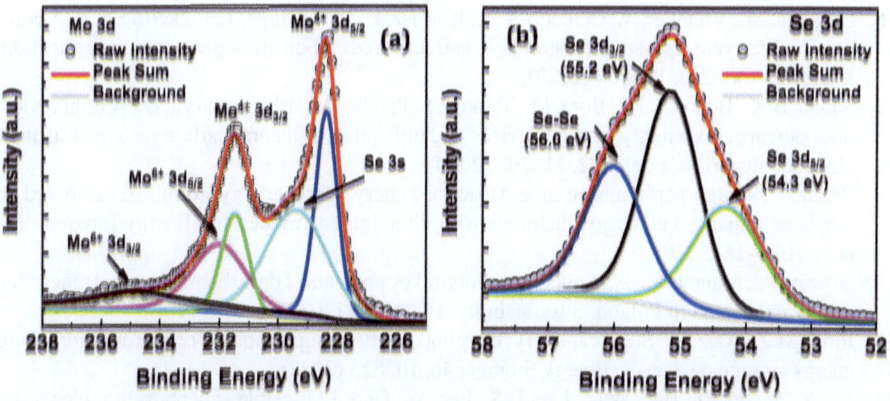

Fig. 4. XPS spectra of $MoSe_2$ (a) Mo 3d region (b) Se 3d region.

4 Conclusion

In conclusion, the $MoSe_2$/rGO nanocomposite was synthesized via the hydrothermal method. The XRD pattern of $MoSe_2$ confirmed its hexagonal, two-dimensional sheet-like structure and the formation of the $MoSe_2$/rGO nanocomposite. The XPS analysis highlights the presence of mixed oxidation states of molybdenum, along with characteristic selenium binding energies. The detection of unreacted selenium and residual MoO_3 indicates the need for further optimization in the synthesis process. The synthesized $MoSe_2$/rGO nanocomposite has potential use as electrode material for supercapacitors.

Acknowledgments. Ms. Monika would like to thank the UGC for providing fellowship support for her research. The authors extend their gratitude to the Dr. Balram Tripathi, S.S. Jain Subodh P.G. College, Jaipur for providing access to the characterization facilities and to the Department of physics, University of Rajasthan, Jaipur for XPS, UV-Visible and their support in this research.

References

1. Muzaffar, A., Ahamed, M.B., Deshmukh, K., Thirumalai, J.: A review on recent advances in hybrid supercapacitors: design, fabrication and applications. Renew. Sustain. Energy Rev. **101**, 123–145 (2019)
2. Tamang, S., Rai, S., Bhujel, R., Bhattacharyya, N.K., Swain, B.P., Biswas, J.: A concise review on GO, rGO and metal oxide/rGO composites: fabrication and their supercapacitor and catalytic applications. J. Alloys Compd. **947**, 169588 (2023)
3. Dissanayake, K., Kularatna-Abeywardana, D.: A review of supercapacitors: materials, technology, challenges, and renewable energy applications. J. Energy Storage. **96**, 112563 (2024)
4. Jalal, N.I., Ibrahim, R.I., Oudah, M.K.: A review on supercapacitors: types and components. J. Phys.: Conf. Ser. **1973**, 012015, IOP Publishing (2021)
5. Lemian, D., Bode, F.: Battery-supercapacitor energy storage systems for electrical vehicles: a review. Energies. **15**(15), 5683 (2022)

6. Cossutta, M., Vretenar, V., Centeno, T.A., Kotrusz, P., McKechnie, J., Pickering, S.J.: A comparative life cycle assessment of graphene and activated carbon in a supercapacitor application. J. Clean. Prod. **242**, 118468 (2020)
7. Saikia, B.K., Benoy, S.M., Bora, M., Tamuly, J., Pandey, M., Bhattacharya, D.: A brief review on supercapacitor energy storage devices and utilization of natural carbon resources as their electrode materials. Fuel. **282**, 118796 (2020)
8. Zhang, F.: A high-performance supercapacitor-battery hybrid energy storage device based on graphene-enhanced electrode materials with ultrahigh energy density. Energy Environ. Sci. **6**(5), 1623–1632 (2013)
9. Nowrot, A., Manowska, A.: Supercapacitors as key enablers of decarbonization and renewable energy expansion in Poland. Sustainability. **16**(1), 216 (2023)
10. Iqbal, M.Z., Aziz, U.: Supercapattery: merging of battery-supercapacitor electrodes for hybrid energy storage devices. J. Energy Storage. **46**, 103823 (2022)
11. Ko, Y., Kwon, M., Bae, W.K., Lee, B.S., Lee, W., Cho, J.: Flexible supercapacitor electrodes based on real metal-like cellulose papers. Nat. Commun. **8**(1), 536 (2017)
12. Kamila, S., Jena, B.K., Basu, S.: In: Sahoo, U. (ed.) Advances in Electrochemical Energy Storage Device: Supercapacitor in Energy Storage, pp. 119–147. Wiley, Hoboken (2021)
13. Lekakou, C., Moudam, O., Markoulidis, F., Andrews, T., Watts, J.F., Reed, G.T.: Carbon-based fibrous EDLC capacitors and supercapacitors. J. Nanotechnol. **2011**, 1–8 (2011)
14. Pershaanaa, M., Bashir, S., Ramesh, S., Ramesh, K.: Every bite of Supercap: a brief review on construction and enhancement of supercapacitor. J. Energy Storage. **50**, 104599 (2022)
15. Taslim, R., Apriwandi, A., Taer, E.: Novel Moringa oleifera leaves 3D porous carbon-based electrode material as a high-performance EDLC supercapacitor. ACS Omega. **7**(41), 36489–36502 (2022)
16. Ahmadi, M.: 2D transition metal dichalcogenide nanomaterials: advances, opportunities, and challenges in multi-functional polymer nanocomposites. J. Mater. Chem. A. **8**(3), 845–883 (2020)
17. Azwar, E.: Transformation of biomass into carbon nanofiber for supercapacitor application—a review. Int. J. Hydrog. Energy. **43**(45), 20811–20821 (2018)
18. Zhao, X., Sánchez, B.M., Dobson, P.J., Grant, P.S.: The role of nanomaterials in redox-based supercapacitors for next generation energy storage devices. Nanoscale. **3**(3), 839–855 (2011)
19. Chatterjee, D.P., Nandi, A.K.: A review on the recent advances in hybrid supercapacitors. J. Mater. Chem. A. **9**(29), 15880–15918 (2021)
20. Dubal, D.P., Ayyad, O., Ruiz, V., Gomez-Romero, P.: Hybrid energy storage: the merging of battery and supercapacitor chemistries. Chem. Soc. Rev. **44**(7), 1777–1790 (2015)
21. Zhao, B.: A high-energy, long cycle-life hybrid supercapacitor based on graphene composite electrodes. Energy Storage Mater. **7**, 32–39 (2017)
22. Chaturvedi, R., Garg, A.: Environment-friendly approach to rGO–TMD composite synthesis for use as a supercapacitor. Bull. Mater. Sci. **47**(3), 220 (2024)
23. Sha, R., Maity, P.C., Rajaji, U., Liu, T.-Y., Bhattacharyya, T.K.: MoSe2 nanostructures and related electrodes for advanced supercapacitor developments. J. Electrochem. Soc. **169**(1), 013503 (2022)
24. Upadhyay, S., Pandey, O.P.: Studies on 2D-molybdenum diselenide ($MoSe_2$) based electrode materials for supercapacitor and batteries: a critical analysis. J. Energy Storage. **40**, 102809 (2021)
25. Tanwar, S., Sharma, A.L.: Aging impact on morphological and electrochemical performance of MoSe2 composite for supercapacitor application. Ceram. Int. **49**(11), 18281–18295 (2023)
26. Hu, Q.: Graphene oxide-drove transformation of NiS/Ni_3S_4 microbars towards Ni_3S_4 polyhedrons for supercapacitor. J. Colloid Interface Sci. **559**, 115–123 (2020)

27. Malik, M.T.U., Sarker, A., Rahat, S.S.M., Shuchi, S.B.: Performance enhancement of graphene/GO/rGO based supercapacitors: a comparative review. Mater. Today Commun. **28**, 102685 (2021)
28. Hu, Y.: All-in-one graphene fiber supercapacitor. Nanoscale. **6**(12), 6448–6451 (2014)
29. Ogata, C.: All-graphene oxide flexible solid-state supercapacitors with enhanced electrochemical performance. ACS Appl. Mater. Interface. **9**(31), 26151–26160 (2017)
30. Robertson, B., Sapna, R., Hegde, V., Hareesh, K.: A comprehensive review on $MoSe_2$ nanostructures with an overview of machine learning techniques for supercapacitor applications. RSC Adv. **14**(51), 37644–37675 (2024)
31. Ahmad, K., Kim, H.: Hydrothermally synthesized $MoSe_2$/rGO composite as electrode modifier for the construction of non-enzymatic urea sensor. Mater. Chem. Phys. **286**, 126206 (2022)
32. Ye, F.: Synthesis of two-dimensional plasmonic molybdenum oxide nanomaterials by femtosecond laser irradiation. Chem. Mater. **33**(12), 4510–4521 (2021)
33. Upadhyay, S., Pandey, O.P.: Synthesis of layered 2H–$MoSe_2$ nanosheets for the high-performance supercapacitor electrode material. J. Alloys Compd. **857**, 157522 (2021)
34. Jeevanandham, G., Vediappan, K., ALOthman, Z.A., Altalhi, T., Sundramoorthy, A.K.: Fabrication of 2D-$MoSe_2$ incorporated NiO nanorods modified electrode for selective detection of glucose in serum samples. Sci. Rep. **11**(1), 13266 (2021)
35. Bhat, K.S., Nagaraja, H.S.: Effect of isoelectronic tungsten doping on molybdenum selenide nanostructures and their graphene hybrids for supercapacitors. Electrochim. Acta. **302**, 459–471 (2019)
36. Dogra, N., Sharma, S.: Hydrothermally obtained $MoSe_2$/reduced graphene oxidebased composite with superior electrocatalytic performance. Int. J. Hydrog. Energy. **55**, 78–87 (2024)
37. Amin, B.G., Masud, J., Nath, M.: Facile one-pot synthesis of $NiCo_2Se_4$-rGO on Ni foam for high performance hybrid supercapacitors. RSC Adv. **9**(65), 37939–37946 (2019)
38. Siroha, P., Khandelwal, V., Singh, D., Ramovatar, Gangwar, J.: Remarkable enhancement in adsorption capacity of methylene blue by hydrothermally processed $MoSe_2$ nanosheets for environmental application. Bull. Mater. Sci. **47**(4), 233 (2024)
39. Shui, J., Bai, B., Jiang, X., Du, P.: High-performance $MoSe_2$/rGO composites based on interface and phase engineering for all-solid-state symmetric supercapacitors. Electrochim. Acta. **469**, 143257 (2023)
40. Jhankal, D., Khan, M.S., Jhankal, K.K., Sachdev, K.: Charge storage kinetics of MoS_2 flower decorated reduced graphene oxide for quasi solid-state symmetric supercapacitor. J. Phys. Chem. Solids. **173**, 111117 (2023)
41. Dong, N.: Optical limiting and theoretical modelling of layered transition metal dichalcogenide nanosheets. Sci. Rep. **5**(1), 1–10 (2015)
42. Luo, X.-L., Chen, Z.-Y., Yang, S.-Y., Xu, Y.-H.: Two-step hydrothermal synthesis of peanut-shaped molybdenum diselenide/bismuth vanadate ($MoSe_2$/$BiVO_4$) with enhanced visible-light photocatalytic activity for the degradation of glyphosate. J. Colloid Interface Sci. **532**, 456–463 (2018)
43. Luan, C.-Y., Xie, S., Ma, C., Wang, S., Kong, Y., Xu, M.: Elucidation of luminescent mechanisms of size-controllable $MoSe_2$ quantum dots. Appl. Phys. Lett. **111**(7), 073105 (2017)
44. Wu, J., Liu, Y., Yao, Y., Shao, Y., Wu, X.: Graphene-like 2H/1T-$MoSe_2$ with superior full spectrum absorption: morphology and phase engineering. J. Alloys Compd. **877**, 160317 (2021)
45. Singh, A., Poddar, D., Thakur, S., Jha, R.: Ternary composite based on $MoSe_2$-rGO/polyaniline as an efficient counter electrode catalyst for dye sensitized solar cells. Mater. Chem. Phys. **273**, 125043 (2021)

46. Eftekhari, A.: Molybdenum diselenide ($MoSe_2$) for energy storage, catalysis, and optoelectronics. Appl. Mater. Today. **8**, 1–17 (2017)
47. Zhu, M.: N-doped one-dimensional carbonaceous backbones supported $MoSe_2$ nanosheets as superior electrodes for energy storage and conversion. Chem. Eng. J. **334**, 2190–2200 (2018)
48. Papanai, G.S., Sahoo, K.R., Gupta, S., Gupta, B.K.: Role of processing parameters in CVD grown crystalline monolayer $MoSe_2$. RSC Adv. **12**(21), 13428–13439 (2022)
49. Azcatl, A.: HfO_2 on UV–O_3 exposed transition metal dichalcogenides: interfacial reactions study. 2D Mater. **2**(1), 014004 (2015)
50. Xie, H., Xiong, X.: A porous molybdenum disulfide and reduced graphene oxide nanocomposite (MoS_2-rGO) with high adsorption capacity for fast and preferential adsorption towards Congo red. J. Environ. Chem. Eng. **5**(1), 1150–1158 (2017)
51. Chen, W.: Tailoring the thickness of $MoSe_2$ layer of the hierarchical double-shelled N-doped carbon@ $MoSe_2$ hollow nanoboxes for efficient and stable hydrogen evolution reaction. J. Catal. **381**, 363–373 (2020)

Boosting Hydrogenation Performance of Magnesium Via 2D Layered MXene and Chromium

Madhu Yadav[1,2], S. N. Dolia[1,2], and Chhagan Lal[1,2](✉)

[1] Department of Physics, University of Rajasthan Jaipur, JLN Marg, Jaipur 302004, Rajasthan, India
clsaini52@gmail.com, clsaini52@uniraj.ac.in

[2] Center for Non-conventional Energy Resources, University of Rajasthan Jaipur, JLN Marg, Jaipur 302004, Rajasthan, India

Abstract. Hydrogen is a pure, high-energy fuel that is readily available worldwide. However, its economic system, which depends on hydrogen, is confronted with difficulties in manufacturing, storage, and application, rendering it a difficult option. Extensive research has been conducted on chemical hydrogen storage materials, which are chemically bound to other elements. The initial research concentrated on metal alloys, which have a low gravimetric hydrogen density of typically less than 2 wt% but absorb hydrogen swiftly. MgH_2 extensive application is hindered by its poor cycle performance, sluggish hydrogen sorption kinetics, and high thermal stability, despite its numerous benefits. Hydrogen is an alternate source of energy that can be utilized either directly or indirectly. Storage hydrogen, thereby providing a more flexible deployment and increased energy density developments facilitate the transformation of surplus renewable energy through. Additionally, investigation is required to facilitate interaction. This investigation employs the ball milling method to generate a nanocomposite composed of Mg + 5 wt % (Ti_3C_2F + Cr). 2.52 wt% of hydrogen can be stored in the nano-composite Mg + 5 wt% (MX + Cr) at a temperature of 320 °C. Thermal behavior and desorption reactions induced by additives are demonstrated by the 11.4 wt% total weight loss. To characterize samples, XRD was implemented. The dehydrating reaction has an activation energy of 168.77 kJ/mol.

Keywords: Storage · Hydrogen · Energy · Magnesium hydride · Kinetics

1 Introduction

Mitigating climate change and transitioning to a low-carbon economy necessitates the development of sustainable and efficient energy storage technologies, with hydrogen serving as a potential carrier [1]. The utilization of alternative energies is required to resolve the harmful emissions from conventional energy sources, in addition to addressing the increasing demand for energy and the transition away from fossil fuels [2]. For several uses, such as power generation, cooking, autos, hydrogen-powered sectors, jet

S. Kumari et al. (Eds.): ICMTP 2024, SPPHY 311, pp. 103–112, 2025.
https://doi.org/10.1007/978-981-96-6795-6_13

jets, and household energy requirements, hydrogen gas is a viable fuel for the future. In addition to being utilized in space initiatives for aerospace transportation, powering international space stations, and supplying water and electricity, it is formerly utilized in exploratory automobiles [3]. The shortcomings of traditional storage techniques are now being addressed by the proposal of solid-state hydrogen storage materials [4]. Metal hydrides like MgH_2, are convenient and secure for storing hydrogen because of their massive potential ability, cost-effectiveness, compact size and excess. However, they are not suitable for commercial applications because of their toxicity, sluggish sorption rate, and extreme expulsion temperature [5]. Many methods have been used to improve magnesium hydride (MgH_2) hydrogenation capacity, including mechanical milling for nano structuring, catalyst addition, and thin-film technology exploration. Particle size is decreased, faults are created, and the site of action is exposed to hydrogen through mechanical milling. It demonstrates the way reducing the metallic bond strength and allowing atomic hydrogen to diffuse inside the structure of metal may enhance performance. Because of their strong attraction for hydrogen, transition metal compounds are thought to be highly efficient [6]. To optimize the hydrogen storage capacity of MgH_2/Mg, a variety of catalysis are being implemented, such as metals, oxides, halogens, covalent substances and carbons [7]. The outstanding efficiency of two-dimensional (2D) materials in hydrogen storage is attributed to their lower dimensions. Materials such as graphene, MXene, hexagonal boron (h-BN), phosphorene, borophene, and transition metal dichalcogenides (TMDs) are included in this category. These materials exhibit distinctive attributes, including a layered construction, large surface area, great mechanical qualities, porosity, low band gap tuning, thermal and electrical performance. Their layered structure also enables them to provide rapid hydrogen transport, and they satisfy the hydrogen intake quantity, strain, and adsorption-desorption temperature of hydrogen requirements of the United States Department of Energy [8–16]. The efficiency of hydrogen storage is greatly improved by alloying, which lowers the rate of desorption temperature, improves absorption/desorption dynamics, and increases periodicity. Compositions including Mg-V, Mg-Ti-Fe, and Mg-La-Ni show great capacity [17]. Verma et al. show that MgH_2 was catalyzed by VS_2 at greater speed than Gr and pristine MgH_2 releasing 2.54 wt% within 10 min and demonstrating exceptional durability and flexibility for up to 25 cycles [18]. Hang et al. show that lamellar structure of Ti_3C_2 effectively separates agglomerated $NaMgH_3$ particles. Its thermal storage density is 2562 kJ/kg, rendering it appropriate for heating storage applications [19]. Liu et al. show that the hydrocarbon storage dynamics capabilities of the two-dimensional $Nb_4C_3T_x$ (MXene) composite are better when applied to MgH_2. It has a lower activation energy for dehydrogenation, reduces the onset temperature, and sheds hydrogen entirely all of which suggest that it has extensive catalytic activity [20]. Chen et al., show that de-hydrogenation initial temperature of the $4MgH_2$-$LiAlH_4$-Ti_3C_2 composite is reduced by 64 K and 274 K, respectively [21]. Polanski et al. show that nanocomposite showed a steady decrease in hydrogen storage capacity following 150 cycles at 325 °C with PCT curves showing a decrease from approximately 5.2 wt.% after 1 cycle to 4.6 wt.% after 150 cycles [22]. Vijay et al. show that rate of absorption increased as the Cr_2O_3 content increased, while the absorption capacity decreased because of the temperature [23]. Pukazhselvan et al. show that 5 wt.% CrO_3 additive decreases the dehydration activating energy of MgH_2 by 68 kJ/mol and decreases the necessary dehydrogenation temperature by 80 °C. The presence of CrO_3 in MgH_2 [24]. Zhu et al. show that at standard temperature, the MgH_2 + Ni@Ti-MX composite takes 5.4 weight percent H_2 in 25 s at 125 °C. And emits 5.2 weight percent H_2 in 15 min at 250 °C. In 5 h, it consumes 4 weight percent H_2 [25].Yuan et al. show that initial dehydriding temperature of MgH_2 was reduced by 114 °C through the addition of 5 wt% S-Ti_3C_2 [26]. In comparison to pristine MgH_2, the system exhibited a 35.3% decreased E_a of hydrogen release and a reduction in the start release temperature to 142 °C [27].

In this study, the ball-milling technique is used. Ball milling is used to create tiny particles that attain variation states.

Furthermore, MXene a 2D material and Chromium might influence the thermal and kinetic properties of magnesium-based hydrogen storage materials. During hydrogenating by XRD investigate the crystalline features of MgH_2 and Mg – 5 wt% (MX + Cr).

2 Sample Preparation

The sample was prepared using the Ball-Milling method at CNCER, Jaipur. To prepare 1 g sample the Chromium and MXene are taken in the 5 wt% ratio of Magnesium. The Magnesium powder (purity 99%), chromium powder (powder 99.5%) is purchased from Sigma -Aldrich and the source of MXene (Ti_3C_2F) is from RRCAT, Indore. Then the 1 g sample was loaded into the Restch PM-100 ball-milling machine. The rpm was taken 250 within a 5-min interval in 2 h for 10 h. This could produce finer particles. After ball-milling the sample was open in glove box in the inert atmosphere. So that no oxidisation take place.

3 Materials and Methods

The X-ray diffractometer utilizes CuKα radiation ($\lambda = 1.5406$ Å) to characterize the structural properties of the prepared samples. The Hydrogen storage experiment was performed at Homemade seivert-type Pressure-Composition-Isotherm (PCT). Enthalpy and entropy of the rehydrogenation process were computed using the Van't Hoff equation. The prepared samples were subjected to thermodynamic analysis using the TGA (STA Model Hitachi TGA-7300) at a varying rate of 5 °C/min, 10 °C/min, and 15 °C/min, while nitrogen (60 ml/min) was flowing. Using the Kissinger equation, the activation energy can be determined by analyzing TGA at various heating rates.

4 Results and Discussion

4.1 XRD Analysis

Figure 1 illustrates the XRD patterns of MgH_2 with catalysts (Chromium and MXene) after ball milling. All diffractograms show the peaks associated to β-MgH_2 and γ-MgH_2 resulting from the initial grinding of MgH_2. Following comes generates the metastable γ-MgH_2 phase. Mechanical milling resulting from processing conditions generated in energy terms. The XRD spectrum suggests that the sample contains a wide range of phases, some of which may have been generated via the ball-milling method. The hydrogenation capabilities of nanocomposite are improved by the creation of additional active surfaces through ball-milling. XRD analysis has demonstrated that ball-milling results in a non-homogeneous reduction in particle size.

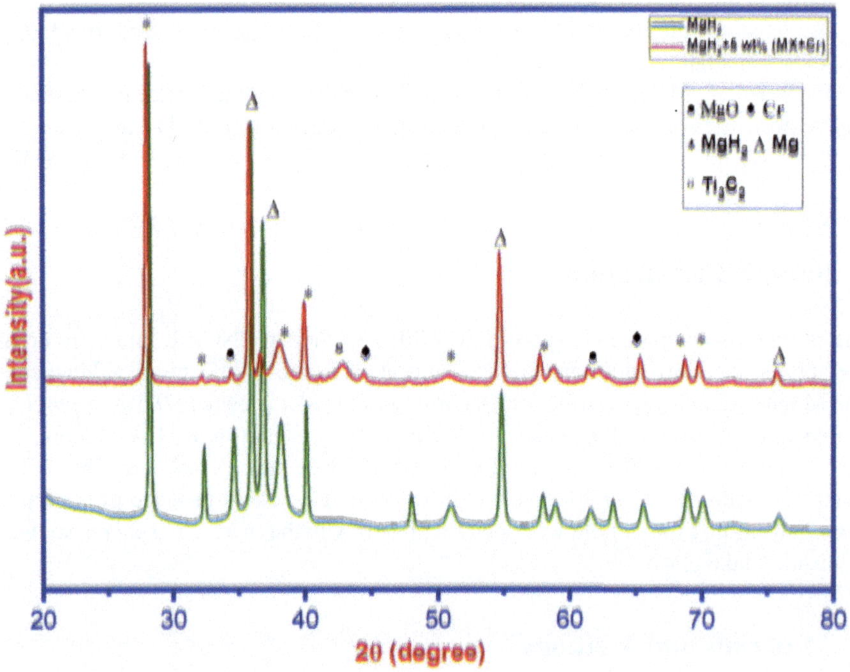

Fig. 1 XRD plot of MgH_2 and Mg + 5 wt% (MX + Cr)

Debye-Scherrer equation was employed to determine the average grain size using X-ray diffraction (XRD) data [28].

$$D = \frac{0.9\,\lambda}{\beta\, cos\theta}$$

where λ, β, and θ represent the wavelength, line widening, and Bragg angle, respectively.

Dislocation density

To determine dislocation density Williamson and Smallman's formula was used [29].

$$\delta = \frac{1}{D^2}$$

Microstrain

Microstrain value is defined by [30]

$$\epsilon = \frac{\beta cos\theta}{4}$$

The addition of catalyst lead to a lower grain size, which enhances the amount of surface area per volume and provides additional sites with activity for chemical reactions. The chemical process proceeds at a faster pace due to the expanded surface that facilitates the interaction of tiny particles. Additionally, the energy limit for the change of phase lowers which improves reaction thermodynamics. Dislocation density is lowered which indicates lowered retaining at imperfections sites, which improves the storage of hydrogen capacity. This results in better long-term cycling stability due to fewer induced reactions (Table 1).

Table 1. The average grain size, microstrain and dislocation density

Sample	Average grain size (D) nm	Dislocation density ($\times 10^{-4}$) nm^{-2}	Microstrain ($\times 10^{-3}$)
MgH_2	46.04	9.19	4.48
Mg + 5 wt%(MX+Cr)	38.76	6.66	3.70

4.2 PCT Analysis

The 1 g sample was loaded into the PCT setup. The activation process was followed for 36 h. The 6 cycles of 6 h were followed at 350 °C at 2 MPa H_2. The measurements were taken at different -different temperatures. The hydrogen storage analysis of Mg and Mg + 5 wt% (MX + Cr) is 5.29 wt%. and 2.53 wt% at 320 °C (Fig. 2).

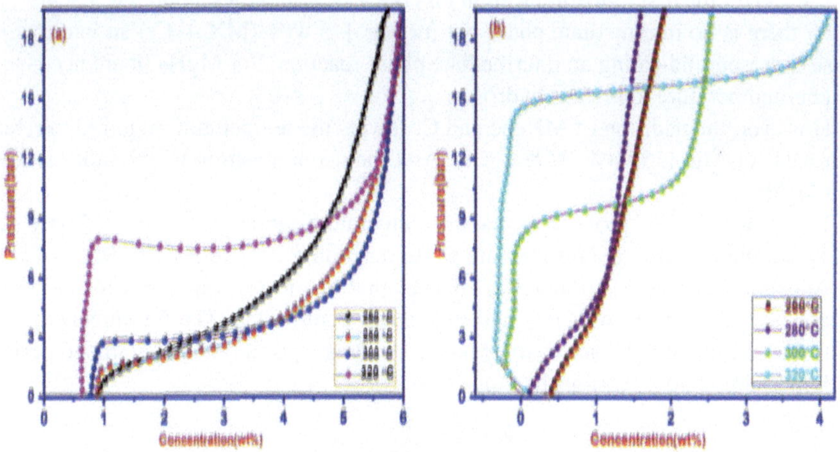

Fig. 2 PCT plot of (a) MgH_2 (b) Mg + 5 wt% (MX + Cr)

The van' t Hoff plot was plotted using PCT plot. The enthalpy and entropy of MgH_2 and Mg + 5 wt% (MX + Cr) is 59.14 kJ/mol, 114.31 J/mol-K and 214.5 kJ/mol,

388.67 J/mol-K. Enthalpy controls the amount of heat absorbed or emitted during hydrogen intake or release processes. Because the additions create more favourable reaction paths, it could require less heat to break bonds in the reactants. The additions most likely help the structural order of the system to be improved during hydrogenation and dehydrogenation (Fig. 3).

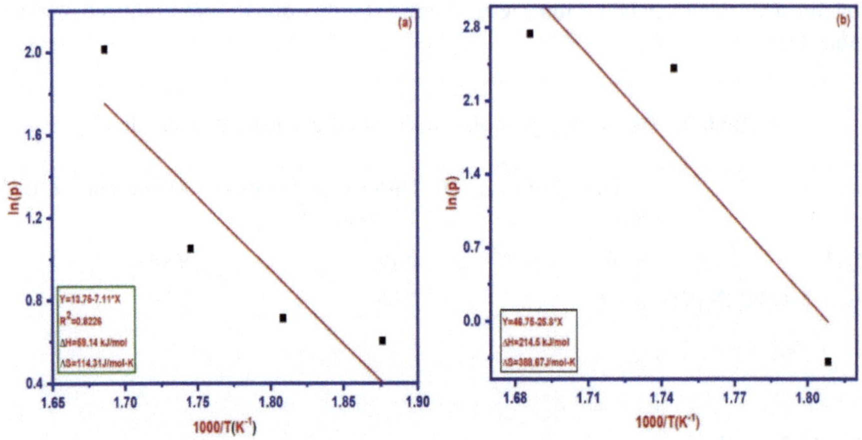

Fig. 3 Van't Hoff plot of (a) MgH_2 (b) Mg + 5 wt% (MX + Cr)

4.3 Thermodynamics

In Fig. 4 TGA plot shows that the weight loss for MgH_2 is 4.3 wt% and for Mg + 5 wt% (MX + Cr) for the first transition weight loss is 8.1 wt% and for second it is 3.3 wt%. For MgH_2 there is no intermediate phase but for Mg + 5 wt% (MX + Cr) an intermediate phase is present indicating an intermediate phase reaction. For MgH_2, there needs to be a higher temperature to release hydrogen.

However, the addition of MXene and Cr lowers the temperature required to release hydrogen for Mg + 5 wt% (MX + Cr). And the decomposition temperature is lower than MgH_2.

In Fig. 4(c, d) DSC plot shows that the temperature decreases for Mg + 5 wt% (MX + Cr) and there is also an intermediate phase reaction for Mg + 5 wt% (MX + Cr).

Activation energy is a theoretical barrier that has to be conquered to initiate the release of hydrogen from MgH_2 and their nano-composites. The Kissinger equation helps one to ascertain the activation energy. Using an appropriate apparatus to perform TGA measurements at various heating rates:

$$lnk = -\frac{E_a}{RT_p} + \alpha$$

where $k = \frac{\beta}{T_p^2}$, β is heating rate, T_p is highest temperature, R is gas constant and E_a is activation energy. The slope is equivalent to $\frac{-E_a}{R}$ when a graph of ln(k) is plotted in against to 1/T.

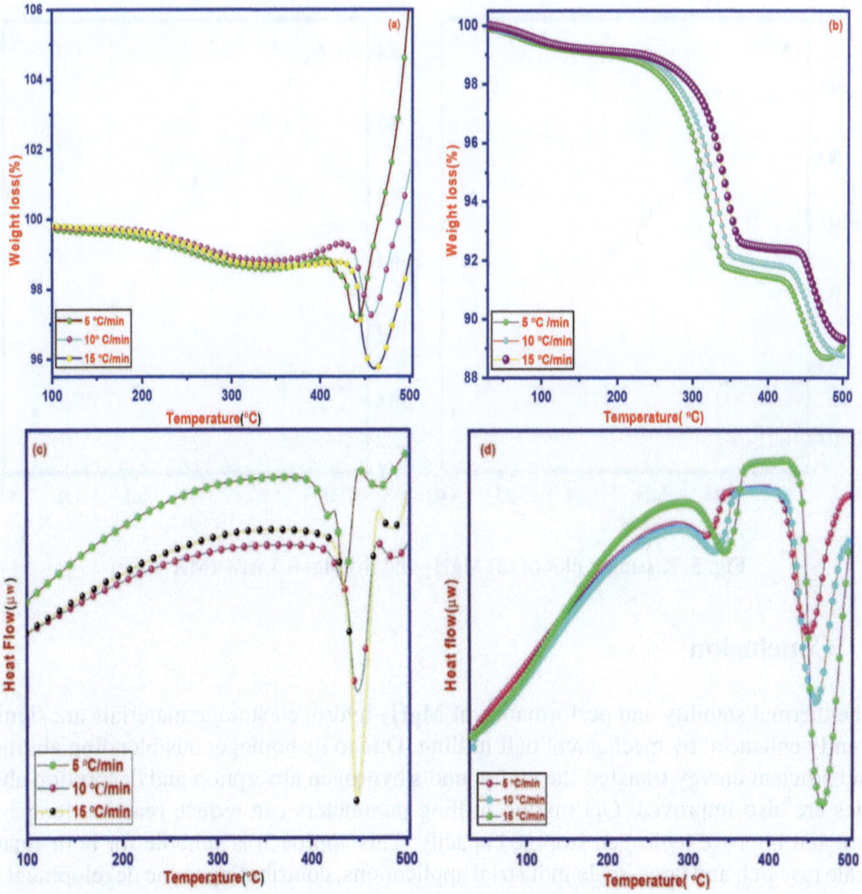

Fig. 4 TGA plot of (a) MgH_2 (b) Mg + 5 wt% (MX + Cr). DSC plot of (c) MgH_2 (d) Mg + 5 wt% (MX + Cr)

Figure 5 shows the Kissinger plot of (a) MgH_2 and (b) Mg + 5 wt% (MX + Cr). The activation energy for MgH_2 is 218.57 kJ/mol, while for Mg + 5 wt% (MX + Cr) is 168.77 kJ/mol. This shows that the activation energy value decreases by 40.8 kJ/mol by the addition of the catalysts (MXene and Cr). Because the desorption rate is also based upon the catalytic effect, particle size, surface area, and nature of the catalyst. The activation energy suggests that the sample has a high level of thermal stability.

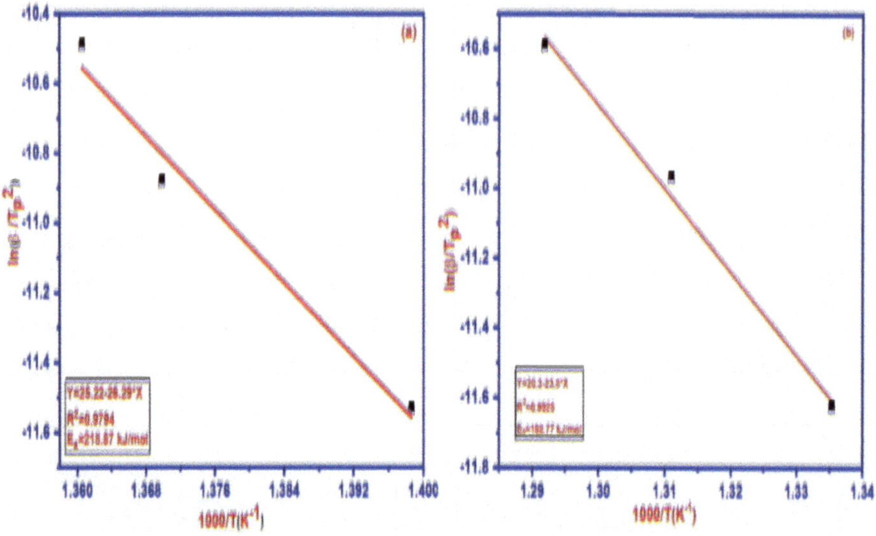

Fig. 5 Kissinger plot of (a) MgH_2 and (b) Mg + 5 wt% (MX + Cr)

5 Conclusion

The thermal stability and performance of MgH_2 hydrogen storage materials are significantly enhanced by mechanical ball milling. Due to its homogenous blending abilities and efficient energy transfer, the compound's hydrogen absorption and desorption abilities are also improved. Optimizing milling parameters can reduce reaction temperature and improve hydrogen storage capacity. This approach is suitable for both small-scale research and large-scale industrial applications, contributing to the development of hydrogen storage systems. Adding the catalysts (MXene and Chromium) to Magnesium hydride leads to a decrease in grain size, dislocation density and microstrain. The hydrogen storage capacity of 2.53 wt% at 320 °C was achieved for Mg + 5 wt% (MX + Cr). The weight loss for Mg + 5 wt% (MX + Cr) for the first transition is 8.1 wt%, for the second transition it is 3.3 wt% and shows an intermediate phase reaction. The activation energy decreases by 40.8 kJ/mol. DSC measurements also show that the MXene and Chromium compounds improve the activation energy.

Acknowledgments. I am grateful to S.S.Jain Subodh P.G. College Jaipur for the characterization. And Dr. Ravindra Jangid from RRCAT, Indore to provide the MXene sample.

Authors Contribution. **Madhu Yadav**- Writing-Original draft preparation, Writing-review editing, Software, Data curation, **Prof. S. N. Dolla** - Methodology, Conceptualization, Validation, and **Dr. Chhagan Lal**-Visualization, Resources, and Supervision of the whole work.

Funding. I want to thank the UGC for JRF.

Declaration of Competing Interest. No competing interests have been declared by the authors.

References

1. Abifarin, J.K., Torres, J.F., Lu, Y.: 2D materials for enabling hydrogen as an energy vector. Nano Energy. **129**, 109997 (2024)
2. Balan, A.P., Puthirath, A.B., Roy, S., Costin, G., Oliveira, E.F., Saadi, M.A.S.R., Sreepal, V., Friedrich, R., Serles, P., Biswas, A., Iyengar, S.A., Chakingal, N., Bhattacharyya, S., Saju, S.K., Pardo, S.C., Sassi, L.M., Filleter, T., Krasheninnikov, A., Galvao, D.S., et al.: Non-van der Waals quasi-2D materials; recent advances in synthesis, emergent properties and applications. Mater. Today. **58**, 164–200 (2022)
3. Bushroa, A.R., Rahbari, R.G., Masjuki, H.H., Muhamad, M.R.: Approximation of crystallite size and microstrain via XRD line broadening analysis in TiSiN thin films. Vacuum. **86**(8), 1107–1112 (2012)
4. Chen, G., Zhang, Y., Cheng, H., Zhu, Y., Li, L., Lin, H.: Effects of two-dimension MXene Ti_3C_2 on hydrogen storage performances of MgH_2 -$LiAlH_4$ composite. Chem. Phys. **522**(March), 178–187 (2019)
5. Dangwal, S., Ikeda, Y., Grabowski, B., Edalati, K.: Machine learning to explore high-entropy alloys with desired enthalpy for room-temperature hydrogen storage: prediction of density functional theory and experimental data. Chem. Eng. J. **493**, 152606 (2024)
6. Gattia, D.M., Jangir, M., Jain, I.P.: Study on nanostructured MgH_2 with Fe and its oxides for hydrogen storage applications. J. Alloys Compd. **801**, 188–191 (2019)
7. Ghotia, S., Kumar, P., Srivastava, A.K.: A review on 2D materials: unveiling next-generation hydrogen storage solutions, advancements and prospects. J. Mater. Sci. **60**, 1071–1097 (2024)
8. Han, G., Lu, Y., Jia, H., Ding, Z., Wu, L., Shi, Y.: Magnesium-based energy materials: progress, challenges, and perspectives. J. Magn. Alloys. **11**(11), 3896–3925 (2023)
9. Hang, Z., Hu, Z., Xiao, X., Jiang, R., Zhang, M.: Enhancing hydrogen storage kinetics and cycling properties of $NaMgH_3$ by 2d transition metal carbide MXene Ti_3C_2. Processes. **9**(10), 1690 (2021). https://doi.org/10.3390/pr9101690
10. Huang, G., Lu, Y., Liu, X., Tang, W., Li, X., Wang, F., Shui, J., Yu, R.: Layered double hydroxide-derived $Mg_2Ni/TiH_{1.5}$ composite catalysts for enhancing hydrogen storage performance of MgH_2. J. Magn. Alloys. **12**(12), 4966–4975 (2023)
11. Jain, I.P.: Hydrogen the fuel for 21st century. Int. J. Hydrog. Energy. **34**(17), 7368–7378 (2019)
12. Jain, I.P., Jain, P., Jain, A.: Novel hydrogen storage materials: a review of lightweight complex hydrides. J. Alloys Compd. **503**(2), 303–339 (2010)
13. Jangir, M., Jain, A., Yamaguchi, S., Ichikawa, T., Lal, C., Jain, I.P.: Catalytic effect of TiF4 in improving hydrogen storage properties of MgH_2. Int. J. Hydrog. Energy. **41**(32), 14178–14183 (2016)
14. Kumar, P., Singh, S., Hashmi, S.A.R., Kim, K.H.: MXenes: emerging 2D materials for hydrogen storage. Nano Energy. **85**, 105989 (2021)
15. Kumar, P., Yu, S., Shahzad, F., Hong, S.M., Kim, Y.H., Koo, C.M.: Ultrahigh electrically and thermally conductive self-aligned graphene/polymer composites using large-area reduced graphene oxides. Carbon. **101**, 120–128 (2016)
16. Li, Y., Guo, Q., Ding, Z., Jiang, H., Yang, H., Du, W., Zheng, Y., Huo, K., Shaw, L.L.: MOFs-based materials for solid-state hydrogen storage: strategies and perspectives. Chem. Eng. J. **485**, 149665 (2024)
17. Liu, Y., Gao, H., Zhu, Y., Li, S., Zhang, J., Li, L.: Excellent catalytic activity of a two-dimensional $Nb_4C_3T_x$ (MXene) on hydrogen storage of MgH_2. Appl. Surf. Sci. **493**, 431–440 (2019)
18. Park, Y.J., Lee, H., Choi, H.L., Tapia, M.C., Chuah, C.Y., Bae, T.H.: Mixed-dimensional nanocomposites based on 2D materials for hydrogen storage and CO_2 capture. npj 2D Mater. Appl. **7**(1), 61 (2023)

19. Polanski, M., Bystrzycki, J., Varin, R.A., Plocinski, T., Pisarek, M.: The effect of chromium (III) oxide (Cr_2O_3) nanopowder on the microstructure and cyclic hydrogen storage behavior of magnesium hydride (MgH_2). J. Alloys Compd. **509**(5), 2386–2391 (2011)
20. Pukazhselvan, D., Çaha, I., de Lemos, C., Mikhalev, S.M., Deepak, F.L., Fagg, D.P.: Understanding the catalysis of chromium trioxide added magnesium hydride for hydrogen storage and Li-ion battery applications. J. Magn. Alloys. **12**(3), 1117–1130 (2024)
21. Sharma, H., Kumar, S., Yadav, J., Prasad, J., Singh, M.: Temperature treatment effect on the physical and optical properties of ZnO thin films. J. Mater. Sci. Mater. Electron. **35**(1), 1–8 (2024)
22. Verma, S.K., Shaz, M.A., Yadav, T.P.: Enhanced hydrogen absorption and desorption properties of MgH_2 with graphene and vanadium disulfide. Int. J. Hydrog. Energy. **48**(56), 21383–22139 (2023)
23. Vijay, R., Sundaresan, R., Maiya, M.P., Murthy, S.S.: Hydrogen storage properties of Mg-Cr_2O_3 nanocomposites: the role of catalyst distribution and grain size. J. Alloys Compd. **424**(1–2), 289–293 (2006)
24. Wang, Y., Wang, Y.: Recent advances in additive-enhanced magnesium hydride for hydrogen storage. Prog. Nat. Sci.: Mater. Int. **27**(1), 41–49 (2017)
25. Wu, Z., Fang, J., Liu, N., Wu, J., Kong, L.: The improvement in hydrogen storage performance of MgH_2 enabled by multilayer Ti_3C_2. Micromachines. **12**(10), 1–14 (2021)
26. Xu, Y., Zhou, Y., Li, Y., Ding, Z.: Research progress and application prospects of solid-state hydrogen storage technology. Molecules. **29**(8), 1767 (2024)
27. Xu, Y., Zhou, Y., Li, Y., Hao, Y., Wu, P., Ding, Z.: Magnesium-based hydrogen storage alloys: advances, strategies, and future outlook for clean energy applications. Molecules. **29**(11), 2525 (2024)
28. Yadav, J., Prasad, J., Sharma, H., Kumar, S., Singh, M.: Effect of Zn and Sn incorporation on the crystallinity of spin-coated $CuInSe_2$ thin films. J. Mater. Sci. Mater. Electron. **33**(24), 19119–19128 (2022)
29. Yuan, Z., Zhang, X., Wu, Y., Guan, S., Zhao, S., Ji, L., Peng, Q., Han, S., Fan, Y., Liu, B.: Effectively enhanced catalytic effect of sulfur doped Ti_3C_2 on the kinetics and cyclic stability of hydrogen storage in MgH_2. J. Magn. Alloys. (2024)
30. Zhu, W., Panda, S., Lu, C., Ma, Z., Khan, D., Dong, J., Sun, F., Xu, H., Zhang, Q., Zou, J.: Using a self-assembled two-dimensional MXene-based catalyst (2D-Ni@Ti_3C_2) to enhance hydrogen storage properties of MgH_2. ACS Appl. Mater. Interfaces. **12**(45), 50333–50343 (2020)

Room Temperature Thermoelectric Properties of β-Zn_4Sb_3 Thermally Evaporated Thin Films

Avinash Kumar and S. K. Tripathi(✉)

Department of Physics, Centre for Advanced Study in Physics, Panjab University, Chandigarh 160014, India

surya@pu.ac.in

Abstract. Main aim of present study is to investigate room temperature (RT) thermoelectric (TE) parameters of β-Zn_4Sb_3 thin films after heat treatment for 6 h at 323 K, 343 K, 363 K and 383 K. TE measurements of electrical conductivity (σ), Seebeck coefficient (S) and power factor (PF) were performed and values were determined at RT. Melt-quench and thermal evaporation routes were used for synthesis and thin film deposition of β-Zn_4Sb_3 respectively. Maximum σ, S, PF values of 9.5×10^3 Sm^{-1},138 μVK^{-1} and 145 μWm^{-1} K^{-2} respectively were achieved at RT after heat treatment of films and an enhancement of 39.85%, 48.38% and 145.76% respectively compared to without heat treated thin film sample. Structural properties of thin films were investigated using X-ray diffraction (XRD) and crystallite sizes were determined. Optical properties of thin films were examined using ultraviolet-visible spectroscopy (UV-Vis) and photoluminescence (PL). Direct band gap energy (E_g) of 1.27 eV was estimated using tauc plot method. Field Emission Scanning Electron Microscope (FESEM) was used to investigate morphological properties and thickness of thin films. Thin film thickness of 327 nm was determined using cross sectional FESEM. Energy Dispersive X-ray (EDX) was used to analyse atomic percentages composition.

Keywords: β-Zn_4Sb_3 thin films · Thermoelectric · Thermal evaporation · Power factor

1 Introduction

Problem of exponentially increasing electrical power demands with limited available fossil fuels and deteriorating environmental conditions focused researchers to find sustainable, durable and emission free power generation alternatives [1]. Thermoelectric materials emerged as potential alternative of fossil fuels and presently available renewable resources for electricity generation due to their potential to convert thermal energy available in form of waste heat directly into electrical energy. Thin films based TE modules have gained large interest for power production and cooling purposes in wearable devices, sensors, chips, spacecrafts due to its miniature size, lightweight, versatility and low maintenance cost [2, 3]. Efficiency of a TE materials is estimated by determining

S. Kumari et al. (Eds.): ICMTP 2024, SPPHY 311, pp. 113–119, 2025.
https://doi.org/10.1007/978-981-96-6795-6_14

term ZT expressed as Eq. (1) in terms of S, σ, thermal conductivity (k) and operating temperature (T). Parameter $S^2\sigma$ in Eq. (1) is collectively known as power factor (PF) [4].

$$ZT = \frac{S^2 \sigma T}{k} \quad (1)$$

β-Zn_4Sb_3 has shown promising TE performance near room temperature range which can replace state of art materials such as Bi_2Te_3 and PbTe based alloys in near future for large scale energy production applications due to its exhibition of high values of conductivity and Seebeck coefficient with ultra-low values of thermal conductivity($<1\ Wm^{-1}\ K^{-1}$). β-Zn_4Sb_3 also exhibit other major requirements such as low cost and environment friendliness due to use of elements which are present in abundance and non-toxic in nature [5]. Several routes are earlier reported to synthesize β-Zn_4Sb_3 such as solid state reaction, mechanical alloying, melt quenching, flux growth [6]. In present study, β-Zn_4Sb_3 powder was obtained with the help of melt quench synthesis method because mixing of elements in melting phase results in low unreacted impurities and better homogeneity. Quenching inhibits formation of different microstructures which results in material with tunable transport properties. Thin film of synthesized material were deposited using thermal evaporation technique.

Post deposition strategies such as nano structuring, doping, band gap engineering, introduction of pores and defects are earlier reported to enhance TE performance of β-Zn_4Sb_3 thin films [5, 6]. Jindal et al. [7, 8] and Singh et al. [9] reported post deposition heat treatment as an effective strategies to enhance power factor of thin films. They reported reorientation of crystallographic planes, improved crystallinity and defects curing due to heat treatment of thin films which resulted in optimized transport properties and enhanced power factor [7–9].

In present work, variation of TE parameters S, σ, PF of thin films of β-Zn_4Sb_3 are investigated at RT after heat treatment for 6 h at 323 K, 343 K, 363 K and 383 K. Additionally, structural, optical, morphological properties and atomic composition are also examined.

2 Experimental Details and Methodology

2.1 Synthesis and Thin Film Deposition of β-Zn_4Sb_3

High purity (5N, Sigma Aldrich) Zn, Sb elements were weighed according to desired stoichiometric composition and loaded in to quartz ampoules which were further sealed under the maintained vacuum pressure of 2×10^{-5} mbar. Ampoules were then transferred to muffle furnace and kept at 1273 K for 10 h for homogeneous mixing and reaction of elements in melting phase. Ampoules were quenched into ice to obtain β-Zn_4Sb_3 ingot. Powdered material was obtained after crushing the ingot into mortar pestle. Thin films were deposited on glass substrate by evaporating powdered β-Zn_4Sb_3 kept into molybdenum (Mo) boat under the same vacuum pressure of through resistive heating.

2.2 Thin Film Samples Characterizations

To investigate structural properties of deposited thin films, XRD spectra were recorded for $2\theta = 20°$ to $80°$ using Rigaku Miniflex-600 diffractometer operated with Cu radiation source (λ = 1.54 Å). Optical characteristics of thin films were examined using UV-Vis instrument from Lab India spectrophotometer and PL instrument from Shimadzu RF-6000 spectrofluoro-photometer. Morphological characteristics & thickness measurements of deposited samples were examined with the help of Hitachi SU 8010 FESEM. Atomic percentages composition of samples were determined using Brukers Xflash 6130 EDX. Electrical conductivity of thin films were determined by performing I-V experiment using Keithley 6517A as voltage source. Seebeck experiments were performed by applying a temperture gradient of ~2 ° C across thin film samples and variation of generated voltage was measured using Keithley 2182 A nanovoltmeter.

3 Results and Discussions

Figure 1 shows the recorded XRD spectra of β-Zn_4Sb_3 thin film sample without post deposition heat treatment (As grown). Intense peak at $2\theta = 44.5°$ corresponds to (330) plane and other minor peaks as indexed in Fig. 1 represents Zn_4Sb_3 phase. No peaks of secondary phases such as ZnSb and elemental impurities phases of Zn, Sb are depicted in XRD spectra. Reference pattern with JCPDS card no. 00-031-0112 was used for indexing β-Zn_4Sb_3 peaks. Crystallite size (D) corresponding to most intense peak at $2\theta = 44.5°$ was determined using Debye-Scherrer formula ($D = \frac{0.9\lambda}{\beta cos\theta}$) where λ, β, θ are X-ray wavelength, full width half maxima and Bragg's angle respectively [3]. Determined D value of 19.66 nm revealed the nano crystalline nature of thin film.

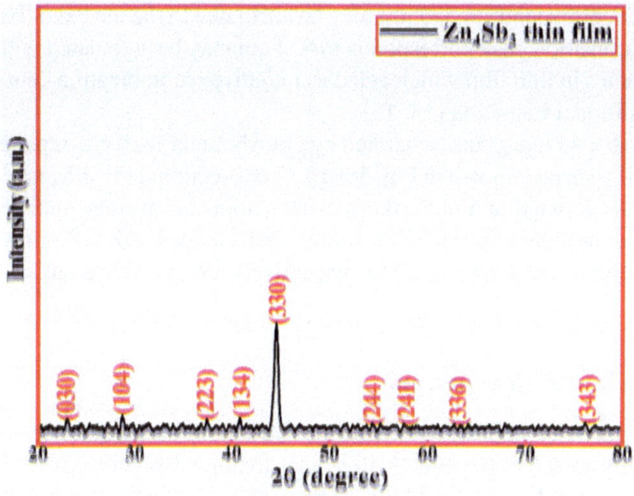

Fig. 1. XRD spectrum of β-Zn_4Sb_3 As grown thin film.

Figure 2 (left) shows the tauc plot extrapolation used to estimate the direct band gap energy. Inset shows recorded absorbance spectra of As grown thin film.

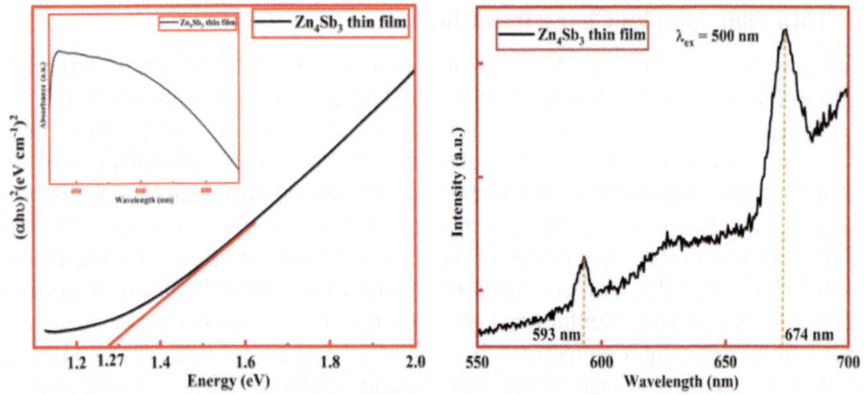

Fig. 2. Tauc plots of As grown thin film to estimate band gap (**left**) and PL spectra of As grown thin film (**right**).

E_g value of 1.27 eV was estimated using tauc Eq. (2) as shown in Fig. 2 (left).

$$\alpha h\upsilon = A(h\upsilon - E_g)^n \tag{2}$$

where α, hυ, E_g and A are the absorption coefficient of absorption, energy of photon, bandgap value and material specific constant respectively A. Value of variable n are 2 and 1/2 for the indirect and direct electronic allowed transitions [10]. Estimated E_g value of 1.27 eV is in close agreement with earlier reported values of 1.29 eV, 1.22 eV, 1.20 eV by Sanchela et al. [11], Sittichai et al. [12], Tahashi et al. [13]. Figure 2 (right) shows PL emission spectra of As grown thin film with excitation wavelength of 500 nm. Emission transition peak corresponds to 593 nm may be attributed to the the band- band transition and emission transition peak corresponds to 674 nm may be associated with presence of defects, vacancies in thin film which acted as radiative recombination centres at surface to provide additional transition [14, 15].

Spherical shaped like grains separated by grain boundaries were depicted on without annealed As film surface shown in Fig. 3(a, b). Cross sectional FESEM revealed uniform deposition of As grown thin film via thermal deposition and average thickness of 327 nm was determined as shown in Fig. 3(c). Atomic percentages of Zn, Sb in As grown thin film found to be 65.12% and 34.88% respectively using EDX analysis as shown in Fig. 3(d).

3.1 Thermoelectric Measurements

Electrical conductivity value of 6.8×10^3 Sm^{-1} was determined at RT for As film which reduced slightly to 6.6×10^3 Sm^{-1} after heat treatment of films at 323 K might be attributed to enhanced charge carrier scattering from defects and grain boundaries [7, 8]. Electrical conductivity values showed uptrend for further heat treatments at 343 K, 363 K, 383 K with a maximum value of 9.5×10^3 Sm^{-1} at RT as shown in Fig. 4 (left). This uptrend may be attributed to reduction in defects & grain boundaries, improved crystallinity and reorientation of crystallographic planes via high temperature heat treatment

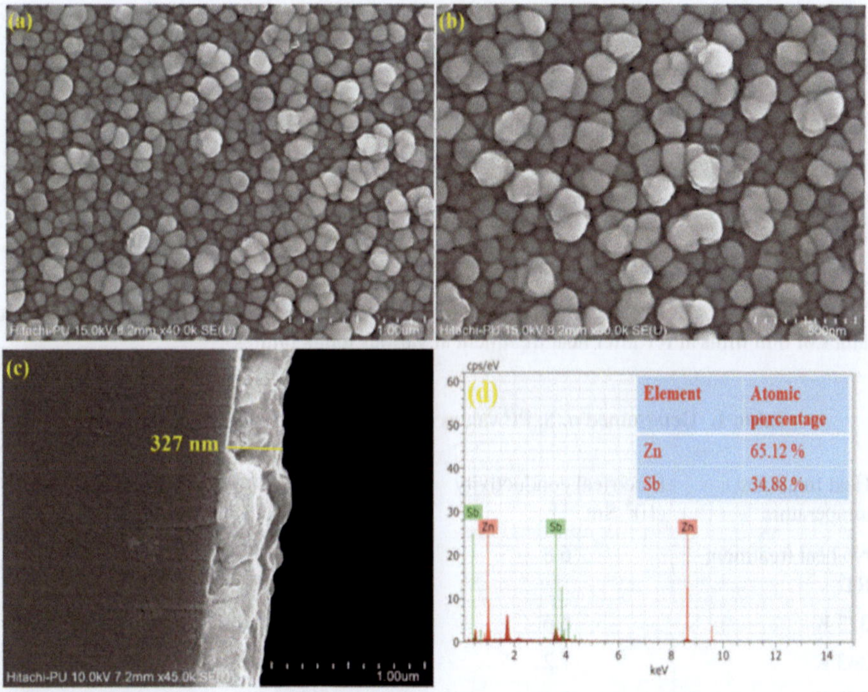

Fig. 3. **(a, b)** FESEM images of as grown thin film **(c)** cross sectional FESEM for thickness determination **(d)** EDX spectra of without annealed As film.

which provided larger mean free path and reduced scattering of carriers [7–9]. Seebeck coefficient of 93 μVK^{-1} was determined for As grown thin film which remained almost unchanged with value of 90 μVK^{-1} after heat treatment at 323 K. S value increased up to highest value of 138 μVK^{-1} after heat treatment at 363 K as shown in Fig. 4 (middle) attributed to energy filtering effects of low energy charge carriers via defects and grain boundaries. S value reduced to 110 μVK^{-1} at RT after heat treatment at 383 K attributed to enhanced electrical conductivity and high diffusion of low energy charge carriers. Power factor values showed similar trend as S values as shown in Fig. 4 (right) which shown the quadratic role of S values in determination of PF values [7–9]. Table 1 shows the determined σ, S, PF values at RT after heat treatment of thin films.

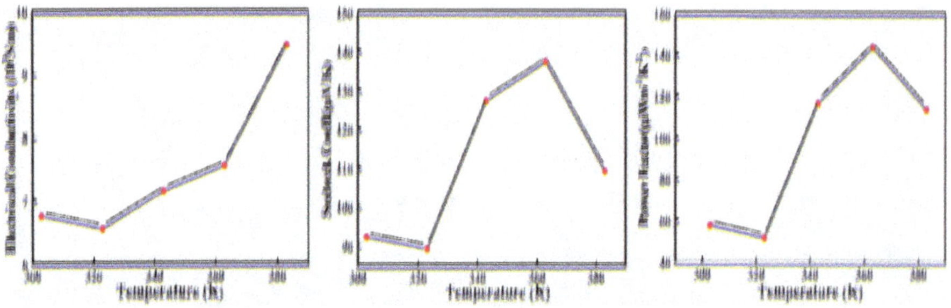

Fig. 4. Variation of (left) electrical conductivity (middle) Seebeck coefficient (right) power factor values of thin films at RT after heat treatment at various temperatures.

Table 1. Determined σ, S, PF values at RT after heat treatment of thin films.

Heat treatment temperature	Electrical conductivity (10^3 Sm^{-1})	Seebeck coefficient (μVK^{-1})	Power factor (μWm^{-1} K^{-2})
No heat treatment-RT	6.8	93	59
323 K	6.6	90	53
343 K	7.2	128	118
363 K	7.6	138	145
383 K	9.5	110	115

4 Conclusions

Thermally evaporated thin films of β-Zn_4Sb_3 were deposited that displayed improved σ, S, PF values of 6.8 $\times 10^3$ Sm^{-1}, 138 μVK^{-1}, 59 μWm^{-1} K^{-2} respectively after heat treatment at various temperatures. Post deposition heat treament strategy proved to be effective to improve TE performance of β-Zn_4Sb_3 significantly in present study which enhanced σ, S, PF by 39.85%, 48.38% and 145.76% respectively. This enhancement in TE parameters is attributed to improved crystallinity, optimized defects & grain boundaries and reorientation of crystallographic planes via high temperatures heat treatment which led to improved charge carrier transportation. Estimated band gap value of 1.27 eV using tauc method is in optimum range for TE power genaration and cooling applications.

Acknowledgement. Mr. Avinash Kumar is thankful to CSIR, New Delhi for providing fellowship during the work.

References

1. Yuan, X., Zhao, Y., Dai, Z.: Typical thermoelectric materials: progress and prospects. Phys. Lett. A. **535**, 130280 (2025)

2. Singh, J., Kaur, H., Singh, G., Tripathi, S.K.: High thermoelectric performance in cubic inorganic halide perovskite material $AgCdX_3$ (X= F and Cl) from first principles. Mater. Today Energy. **21**, 100820 (2021)
3. Singh, S., Sidhu, J., Tripathi, S.K.: High thermopower in (00l)-oriented nanocrystalline Bi-Sb-Te thin films produced by one-step thermal evaporation process. Vacuum. **165**, 12–18 (2019)
4. Singh, S., Jindal, S., Tripathi, S.K.: High Seebeck coefficient in thermally evaporated Sb-In co-alloyed bismuth telluride thin film. J. Appl. Phys. **127**(5), 055103 (2020)
5. Yang, J., Liu, G., Shi, Z., Lin, J., Ma, X., Xu, Z., Qiao, G.: An insight into β-Zn_4Sb_3 from its crystal structure, thermoelectric performance, thermal stability and graded material. Mater. Today Energy. **3**, 72–83 (2017)
6. Zou, T., Xie, W., Feng, J., Qin, X., Weidenkaff, A.: Recent developments in β-Zn_4Sb_3 based thermoelectric compounds. J. Nanomater. **16**(1) (2016)
7. Jindal, S., Singh, S., Saini, G.S.S., Tripathi, S.K.: Enhanced thermopower in (013)-oriented silver selenide films produced by thermal annealing. Appl. Phys. A. **126**, 1–14 (2020)
8. Jindal, S., Singh, S., Saini, G.S.S., Tripathi, S.K.: Optimization of thermoelectric power factor of (013)-oriented Ag_2Se films via thermal annealing. Mater. Res. Bull. **145**, 111525 (2022)
9. Singh, S., Singh, J., Kaushal, J., Tripathi, S.K.: Effects of annealing on the thermoelectric properties of nanocrystalline $Bi_{1.2}Sb_{0.8}Te_3$ thin films prepared by thermal evaporation. Appl. Phys. A. **125**, 1–13 (2019)
10. Manyani, N., Siwatch, P., Rana, S., Sharma, K., Tripathi, S.K.: Study of electrochemical behaviour of binder-free nickel metal-organic framework derived by benzene-1, 3, 5-tricarboxylic acid for supercapacitor electrode. Mater. Res. Bull. **165**, 112320 (2023)
11. Sanchela, A.V., Tomy, C.V., Thakur, A.D.: Effect of Sb deficiency on the thermoelectric properties of Zn_4Sb_3. Solid State Commun. **218**, 49–53 (2015)
12. Sitthichai, S., Thongtem, T., Thongtem, S., Suriwong, T.: One-step synthesis of Zn_4Sb_3 nanocrystals and Zn_4Sb_3–ZnSb composites. Superlattice. Microst. **64**, 433–438 (2013)
13. Tahashi, M., Goto, H.: Enhancement of electrical conductivity of Zn_4Sb_3 by addition of ZnO particles. AIP Adv. **13**(1), 015308 (2023)
14. Kumar, A., Singh, J., Tripathi, S.K.: High thermoelectric coefficients of β-Zn_4Sb_3 thin films at room temperature via chemical etching. Mater. Res. Bull. **184**, 113281 (2025)
15. Kumar, A., Manyani, N., Tripathi, S.K.: Synthesis and characterization of thermoelectric material Bi_2Te_3—A potential alternative for power generation. In: International Conference on Nanotechnology: Opportunities and Challenges, pp. 287–292. Springer Nature Singapore, Singapore (2022)

Investigation of Structural, Electronic and Optical Properties of $EASnBr_3$ Based Perovskites

Anusha Dubey[1], Naincy Pandit[2], Yogendra Kumar Saraswat[3], and Ajay Singh Verma[4,5](✉)

[1] Department of Physics, Noida Institute of Engineering and Technology, Greater Noida 201306, India
[2] Department of Physics, School of Allied Sciences, Dev Bhoomi Uttarakhand University, Dehradun 248007, India
[3] Department of Chemistry, B S A College, Mathura 281004, India
[4] Division of Research and Innovation, School of Applied and Life Sciences, Uttaranchal University, Dehradun 248007, India
ajay_phy@rediffmail.com
[5] University Centre for Research and Development, Department of Physics, Chandigarh University, Mohali 140143, Punjab, India

Abstract. A variety of solar systems can include hybrid halide perovskites. These types of perovskites have lower toxicity, more structural stability and show improvement in the bandgap as compared to ethyl ammonium-based hybrid perovskites. The structural, electrical, optical and thermoelectric properties of Ethyl ammonium Tin Bromide have been evaluated with the help of the software density functional theory (DFT), in particular WIEN2K simulation code. The use of the exchange-correlation potential KTB-mBJ makes it possible to determine the optimized structure, density of states, and band structure of EASnBr3 which is quite evidently a semiconductor material that has a direct bandgap and a very high absorption coefficient in the Ultraviolet to Visible range. These findings help to tackle the environmental issues as it makes sure that it is an appropriate and cheap material to be used in solar panels.

Keywords: Density Functional Theory · WIEN2k · Photovoltaic material · Optoelectronic

1 Introduction

Renewable energy sources represent the inevitable path for energy consumption in the twenty-first century, given the impending depletion of finite fossil fuel resources. The Sun, being a virtually limitless energy reservoir, is anticipated to play a central role in this shift, with photovoltaic (PV) devices converting sunlight into electricity gaining prominence as a primary renewable energy source [1, 2]. Perovskite solar cells have rapidly risen to the forefront among emerging photovoltaic technologies, presenting

S. Kumari et al. (Eds.): ICMTP 2024, SPPHY 311, pp. 120–125, 2025.
https://doi.org/10.1007/978-981-96-6795-6_15

themselves as promising candidates for electricity generation [3]. These solar cells boast a noteworthy track record, achieving a high conversion efficiency of 25.17% in the field of photovoltaic research, coupled with the advantage of being cost-effective [4]. The general structure of hybrid halide perovskites is ABX_3, Where A is a monovalent cation such as MA^+, EA^+, and FA^+, B is the divalent cation like Pb^{2+}, Sn^{2+}, and Ge^{2+}, and X is the halide namely I^-, Br^-, and Cl^- [5–8]. Dubey et al. [3] utilized the WIEN2K code to delve the various properties of the $EASnCl_3$. Notably, $EASnCl_3$ exhibits direct band gap semiconductor behavior with a band gap of 3.006 eV, making it promising for photovoltaic applications, particularly in UV and visible light regions.

Taking into account the literature available in this study, our objective is to identify a suitable lead-free perovskite PV material by replacing lead with Tin. We anticipate that the resulting material, $CH_3CH_2NH_3SnBr_3$ or $EASnBr_3$, will possess all the desired requirements for efficient PV applications.

2 Computational Details

In modern science, density functional theory (DFT) is arguably the most widely accepted approach to ab initio calculations, especially with respect to ground-state properties of molecules and solids [9]. We applied this established approach to calculate the structural, electronic, optical, and thermoelectric parameters of EASnBr3. The computations were performed using the WIEN2K package, applying the Full-potential linearized augmented plane wave (FP-LAPW) method within DFT. To calculate the atomic positions, lattice constant (a), bulk modulus (B), and minimum energy (E_0), an optimized structure was taken. An energy difference of −6.0 Ry was taken for the distinction between the core and valance electron states [11]. We used the GGA-WC approximation and followed the convergence criterion (RMT × Kmax = 3), where RMT is the smallest value of the Muffin tin (MT) radii and Kmax is the cut-off plane wave vector, for the optimization process [12]. We use the KTB-mBJ potential to obtain the electronic, optical, and thermodynamic properties of the crystal. For calculating the properties of $EASnBr_3$, we use 700 k points. The optical parameters of $EASnBr_3$ were determined by considering an extensive 10,000 k points within the irreducible Brillouin zone.

3 Results and Discussion

3.1 Structural Properties

Structural parameters were computed using LDA exchange-correlation potential through an optimization method. The obtained lattice constants were then fitted into Birch-Murnaghan's equation of state. Figure 1 demonstrates the correlation between energy (E) and per unit cell volume (V), and Table 1 represents the optimized parameters of material:

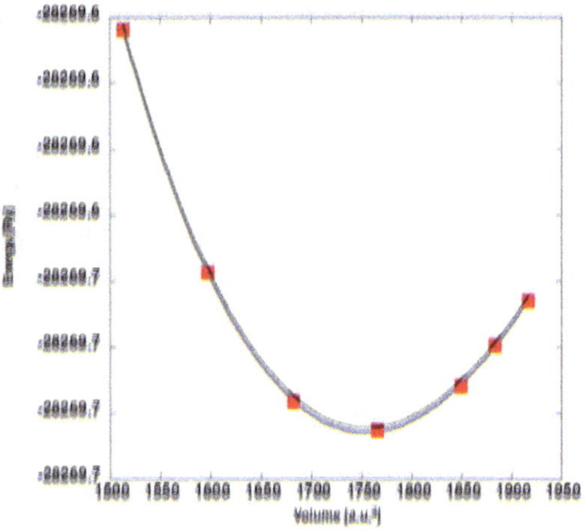

Fig. 1. Optimization curve of $EASnBr_3$.

Table 1. Lattice parameter of the optimized structure of $EASnBr_3$.

Material	Lattice constant a (Å)	Equilibrium volume V_0 (Å^3)	Bulk modulus B (GPa)	Pressure derivative of bulk modulus B'
$CH_3CH_2NH_3SnBr_3$	6.409	1776.8220	43.6025	3.71

3.2 Electronic Properties

Figure 2 shows the electronic band structure as well as the density of states (DOS) for this material in the IBZ. This plot gives some insights in regards to the type of conduction that takes place within the material as well as the type of conduction band and the band gap. The Fermi level within the band gap of EASnBr3 is free, which confirms the fact that the compound is naturally a semiconductor. In addition, both Valance band maxima (VBM) and conduction band minima (CBM) are situated at the symmetry point R which gives $EASnBr_3$ a direct band gap of 2.093 eV. The observed band gap is a result of the formation of the CB and VB from the combination of Sn-5p, Sn-5s and Br-4p orbitals confirms from the DOS of the material.

3.3 Optical Properties

The optical properties of the materials, which include the absorption coefficients, refractive index, extinction coefficient, optical conductivity, reflectivity, and electronic loss energy function, are closely connected to the dielectric function and are important to finding the PV characteristics of the material. Figure 3 depicts the absorption coefficient $\alpha(\omega)$ of $EASnBr_3$ as a function of photon energy. The $\alpha(\omega)$ describes the ability of a substance to absorb energy. Figure 3 shows when the energy of photons passes the band gap value of the materials, the value of $\alpha(\omega)$ starts increasing. When the photon energy

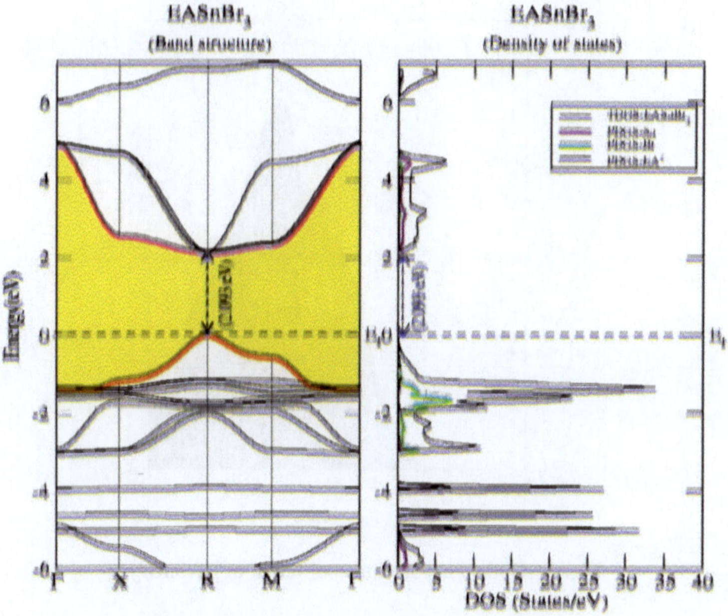

Fig. 2. Band structure and DOS of EASnBr3.

is 4.56 eV the maximum value of $\alpha(\omega)$ 49.93 × 10^4 cm^{-1} for $EASnBr_3$. The presence of direct transitions is more likely when the absorption coefficient value exceeds 10^4 cm^{-1} for the material. It is possible to state that material have good absorption coefficients for use in Solar cell technology.

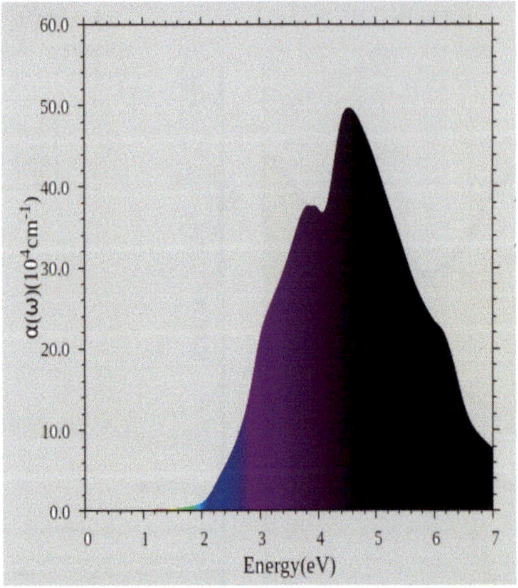

Fig. 3. Optical absorption coefficient of $EASnBr_3$.

4 Conclusion

We have investigated the physical properties of $EASnBr_3$ for photovoltaic application. The results show that the compound is dynamically and mechanically stable. This compound possesses the direct band gap nature. The optical absorption coefficient of this compound shows the good optical absorption in visible region makes it suitable for photovoltaic application.

References

1. Pandit, N., Singh, R., Joshi, T.K., Shukla, A., Kamlesh, P.K., Dubey, A., Kumar, T., Chauhan, M.S., Verma, A.S.: Investigation of a potential photovoltaic absorber based on first principles spectroscopic screening of chalcogenide perovskites: $CaZrX_3$ (X=S, Se). J. Comput. Electron. **24**, 1–19 (2025)
2. Pandit, N., Singh, R., Kamlesh, P.K., Kumar, N., Sharma, P., Kumari, S., Kumar, T., Al-Qaisi, S., Verma, A.S.: Study of chalcogenide-based metal perovskites bazrx3 (x= s and se): Dft insight into fundamental properties for sustainable energy generation using amps-1d. J. Comput. Electron. **23**, 1014–1028 (2024)
3. Dubey, A., Mishra, K., Srivastava, R., Kumari, A., Jangra, P., Joshi, T., Choudhary, B., Verma, A.: Structural, optoelectronic, thermal and transport properties of hybrid perovskite ($EAGeCl_3$). Mater. J. Ovonic Res. **20**, 381–394 (2024)
4. Shukla, A., Sharma, V.K., Gupta, S.K., Verma, A.S.: Computational determination of the physical-thermoelectric parameters of tin-based organomatallic halide perovskites ($CH_3NH_3SnX_3$, X= Br and I): emerging materials for optoelectronic devices. Mater. Chem. Phys. **253**, 123389 (2020)

5. Joshi, T.K., Pravesh, Sharma, G., Verma, A.S.: Investigation of structural, electronic, optical and thermoelectric properties of ethylammonium tin iodide $CH_3CH_2NH_3SnI_3$: an appropriate hybrid material for photovoltaic application. Mater. Sci. Semicond. Process. **115**, 105111 (2020)
6. Zhang, Z., Liu, D., Wu, K.: First-principles study of structural stability, electronic and optical properties of GA-doped $MAPbI_3$. Spectrochim. Acta A Mol. Biomol. Spectrosc. **226**, 117638 (2020)
7. He, C., Zha, G., Deng, C., An, Y., Mao, R., Liu, Y., Chen, Z.: Refractive index dispersion of organic–inorganic hybrid halide perovskite $CH_3NH_3PbX_3$ (X= Cl, Br, I) single crystals. Cryst. Res. Technol. **54**, 1900011 (2019)
8. Liu, D., Li, Q., Wu, K.: Ethylammonium as an alternative cation for efficient perovskite solar cells from first-principles calculations. RSC Adv. **9**, 7356–7361 (2019)
9. Blaha, P., Schwarz, K., Madsen, G.K.H., Laskowsk, R., Tran, F., Marks, L.D.: WIEN2K: an augmented plane wave plus local orbitals program for calculating properties of solid. J. Chem. Phys. **152** (2020)
10. Singh, D.J., Nordstrom, L.: Planewaves, Pseudopotentials, and the LAPW Method. Springer Science and Business Media, Berlin (2006)
11. Perdew, J.P., Ruzsinszky, A., Csonka, G.I., Vydrov, O.A., Scuseria, G.E., Constantin, L.A., Zhou, X., Burke, K.: Restoring the density gradient expansion for exchange in solids and surfaces. Phys. Rev. Lett. **100**, 136406 (2008)
12. Wu, Z., Cohen, R.E.: More accurate generalized gradient approximation for solids. Phys. Rev. B. **73**, 235116 (2006)

A DFT Study of Optoelectronic Properties of New Lead Free Single Perovskite $LiSrBr_3$ for Energy Applications

Anuradha(✉), Deepak K. Meena, Monika Rani, Sunil Kumawat, and Gunjan Arora

Department of Physics, University College of Science, Mohanlal Sukhadia University, Udaipur 313001, Rajasthan, India
anuradhabarala82@gmail.com

Abstract. To explore the structural and optoelectronic properties of $LiSrBr_3$ halide perovskite, generalised gradient potential is incorporated within the density functional theory framework. The electronic band structure and density of states reveals the p-type semiconducting nature of $LiSrBr_3$ sample with an indirect band gap value of 2.96 eV. Optical properties of the present perovskite are studied in terms of dielectric constant, absorption coefficient, reflectivity and refractivity, which strongly support the bandgap value of material. The multiple optical absorption peaks of $LiSrBr_3$ are lying in the UV region which reveals its applications in optoelectronic and high-power operation devices.

Keywords: Perovskite halide · Density functional theory · Optical properties · Band structure

1 Introduction

Perovskite halides show greater device performance due to their important physical characteristics like strong carrier mobility, absorption, an adjustable band gap, and luminous efficiency. Halide perovskites belong to a class of materials that are used in optoelectronic devices. In ideal cubic structure with chemical formula ABX_3, B and X form a corner shared octahedral, BX_6 in three dimensions where, the A-site cation is present in the cage, which is made by the octahedral and surrounded by 12 evenly spaced anions [1–3]. Metal halide perovskites, AMX_3 where A/M = rare-earth/alkali-earth, X = Br, Cl, F, I show promising optical properties. The direct/indirect band gap characteristics of these compounds can be controlled by various chemical adjustments [4]. For the study of potential applications of $Li(Be,Mg)F_3$ perovskite compounds in optoelectronic and UV protection devices, *ab-inito* calculations have been utilized to evaluate the electronic structure and optical absorption characteristics of these compounds [5, 6]. Kaleemullah et al. [7] have discussed optoelectronic properties of AMX_3 perovskites. Some other previously reported work in the field of optoelectronic applications [8–12]. Perdew-Burke-Ernzerhof (PBE) methodology have been precisely applied in the current work for the study of structure of energy bands, density of states (DOS) and optical properties of $LaSiBr_3$ perovskite.

S. Kumari et al. (Eds.): ICMTP 2024, SPPHY 311, pp. 126–130, 2025.
https://doi.org/10.1007/978-981-96-6795-6_16

2 Methodology

The optical and electronic properties of $LiSrBr_3$ compound are deduced by implementing full-potential linearized augmented plane wave approach [13]. DFT calculations for $LiSrBr_3$ sample are performed by using PBE exchange-correlation approximation [14]. Throughout the computations, the cut-off parameter, $R_{MT} \times k_{max} = 7.0$, $G_{max} = 12$, and $l_{max} = 10$ were maintained constant to achieve the basis sets convergence. We have used crystal structure and lattice parameter information from previously reported data [7].

3 Results and Discussions

Electronic configurations of Li, Sr and Br are taken as $1s^22s^1$, $1s^22s^2\ 2p^63s^2\ 3p^64s^23d^{10}4p^65s^2$ and $1s^22s^2\ 2p^63s^2\ 3p^64s^23d^{10}4p^5$, respectively. DOS curves (shown in Fig. 1) gives the facts about the influences of the atoms and states in the formation of band structure (BS) and electronic band transitions. Total DOS of $LiSrBr_3$ depict that the valence bands mainly occur due to the impact of the 4p state of the Br and in the conduction band the involvement of 5s state of Sr, 2s state of Li and 4p state of Br dominate.

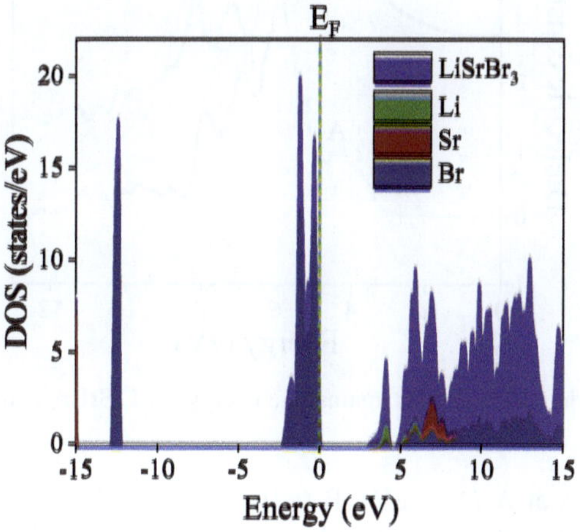

Fig. 1. Density of state curves for $LiSrBr_3$ computed using PBE scheme.

Computed electronic BS is shown in Fig. 2. From this figure we observe that Fermi level (E_F) is very close to the valence band which represent the p-type semi-conducting nature of this material. The BS below the E_F is constructed due to the major contribution of the 4p state of the Br anion and above the E_F 2s electronic state of the Li and 5s orbitals of the Sr cations also show significant contribution. The bandgap (BG) is found as 2.96 eV with indirect nature because the maxima of valence band lies on R point and the minima of the conduction band lies at the Γ point (Fig. 3).

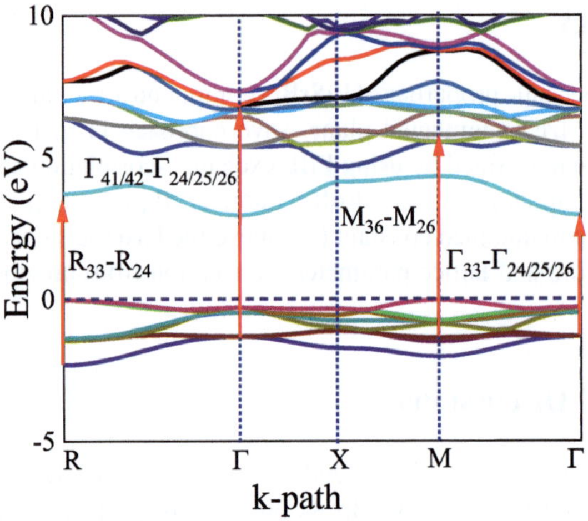

Fig. 2. Band structure of $LiSrBr_3$ deduced using PBE scheme.

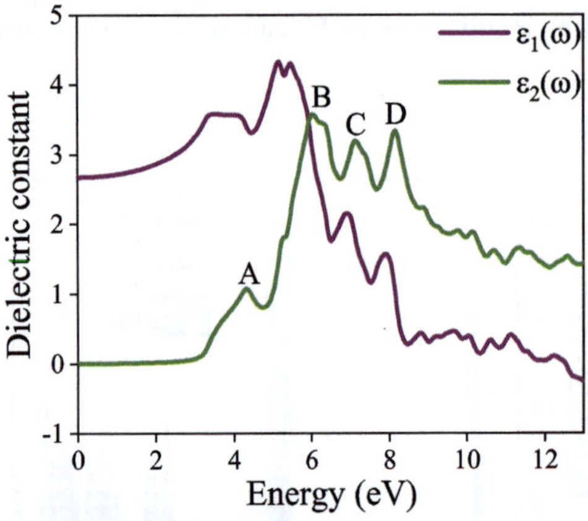

Fig. 3. Dielectric constant plots against the energy for $LiSrBr_3$ using PBE scheme.

Four major peaks at A (4.25 eV), B (6.04 eV), C (7.12 eV) and D (8.13 eV) are observed in the spectra of ε_2 (ω). Band transitions corresponding to A, B, C and D peaks are $\Gamma_{24/25/26}$ to Γ_{33}, R_{24} to R_{33}, M_{26} to M_{36} and $\Gamma_{24/25/26}$ to $\Gamma_{41/42}$. These band transitions occur due to the movement of electrons from the 4p state of Br to the 2s and 5s states of Li and Sr. Initially, corresponding to 0–3.07 eV photon energy, no absorption peak is observed in the absorption spectra indicating a transparent behavior in this energy range. After this energy absorption peaks are observed in the UV region (Fig. 4). The motion of charge in materials explained by band theory and BG between 0 and 3 eV classified as semiconducting nature, which strongly correlated to the electrical conductivity, thermal conductivity and optical properties. As we shown in Fig. 4, the absorption edge starting to increase around 3 eV. The high BG suggests the high energy required to jump electron VB to CB in material. In our case, UV-region energy required for electron motion from VB to CB. These material are shows potential application in high-power operation devices.

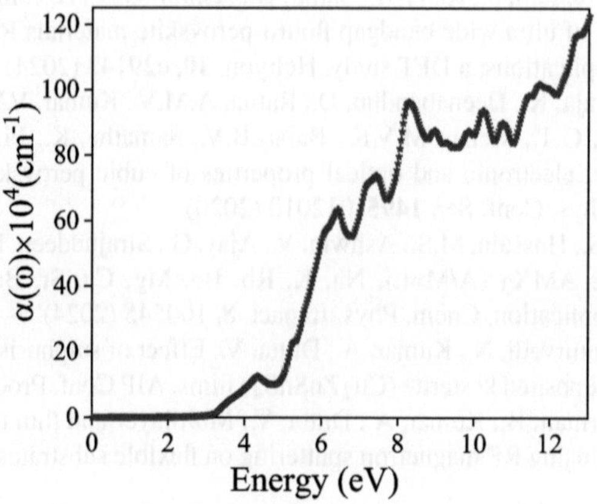

Fig. 4. Absorption coefficient with respect to energy for $LiSrBr_3$ calculated using PBE approach.

4 Conclusions

In the current study, PBE potential is used to investigate the optoelectronic properties of lead-free halide perovskite, $LiSrBr_3$. It has cubic crystal structure with Pm-3m (221) space group. Based on the band structure designs, $LiSrBr_3$ perovskite show p-type semiconducting nature with indirect BG of 2.96 eV. Suitable BG, reasonably excellent absorption spectra and environmental friendly nature of lead-free perovskite $LiSrBr_3$ is recommended for the application of optoelectronic devices.

Acknowledgments. We are grateful for the financial support from research project namely, RUSA 2.0, MHRD, New Delhi (No.F.30(16)SPD/RUSA/2016/178) sanctioned to GA. Anuradha would like to thank University Grants Commission (UGC), New Delhi for the award of junior research fellowship (No. 201610150226).

References

1. Gladkikh, V., Kim, D.Y., Hajibabaei, A., Jana, A., Myung, C.W., Kim, K.S.: Machine learning for predicting the band gaps of ABX_3 perovkites from elemental properties. J. Phys. Chem. C. **124**, 8905–8918 (2020)
2. Aleksandrov, K.S., Bartolom, J.: Structural distortions in families of perovskite-like crystals. Ph. Transit. **74**, 255–335 (2009)
3. Anuradha, Meena, D.K., Rani, M., Kumawat, S., Gurjar, V.K., Verma, A.S., Arora, G.: DFT study of structural, electronic, optical and elastic properties of lead-free Bi based perovskites $X_2NaBiCl_6$ (X= K and Rb) with pressure. Mater. Today Commun. **41**, 110433 (2024)

4. Kaleemullah, N.S., Hussain, M.S., Ashwin, V., Ajay, G., Sirajuddeen, M.M.S.: Novel ternary halide perovskite AMX_3 (A/M= Li, Na, K, Rb, Be, Mg, Ca, Sr, Ba, X= Br, Cl, F, I) for optoelectronic applications. Chem. Phys. Impact. **8**, 100545 (2024)
5. Shahzad, M.K., Hussain, S., Riaz, M., Sattar, H., Ashraf, G.A., Azeem, W., Ali, S.M., Alam, M.: Investigation of ultra wide bandgap flouro-perovskite materials $RBeF_3$ (RK and Li) for smart window applications: a DFT study. Heliyon. **10**, e29143 (2024)
6. Babu, K.E., Neeraja, K., Deenabandhu, D., Ratna, A.M.V., Kumar, V.V., Kumari, K.B., Taddesse, P., Aregai, G.T., Mehar, M.V.K., Babu, B.V., Samatha, K.: First-principles study of structural, elastic, electronic and optical properties of cubic perovskite $LiMgF_3$ for novel applications. J. Phys. Conf. Ser. **1495**, 012010 (2020)
7. Kaleemullah, N.S., Hussain, M.S., Ashwin, V., Ajay, G., Sirajuddeen, M.M.S.: Novel ternary halide perovskite AMX_3 (A/M=Li, Na, K, Rb, Be, Mg, Ca, Sr, Ba, X=Br, Cl, F, I) for optoelectronic application. Chem. Phys. Impact. **8**, 100545 (2024)
8. Swami, S.K., Chaturvedi, N., Kumar, A., Dutta, V.: Effect of sulphurisation on the activation energy of spray deposited kesterite (Cu_2ZnSnS_4) films. AIP Conf. Proc. **1731**, 120001 (2016)
9. Swami, S.K., Barman, B., Kumar, A., Dutta, V.: Multilayer and thin transparent conducting oxide fabrication using RF magnetron sputtering on flexible substrates. Org. Mater. **5**, 59–65 (2023)
10. Swami, S.K., Chaturvedi, N., Kumar, A., Dutta, V.: Effect of electric field on spray deposited zinc sulphide films. AIP Conf. Proc. **1665**, 080075 (2015)
11. Swami, S.K., Chaturvedi, N., Kumar, A., Dutta, V.: Effect of zinc precursor on Cu_2ZnSnS_4 nanoparticles synthesized by the solvothermal method and its application in dye-sensitized solar cells as the counter electrode. Mater. Today Energy. **9**, 377–382 (2018)
12. Kumar, A., Dalal, D., Singh, A., Kumar, D., Narwal, V., Swami, S.K., Chaturvedi, N., Kumar, M.: Study of deposition temperature effect on spray-deposited copper oxide thin films and its schottky diodes. Phys. Scr. **99**, 075927 (2024)
13. Petersen, M., Wagner, F., Hufnagel, L., Scheffler, M., Blaha, P., Schwarz, K.: Improving the efficiency of FP-LAPW calculations. Comput. Phys. Commun. **126**, 294–309 (2000)
14. Perdew, J.P., Burke, K., Ernzerhof, M.: Generalized gradient approximation made simple. Phys. Rev. Lett. **77**, 3865–3868 (1996)

Effect of Synthesis Strategy on the Thermoelectric Properties of $Bi_{0.5}Sb_{1.5}Te_3$

Ranu Bhatt[1,2](✉), Rishikesh Kumar[1,3], Shovit Bhattacharya[1,2], and Ajay Singh[1,2]

[1] Technical Physics Division, Bhabha Atomic Research Centre, Mumbai 400085, India
rbahtt@barc.gov.in
[2] Homi Bhabha National Institute, Mumbai 400094, India
[3] Department of Mechanical Engineering, Indian Institute of Technology, Patna 801103, India

Abstract. Effect of the synthesis methods and process parameters on the thermoelectrics performance of $Bi_{0.5}Sb_{1.5}Te_3$ material were investigated, essential for thermoelectric generators (TEGs). We systematically explore three synthesis techniques: (i) mechanical alloying at 500 RPM (BM500) and 300 RPM (BM300), (ii) melt growth (MG), and (iii) a hybrid approach combining mechanical alloying followed by melt growth (BM500-MG). Positive Seebeck coefficient (α) values confirm *p*-type conductivity across all methods. The BM500-MG sample achieves the highest power factor ($\alpha^2\sigma$) of ~6.5 W/m·K^2 and a reduced thermal conductivity (κ) of ~1.36 W/mK, resulting in a maximum dimensionless figure-of-merit (zT) of ~1.4 at 300 K. Theoretical, COMSOL simulations based on experimental parameters predict a maximum power output of 250 mW and a conversion efficiency (η) ~ 12% for MG-BM500 sample.

Keywords: Thermoelectricity · Energy materials · Microstructures · COMSOL

1 Introduction

Bismuth Antimony Telluride is a well-established thermoelectric material with highly anisotropic transport behaviour, thus requires a careful optimization for thermoelectric application. Over single crystal materials, polycrystalline materials show an advantage in terms of better $\alpha^2\sigma$, where α is Seebeck coefficient, σ is electrical conductivity. The texturing effect has been extensively investigated by various research group using various hot deformation techniques. Synthesis methods such as mechanical alloying, nanostructuring, and melt spinning are advantageous in tailoring the phonon scattering in material resulting in lower κ. In order to achieve higher zT ($=\alpha^2\sigma.T/\kappa$), all the three parameters (α, σ, and κ) must be synergistically tuned using combination of methodologies [1, 2].

2 Experimental Procedures

The stochiometric $Bi_{0.5}Sb_{1.5}Te_3$ composition was synthesized using three approaches; (i) mechanical alloying using ball-mill at 500 RPM (BM500) and 300 RPM (BM300), (ii) vacuum melting of elemental chunk in rocking furnace (MG), and (iii) hybrid approach

S. Kumari et al. (Eds.): ICMTP 2024, SPPHY 311, pp. 131–135, 2025.
https://doi.org/10.1007/978-981-96-6795-6_17

where BM500 powder was vacuum melt in quartz ampoule under rocking condition (MG-BM500). All samples were densified under vacuum at 450 °C with an applied load of 50 kg. Thermal transport properties, including the α, ρ, and κ, were measured using Linseis, LSR and LFA setups, while the theoretical power output was simulated Finite element COMSOL Multiphysics software [3].

3 Results and Discussion

The XRD patterns of the synthesized samples, analysed using Rietveld refinement, (data shown in Table 1), confirmed the formation of a single-phase rhombohedral structure (space group R-3m), with polycrystalline $Bi_{0.5}Sb_{1.5}Te_3$ as the dominant phase, Fig. 1. It was observed that Bi exhibited higher site occupancy at lower milling speeds, attributed to insufficient atomic diffusion. On increasing the milling speed, the stabilization of Bi occupancy was noted; however, this resulted in a reduced Te_2 site occupancy, indicating an increased likelihood of Bi/Sb antisite defect formation. The MG approach revealed significant variations in Bi, Sb, Te_1, and Te_2 occupancies, suggesting non-homogeneous mixing at the atomic level. This non-uniformity is likely due to the larger particle size of the elemental materials used during synthesis. Vacuum melting the BM-500 powder under similar conditions, restored stoichiometric elemental occupancy, though a persistent reduction in Te_2 site occupancy was observed, which can be attributed to Te high volatility. For comparison the XRD parameters of commercial $Bi_{0.5}Sb_{1.5}Te_3$ is also given in Table 1. The microstructure of the material analyzed using scanning electron microscopy (SEM), Fig. 2(a–c) reveals granular morphology in case of BM500 sample, however, the melt grown samples show micron-scale layered features. As compared to MG sample, finer and randomly oriented layer structures are observed in MG-BM500 sample. BM500 powder sample may contain internal defects due high milling speed which get annihilated during high temperature melting and subsequent controlled cooling. This facilitates atomic rearrangement into the lattice, reducing stresses, and improved densification.

Table 1. Rietveld refined XRD parameters of the samples.

Sample	Atom Occupancies				Lattice Parameters			χ2
	Bi	Sb	Te1	Te2	a=b	c	Space group	
BM300	0.065	0.129	0.096	0.194	4.29	30.466	R-3m	0.138
HEBM500	0.048	0.129	0.095	0.176	4.295	30.477	R-3m	0.224
MG	0.18	0.099	0.175	0.363	4.272	30.33	R-3m	4.21
MG-BM500	0.048	0.129	0.096	0.178	4.295	30.477	R-3m	0.257
Commercial	0.047	0.129	0.092	0.195	4.282	30.404	R-3m	0.145

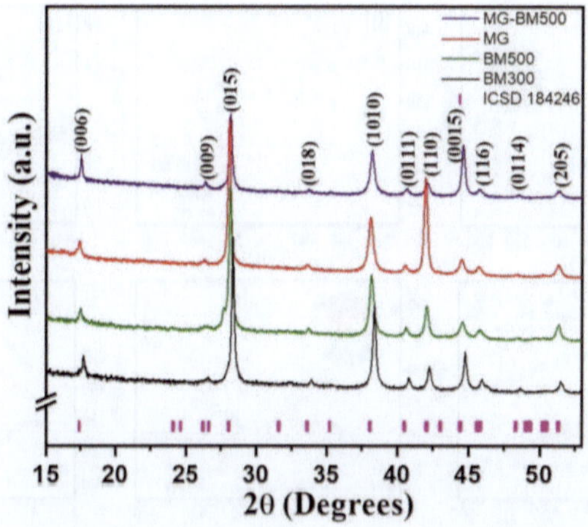

Fig. 1. Powder XRD pattern of powdered sample.

Fig. 2. SEM images captured on fractured surface of samples.

The effect of microstructure is evident in the temperature dependent thermal transport data, Fig. 3(a–d). Sample BM300 shows the minimum ρ ~ 4.5 μΩ-m, which increases significantly (~23 μΩ-m) at higher milling speed, due to finer grains and defects. Vacuum melting reduces ρ by ~68% (~6 μΩ-m) in MG and MG-BM500 samples, with different temperature response above 400 K. All the samples show positive α value varying between 220 and 80 μV/K (*p*-type) near room temperature. Hybrid method significantly enhances the $\alpha^2\sigma$ value from 2.5 mW/m-K^2 (BM500) to 6.5 mW/m-K^2 (MG-BM500). The minimum κ_t (~1.2 W/m-K) is observed in BM500 sample due higher defect density which increases to ~1.36 W/m-K in MG-BM500 samples, attributed to defect quenching. The temperature dependent zT (Fig. 3(d)) shows superior performance in MG-BM500 sample, achieving a ~75% enhanced zT ~ 1.4, at 300 K. The engineering zT value for this material is ~0.8 at temperature gradient (ΔT) of 300 K (T_H = 600 K), estimated using relation [4]; $\left(zT_{\text{eng}} = \frac{\left(\int_{T_c}^{T_H} \alpha(T)d(T)\right)^2}{\int_{T_c}^{T_H} \rho(T)d(T)\cdot\int_{T_c}^{T_H} \kappa(T)}\Delta T\right)$.

The open circuit voltage (V_{oc}), power output (P_{out}) and conversion efficiency (η) of synthesized material was simulated using COMSOL Multiphysics software for single leg thermoelectric generators (TEGs) with dimension 2 mm × 2 mm × 5 mm (inset

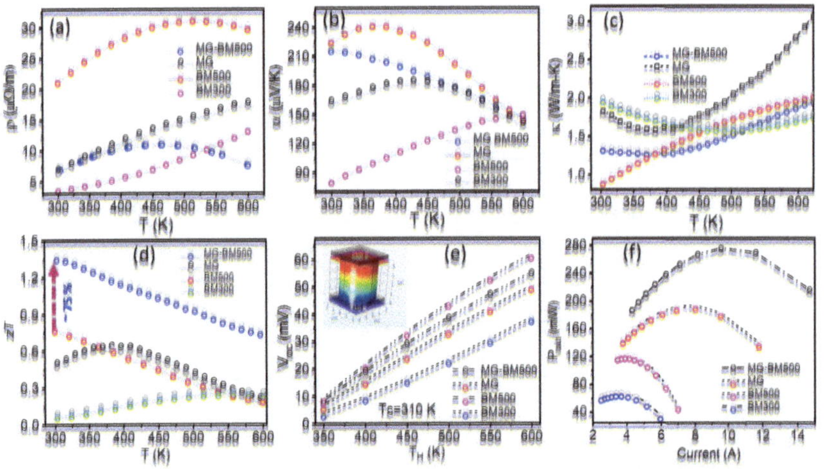

Fig. 3. Temperature dependence of (a) ρ, (b) α, (c) κ, (d) zT, and (e) COMSOL simulated open circuit voltage. (f) shows P_{out} as a function of current.

of Fig. 3(e)). For simulation, TEG was kept at T_H of 600 K, maintaining ΔT ~ 290 K. For power output calculation, the integrated value of the experimental thermoelectric parameters was considered [3]. Among the three-synthesis method, the MG-BM500 sample yielded the maximum P_{out} ~ 250 mW (Fig. 3(f)), and the η ~ 12% with a simulated heat input of 2.1 W/m-K. Achieving this η experimentally requires optimized interface diffusion. The study shows that combining vacuum melting with ball milling is a promising approach for $Bi_{0.5}Sb_{1.5}Te_3$ thermoelectric materials.

4 Conclusions

Microstructure tailoring effectively decouple the thermoelectric parameters, resulting in enhanced zT value. Grain size of starting material has significant effect on the morphology of the synthesized material. Hybrid synthesis approach resulted in high zT (~1.4) value, with effective zT_{eng} value ~0.8 at ΔT of 300 K in MG-BM500 sample. COMSOL simulation shows a maximum power generation capability of ~250 mW in single leg TEG with η ~ 12%, designed using this material.

References

1. Kim, T.-S., Chun, B.-S.: Microstructure and thermoelectric properties of n- and p-type Bi_2Te_3 alloys by rapid solidification processes. J. Alloys Comp. **437**, 225 (2007)
2. Tan, C., Tan, X., Yu, B., Liu, G.-Q., Wang, H., Luo, G., Xu, J., Wu, Q., Liang, B., Jiang, J.: Synergistically optimized thermoelectric performance in $Bi_{0.48}Sb_{1.52}Te_3$ by hot deformation and Cu doping. ACS Appl. Energy Mater. **2**, 6714 (2019)

3. Doraghi, Q., Jouhara, H.: Thermoelectric generator efficiency: an experimental and computational approach to analyzing thermoelectric generator performance. Thermal Sci. Eng. Prog. **55**, 102884 (2024)
4. Kim, H.S., Liu, W., Chen, G., Chu, C.W., Ren, Z.: Relationship between thermoelectric figure of merit and energy conversion efficiency. Proc Natl. Acad. Sci U S A. **112**(27), 8205–8210 (2015)

Electron Density Mapping and Optical Characteristics of $BaAl_2O_4$ Synthesized via Sol–Gel Method

Mamta[1], Neha Singh[1](✉), Seema Verma[1,2], Anita[1], Parvez Ahmed Alvi[1], and Aakansha[1](✉)

[1] Department of Physical Sciences, Banasthali Vidyapith, Jaipur 304022, Rajasthan, India
{nehasingh,aakansha}@banasthali.in

[2] Department of Computer Science Engineering, Vivekanand Global University, Jaipur 303012, Rajasthan, India

Abstract. Nowadays, $BaAl_2O_4$ grab a huge attention towards the technological applications such as optoelectronic devices, catalyst, sensors etc. Therefore, in this study, the $BaAl_2O_4$ nanomaterial was synthesized through a sol–gel chemical technique. The structural properties, functional group and optical characteristics of products were evaluated by X-ray diffraction (XRD), FTIR, Raman spectroscopy, UV–vis spectroscopy and photoluminescence (PL) analysis. The experimental result shows that $BaAl_2O_4$ nanomaterial exhibits hexagonal spinel structure without other impurity phases confirmed by Rietveld refinement using Fullprof Software. The extended functions of Fullprof Software were used to simulate the crystal structure and electron density of the material. The Williamson Hall method and the modified Debye–Scherrer relation were used to determine the crystallite size and microstrain. The error in these measurements has reduced with the use of the modified Debye–Scherrer relation. Further, the band gap, 3.4 eV was estimated through the *Tauc* analysis done by UV–vis spectroscopy. Vibrational modes F_{2g}, E_g and A_{1g} was observed in the Raman spectra which belongs to four discrete peaks observed at $242\ cm^{-1}$, $409\ cm^{-1}$, $574\ cm^{-1}$, $818\ cm^{-1}$. The acquired band gap, high PL intensity and decay time make this material suitable for the range of photonic applications.

Keywords: Barium Aluminate · Rietveld Refinement · Electron Density

1 Introduction

Numerous researchers have examined the physical characteristics of materials based on spinels due to their extensive useful applications, including those in photocatalysts, pigments, sensors, magnetic, optical and electrode materials [1–8]. Among spinel compounds, aluminate spinels with the chemical formula MAl_2O_4, where M represents a divalent metal cation (Mg, Co, Ni, Zn, Ba, etc.), have attracted great attention for various applications owing to their high chemical and thermal stabilities. It has long been

S. Kumari et al. (Eds.): ICMTP 2024, SPPHY 311, pp. 136–148, 2025.
https://doi.org/10.1007/978-981-96-6795-6_18

known choosing M atom affects the crystallinity and morphology, which exert a significant influence on their structural and optical properties [9–11]. Based on M, spinel aluminates majorly have cubic symmetry (M = Mg, Co, Ni, Zn, etc.) [12–15] while few of them exhibits hexagonal symmetry (M = Ca, Ba) as well [16, 17]. One of the many versatile materials is barium aluminate ($BaAl_2O_4$), which may be employed as sensors, afterglow phosphors, catalyst support, ultraviolet (UV) photoelectronic devices and light-cumulative fluorescent materials due to its wide band gap (~5.8 eV), high fluorescence efficiency, along with thermal and chemical stability [18–21]. $BaAl_2O_4$ exhibits hexagonal crystal unit cell at room temperature (RT), in $P6_3$ (space group no. 173) with lattice parameters $a = b = 10.4490$ Å and $c = 8.7930$ Å [10–12] whereas $P6_322$ (space group no. 182) with lattice constants $a = b = 5.2211$ Å and $c = 8.7898$ Å with $\alpha = \beta = 90^o$ and $\gamma = 120^o$ [22], depending on the synthesis method. Martina et al [23] have reported the $P6_3$ hexagonal symmetry where the $BaAl_2O_4$ was synthesized by hydrothermal method. However, Bele et al [24] have reported that, $BaAl_2O_4$ synthesized by Sol–gel route with $P6_322$ hexagonal symmetry. $BaAl_2O_4$ with $P6_3$ symmetry is composed of Ba^{2+} ions arranged into networks created by corner-sharing [AlO_4] tetrahedral. There are two inequivalent Ba^{2+} sites Ba1 and Ba2 where the first Ba1 site is bonded in octahedral symmetry coordinated with eight O^{2-} atoms whereas Ba2 site is bonded to nine O^{2-} atoms. The $P6_322$ space group, which contains 56 atoms in its unit cell and a structural topology akin to the $P6_3$ space group and the broken symmetry phase, characterize the hexagonal crystal structure of $BaAl_2O_4$. All Ba and Al atoms in $BaAl_2O_4$ with $P6_322$ hexagonal space group are symmetrically equivalent. Many authors have worked on the $BaAl_2O_4$ material and tune its structural and optical properties with substituting it with either at Ba site or Al site. With rare earth (Pr, Nd, Sm, Gd, Dy and Er) substitution at Ba site the material reported as a phosphor material and becomes suitable for advance light emitting device. V. M. Maphiri et al [25] have reported the Cu^{2+} substituted $BaAl_2O_4$, influence the bluish emission. A.H. Wako et al [26] have reported the rare earth element Eu^{2+} and Nd^{3+} doped $BaAl_2O_4$, prepared by a combustion process at 500 °C. The obtained band gaps (E_g) were 5.9 eV and 4.43 eV corresponding to Eu and Nd substitution respectively. Besides, the high decay time and luminescence intensity of $BaAl_2O_4$:Eu^{2+}, Nd^{3+} makes it suitable for plasma display panel applications. Additionally, according to Ganesh Kumar et al [27], Ce/Li based $BaAl_2O_4$ yellow phosphors were found to generate yellow light (589 nm) when stimulated by 320 nm radiation.

According to the reported literature, it has observed that the deep understanding on the structure and the optical properties of the polycrystalline $BaAl_2O_4$ material is required to make it more suitable in the pure form as well. Therefore, we believe that there is still a large room for essential progressions. Hence, in this novel work, a comprehensive experimental investigation of $BaAl_2O_4$ was carried out towards the structural properties which consists of the XRD analysis along with electron density mapping (confirms the number of electrons per unit volume in the material). Along with XRD and electron density, the Fourier transform infrared (FTIR) spectroscopy and Raman spectroscopy were also carried out including the optical properties of the synthesized material. These properties exploration will give a broad understanding towards the miniaturization of these properties for the required technological applications.

2 Experimental Techniques

2.1 Material Synthesis

This synthesis of nanoparticles of $BaAl_2O_4$ was carried out through a sol–gel chemical method. Where, barium nitrate pentahydrate ($Ba(NO_3)_2.5H_2O$) aluminium nitrate nano-hydrate ($Al(NO_3)_2.9H_2O$), citric acid monohydrate ($C_6H_8O_7.H_2O$), ethylene glycol ($C_2H_6O_2$) and deionized water were taken as precursors and all of them have high purity >99%. 1 mole of barium nitrate, 2 moles of aluminium nitrate and 2 moles of citric acid were dissolved in deionized water. Precursor mixture is stirred at 50 °C until a homogeneous solution is achieved, with the pH of the solution being 5. Ethylene glycol was added to the solution for the gel formation after a few minutes. Following gelation, the solution's temperature is raised to 160 °C and maintained there for about 4 h until it turns into a granular substance. The granule of dry material was crushed for 30 min to a fine powder using an agate mortar-pestle. Final sintering of the power was done at 850 °C for 4 h following intermediate grinding.

2.2 Characterization

The synthesized material underwent various characterizations such as XRD pattern was collected through X-Ray Diffractometer made of Bruker D8 Advance, made in Germany. Using a JASCO FTIR 4600 spectrophotometer, the Fourier transform infrared (FTIR) spectra was obtained at room temperature within the wavenumber range of 500 to 4000 cm^{-1}. A 532 nm laser-equipped Thermo Scientific DXRxi Raman imaging microscope was used to record the Raman spectra of $BaAl_2O_4$. An absorbance spectrum was collected through the UV–vis made of Perkin-Elmer, Singapore. The photoluminescence (PL) excitation and emission spectra along with decay curve of the prepared material were recorded using the fluoromax-plus spectrofluorometer, which was fitted with a 150 W xenon flash lamp.

3 Results and Discussions

3.1 Structural Analysis

The XRD pattern of $BaAl_2O_4$ identifies the compound's crystalline structure and phase purity. Material was subjected to in-depth investigation for these observations. Rietveld refinement was performed with space group $P6_322$ (No. 182) in the Fullprof program to verify the phase purity. Since the material was synthesized using sol–gel and was previously reported by Xie et al. [22], the structure belongs to space group P6322. Table 1 expresses the atomic position of $BaAl_2O_4$. The material exhibits hexagonal structure with lattice parameters $a = b \neq c$ and angles $\alpha = \beta = 90°$ and $\gamma = 120°$. Figure 1(a) displays the improved and refined XRD pattern, with *hkl* planes corresponding to XRD peaks. After performing a couple of iterations while varying several parameters, the obtained value of $\chi^2 = 2.7$, indicating an ideal match between the observed and experimental data. The blue solid line represents the difference between observed and experimental data, shown in Fig. 1(a). The parameters collected from the refinement are as follows:

(i) crystal lattice parameter, $a = b = 5.2254(5)$ Å, $c = 8.8030(3)$ Å (ii) unit cell volume = 208.163 Å^3 (iii) reliability factors, $R_p = 4.87$ $R_{wp} = 6.57$ and $R_{exp} = 3.92$, $R_f = 1.63$ and $R_{Bragg} = 1.74$.

Table 1. The atomic position of $BaAl_2O_4$ with space group $P6_322$ with hexagonal symmetry.

Atoms	Wyckoff position	Atomic position (x, y, z)
Ba	2b	(0, 0, 1/4)
Al	4f	(1/3, 2/3, z)
O1	2c	(1/3, 2/3, 1/4)
O2	6g	(x, 0, 0)

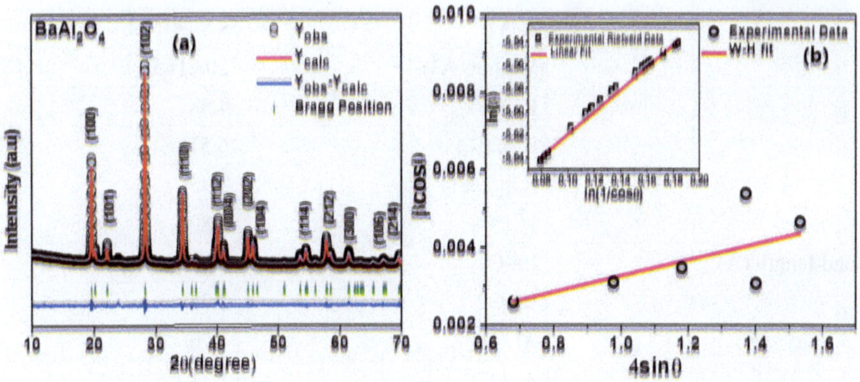

Fig. 1. (a) The Rietveld refined XRD pattern displays the observed and experimental data along with matching between them and Bragg's positions along with (b) W-H plot of $BaAl_2O_4$.

The crystallite size and microstrain were calculated using the linear fitting of the Williamson Hall (WH) plot [28] as represented in Fig. 1(b) by using the following Eq. 1:

$$\beta cos\theta = \frac{0.9\,\lambda}{S} + \epsilon.sin\theta \tag{1}$$

where λ is the wavelength of CuKα = 1.5406 Å, β is full width at half maximum of the peaks for different hkl planes, S is the crystallite size, 0.9, is Scherrer's constant for circular shape; θ is the Bragg's angle and ϵ is the lattice strain. The intercept and slope of this linearly fitted curve yielded crystallite size and strain, respectively. The obtained crystallite size and microstrain are shown in Table 2. Moreover, the average crystallite size was also estimated through the modified Debye–Scherrer equation to reduce the source of errors in calculation of crystallite size. The least squares method, which takes into consideration all peaks to statistically reduce mistakes, is used to compute the average crystallite size. The modified Scherrer plot of ln(β) versus ln(1/Cosθ) are shown in inset of Fig. 1(b). The crystallite size is determined by taking the intercept of a least squares fitted line that passes through each potential peak. The Rietveld methodology, modified Scherrer plot and Williamson–Hall plot provide different average crystallite values. The reason for this is that every method approaches peak profile analysis in a different way [30]. The Scherrer relation $\epsilon = \beta\cos\theta/4$ is used to get the micro-strain (ε) [29]. Further, the dislocation density, σ was determined using the relation $\sigma = \frac{1}{S^2}$, where the S was the crystallite size obtained by the Modified Scherrer equation and the evaluated value is $3.14 \times 10^{15}\ m^{-2}$. The calculated tolerance factor (T) for the $BaAl_2O_4$ hexagonal spinel aluminate was found to be 0.7416 and noted in Table 2.

Table 2. Summary of obtained parameter of $BaAl_2O_4$

Parameters	$BaAl_2O_4$	
Rietveld parameters	Space group	$P6_322$
	$a = b$(Å)	5.2254
	c (Å)	8.8030
	Volume(Å^3)	208.165
	R_p (%)	4.88
	R_{wp} (%)	6.57
	R_{exp} (%)	3.93
	χ^2 (%)	2.80
Bond-length (Å)	Ba-O1	3.01
	Ba-O2	2.95
	Al-O1	1.70
	Al-O2	1.71
	Al-Ba	4.04
Bond-angle (degree)	O1-Ba-O2	61.37
	O2-Al-O2	61.20
	Ba-Al-Ba	80.50
	Ba-O1-Ba	121.30
	Al-O2-Al	86.44
Crystalline size (nm)	W-H plot	49.85
	Modified Scherrer	17.83
Strain	W-H plot	3.44×10^{-4}
	Scherrer	7.45×10^{-3}
Tolerance factor	T	0.74

After refinement, an additional feature of Fullprof software like VESTA was used to mimic the 3-D crystal structure of $BaAl_2O_4$ (shown in Fig. 2). Left side image represents the ball and stick frame of Ba (green sphere), Al (Blue sphere) and O1/O2 (red/yellow sphere) in the mentioned space. However, the right-side image shows the polyhedron 3D structure of the similar ball and stick frame. These images illustrate the bonds, bond angle and coordination among all the cations (Ba^{2+}, Al^{3+}) and anions (O^{2-}). The extracted bond length and bond angles are mentioned in Table 2.

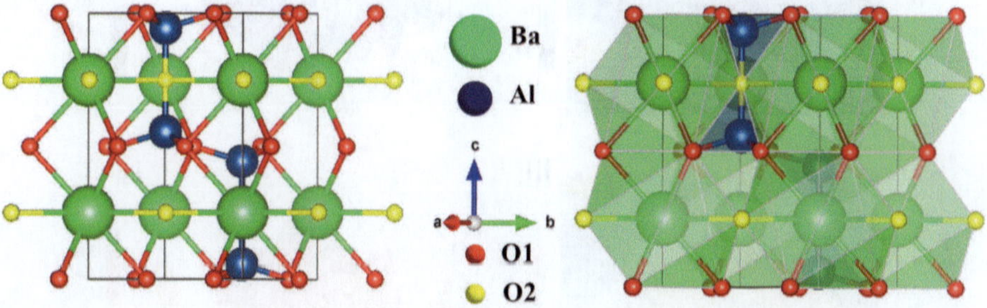

Fig. 2. Simulated 3-D crystal structure of $BaAl_2O_4$

3.2 Electron Density

The extended program of Fullprof, i.e. GFourier program was used to map the electron density distribution. The complete unit cell's electron density is obtained by using the geometrical structural factor's Fourier transform [6]. This electron density contour represents electron density around the respective atom in the material. Figure 3(i–iv) demonstrates the 2D contour in the x-y and y-z plane along with the corresponding 3D contour. The plot in the y-z plane represents the Ba atoms surrounded with oxygen atoms. The mapping shows the electron density distribution in the colour fill manner where red colour represents the Ba atom itself and the variation in colour from red to blue indicating the decrease in electron density around the Ba atom. The corresponding 3D contour shows the peaks representing Ba atoms. The highest value of electron density for Ba^{2+} atoms is 27.0.

Similarly, the 2D contour in the x-y plane displays the Al atoms in the centre of the plot while at the corners the Ba atom is displayed. The electron density surrounding the Al atoms is also represented with colour variation. The 3D contour of Al atoms confirms the peaks associated with the Al atoms. Whereas the peaks in the corner indicate the presence of Ba atoms. The obtained value of electron density for Al atom is 2.4. The high value of electron density of Ba as compared to the Al is attributed to the larger ionic radii of Ba^{2+} (1.47 Å) as well as the coordination of Ba^{2+} and Al^{3+} with the O^{2-} in $BaAl_2O_4$. Moreover, the electronic configuration also explains the electron density around Ba and Al atom. The electronic configuration of Ba^{2+} atom is [Xe] $6s^2$ and Al^{3+} has [He] $2s^2$ $2p^4$. The bonds among Ba, Al and Oxygen ions are expressed in Fig. 3(v, vi) in the y-z and x-y planes respectively. These simulated crystal structures are displaying the presence of Ba and Al atoms in the mentioned plane against the plotted electron density distribution.

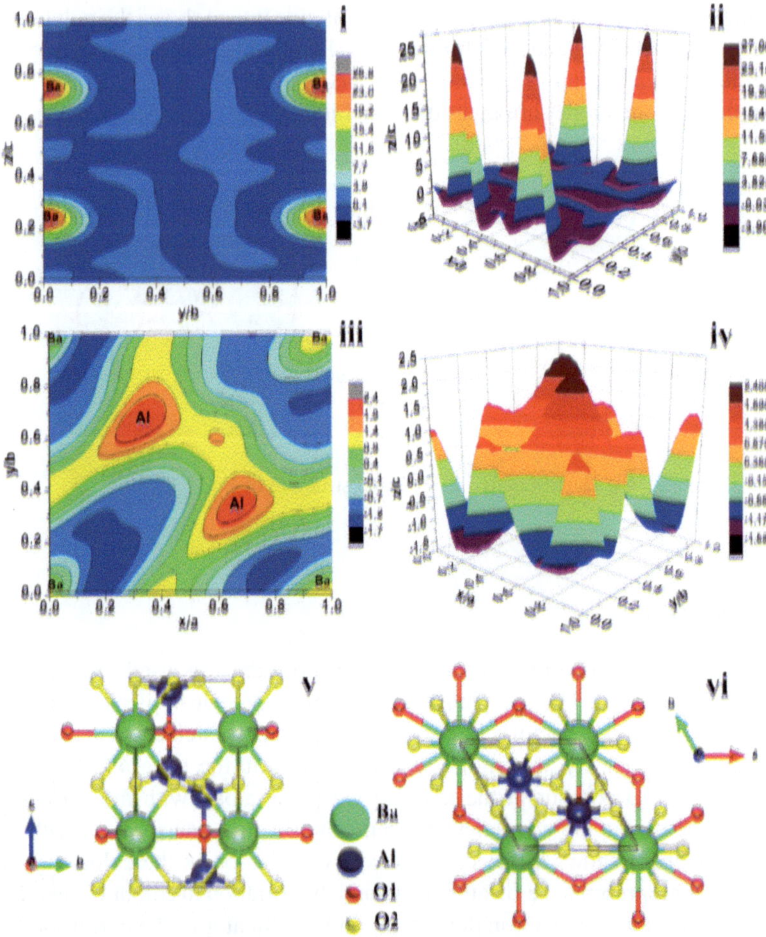

Fig. 3. (i–iv) Electron density 2D and 3D contour, (v) and (vi) simulated crystal structure of $BaAl_2O_4$ in y-z and x-y plane.

3.3 Fourier-Transform Infrared Spectroscopy

Figure 4 displays the $BaAl_2O_4$ FTIR spectrum. The allocated peaks in the spectrum at 627, 802, 837 cm^{-1}, are designated to the stretching vibrations of Al^{3+}-O^{2-}, Ba^{2+}-O^{2-} and Ba^{2+}-O^{2-}-Al^{3+}, respectively [18]. The enlarged view of peak ~802 cm^{-1} and 837 cm^{-1} is displayed in the inset of Fig. 4. However, the peak detected at wavenumber 1454 cm^{-1} is associated to the stretching vibration of tetrahedral unit AlO_4 of $BaAl_2O_4$, which confirms the phase formation of the $BaAl_2O_4$.

3.4 Raman Spectra Analysis

Figure 5 shows the Raman spectra of the synthesized $BaAl_2O_4$ which is recorded in the range of 150 cm^{-1} to 1000 cm^{-1}. The Raman spectra exhibit 4 distinct peaks having

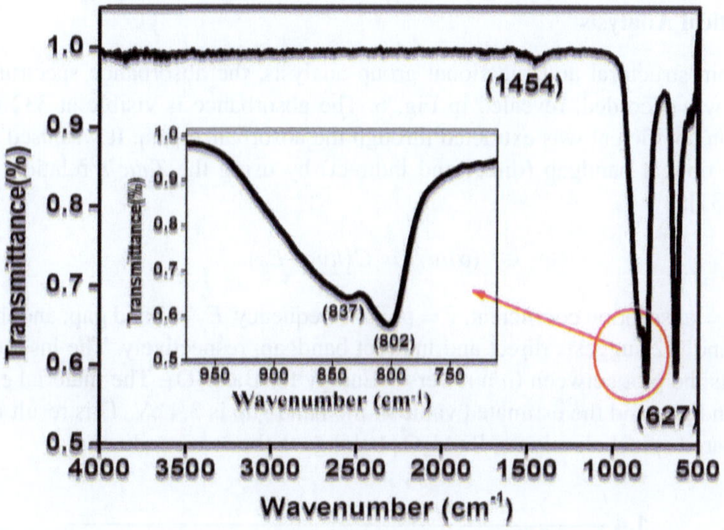

Fig. 4. FTIR spectrum of $BaAl_2O_4$, where inset represents the enlarged view of transmittance ~800 cm^{-1}.

vibrational modes at 242 cm^{-1} (F_{2g}), 409 cm^{-1} (E_g), 574 cm^{-1} (F_{2g}) and 818 cm^{-1} (A_{1g}). In the crystal structure, the translation of the tetrahedron, which is usually filled by Ba^{2+} ions, is attributed to the Raman mode at around 242 cm^{-1}. The mode observed at ~409 cm^{-1}, ascribed to the asymmetric bending motion between Ba^{2+} and O^{2-} ion in the tetrahedral sites. The symmetric bending of O^{2-} with respect to the Al^{3+} ions is responsible for the peak visible at ~574 cm^{-1}. An A_{1g} mode peak at around 818 cm^{-1} was identified, and it was linked to a stretching vibration of Ba^{2+}-O^{2-} [27].

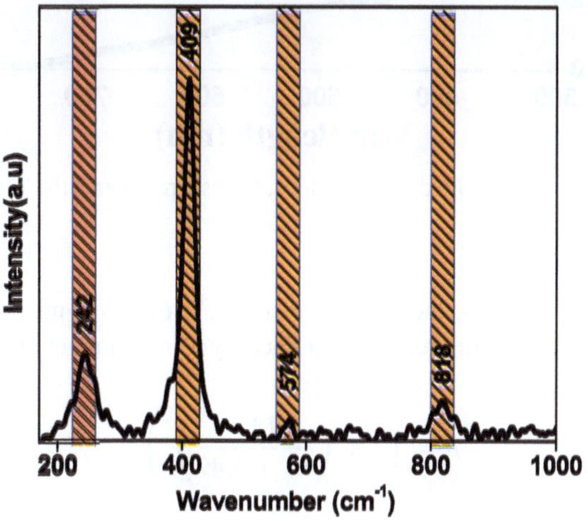

Fig. 5. Raman Spectra of $BaAl_2O_4$.

3.5 Optical Analysis

Apart from structural and functional group analysis, the absorbance spectrum of the material was recorded, revealed in Fig. 6. The absorbance is visible at 332 nm. The absorption coefficient was extracted through the absorbance data. It was used to determine the optical bandgap (direct and indirect) by using the *Tauc's* relation given as follows [31]:

$$(\alpha h\nu)^n = C(h\nu - E_g) \tag{2}$$

where, α = absorption coefficient, ν = photon frequency, E_g = band gap; and the n with value 2 and 1/2 suggests direct and indirect bandgap, respectively. The inset of Fig. 6 represents the plot between $(ah\nu)^2$ verses Energy for $BaAl_2O_4$. The material exhibits a direct band gap and the estimated value of the band gap is 3.4 eV. This result indicates that prepared spinel aluminate, $BaAl_2O_4$ belongs to the semiconductor.

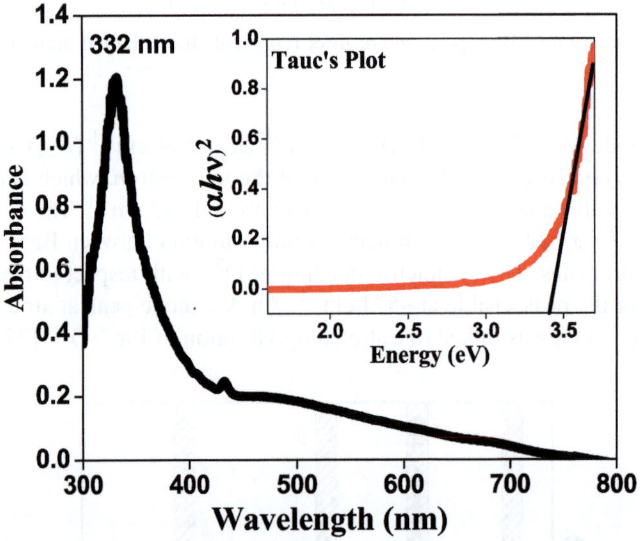

Fig. 6. Absorbance variation with the wavelength and inset represents the *Tauc* plot for $BaAl_2O_4$ spinel aluminate.

In addition to band gap, the refractive index, μ of the synthesised material was calculated using the following relation obtained by Vandamme et al. [32]:

$$\mu = \left(1 + \left(\frac{13.6}{E_g + 3.47}\right)^2\right)^{\frac{1}{2}} \tag{3}$$

where, E_g is the band gap. The calculated value of refractive index is 2.25.

Photoluminescence (PL) emission spectra of $BaAl_2O_4$ is displayed in Fig. 7, where the excitation wavelength is 367 nm which is shown in inset of Fig. 7. A Gaussian fit was used to deconvolute the PL spectra between 450 and 750 nm, resulting in several emission peaks at 470 nm (2.64 eV), 481 nm (2.58 eV), 500 nm (2.48 eV) and 542 nm (2.92 eV). This observed high luminescence intensity indicates less impurities and defects present on the surface of the grown material. This is an important mechanism for nanomaterials' nonradiative relaxation. Figure 7 shows strong blue emission peaks at 470 nm and 481 nm, green emission at 500 nm and 542 nm. Blue emission observed due to surface defects and confined particle size in prepared material which is in good agreement with XRD data. Moreover, the energy levels involved in electron transitions match the energy of the excitation light, thus resulting in efficient photoluminescence spectra and the band gap (3.4 eV) in $BaAl_2O_4$ align well with excitation energy. The rise of singly ionized oxygen vacancies causes the green emission to become more noticeable.

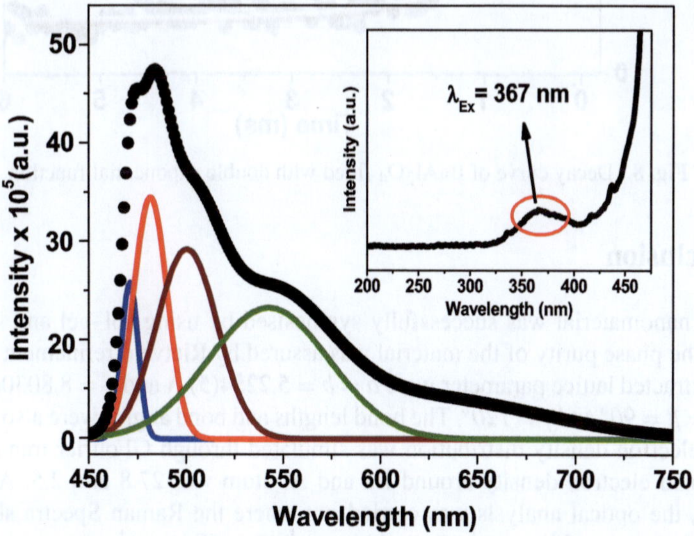

Fig. 7. PL spectra of $BaAl_2O_4$ showing emission in the wavelength range of 450–750 nm. Inset shows the excitation wavelength plot.

In addition to emission spectra, Fig. 8 illustrates the decay curves for $BaAl_2O_4$. The decay curve was fitted by double-exponential Eq. 4, given as follows [26]:

$$I = A_1 \exp\left(-\frac{t}{\tau_1}\right) + A_2 \exp\left(-\frac{t}{\tau_2}\right) \tag{4}$$

$$\tau = \frac{(A_1\tau_1^2 + A_2\tau_2^2)}{(A_1\tau_1 + A_2\tau_2)} \tag{5}$$

The average decay times (τ) of $BaAl_2O_4$ can be calculated by the following Eq. 5. Where I is the luminescence intensity; A_1 and A_2 are the constants; t is the time, and τ_1 and τ_2 are the decay times for exponential components. According to these parameters, the calculated decay time is 0.702 ms. This high value of decay time is in accordance with the high photoluminescence intensity.

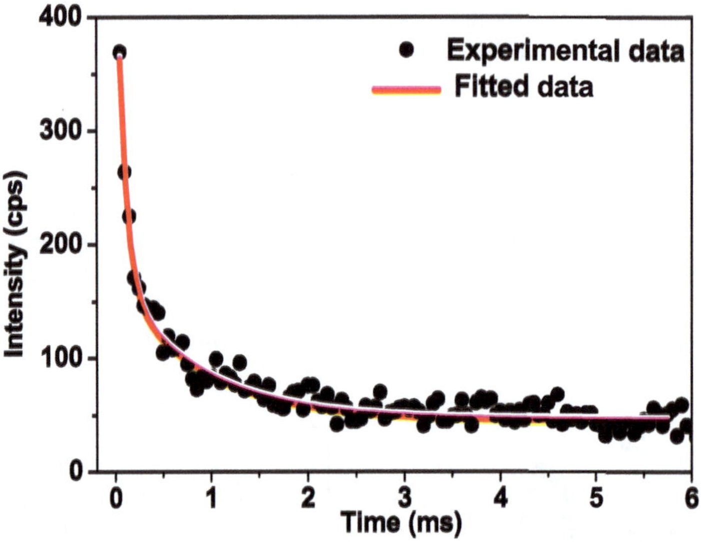

Fig. 8. Decay curve of $BaAl_2O_4$ fitted with double exponential function.

4 Conclusion

$BaAl_2O_4$ nanomaterial was successfully synthesised by using sol–gel and sintered at 850 °C. The phase purity of the material was assured by Rietveld refinement technique and the extracted lattice parameter were $a = b = 5.2254(5)$ Å and $c = 8.8030(3)$ Å with angles $\alpha = \beta = 90°$ and $\gamma = 120°$. The bond lengths and bond angles were also extracted. Besides, electron density distribution was simulated through GFourier transformation and obtained electron density around Ba and Al atom was 27.8 and 2.5. Along with structural, the optical analysis was carried out where the Raman Spectra showed the vibrational modes at 242 cm^{-1} (F_{2g}), 409 cm^{-1} (E_g), 574 cm^{-1} (F_{2g}) and 818 cm^{-1} (A_{1g}). The measured decay time of 0.702 ms confirms enhanced luminescence efficiency, making it suitable for photonic applications.

Acknowledgement. The authors acknowledge the Department of Physical Science, Banasthali Vidyapith for providing instrumental facility such as XRD, Raman, UV–vis and PL spectroscopy. Authors would also like to acknowledge the Department of Biotechnology, Banasthali Vidyapith for providing FTIR facility.

Contributor Role Taxonomy (CRediT) **Mamta:** Sample preparation, Characterizations, Conceptualization, Software, Writing - Original draft preparation.

Neha Singh: Data curation, Visualization, Reviewing and Editing, Supervision, Validation.

Seema Verma: Data curation.

Anita: Characterizations, Data curation.

P. A. Alvi: Software, Visualization.

Aakansha: Conceptualization, Data curation, Software, Visualization, Validation, Investigation, Reviewing and Editing, Supervision.

Declaration of Competing Interest. The authors state that none of the work presented in this study may have been influenced by any known conflicting financial interests or personal ties.

References

1. Tatarchuk, T., Al-Najar, B., Bououdina, M., Ahmed, M.A.: Catalytic and photocatalytic properties of oxide spinels. Handbook Ecomater. **3**, 1701–1750 (2019)
2. Dippong, T., Mereu, R.A.: Effect of La^{3+} on thermal, structural and morphological properties of Zn-Co ferrite spinel-based pigments. Ceram. Int. **50**(7), 10314–10324 (2024)
3. Jarusheh, H.S., Yusuf, A., Banat, F., Haija, M.A., Palmisano, G.: Integrated photocatalytic technologies in water treatment using ferrites nanoparticles. J. Environ. Chem. Eng. **10**(5), 108204 (2022)
4. Zhang, Y., Liu, Y., Wang, L., Zhou, H., Meng, W., Li, Y., He, Z., Dai, L.: A mixed-potential type NH_3 sensors based on spinel Zn_2SnO_4 sensing electrode. Sens. Actuators B: Chem. **367**, 132154 (2022)
5. Lafont, U., Locati, C., Kelderv, E.M.: Nanopowders of spinel-type electrode materials for Li-ion batteries. Solid State Ionics. **177**(35–36), 3023–3029 (2006)
6. Dubey, K., Dubey, S., Sahu, V., Parry, R.A., Modi, A., Gaur, N.K.: Structural, optical and magnetic properties of $CoFe_2O_4$ nanoparticle synthesized by ultrasonication-assisted sol–gel technique. Appl. Phy. A. **128**(7), 560 (2022)
7. Lu, Q., Wei, Z., Wu, X., Huang, S., Ding, M., Ma, J.: Electronic structure and optical properties of spinel structure $Zn_{1-x}Ni_xAl_2O_4$ nanopowders synthesized by sol–gel method. Chem. Phys. Lett. **772**, 138582 (2021)
8. Choudhary, B.L., Kumari, N., Kumari, J., Kumar, A., Dolia, S.N.: Relaxation mechanism in $Ni_{0.5}Zn_{0.5}Fe_2O_4$ nanocrystalline ferrite at a lower temperature. Mater. Lett. **304**, 130731 (2021)
9. Halder, R., Bandyopadhyay, S.: Effect of cation ratio on microstructure and optical absorbance of magnesium aluminate spinel. Mater. Res. Express. **7**(1), 015065 (2020)
10. Mahi, K., Ouari, K.E., Mostefa, R.: Synthesis of nanoscale spinel aluminates powder and studies of their optical and structural properties. Phys. Scr. **99**(9), 095982 (2024)
11. Kaur, P., Rani, S.: Effect of sintering temperature on structural and optical properties of magnesium aluminate spinel. J. Opt. **202352**(4), 2366–2374 (2023)
12. Mathur, S., Veith, M., Haas, M., Shen, H., Lecerf, N., Huch, V., Hüfner, S., Haberkorn, R., Beck, H.P., Jilavi, M.: Single-source sol-gel synthesis of nanocrystalline $ZnAl_2O_4$: structural and optical properties. J. Am. Ceram. Soc. **84**(9), 1921–1928 (2001)
13. Duan, X., Pan, M., Yu, F., Yuan, D.: Synthesis, structure and optical properties of $CoAl_2O_4$ spinel nanocrystals. J. Alloys Compd. **509**(3), 1079–1083 (2011)
14. Zurita-Méndez, N.N., Espinosa-Medina, M.A., Carbajal-De la Torre, G.: Vacuum sol-gel synthesis and characterization of $NiAl_2O_4$ nanoparticles. Microsc. Acta. **30**(2), 71–79 (2021)
15. Hosseini, S.M.: Structural, electronic and optical properties of spinel $MgAl_2O_4$ oxide. Phys. Status Solidi. B Basic Res. **245**(12), 2800–2807 (2008)
16. Janáková, S., Salavcová, L., Renaudin, G., Filinchuk, Y., Boyer, D., Boutinaud, P.: Preparation and structural investigations of sol–gel derived Eu^{3+}-doped $CaAl_2O_4$. J. J. Phys. Chem. Solids. **68**(5–6), 1147–1151 (2007)
17. Larsson, A.K., Withers, R.L., Perez-Mato, J.M., Gerald, J.F., Saines, P.J., Kennedy, B.J., Liu, Y.: On the microstructure and symmetry of apparently hexagonal $BaAl_2O_4$. Solid State Chem. **181**(8), 1816–1823 (2008)
18. Wang, S., Gao, H., Fang, L., Wei, Y., Li, Y., Lei, L.Z.: Synthesis and characterization of $BaAl_2O_4$ catalyst and its photocatalytic activity towards degradation of methylene blue dye. Phys. Chem. **233**(8), 1161–1181 (2019)

19. Mumanga, T.J., Díaz-Torres, L.A., Gómez-Solís, C.: Nd^{3+} doped $BaAl_2O_4$ for enhanced photocatalytic degradation of methylene blue. Mater. Lett. **292**, 129664 (2021)
20. Nagarasanakote, J.S., Janardhana, D., Erasmus, L.J.B., Coetsee, E., Motaung, D.E., Swart, H.C.: Influence of annealing temperature on persistent luminescence in $BaAl_2O_4$:Eu^{2+}/Eu^{3+} nanocrystals and its application for latent fingerprint detection. Dalton Trans. **53**(40), 16557–16576 (2024)
21. Cai, Y., Wang, X., Huang, Y.M.: Photoluminescence and afterglow of Tb^{3+} doped $BaAl_2O_4$. Results Phys. **60**, 107656 (2024)
22. Xie, C., Zeng, Q., Dong, D., Gao, S., Cai, Y., Oganov, A.R.: First-principles calculations of the dielectric and vibrational properties of ferroelectric and paraelectric $BaAl_2O_4$. Phy. Lett. A. **378**(26–27), 1867–1870 (2014)
23. Vrankic, M., Saric, A., Bosnar, S., Barisic, D., Pajic, D., Lützenkirchen-Hecht, D., Badovinacet, I.J.: Structural behavior and spin-state features of $BaAl_2O_4$ scaled through tuned Co^{3+} doping. Inorg. Chem. **60**(12), 8475–8488 (2021)
24. Bele, A., Mhlongo, M.R., Koao, L.F., Motaung, T.E., Malevu, T.D., Hlatshwayo, T.T., Mpelane, S., Mlambo, M., Motloung, S.V.: Effects of varying Sm^{3+} concentration on the structure, morphology and photoluminescence properties of the $BaAl_2O_4$/$CaAl_2O_4$/$Ca_4Al_6O_{13}$/$Ca_3Al_2O_6$: x% Sm^{3+}(x= 0-1.9) mixed phases using citrate sol-gel method. Heliyon. **8**(12), e12573 (2022)
25. Maphiri, V.M., Mhlongo, M.R., Hlatshwayo, T.T., Motaung, T.E., Koao, L.F., Motloung, S.V.: Citrate sol-gel synthesis of $BaAl_2O_4$: x% Cu^{2+} ($0 \leq x \leq 1$) nano-phosphors: structural, morphological and photoluminescence properties. Opt. Mater. **109**, 110244 (2020)
26. Wako, A.H., Dejene, F.B., Swart, H.C.: Combustion synthesis, characterization and luminescence properties of barium aluminate phosphor. J. Rare Earth. **32**(9), 806–811 (2014)
27. Ganesh Kumar, K., Balaji Bhargav, P., Aravinth, K., Balaji, C.: Enhanced photoluminescence properties of $BaAl_2O_4$: Ce^{3+}/Li^+ yellow phosphors. J. Mater. Sci. Mater. Electron. **33**(19), 15323–15332 (2022)
28. Arya, J., Leel, N.S., Aakansha, Quraishi, A.M., Hashmi, S.Z., Kumar, S., Choudhary, B.L., Dalela, S., Hasan, Z., Singh, J., Alvi, P.A.: Enhancement of dielectric response and optical bandgap tuning with improving surface morphology of recyclable polyvinyl alcohol-CeO_2 polymer nanocomposites. Phys. Status Solidi. A. **221**(4), 2300742 (2024)
29. Pandey, D.K., Modi, A., Pandey, P., Gaur, N.K.: Variable excitation wavelength photoluminescence response and optical absorption in $BiFeO_3$ nanostructures. J. Mater. Sci. Mater. Electron. **28**, 17245–17253 (2017)
30. Paswan, S.K., Kumari, S., Kar, M., Singh, A., Pathak, H., Borah, J.P., Kumar, L.: Optimization of structure-property relationships in nickel ferrite nanoparticles annealed at different temperature. J. Phys. Chem. Solids. **151**, 109928 (2021)
31. Viezbicke, B.D., Patel, S., Davis, B.E., Birnie III, P.D.: Evaluation of the Tauc method for optical absorption edge determination: ZnO thin films as a model system. Phys. Status Solidi. B. **252**(8), 1700–1710 (2015)
32. Yuliantika, D., Taufiq, A., Hidayat, A., Hidayat, N.S., Soontaranon, S.: Exploring structural properties of cobalt ferrite nanoparticles from natural sand. IOP Conf. Ser. Mater. Sci. Eng. **515**, 012047 (2019)

Investigation of Structural, Electronic and Optical Properties of CoX_3 (X = As, P, Sb) Using FP-LAPW Method

Bhanu Pratap Singh[1], Shradha Kalkhundiya[1], Yashaswi Soni[2], Naincy Pandit[3], Vibhav Kumar Saraswat[4], and Ajay Singh Verma[1,5](✉)

[1] Department of Physics, School of Applied and Life Science, Uttaranchal University, Dehradun 248007, India
ajay_phy@rediffmail.com

[2] Department of Physics, Rajeev Gandhi Govt. P.G. College, Mandsaur 458001, Madhya Pradesh, India

[3] Department of Physics, School of Allied Science, Dev Bhoomi Uttarakhand University, Dehradun 247232, India

[4] Department of Physics, Army Cadet College, Indian Military Academy, Dehradun 248007, India

[5] University Centre for Research and Development, Department of Physics, Chandigarh University, Mohali 140143, Punjab, India

Abstract. Skutterudite is a material studied widely for its conversion of thermal energy into electrical energy, and here, we have investigated structural, electronic and optical properties of CoX_3 (X = As, P, Sb) using density functional theory (DFT) and a full potential linearised augmented plane wave (FP-LAPW) technique. CoP_3 presents a narrow bandgap of 0.524 eV when Modified Becke-Johnson potential (TB-mBJ) is used for electronic band structure calculations. The transport properties calculations give a possibility to construct thermoelectric material with optimised performance and present high values of the figure of merit ZT.

Keywords: Density function theory · Skutterudite · FP-LAPW · TB-mBj · Thermoelectric

1 Introduction

The pace with which the world population is increasing is getting harder day by day to meet the needs of everyone from our existing resource pools. The resource pools are broadly classified into two categories, mainly renewable and non-renewable resources [1, 2]. Scientists have started considering renewable resources as an alternative to non-renewable resource problems [3]. Forty developed, and seventy-three developing nations consider Renewable energy economically beneficial and statistically beneficial [4]. To solve this problem [5], scientists and researchers all over the world have started deducing different methods and ways to use the waste heat energy and convert it into other various forms so that it can be used in different forms to meet consumer demand.

S. Kumari et al. (Eds.): ICMTP 2024, SPPHY 311, pp. 149–154, 2025.
https://doi.org/10.1007/978-981-96-6795-6_19

About 30 years ago, thermo-electrics were at the peak of their potential, stratifying the need for solid-state refrigeration needs. Around the 1950s–60s, materials were broadly optimized for their thermoelectric application. One of the methods implemented is to use the thermoelectric material to construct a thermoelectric generator. Thermoelectric materials can transform thermal energy into electrical energy. This is done by using the temperature gradient between two ends of the device, which results in the diffusion of the charge carrier. The flow of charges between the hot and the cold end finally creates a potential difference. Thermoelectric generators do not use any kind of moving part, which rules out the possibility of it being a bulky device. It does not require any kind of cooling or fuel to operate, which makes it even better than the conventional generators used for energy production. High-efficiency thermoelectric generators are produced every day, with an efficiency rate of 33–37% in a higher temperature range [6]. The dimensionless figure of merit ZT is used to determine if a given thermoelectric material is an effective heat-to-electricity converter. There is a significant potential to turn the waste heat energy produced by heat engines into electrical energy using a mineral called skutterudite [7–9]. Skutterudites with high-value figures of merit are among the finest thermoelectric materials in the 600–700 °C temperature range. The enhancement of ZT can be done by improving the see beck coefficient and electrical conductivity and by reducing the values of the heat capacity [10]. Different methods have been implemented to improve the performance, and new fabrication methods have been introduced to produce high-performing skutterudites with ease [11–13]. There are still a lot of future possibilities regarding the betterment of the material not only in terms of performance but also in terms of its large-scale production technique. Serrano et al. [7] have fabricated nanostructured thermoelectric skutterudite-based materials. This cost-effective method operates at a reduced processing temperature. Here, they used a cold sintering process (CSP) to form a large group of $CoSb_3$-based thermoelectric materials. Li et al. [11] raised skutterudites' filling fraction limitation (FFL), allowing for a high ZT value. Rodrigues et al. [14] have evaluated the electrical and structural characteristics of skutterudite-based thermoelectric using the X-ray absorption technique. The conclusion that was drawn from the evaluation was that $CoSb_3$ is a body-centred cubic crystal. Lin et al. [15] added a minor quantity of Fe to the pure CoP_3 skutterudite nanoneedles; they increased water splitting in an alkaline environment, hydrogen evolution reaction (HER), and oxygen evolution reaction (OER). Mahi et al. [19] obtained under different pressures and temperatures $CoAs_3$ is capable of applications in energy harvesting and optoelectronic systems.

2 Methodology

To proceed in our desired direction of work, we will be using wien2k computational code for the computation calculation and to predict and explain the properties of real materials within the most popular approach, the first principles method [16]. The first principal calculation methods refer to the calculations that are being done using the basic laws of physics, which deal with the fundamental physical quantities which can successfully describe the behaviour of material or crystal structure as a reduced dimensional system of electron correlation. Using the Wien2k code, the density functional theory is implemented for solving the many-electron wave functions to predict as well as explore the properties related to the electronic ground state of the system. This many-electron system can be transformed into a self-contained field of one particle system using mathematical methods and representations. The ground state total energy can be used to calculate numerous crystal properties like structural, electronic, optical, and elastic, as a function of chemical composition. The Wien2k code is based on the augmented plane wave method (FP-LAPW), which is the most significant approach to addressing the Kohn-Sham problem. In this method, atoms are regarded as spheres the size of muffin tins, with spherically symmetric potentials and spherically symmetric wave functions within the atomic sphere. An atomic partial wave augments each plane wave, which is the wave function that corresponds to the interstitial region.

3 Results and Discussions

3.1 Structural Properties

DFT is used to analyze the structural properties of binary skutterudite CoX_3 (As, P, and Sb). WC-GGA potential is used to optimize the compound CoX_3 (As, P, and Sb), properties calculated using the Birch-Murnaghan equation of the states. CoX_3 (As, P, and Sb) belongs to the 204_Im-3 space group. By using the full potential approach in this research work, the lattice constant of CoX_3 (As, P, and Sb) are 8.334 Å, 7.806 Å, and 9.454 Å, respectively. CoP_3 has the lowest lattice constant out of all due to reduced interatomic distance. Figure 1(a–c) illustrates the variation in lattice constant and unit cell resulting due to variation of X compound (Table 1).

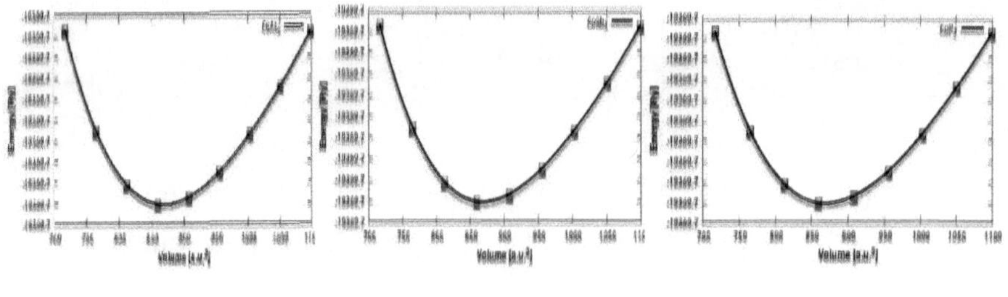

(a) (b) (c)

Fig. 1. Geometry optimization (E vs V) (a) $CoAs_3$, (b) $CoSb_3$, (c) CoP_3.

3.2 Electronic Properties

To determine the electronic properties and band gap of the skutterudite structure, the FP-LAPW was used. From Fig. 2(a–c), CoX_3 (As, P, and Sb) exhibit metallic character with the direct band gap of 0 eV with the transition at Γ position of k-point. From Fig. 3(a–c), it is observed that there is no overlapping, and the bond is ionic between the elements.

Table 1. Lattice parameter of the optimized structure of CoX_3 (X = As, P, Sb).

Compound	Lattice constant (a)	Bulk modulus (GPa)	Derivative of bulk modulus	Ground state volume (V_o)	Ground state energy (E_o)
$CoAs_3$	8.3344	133.9087	4.0336	1857.8503	−65399.5051
CoP_3	7.8059	165.1355	4.3106	1529.1913	−19350.0568
$CoSb_3$	9.4540	87.9963	4.9437	1560.3356	−166754.9930

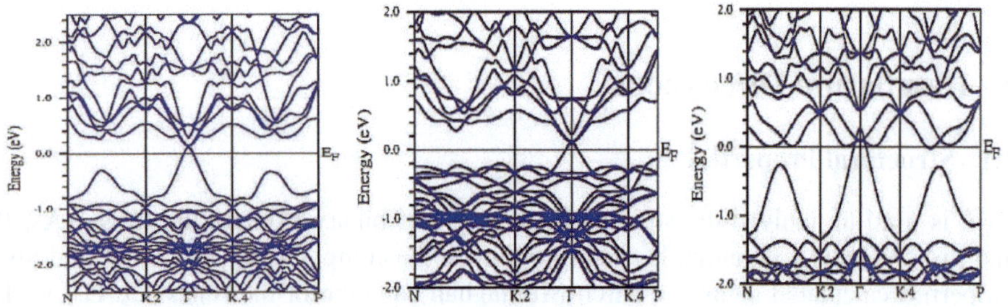

(a) (b) (c)

Fig. 2. Energy band diagram of (a) $CoAs_3$, (b) $CoSb_3$, (c) CoP_3.

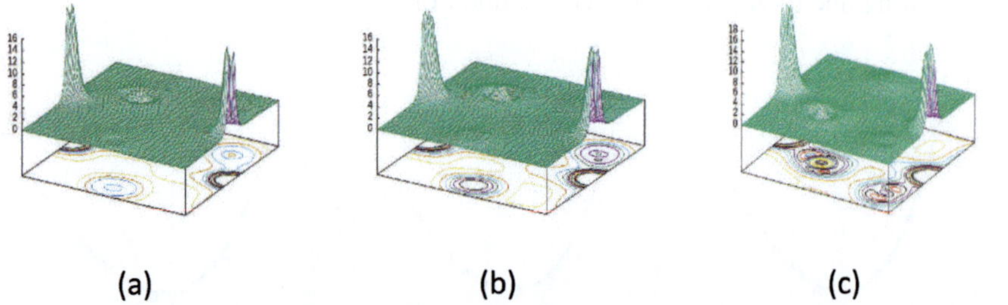

(a) (b) (c)

Fig. 3. Electron-density diagram of (a) $CoAs_3$, (b) $CoSb_3$, (c) CoP_3.

3.3 Optical Properties

To describe the interaction between the photons and materials, optical properties play a vital role. Not only do they describe the information about materials' structure but also electronic dissipation and energy dissipation. Absorption coefficient and reflective index are thoroughly investigated, these properties are arising due to interaction between materials and photons.

The absorption coefficient is one of the vital parameters for optimizing the efficiency of solar energy conversation. Figure 4(b) shows each compound displays an exclusive absorption line, signifying variances in their electronic structures. Around low energies around 2.49 eV, have relatively small bandgaps to allow visible range light. The highest absorption edge is obtained for $CoAs_3$ where $CoSb_3$ has the lowest absorption intensity. Surface characteristics are concluded by optical reflectivity; the surface properties of the solids can be determined by the light reflected by the surface. From Fig. 4(a), all three compounds display relatively low reflectivity. All the compounds illustrate the gradual increase in reflectivity towards the UV region.

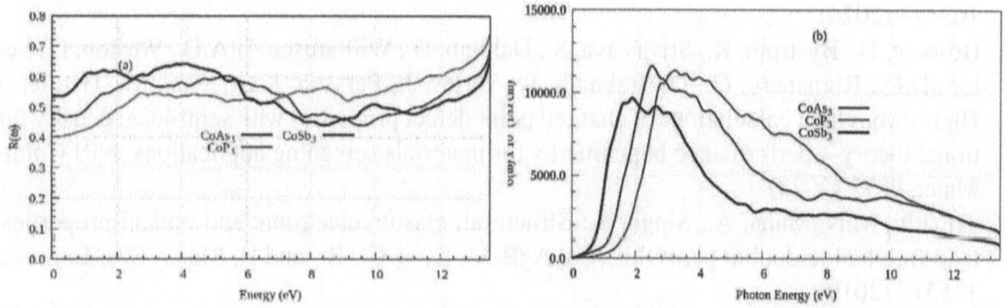

Fig. 4. (a) Reflectivity (b) Absorption Coefficient of CoX_3 (X = As, P, Sb).

4 Conclusion

We used DFT to analyse the structural, electronic and optical properties of the binary compound CoX_3 (X = As, P, Sb), the study revealed that CoP_3 exhibits higher bulk modulus and $CoSb_3$ has the lowest bulk modulus among the compounds under study. The electronic properties indicate the materials are showing metallic behaviour with a direct band gap. In contrast, the optical properties indicate that $CoAs_3$ has the greatest absorption among the compounds. Overall, this study provides a comprehensive understanding of the structural, electronic, and optical characteristics of CoX_3 (As, P, Sb) and paves the way for further investigations into their potential applications in energy harvesting, optoelectronic system, supercapacitor electrode material, energy storage and conversion [19, 20].

References

1. Chu, L., Ahmad, W., Liu, W., Yang, J., Zhang, R., Sun, Y., Yang, J., Li, X.: Lead-free halide double perovskite materials: a new superstar toward green and stable optoelectronic applications. Nano-Micro Lett. **11**, 16 (2019)
2. Lu, R., Liu, Y., Zhang, J., Zhao, D., Guo, X., Li, C.: Highly efficient (200) oriented $MAPbI_3$ perovskite solar cells. Chem. Eng. J. **433**, 133845 (2021)
3. Cucco, B., Pedesseau, L., Katan, C., Even, J., Kepenekian, M., Volonakis, G.: Silver–Bismuth halide double salts for lead-free photovoltaics: insights from symmetry-based modeling. Sol. RRL. **6**, 2200718 (2022)
4. Chen, H., Zhang, C., Liu, Z., Gong, J., Zhang, M., Wu, Y., Chen, Y., Chen, H.: Optoelectronic properties of Rb-doped inorganic double perovskite $Cs_2AgBiBr_6$. Chem. Phys. Lett. **771**, 138501 (2021)

5. Ikram, M., Malik, R., Raees, R., Imran, M., Wang, F., Ali, S., Khan, M., Khan, Q., Maqbool, M.: Recent advancements and future insight of lead-free non-toxic perovskite solar cells for sustainable and clean energy production: a review. Sustain. Energy Technol. Assess. **53**, 102433 (2022)
6. Guo, S., Chen, C., Zhang, Z., Cai, Y., Wang, S., Gao, S., Chen, W., Abduryim, E., Dong, C., Guan, X., Liu, Y., Lu, P.: Theoretical investigations of double perovskite Rb_2AgBiX_6 (X = Cl, Br, I) for solar cell and photocatalysis applications. Mater. Sci. Semicond. Process. **177**, 108373 (2024)
7. Broberg, D., Bystrom, K., Srivastava, S., Dahliah, D., Williamson, B.A.D., Weston, L., Scanlon, D.O., Rignanese, G., Dwaraknath, S., Varley, J., Persson, K.A., Asta, M., Hautier, G.: High-throughput calculations of charged point defect properties with semi-local density functional theory—performance benchmarks for materials screening applications. NPJ Comput. Mater. **9**, 72 (2023)
8. Tripathi, M.N., Saha, A., Singh, S.: Structural, elastic, electronic and optical properties of lead-free halide double perovskite Cs_2AgBiX_6 (X = Cl, Br, and I). Mater. Res. Express. **6**, 115517 (2019)
9. Perdew, J.P., Burke, K., Ernzerhof, M.: Generalized gradient approximation made simple. Phys. Rev. Lett. **77**, 3865 (1996)
10. Okoye, C.M.I.: Theoretical study of the electronic structure, chemical bonding and optical properties of $KNbO_3$ in the paraelectric cubic phase. J. Phys.: Condens. Matter. **15**, 5945 (2003)
11. Sharma, S., Verma, A.S., Jindal, V.K.: Ab initio studies of structural, electronic, optical, elastic and thermal properties of silver gallium dichalcogenides ($AgGaX_2$: X= S, Se, Te). Mater. Res. Bull. **53**, 218–233 (2014)
12. Murnaghan, F.D.: The compressibility of media under extreme pressures. Proc. Natl. Acad. Sci. U. S. A. **30**, 244–247 (1944)
13. Bakeer, A.M., Mahmoud, N.T., Shaltaf, R.: Coupled insights into structural, electronic, and optical properties of $GeTMS_3$ (TM= Sc, Fe, Zn) sulfide perovskite across different phases: ab initio study. Int. J. Energy Res. **2024**, 2316636 (2024)
14. Liu, X., Xie, B., Duan, C., Wang, Z., Fan, B., Zhang, K., Lin, B., Colberts, F.J.M., Ma, W., Janssen, R.A.J., Huang, F., Cao, Y.: A high dielectric constant non-fullerene acceptor for efficient bulk-heterojunction organic solar cells. J. Mater. Chem. A. **6**, 395–403 (2018)
15. Shahzad, M.K., Zhang, Y., Cui, L., Liu, L., Butt, M.K., Li, H.: Dispersing upconversion nanocrystals in PMMA microfiber: a novel methodology for temperature sensing. RSC Adv. **8**, 19362–19368 (2018)
16. Ilyas, I., Zafar, A.A., Ullah, H.M.N., Rizwan, M., Usman, Z., Gillani, S.: Pressure-induced elastic, mechanical and optoelectronic response of $RbCdF_3$: a comprehensive computational approach. J. Phys. Chem. Solids. **165**, 110642 (2022)
17. Wang, L.J., Kuzmich, A., Dogariu, A.: Gain-assisted superluminal light propagation. Nature. **406**, 277–279 (2000)
18. Mugnai, D., Ranfagni, A., Ruggeri, R.: Observation of superluminal behaviours in wave propagation. Phys. Rev. Lett. **84**, 4830–4833 (2000)
19. Mahi, F., Meghoufel, Z.F., Mostefa, Z., Kara, I., Boukra, A.: Impact of lattice strain on the electronic and thermoelectric properties of CoAs3 Skutterudite material. Phys. Status Solidi B. **262**(2), 2400339 (2025)
20. Jiang, J., Li, Z., He, X., Hu, Y., Li, F., Huang, P., Wang, C.: Novel skutterudite CoP_3–based asymmetric supercapacitor with super high energy density. Small. **16**(31), 2000180 (2020)

Microstructural, Morphological, Optical and Electrical Characteristics of Cobalt Doped Zinc Oxide Thin Films Prepared by Sol-Gel Dip Coating Technique

Mahitha Mohan(✉), K. Thangavel, and V. Balaprakash

Hindusthan College of Arts & Science, Coimbatore 641028, Tamilnadu, India
malumahitha09@gmail.com

Abstract. Bulk specimens of Cobalt-doped Zinc Oxide (CZO) thin films were fabricated using the Sol-Gel dip coating technique, with diverging Cobalt (Co) proportions. It was found that Co was successfully incorporated into the of ZnO's crystal lattice, and diffraction pattern analysis affirmed the presence of a hexagonal wurtzite crystalline arrangement. The obtained particle size were varied between 21.871 and 31.626 nm. Optical properties of CZO thin films were investigated through absorption curve, showing a noticeable reduction in the forbidden energy gap with increased doping proportion and annealing temperature. The forbidden energy gap ranged from 3.20 eV to 3.31 eV, with the minimum electronic band gap (3.20 eV) observed for a thin film of CZO doped with 1 at. %, treated at 450 °C. The pure ZnO thin film displayed typical diamagnetic behaviour and as the Co doping increases the material approaches ferromagnetism. These findings indicate that CZO thin films are ideal for transparent conducting oxides (TCOs) for solar cells, while their ferromagnetic properties make them promising for sensor applications.

Keywords: Sol-Gel · Dip coating · Forbidden energy gap · Optoelectronics

1 Introduction

Recently, wide energy gap possessed by group II-VI semiconductors attracted significant attention from various scientific teams. Zinc Oxide (ZnO) a transparent broad-bandgap material with a 3.37 eV energy gap coupled with a significant electron-hole binding energy quantified as 60 meV, has been widely researched due to its diverse technological implementations [1, 2]. It shows notable applications in optoelectronics, photocatalysis, Light Emitting Diodes (LED's), Ultraviolet (UV) photodetectors, and Gas Sensors. ZnO has the potential to replace another wide energy gap semiconductor, GaN, widely utilized in green, blue, and UV light-emission devices. For the advancement of spintronic and optoelectronic devices, metals in the transition series, like Iron (Fe), Cobalt (Co), and Nickel (Ni) known for their exceptional physical properties and large magnetic moments, are essential. These transition metals exhibit magnetic properties owing to their partially

S. Kumari et al. (Eds.): ICMTP 2024, SPPHY 311, pp. 155–168, 2025.
https://doi.org/10.1007/978-981-96-6795-6_20

filled d states, making them crucial for spintronic-based quantum computers. Despite this, there is limited research on the optical characteristics of Fe, Co, and Ni-doped ZnO. The ability to modify the band gap is crucial for fabricating quantum well devices for specific applications. Doping ZnO with Fe, Co, or Ni can alter its band gap, while unpaired electrons in the 3d partially occupied d states of these transition series metals contribute to magnetic activity [3]. Introducing 3d transition metal ions like Fe and Ni into the ZnO matrix can alter the Fermi level, thus tailoring the band gap. Transition metal ions such as Fe, Co, and Manganese (Mn) produce magnetic characteristics when doped into ZnO, making it suitable for spintronic applications. The current research advances the field by addressing gaps in understanding the impact of Co doping on the structural, surface, and optical characteristics of ZnO thin films [4]. This study emphasizes the sol-gel dip-coating process for depositing Co-doped ZnO (CZO) thin films on glass substrates, a method gaining popularity for its simplicity, cost-effectiveness, and doping versatility. While prior studies have reported variations in the microstructural, electronic, and optical characteristics of ZnO linked to intrinsic and extrinsic defects, this research uniquely investigates the solubility and behaviour of Co within the ZnO matrix. Co is particularly significant due to its high solubility and its ability to modify the configuration, microscopic structure, electrical, and optical properties of ZnO. All these points towards the ability of CZO thin films as transparent conducting oxides (TCO's) in solar cells [5, 6].

This research contributes by exploring impurity phases in Co-doped ZnO, which are reported to emerge even at lower Co concentrations. Sol-gel dip coating is highlighted for its advantages, including low vacuum requirements, safety, minimal waste production, and the ability to coat large surface areas with uniform films. Compared to other methods, dip coating offers superior mechanical, chemical, and thermal resilience, along with excellent transparency and resistance to photodegradation. Moreover, it enables precise control over the nano-structure of ZnO, which is increasingly important for size- and shape-dependent applications in optoelectronic devices. In this context, the present study focuses on the production and characterization of ZnO and Co-doped ZnO thin films, which show promise as alternatives to Indium Tin Oxide (ITO) films in optoelectronic applications [7]. By contextualizing the novelty of Co doping in ZnO and utilizing the sol-gel dip-coating process, this research advances existing studies and provides valuable insights into the potential of these materials for future technological innovations [8, 9].

2 Experiment Specifications

2.1 Thin Film Deposition

The fabrication of pure ZnO and Cobalt doped ZnO (CZO)nanostructured thin films incorporates HIMEDIA® make at pathology grade, non-interleaved fine polished edges, plain glass slides having dimensions of 25.4 mm × 76.2 mm (1″ × 3″), thickness of 1.0 mm-1.2 mm with comer of 90° were used. ZnO and CZO thin films, applied onto glass surfaces employing the sol-gel dip coating approach. Figure 1 illustrates the process flow for preparing the ZnO thin films. Zinc acetate dihydrate [Zn $(CH_3COO)_2\cdot 2H_2O$] and Cobalt acetate tetrahydrate [Co $(CH_3COO)_2\cdot 4H_2O$] were used as the precursor and dopant sources. Ethanol served as the solvent and mono ethanol amine (MEA) as stabilizer. The proportion of metal ion was set at 0.5 mol/L, and molar proportion of MEA to metal ions was sustained at 1.0. The Co doping ratio varied from 0 at. % to 8 at. %. After stirring for 2 h at 60 °C to form a transparent and consistent solution, which left to rest at typical ambient temperature over a 24 h period. Glass substrates were sanitized using an ultrasonic cleaner using acetone for 40 min, followed by rinsing with demineralized water and drying. The solution was then deposited towards the glass substrates by dip coating process (Fig. 1). The films heated around 10 min at 200 °C to dispose of the solvent by evaporation and remove organic residues after each dip coating process. After 30 cycles of dip coating and pre-heating cycles, the films annealed at

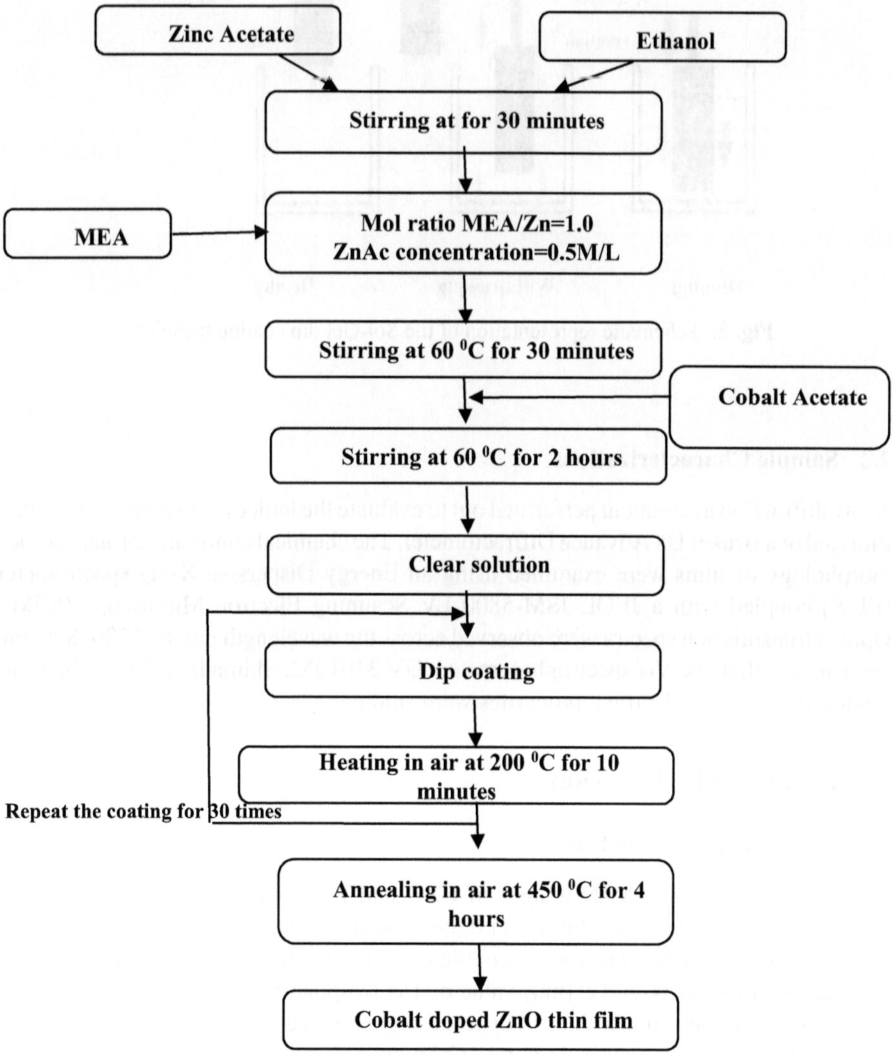

Fig. 1. Flowchart for the development of ZnO and CZO thin films preparation.

450 °C for 4 h. Post annealing is indispensable for achieving high-performance thin films tailored to specific applications. It is an essential step to enhance their structural, electrical, optical, and mechanical properties [10, 11] (Fig. 2). The thin film deposition conditions are specified in Table 1.

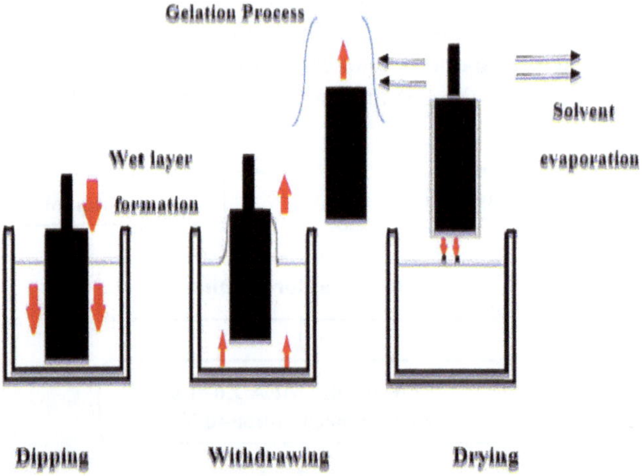

Fig. 2. Schematic representation of the Sol-Gel dip coating technique.

2.2 Sample Characterizations

X-ray diffraction assessment performed out to evaluate the lattice arrangement of the thin films aid of a Bruker D8 Advance Diffractometer. The chemical configuration and surface morphology of films were examined using an Energy Dispersive X-ray spectrometer (EDX) coupled with a JEOL JSM-5800 LV Scanning Electron Microscope (SEM). Optical transmission spectra were observed across the wavelength range of 200–800 nm, measured with a UV-Vis spectrophotometer (UV-3101 PC, Shimadzu). Using the four-point probe method electrical properties were studied.

3 Results and Discussion

3.1 Microstructural Analysis

Figure 3 represents the XRD images of CZO and ZnO thin films, and its outcomes demonstrate those samples exhibit a wurtzite structure with a polycrystalline hexagonal phase (JCPDS 36-1451). The texture coefficient (TC) was used to evaluate the preferred orientation of the films. A TC (hkl) value of 1 corresponds to randomly oriented crystallites, while values greater than 1 suggest a higher concentration of crystallites along specific (hkl) crystallographic direction. Calculated TC shows that all films have their maximal TC values (TC = 1.5) in the (002) plane, indicating preferential growth in the direction of the c-axis, as illustrated below. For all the thin films, the (002) peak is dominant, indicating that Co doping and annealing temperature promote growth along the crystallographic c-axis of ZnO. The films with 1 at. % and 8 at. % cobalt shows the highest intensity of the (002) crystalline peak. The mean crystallographic domain size was calculated by assessing the Full Width at Half Maximum (FWHM) of the main crystalline zenith employing the Debye-Scherrer formula [12, 13]. The observed nanocrystal size varies from 21.871 nm to 27.561 nm for the 0 at. % to 8 at. % CZO thin films, for 1 at. % CZO thin film annealed at 450 °C exhibiting smallest crystallite size (21.871 nm). Doping with Co did not modify the hexagonal wurtzite orientation of the ZnO films, and no secondary phases or Co impurities were observed, indicating that the Co ions introduced by doping were successfully substituted into the ZnO framework. The nano crystallite size estimated using Scherrer's formula.

$$d = \frac{k\lambda}{\beta Cos\theta} \tag{1}$$

Where, d is the average crystalline size, λ is the wavelength, k is a constant (0.9), β is the angular width of the diffraction peak at FWHM on the 2θ and θ is Bragg's angle. The (002) diffraction peaks employed to estimate the crystallite size. The computed average crystalline dimensions of ZnO thin films are presented below.

Table 1. Thin film deposition conditions

Parameter	Deposition conditions
Deposition Temperature	Room temperature (300 k)
Concentration of Zn $(CH_3COO)_2.2H_2O$	0.5 M/L
Doping percentage of Co	0, 0.5, 1, 1.5 and 2 at. %
Dipping and withdrawal speed of substrate	20 mm/min

The average crystalline size changes based as it pertains to the doping concentration and is derived from the FWHM of X-ray diffraction peaks. ZnO film demonstrates a comparatively more substantial crystalline size, indicating well-formed crystallites. A slight decrease in the crystalline size was observed due to the introduction of a small amount of Co, which could cause some lattice distortion, affecting crystal growth. Further reduction in the crystalline size compared to 0.5 at% doping. This suggests that Co doping continues to inhibit crystal growth. A notable increase in crystalline size was observed. This could indicate that at 2 at% doping, Co might act as a nucleation centre, enhancing crystal growth, or that the doping concentration is optimal for crystallite formation. A decrease in crystalline size is observed once again, similar to the behaviour seen at 1 at% doping. This could be due to the saturation of Co atoms in the lattice, causing distortions. At 8 at%, the crystalline size is almost the same as at 4 at%, suggesting that high doping concentrations (beyond a certain point) do not significantly influence crystallite growth. The FWHM is also relatively high, indicating smaller crystallites [14] (Table 2).

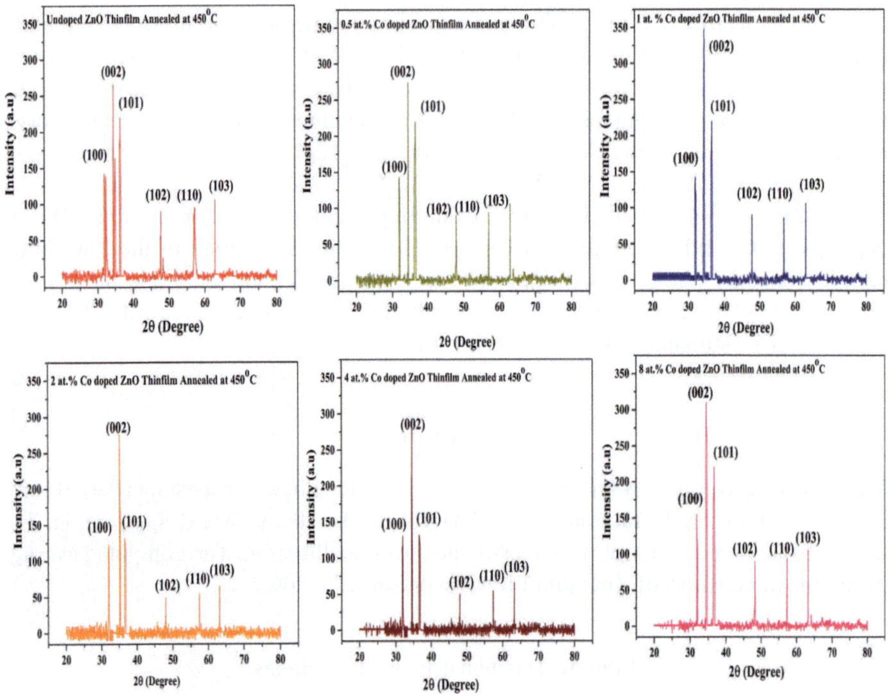

Fig. 3. X-ray diffraction image of ZnO and CZO films

Table 2. Mean crystalline dimension of ZnO and CZO thin films.

Sl no.	Doping concentration (at.%)	Diffraction angle (degree)	Cos (rad)	FWHM (rad)	Average crystalline size (nm)
1	Undoped	17.261	0.954961	0.0055	27.561
2	0.5	17.474	0.953854	0.00612	24.797
3	1	17.236	0.9550939	0.00693	21.871
4	2	17.210	0.9552257	0.00480	31.571
5	4	17.200	0.9552794	0.00705	21.494
6	8	17.552	0.9534425	0.00738	21.885

3.2 Surface Configuration

Figure 4 clearly depicts the typical EDAX spectrum of ZnO and CZO thin film. It exhibits the crests associated with Zinc (Zn) and Oxygen (O) molecules. Figure 4 shows compositional investigation of $ZnOCo_5$ samples, which shows the existence of Zn, O then Co elements alone, which designates that the nanostructured thin film contains Zinc, Oxygen and Cobalt ions. To verify the existence of Co and assess the cleanliness of the films, an EDX measurement was performed. Figure 4 displays the EDX spectrum for both undoped and 8 at. % CZO films, confirming the existence of cobalt in the ZnO matrix. The EDX results indicate impurity introduction of Co into the films.

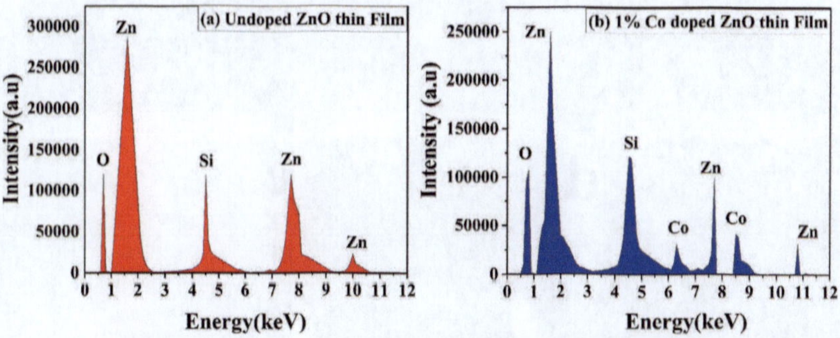

Fig. 4. EDAX image of ZnO thin film, and CZO thin film.

The SEM images in Fig. 5 reveal that the ZnO thin film exhibits a polished, consistent tiny ZnO crystals. In contrast, the CZO films exhibit asymmetrical, fibre such as stripes and a creased network structure, with this morphology being more pronounced and denser in the 4% and 8% CZO films. Generally, sol-gel ZnO thin films display a network-like structure with wrinkles and fibre-like stripes on their surface. The inclusion of Co ions into the ZnO thin films altered the surface structure derived from a granular texture into a crumpled network, likely due to the introduction of contaminants and subsequent Microstructural imperfections in the ZnO crystal. This study suggests that the Co dopant influences nucleation growth in various directions, affecting the overall surface configuration of the CZO films [15].

Increased doping levels and heat treatment temperatures can cause grain clustering, leading to the formation of larger grains. However, as the concentration of Co ions increases, the extent of agglomeration tends to decrease. The agglomeration is attributed to magnetic interactions among the particles. On the nanometre scale, particles possess a high surface-to-volume proportion, which increases surface tension at the interface and contributes to agglomeration. Additionally, the accumulation of Co^{2+} ions at the inter-grain boundaries applies tensile stress and pressure on the grains, which reduces their size. The results confirm successful Co doping in the films, which clearly exhibit a hexagonal shape. In contrast, undoped films show a spongy microstructure. Furthermore, as the annealing temperature rises, the intermolecular spacing reduces, leading to an increase in conductivity [16].

3.3 Optical Properties

The optical absorption spectrum of undoped and CZO thin films heat-treated at different thermal conditions analyzed using UV-Vis Spectroscopy. This plot illustrates the absorbance spectral profiles of ZnO thin films doped across varying levels of concentration (Fig. 6).

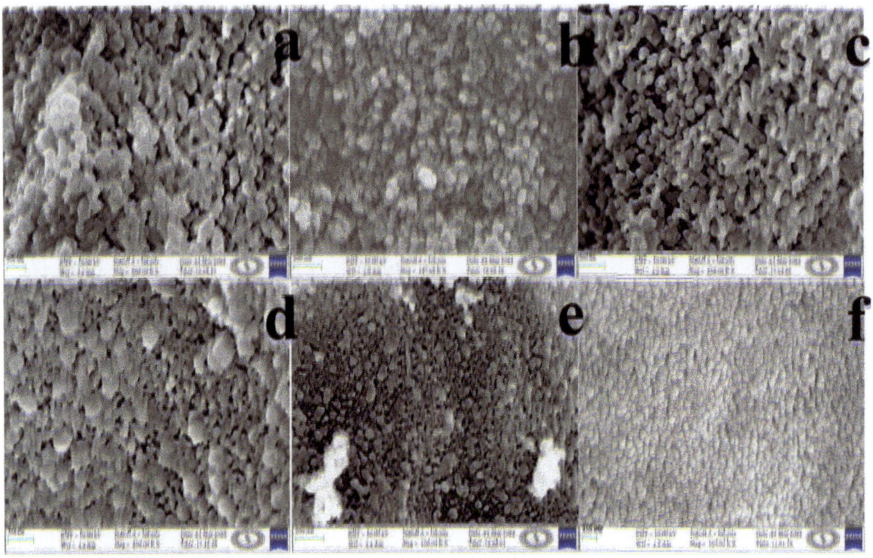

Fig. 5. SEM images of (a) pure ZnO thin film, (b) 0.5 at. %, (c) 1 at. %, (d) 2 at. %, (e) 4 at. % and (f) 8 at. % CZO thin film.

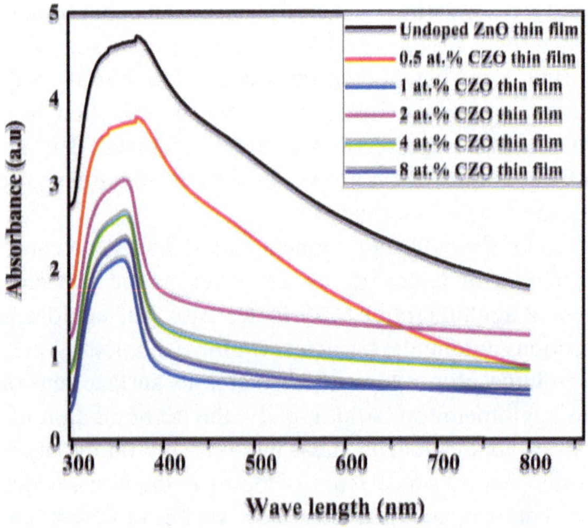

Fig. 6. Absorbance Spectra of ZnO and CZO thin film.

All the spectra exhibit a peak in the absorbance around 300-380 nm, which is typical for ZnO due to its wide bandgap. A region of reduced absorption can be observed between 300 nm and 800 nm [17]. This indicates that, fabricated films are suitable as TCOs. This peak corresponds to the absorption of ZnO, indicating the onset of electronic transformations from the outer electron band to the conduction band. The absorbance ranges between 2.5 a.u and 4.492 a.u. At shorter wavelengths (below ~350 nm), all films exhibit low transmittance. This is likely due to absorption in the ultraviolet ddomains. As the wavelength intensifies toward visible range (~400 to 800 nm), the absorbance decreases and transmittance increases. ZnO films doped with high concentration show lower absorbance in the visible to near-infrared range. The undoped film shows the highest absorbance across the spectrum. Doping with higher concentration reduces the optical absorbances and improves the optical transparency of ZnO thin films. Higher doping concentration likely reduce defects or improve crystallinity, leading to less light scattering and lower absorbance. This type of analysis can be important in optimizing the annealing conditions for ZnO thin films to achieve desired optical properties for applications in photovoltaics, sensors and optoelectronic devices. When a photon beam with energy hv > Eg strikes a semiconductor, a certain amount of absorption occurs, which is determined by the material's properties. Photon wavelength and sample thickness is influenced by the wavelength of the photon and the sample thickness. The band gap energy (Eg) is determined from the absorption spectra with Tauc's relation [18] (Fig. 7),

$$(\infty h\upsilon)^2 = k(h\upsilon - Eg) \tag{2}$$

$$\alpha = C(h\upsilon - Eg)^{1/2} \tag{3}$$

It was noted that the optical absorbance was minimal and the transmittance of all samples was approximately 90%. This indicates that the fabricated films are suitable for solar cell applications. The band gap of the thin films annealed at different temperatures ranges from 3.20 eV to 3.31 eV. The reduction of the band gap attributed to the sp-d exchange coupling between the band electrons and the constrained states 'd' electrons of Co^{2+} substituting Zn^{2+}. The observed growth in forbidden gap is related to the Moss-Baurstein effect [19].

Table 3 reveals that increasing Co doping concentration slightly decreases the bandgap energy for ZnO, attributed to improved crystallinity and reduced defect density. With 0.5 at.% Co doping, the bandgap narrows further due to Co-induced localized states enhancing lattice integration and reducing defect-related states. For 1 at.% Co doping, the bandgap reaches a minimum of 3.20 eV, indicating optimal doping for narrowing the bandgap. At 2 at.% Co doping, the bandgap initially increases to 3.29 eV due to the Moss-Burstein effect, where higher carrier concentration raises the Fermi level, shifting the absorption edge. However, with increased annealing temperature, the bandgap decreases as crystallinity improves and the Moss-Burstein effect diminishes. For 4 at.% Co doping, the bandgap consistently decreases to 3.25 eV, as the Moss-Burstein effect becomes negligible. At 8 at.% Co doping, the bandgap further reduces to 3.23 eV due to more Co-related localized states modifying the conduction band. The undoped ZnO bandgap is 3.31 eV, decreasing slightly with annealing temperature. Doping introduces a nuanced interplay of factors, including the Moss-Burstein effect and localized states, significantly impacting the electronic structure and bandgap of CZO thin films.

3.4 Electrical Properties

Analysing the electrical properties of semiconductor films is crucial for understanding their conductive behaviour. In CZO thin films, electrical conduction arises from the

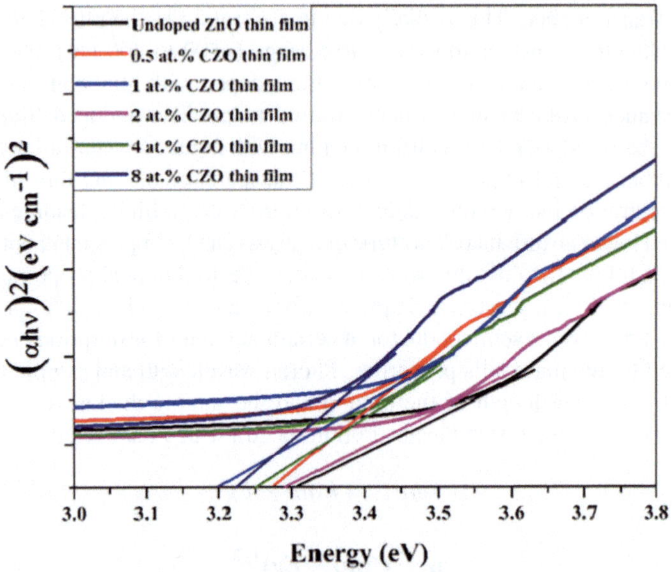

Fig. 7. Band gap of ZnO and CZO thin films.

Table 3. Forbidden energy gap of ZnO and CZO thin films.

Sl no.	Doping concentration (at. %)	Bandgap (eV)
1	0	3.31
2	0.5	3.28
3	1	3.20
4	2	3.29
5	4	3.25
6	8	3.23

bonding of atoms and is highly dependent on temperature. Doping with Co introduces impurity energy levels within the forbidden energy gap of the ZnO. These extrinsic energy levels modify the electrical conductivity by enabling thermal excitation of charge carriers. In this section, the electrical resistivity of CZO thin films was measured using the four-point probe method.

CZO thin films are suitable options for uses such as TCOs, particularly in the context of solar cells and solid -state display devices. For such applications, electrical resistivity is a crucial parameter. The influence of cobalt doping concentration (0 at. % to 8 at. %) on the resistivity of ZnO and CZO thin films was illustrated in Fig. 8.

The electrical conductivity of ZnO is primarily influenced by free electrons generated from intermediate zinc atoms and oxygen vacancies. It was observed that the

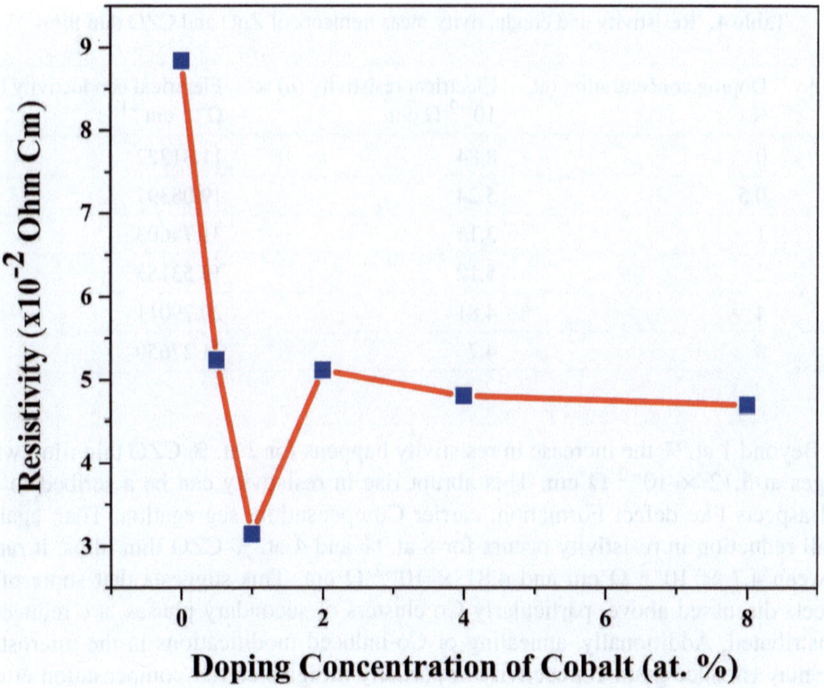

Fig. 8. Effect of Cobalt doping in the electrical resistivity of ZnO thin films.

resistivity of cobalt-doped films is lower than that of undoped films at optimal doping concentrations. At lower doping levels, cobalt acts as an effective electrical dopant, enhancing conductivity. However, at higher concentrations, it behaves as an impurity, contributing to increased resistivity. This is attributed to microstructural and electrical defects. Subsequently, it has been established that resistivity of doped films is inversely proportional to dominance of the (002) crystal orientation, as shown in Table 4. These findings underline the key role of optimizing cobalt doping concentration for achieving superior electrical properties in CZO thin films.

3.4.1 Influence of Doping Level on Resistivity

For all annealing temperatures, the resistivity of ZnO thin films decreases with increasing Co doping concentration up to 1 at. %, after which it increases with further doping. This indicates an optimal doping concentration (1 at. %) at which the resistivity is minimized. The minimum resistivity obtained is 3.15×10^{-2} Ω cm.

The decrease in resistivity with doping suggests that Co doping introduces additional charge carriers, likely due to the substitution of Co ions (Co^{2+}) at Zn sites in the ZnO lattice. This substitution can create shallow donor levels near the conduction band, enhancing carrier concentration.

Table 4. Resistivity and conductivity measurements of ZnO and CZO thin films.

Sl. no	Doping concentration (at. %)	Electrical resistivity (ρ) × 10^{-2} Ω cm	Electrical conductivity (σ) Ω^{-1} cm^{-1}
1	0	8.84	11.31222
2	0.5	5.24	19.08397
3	1	3.15	31.74603
4	2	5.12	19.53155
5	4	4.81	20.79011
6	8	4.7	21.27659

Beyond 1 at. % the increase in resistivity happens for 2 at. % CZO thin film, which ranges at 5.12 × 10^{-2} Ω cm. This abrupt rise in resistivity can be a scribed to several aspects like defect Formation, carrier Compensation, segregation. Then again, a small reduction in resistivity occurs for 8 at. % and 4 at. % CZO thin films, it ranges between 4.7 × 10^{-2} Ω cm and 4.81 × 10^{-2} Ω cm. This suggests that some of the defects discussed above, particularly Co clusters or secondary phases, are reduced or redistributed. Additionally, annealing or Co-induced modifications in the microstructure may enhance grain connectivity or partially mitigate carrier compensation effects, resulting in improved electrical conductivity.

The lowest resistivity is observed at 1 at. % cobalt doping. This indicates that annealing temperature and doping concentrations can significantly improve the conductivity of CZO films, which is essential for their use in TCO applications. However, higher doping levels result in higher resistivity, likely due to defects or reduced carrier mobility. The results demonstrate that Co doping and post-deposition annealing temperature are critical parameters for tailoring the electrical attributes of ZnO thin films. Such doped ZnO films incorporating optimized resistivity could be suitable for applications in optoelectronic devices, transparent conductive oxides, and sensors [20–23].

4 Conclusion

The study presents a comprehensive investigation into the structural, optical, and electrical properties of CZO thin films, deposited by the sol-gel dip-coating method. XRD analysis confirmed a hexagonal wurtzite structure for all films, with a preferential c-axis orientation along the (002) plane. The crystallite size decreased with Co doping, reaching a minimum at 1 at. % Co (21.87 nm), due to lattice distortion caused by Co substitution. Beyond this, crystallite size varied due to saturation effects and nucleation centre formation. EDX confirmed successful incorporation of Co ions without significant impurities. SEM images revealed a transition from smooth granular surfaces in undoped ZnO films to crumpled, fibrous structures in doped films, particularly pronounced at 4 at. % and 8 at. % Co. UV-Vis spectroscopy showed low absorbance and high transparency (~90%) in the visible range as well as a decrease in the optical bandgap with increasing Co concentration, reaching a minimum of 3.20 eV at 1 at. % Co doping. This bandgap narrowing is attributed to sp-d exchange interactions and reduced defect density. Resistivity decreased with increasing Co doping up to 1 at. % (minimum resistivity of 3.15×10^{-2} Ω·cm), enhancing conductivity through additional charge carriers from Co^{2+} substitution. Higher doping levels increased resistivity due to defect formation and carrier compensation effects, although resistivity slightly declined again at 8 at. % doping. CZO thin films exhibit excellent potential as TCOs, suitable for applications in solar cells, optoelectronics, and spintronic devices. Optimal Co doping (1 at. %) enhances the balance between structural integrity, optical transparency, and electrical conductivity. The findings underline the critical role of controlled Co doping in tailoring the multifunctional properties of ZnO thin films for different applications.

References

1. Khantoula, R., Sebaisa, M., Rahalb, B., Boudinea, B., Halimia, O.: Microstructural and Optical Properties of ZnO and Co Doped ZnO Thin Films Prepared by Sol-Gel. (2018) https://doi.org/10.12693/APhysPolA.133.114
2. Thangavel, K., Roshini, T., Balaprakash, V., Gowrisankar, P., Sudha, S., Mohan, M.: Microstructural, morphological and antibacterial properties of ZnO nanofibers fabricated by electrospinning technique. Mater. Today: Proc. **33**, 2160–2166 (2020)
3. Jamal, M., Nishat, S.S., Sharif, A.: Effects of transition metal (Fe, Co & Ni) doping on microstructural, electronic and optical properties of Cuo: Dft + U Study. Chem. Phys. **545**(5), 111160 (2021)
4. Awale, M.B., Kathwate, L.H., Lokhande, S.D.: Synthesis, microstructural and optical properties of Zn1-xCoxO of thin films by spray pyrolysis method. IOSR-JAP. 11(3 Ser. II) e-ISSN.: 2278-4861 (2019)
5. Swami, S.K., Kumar, N., Radu, D.R., Cho, S.W., Lee, J.: Lithium incorporation into TiO_2 photoanode for performance enhancement of dye-sensitized solar cells. ACS Appl. Energy Mater. **6**(16), 8599–8606 (2023). https://doi.org/10.1021/acsaem.3c01579
6. Kumar, A., Singh, A., Kumar, D., Garg, A., Dutta, V., Swami, S.K., Chaturvedi, N., Kumar, M.: Spray deposited carbon nanotube embedded ZnO as an electrons transport layer in inverted organic solar cells. Hybrid Adv. **4**, 100088 (2023). https://doi.org/10.1016/j.hybadv.2023.100088
7. Swami, S.K., Chaturvedi, N., Kumar, A., Dutta, V.: Effects of electric field during deposition on spray deposited indium-doped zinc oxide films. Progress Photovoltaics: Res. Appl. **24**(1), 74–82 (2015). https://doi.org/10.1002/pip.2649
8. Afre, R.A., Sharma, N., Sharon, M., Sharon, M.: Transparent conducting oxide films for various applications: a review. Rev. Adv. Mater. Sci. **53**(1), 79–89 (2018). https://doi.org/10.1515/rams-2018-0006
9. Mahitha, M., Thangavel, K., Balaprakash, V.: Characterization of CZO thin films prepared by Sol – Gel dip coating technique with varied concentrations and annealed at different temperatures. Chem. Phys. **573**, 111981 (2023). https://doi.org/10.1016/j.chemphys.2023.111981, 0301-0104
10. Singh, A., Dinesh Kumar, P.K., Khanna, B.C., Joshi, M.K.: Effect of post annealing temperature on structural and optical properties of ZnCdO thin films deposited by sol–gel method. Appl. Surf. Sci. **258**(5), 1881–1887 (2011). https://doi.org/10.1016/j.apsusc.2011.10.096
11. Singh, A., Kumar, D., Khanna, P.K., Kumar, M., Prasad, B.: Post annealing effect on structural and optical properties of ZnO thin films derived by sol–gel route. J. Mater. Sci.: Mater. Electron. **24**, 4607–4613 (2013). https://doi.org/10.1007/s10854-013-1451-4

12. Swami, S.K., Barman, B., Kumar, A., Dutta, V.: Multilayer and thin transparent conducting oxide fabrication using RF magnetron sputtering on flexible substrates. Organ. Mater. **5**, 59–65 (2023). https://doi.org/10.1055/s-0042-1757982
13. Singh, A., Kumar, D., Khanna, P.K., Kumar, A., Kumar, M., Kumar, M.: Anomalous behavior in ZnMgO thin films deposited by sol–gel method. Thin Solid Films. **519**(17), 5826–5830 (2011). https://doi.org/10.1016/j.tsf.2010.12.188
14. Singh, A., Kumar, D., Khanna, P.K., Kumar, M., Prasad, B.: Phase segregation limit in ZnCdO thin films deposited by sol–gel method: a study of structural, optical and electrical properties. ECS J. Solid State Sci. Technol. **2**(9), Q136 (2013). https://doi.org/10.1149/2.001309jss
15. Lghazi, Y., Bahar, J., Youbi, B., Himi, M.A., Elhaimer, C., Elouadrhiri, A., Bimaghra, I., Ouknin, M., Majidi, L.: Nucleation/growth and optical proprieties of Co-doped ZnO electrodeposited on ITO substrate. Biointerface Res. Appl. Chem. **12**(5), 6776–6787 (2021). https://doi.org/10.33263/BRIAC125.67766787
16. Varudkar, H.A., Kathwate, L.H., Awale, M.B., Lokhande, S.D., Umadevi, G., Dargad, J.S., Mote, V.D.: Structural, morphological, and gas sensing properties of Co-doped ZnO nanoparticles. J. Aust. Ceram. Soc. **58**, 793–802 (2022). https://doi.org/10.1007/s41779-022-00726-1
17. Balachandra Kumar, K., Raji, P.: Synthesis and characterization of nano zinc oxide by sol gel spin coating. Recent Res. Sci. Technol., 48-52 (2011) issn: 2076-5061, www.recent-science.com
18. Viezbicke, B.D., Patel, S., Davis, B.E., Birnie, D.P.: Evaluation of the Tauc method for optical absorption edge determination: ZnO thin films as a model system. Basic Solid State Phys. **252**(8), 1700–1710 (2015). https://doi.org/10.1002/pssb.201552007
19. Zarhri, Z., Cano, A.D., Oubram, O., Ziat, Y., Bassam, A.: Optical measurements and Burstein Moss effect in optical properties of Nb-doped $BaSnO_3$ perovskite. Micro Nanostructures. **166**, 207223 (2022). https://doi.org/10.1016/j.micrna.2022.207223
20. Singh, A., Joshi, B.C., Kumar, D., Khanna, P.K., Kumar, M.: Effect of band gap offset on ZnO (Cd, Mg)/ZnO:P multi quantum well light emitting diodes: a simulation study. Energy Environ. Focus. **3**, 185–188 (2014). https://doi.org/10.1166/eef.2014.1095
21. Singh, A., Kumar, D., Khanna, P.K., Kumar, M.: Reduction in point defects of sol–gel derived ZnO thin films with oxygen ambient. Mater. Lett. **183**, 365–368 (2016). https://doi.org/10.1016/j.matlet.2016.07.055
22. Singh, A., Saini, S., Kumar, D., Khanna, P.K., Kumar, M., Prasad, B.: Influence of doping precursor on band gap and morphologies of sol-gel deposited ZnMgO thin films. Phys. Status Solidi. **11**(9-10), 1488–1492 (2014). https://doi.org/10.1002/pssc.201300593
23. Singh, A., Vij, A., Kumar, D., Khanna, P.K., Kumar, M., Gautam, S., Chae, K.H.: Investigation of phase segregation in sol–gel derived ZnMgO thin films. Semicond. Sci. Technol. **28**(2), 025004 (2013). https://doi.org/10.1088/0268-1242/28/2/025004

DFT Study of $Rb_2AgBiCl_6$'s Optoelectronic, Thermoelectric, and Structural Stability: A Leading Perovskite for Solar Cell

Karishma Sharma[1], Hansraj Karwasara[1], Nidhi Choudhary[2], Kishor Kumar[1], Amit Soni[2(✉)], and Jagrati Sahariya[3(✉)]

[1] Department of Physics, Manipal University Jaipur, Jaipur 303007, Rajasthan, India
[2] Department of Electrical Engineering, Manipal University Jaipur, Jaipur 303007, Rajasthan, India
amit.soni@jaipur.manipal.edu
[3] Department of Physics, National Institute of Technology, Srinagar 246174, Uttarakhand, India
jagrati.sahariya@gmail.com

Abstract. This study examined the structural, electrical, optical, and thermoelectric characteristics of a cubic-phase material, $Rb_2AgBiCl_6$, with a bandgap of 1.895 eV, using density functional theory (DFT) with the Perdew-Burke-Ernzerhof (PBE) exchange correlation potential. The octahedral, Goldschmidt tolerance factor and new tolerance factor depicts structure stability of the perovskite. The electronic structure analysis reveals that the material possesses an indirect bandgap, making it suitable for applications in optoelectronic devices. Its potential for photonics applications is highlighted by optical characteristics including dielectric constant, optical conductivity, and energy loss function calculations. Additionally, the Seebeck coefficient is used to estimate the figure of merit (ZT), and its utility in thermoelectric devices is demonstrated by electrical and thermal conductivity at various temperatures. The combination of a suitable bandgap, stable cubic structure, strong optical absorption and appropriate ZT value suggests that this material could be a promising candidate optoelectronic and thermoelectric applications.

Keywords: Density Functional Theory · FP-LAPW Calculations · Perovskite · Optoelectronics · Thermoelectric

1 Introduction

As the population increasing demand of energy is also increasing in the world and most of the demand fulfills with the use of fossil fuels so we need sustainable approach to mitigate these kinds of problems. The renewable energy is the game changer in the alternate of traditional energy. The shift to renewable energy sources, which are essential for lowering reliance on fossil fuels and limiting environmental harm, is one important component of this. Among these, solar energy stands out as a promising solution due to its abundance, scalability, and low environmental impact [1–3]. Using solar thermal

S. Kumari et al. (Eds.): ICMTP 2024, SPPHY 311, pp. 169–177, 2025.
https://doi.org/10.1007/978-981-96-6795-6_21

systems or photovoltaic panels, solar power captures the energy of the sun to produce clean, sustainable electricity. Solar energy is becoming more widely available due to technological advancements and declining costs, making it a key component of the global movement toward a resilient and sustainable energy future [4, 5]. According to research by Sanjay Kumar Swami et al., sulfurized Cu_2ZnSnS_4 is better suited for use in solar cells [6]. In this series Anuj Kumar et al. suggested that the deposited films' energy band gap expanded from 1.67 to 2.85 eV as the deposition temperature escalated, according to the UV–vis-NIR spectrophotometer for Cu_2O [7].

Double halide perovskites (DHP) turned out to be a promising material for a variety of uses, particularly in the field of solar energy [8]. In the previous several years, notable advancements have been made in understanding and improving the performance of these materials. A remarkable benefit of double halide perovskites is their capacity to efficiently absorb a broad spectrum of the sun's light, which makes them ideal for solar cell applications. The incorporation of double halide compositions has been shown to improve moisture resistance and reduce phase segregation, which can lead to performance degradation over time [9–11]. Beyond solar applications, double halide perovskites have shown potential in other fields, including light-emitting devices, lasers, and photodetectors, due to their excellent optoelectronic properties [12–14].

Recently, many research groups have been studied DHPs such as McClure et al. synthesized Cs_2AgBiX_6 (X = Cl, Br) compounds and highlighted their importance in use of alternate of halide based double perovskite [15]. Additionally, Zhang et al. investigated how pressure affected $Cs_2AgBiCl_6$ and discovered that performance of this compound improves in high pressure [16]. Few months back, Guo and co-authors have been theoretically investigated mechanical and optoelectronic properties of Rb_2AgBiX_6 (X = Br, Cl, I) has demonstrated the mechanical stability and use of these materials in optoelectronic devices [17].

The structural, electrical, optical, and thermoelectric characteristics of $Rb_2AgBiCl_6$ have been computed in this work using PBE-GGA within density functional theory. This chemical demonstrates significant potential for thermoelectric applications, as revealed by our preliminary investigation of its thermoelectric properties at temperatures of 400 K and 800 K. The material has a desired electrical conductivity, optical conductivity, and Seebeck coefficient—all of which are necessary for high thermoelectric efficiency. These results imply that it is appropriate for energy conversion technologies like solid-state cooling systems and waste heat recovery.

2 Computational Details

The structural, optoelectronic, and thermoelectric characteristics of the $Rb_2AgBiCl_6$ compound were investigated using the Wien2k code [16] and the full potential augmented plane wave (FP-LAPW) method. All calculations for electronic and structural were performed by PBE-GGA functional [18]. A thick k-point mesh with 1000 points in the Brillouin zone was utilized to ensure accurate results. The maximal reciprocal lattice vector ($R_{MT} \times K_{max}$) is set to 7 and the muffin-tin radius is multiplied by the geometric mean of the MT radii (G_{MT}) of 12. For this work MT radii of various elements were taken as 2.5 for Rb, Ag, and Bi as well as 2.07 for Cl.

3 Result and Discussion

3.1 Structural Properties

As seen in Fig. 1, the compound $Rb_2AgBiCl_6$ crystallizes in a face-centered cubic structure that is a member of space group Fm-3m. The value of the lattice parameter, which is optimized using volume and energy optimization using PBE-GGA, is a = b = c = 11.083 Å. This optimized value of lattice parameter is taken for further calculation of optoelectronic and thermoelectric properties.

Fig. 1. Crystal structure of $Rb_2AgBiCl_6$ using PBE-GGA

The dimensional stability of double perovskite $A_2BB'X_6$ is largely studied by calculation of octahedral factor (μ) and Goldschmidt tolerance factor (t). Further, some researchers also invented new tolerance factor (τ) to reexamine the calculated stability as some structures were not in the proposed agreement.

Using the below formulas, we have determined the octahedral factor, Goldschmidt tolerance factor, and new tolerance factor:

$$\mu = \frac{r_B}{r_{Cl}} \tag{1}$$

$$t = \frac{r_{Rb} + r_{Cl}}{\sqrt{2}(r_B + r_{Cl})} \tag{2}$$

$$\tau = \frac{r_{Cl}}{r_B} - n_{Rb}\left[n_{Rb} - \frac{\left(\frac{r_{Rb}}{r_B}\right)}{\ln\left(\frac{r_{Rb}}{r_B}\right)}\right] \tag{3}$$

Where r belongs to Shannon's ionic radii [19], n is the oxidation state and r_B is the average of r_{Ag} and r_{Bi}.

The values of μ, t and τ are in the desired range i.e. $0.415 < \mu < 0.895$, $0.813 < t < 1.107$ and $\tau < 4.18$, included in Table 1.

Table 1. Calculated parameters for the compound $Rb_2AgBiCl_6$ through PBE-GGA

Parameter	Value
a	11.083 Å
r_B	1.09 m
μ	0.60
t	0.86
τ	4.15
E_g	1.895 eV
$\varepsilon_1(0)$	3.70
S (electrons) at 400 K	2.32 mV/K
S (holes) at 400 K	2.36 mV/K
S (electrons) at 800 K	1.16 mV/K
S (holes) at 800 K	1.18 mV/K

3.2 Electronic Properties

Utilizing density of state (DOS) and band structure (BS) spectra obtained from the PBE-GGA correlational potential within DFT, we have examined the electronic behavior of the perovskite $Rb_2AgBiCl_6$ in this section. Figure 2(a) illustrates the compound's band structure. It demonstrate that the valence band maxima (VBM) and conduction band minima (CBM) are situated at different high symmetry positions, namely at X and L, respectively suggests that the material has indirect band gap and its value is 1.895 eV which is calculated via BS. It can be used in photovoltaic applications as a lead-free substitute because of its suitable band gap and eco-friendly makeup. It is commonly recognized that the indirect nature of band gaps results in a substantial amount of incident energy being wasted while the electron travels through phonons from the valence band to the conduction band.

Figure 2(b) depicts the total and partial DOS of the compound $Rb_2AgBiCl_6$, which confirms the contribution of particular atoms and orbitals. The energy bands arise in valence band are due to major contribution of Ag-d state and minor contribution of Cl-p and Bi-s states, whereas energy bands in conduction band are due to Bi-p orbital and minor contribution of Rb-s, Ag-s and Cl-p orbitals. The material $Rb_2AgBiCl_6$ has an indirect band gap of 1.895 eV, which directly affects its electrical characteristics. Although this restricts intrinsic conductivity and direct optical transitions, it increases carrier lifetimes, which makes it ideal for photovoltaic applications. Potential enhancements, like doping or band structure engineering, could increase its optoelectronic efficiency and expand its use in sustainable energy systems.

3.3 Optical Properties

The dielectric tensor, optical conductivity, and energy loss function were used to study the material's optical response within the range of energy 0–8 eV. The dielectric constant

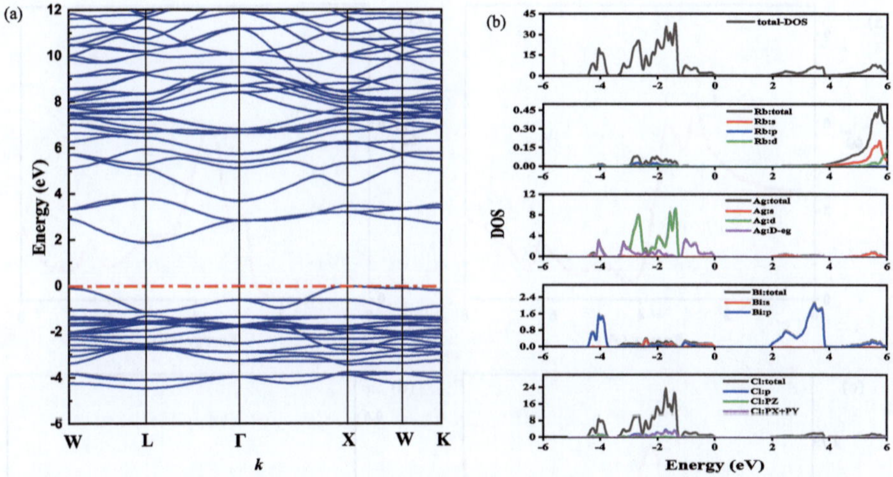

Fig. 2. (a) Band Structure (b) Density of State Spectra of $Rb_2AgBiCl_6$ using PBE-GGA

is written as following equation.

$$\varepsilon(\omega) = \varepsilon_1(\omega) + i\varepsilon_2(\omega) \tag{5}$$

where the dielectric tensor's real and imaginary parts are denoted by ε_1 and ε_2, respectively.

Similarly, optical conductivity σ (ω) and reflectivity R (ω) can be calculated with following equations

$$\sigma(\omega) = \frac{\omega}{4\pi}\varepsilon_2(\omega) \tag{6}$$

$$R(\omega) = \frac{(n-1)^2 + k^2}{(n+1)^2 + k^2} \tag{7}$$

Figure 3(a) shows the real dielectric tensor spectra and its value at 0 eV, i.e. $\varepsilon_1(0)$, is 3.70 which is obtained from the spectra. Whereas, the imaginary part of the dielectric tensor spectrum, which begins with the energy value1.895 eV (Band gap), is shown in Fig. 3(b) and attains maximum value 6.03 at 3.65 eV.

The compound's optical conductivity behaviour is displayed in Fig. 3(c) which confirms that the compound has desirable conductivity value in the visible energy range. The peaks arise in the spectra are due to charge transfer transition between various orbitals where maximum peak value 2.67 arises at 3.72 eV. Further, electron energy loss function (ELF) is another important parameter for optical characteristics, studied at different energy range which is plotted in Fig. 3(d). ELF reveals about scattering of electrons and plasma frequency which highlights the energy loss due to different interactions. The plot reveals that minimum energy loss in visible energy range.

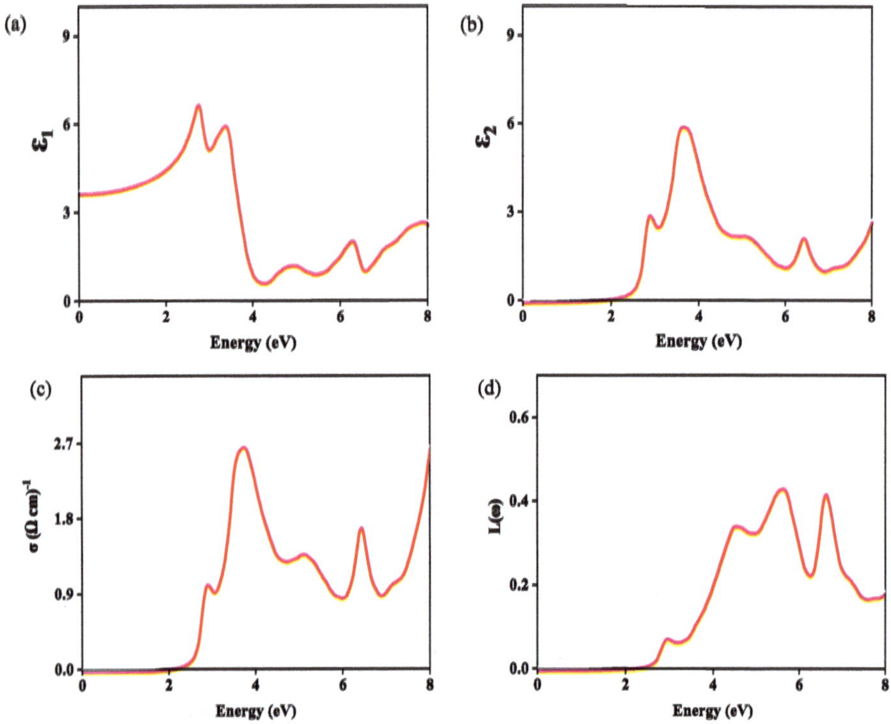

Fig. 3 (a) Real (b) Imaginary part of dielectric tensor of $Rb_2AgBiCl_6$ using PBE-GGA. (c) Optical conductivity (d) Energy loss factor of $Rb_2AgBiCl_6$ using PBE-GGA

3.4 Thermoelectric Properties

Using the Seebeck coefficient, thermal conductivity, and electric conductivity, we have calculated the figure of merit (FOM) in order to investigate the thermoelectric response of the compound $Rb_2AgBiCl_6$. These properties have been examined at two distinct temperatures, 400 K and 800 K. The Seebeck coefficient is studied to know about dominant carrier charge in the system, displayed in the Fig. 4(a). We can see from the figure that the values for electrons (holes) are 2.32 mV/K (2.36 mV/K) and 1.16 mV/K (1.18 mV/K) at 400 K and 800 K respectively. It implies that the compound's primary charge carrier is holes. The stimulation of charge carriers across the valence and conduction bands causes the Seebeck coefficient to drop with increasing temperature.

Another crucial factor in determining the material's thermoelectric behavior is its electrical conductivity (σ), which is shown in Fig. 4(b). For 400 K, the electrical conductivity of electrons (holes) is at its maximum of 1.46 × 106 (1.21 × 106) at chemical potential 2.43 eV (−2.85 eV), while for 800 K, it drops to 1.35 × 106 (1.06 × 106) at chemical potential 2.45 eV (−2.83 eV). In a similar manner, Fig. 4(c) displays the predicted electric thermal conductivity. The maximum value of thermal conductivity (k_e) for electrons (holes) is 12.81 (10.16) at chemical potential 2.45 eV (−2.76 eV) for 400 K which increases to 19.15 (15.84) at chemical potential 2.23 eV (−1.63 eV) for

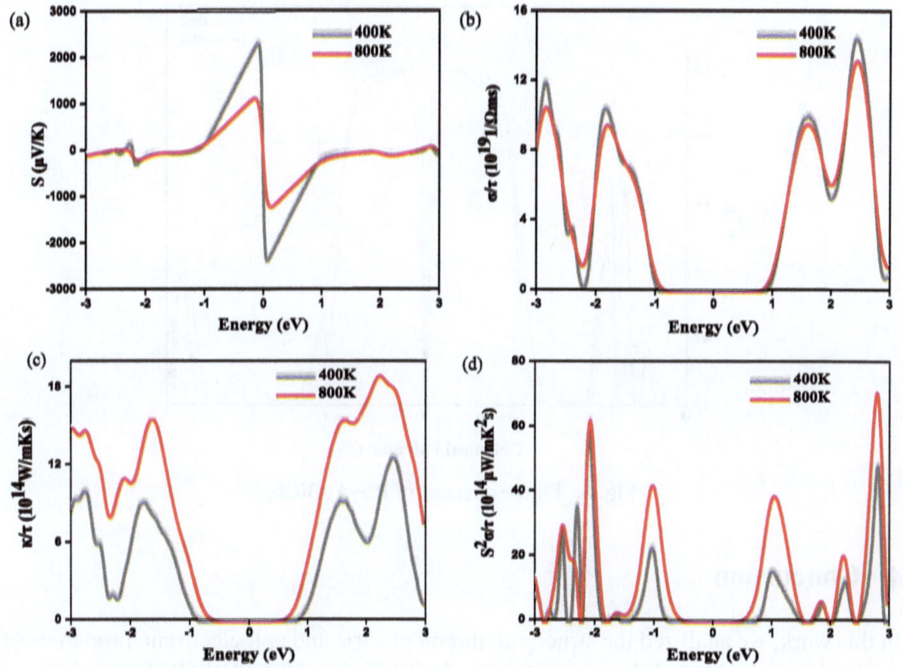

Fig. 4. (a) Seebeck coefficient (b) Electrical conductivity of $Rb_2AgBiCl_6$. (c) Thermal conductivity (b) Power factor of $Rb_2AgBiCl_6$

800 K. Both electrical and thermal conductivity decreases as temperature increase can be explained by Wiedemann-Franz law (k_e/σ is proportional to temperature).

Further, power factor (PF) is calculated with the multiplication of electrical conductivity (σ) and square of Seebeck coefficient (S), i.e. $PF = \sigma S^2$, included in Fig. 4(d). The PF increases as Seebeck coefficient increase and value of PF is higher for electrons because electron are dominant charge carrier in Seebeck coefficient. The highest peaks for electrons (holes) are 48.98 (59.95) at chemical potemtial 2.80 eV (−2.08 eV) and 71.70 (63.36) at chemical potential 2.79 eV (−2.07 eV) for 400 K and 800 K respectively (Fig. 5).

The figure of merit (ZT) can be calculated using the formula below:

$$ZT = \frac{\sigma T S^2}{K} \tag{8}$$

The temperature is T, the electric conductivity is represented by G, the thermal conductivity is K, and the Seebeck coefficient is S. With increasing temperature, the computed values of ZT approach unity.

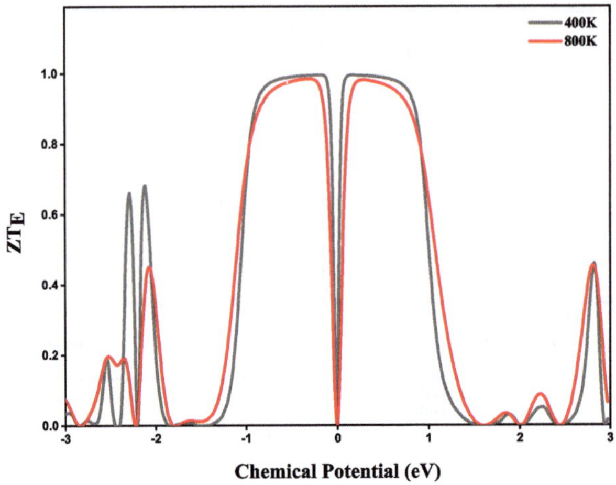

Fig. 5. Figure of merit of $Rb_2AgBiCl_6$

4 Conclusion

In this work, we analyzed the structural, thermoelectric and optoelectronic properties of the lead free double halide perovskite $Rb_2AgBiCl_6$ using the PBE-GGA correlational function in the Wien2K code. The calculated octahedral and tolerance factor using Shannon's radii shows the structure stability of the compound. The electronic band structure reveals the semiconductor behavior and indirect band gap nature of the material while optical properties gives the suitable values in visible energy range which shows its usefulness in optoelectronic devices. Further, we have also studied the thermoelectric properties of the compound and different temperatures which suggests that $Rb_2AgBiCl_6$ is the promising candidate for thermoelectric applications.

Acknowledgement. All the authors are thankful to CSIR, New Delhi for giving research grant vide 03/1502/23/EMR-II and also indebted to Manipal University Jaipur for research facility. Contribution of Prof. Peter Blaha for Wien2k software is also appreciated.

References

1. Mekhilef, S., Saidur, R., Safari, A.: A review on solar energy use in industries. Renew. Sust. Energ. Rev. **15**, 1777–1790 (2011)
2. Kannan, N., Vakeesan, D.: Solar energy for future world: a review. Renew. Sust. Energ. Rev. **62**, 1092–1105 (2016)
3. Kabir, E., Kumar, P., Kumar, S., Adelodun, A.A., Kim, K.-H.: Solar energy: potential and future prospects. Renew. Sust. Energ. Rev. **82**, 894–900 (2018)
4. Hayat, M.B., Ali, D., Monyake, K.C., Alagha, L., Ahmed, N.: Solar energy—A look into power generation, challenges, and a solar-powered future. Int. J. Energy Res. **43**, 1049–1067 (2019)

5. Rabaia, M.K.H., Abdelkareem, M.A., Sayed, E.T., Elsaid, K., Chae, K.-J., Wilberforce, T., Olabi, A.G.: Environmental impacts of solar energy systems: a review. Sci. Total Environ. **754**, 141989 (2021)
6. Swami, S.K., Chaturvedi, N., Kumar, A., Dutta, V.: Effect of sulphurisation on the activation energy of spray deposited kesterite (Cu2ZnSnS4) films. AIP Conf. Proc. **1731**, 120001 (2016)
7. Kumar, A., Dalal, D., Singh, A., Kumar, D., Narwal, V., Swami, S.K., Chaturvedi, N., Kumar, M.: Study of deposition temperature effect on spray-deposited copper oxide thin films and its schottky diodes. Phys. Scr. **99**, 075927 (2024)
8. Ji, F., Boschloo, G., Wang, F., Gao, F.: Challenges and progress in Lead-free halide double perovskite solar cells. Sol. RRL. **7**, 2201112 (2023)
9. Meyer, E., Mutukwa, D., Zingwe, N., Taziwa, R.: Lead-free halide double perovskites: a review of the structural, optical, and stability properties as well as their viability to replace Lead halide perovskites. Metals. **8**, 667 (2018)
10. Heidari Gourji, F., Velauthapillai, D.: A review on Cs-based Pb-free double halide perovskites: from theoretical and experimental studies to doping and applications. Molecules. **26**, 2010 (2021)
11. Igbari, F., Wang, Z.-K., Liao, L.-S.: Progress of Lead-free halide double perovskites. Adv. Energy Mater. **9**, 1803150 (2019)
12. Li, X., Shi, J., Chen, J., Tan, Z., Lei, H.: Lead-free halide double perovskite for high-performance photodetectors: progress and perspective. Materials. **16**, 4490 (2023)
13. Tang, H., Xu, Y., Hu, X., Hu, Q., Chen, T., Jiang, W., Wang, L., Jiang, W.: Lead-free halide double perovskite nanocrystals for light-emitting applications: strategies for boosting efficiency and stability. Adv. Sci. **8**, 2004118 (2021)
14. Lei, L., Dong, Q., Gundogdu, K., So, F.: Metal halide perovskites for laser applications. Adv. Funct. Mater. **31**, 2010144 (2021)
15. McClure, E.T., Ball, M.R., Windl, W., Woodward, P.M.: Cs2AgBiX6 (X = Br, Cl): new visible light absorbing, Lead-free halide perovskite semiconductors. Chem. Mater. **28**, 1348–1354 (2016)
16. Zhang, L., Fang, Y., Sui, L., Yan, J., Wang, K., Yuan, K., Mao, W.L., Zou, B.: Tuning emission and electron–phonon coupling in Lead-free halide double perovskite Cs2AgBiCl6 under pressure. ACS Energy Lett. **4**, 2975–2982 (2019)
17. Guo, S., Chen, C., Zhang, Z., Cai, Y., Wang, S., Gao, S., Chen, W., Abduryim, E., Dong, C., Guan, X., Liu, Y., Lu, P.: Theoretical investigations of double perovskite Rb2AgBiX6 (X = Cl, Br, I) for solar cell and photocatalysis applications. Mater. Sci. Semicond. Process. **177**, 108373 (2024)
18. Haas, P., Tran, F., Blaha, P., Schwarz, K., Laskowski, R.: Insight into the performance of GGA functionals for solid-state calculations. Phys. Rev. B. **80**, 195109 (2009)
19. Shannon, R.D.: Revised effective ionic radii and systematic studies of interatomic distances in halides and chalcogenides. Acta Cryst A. **32**, 751–767 (1976)

Effect of Valence Band Mixing on Density of States and Absorption Coefficient of CdSe/ZnSe Quantum Dots

Prateek Uniyal(✉), Saral Kumar Gupta, and C. M. S. Negi

Department of Physical Sciences, Banasthali Vidyapith, Jaipur 304022, Rajasthan, India
prateekuniyal1983@gmail.com

Abstract. This study explores the impact of quantum dot (QD) height and material concentration on the density of states (DOS) and absorption coefficient of CdSe/ZnSe QD. A multi-band envelope (k·p) approximation computes energy levels and wave functions, incorporating strain effects via deformation potentials. A reduction in QD height from 7 nm to 6 nm results in a blue shift of approximately 7.3 meV in the absorption spectrum and an increase in absorption values by 30×10^3 cm^{-1}, attributable to the quantum confinement effect. Additionally, increasing strain values from 0.72 to 2.12% significantly enhances the absorption coefficient by 9×10^4 cm^{-1} and induces a shift towards higher energy in the peak. This work highlights the critical role of QD geometry and strain engineering in fine-tuning the optical properties of QDs, offering valuable insights for optimizing QDs in next-generation optoelectronic devices.

Keywords: Quantum dots · Strain · Density of States · Dipole matrix element

1 Introduction

Several functional materials are approaching their performance limits because of intrinsic trade-offs between key physical characteristics. Engineering some novel materials that have different compositions in terms of grain size and constituents shows better performances. Growing energy demands in the modern period have spurred huge technological improvements, and the concurrent, ever-impending semiconductor industry constraints have made consistent developments in new functional materials. Materials having dimensions in nano size, especially like QDs, have attracted significant attention in the last few years due to their beneficial characteristics for both basic research and practical uses. Low-dimensional structures exhibit unique optical and electronic properties, including enhanced nonlinearity, due to the quantum confinement effect. In this effect, the continuous energy levels of bulk materials become discrete with larger separations, and the band gap between the valence and conduction bands increases [1]. It is noteworthy that due to this effect, the nonlinear optical properties of QDs, such as third-order susceptibility and nonlinear absorption coefficient, are enhanced by several orders of magnitude. This enhancement arises because the interaction between excited electron-hole pairs is stronger in confined systems [2–11]. Owing to their superior nonlinear

S. Kumari et al. (Eds.): ICMTP 2024, SPPHY 311, pp. 178–188, 2025.
https://doi.org/10.1007/978-981-96-6795-6_22

optical properties, QDs are highly suitable for applications in optical signal processing, switching, photodetection, bioimaging, displays, solar cells, multiphoton imaging, modulators, and ultra-compact nonlinear optical elements [11–15]. In semiconductors, the valence band consists of heavy-hole, light-hole, and split-off bands, which are typically closely spaced in energy in bulk materials. However, in low-dimensional semiconductor systems, the strain effect arising from semiconductor heterojunctions, particularly when using the Stranski-Krastanov (SK) fabrication method, plays a significant role. Strain develops due to the lattice mismatch between a film and its substrate, where the epitaxial deposition of a lattice-mismatched material on a substrate results in the formation of nanoscale islands. These islands retain substantial residual strain, which significantly influences the material's optoelectronic properties. Self-assembled QDs (SAQDs) formed via the SK approach exhibit structural anisotropy due to limited control over the formation process during fabrication. Strain induces mixing of the valence bands due to reduced symmetry, spin-orbit coupling, and quantum confinement of carriers. Valence-band mixing causes the wave functions to exhibit characteristics of both heavy-hole and light-hole states. Additionally, valence-band mixing increases the overlap of wave functions between the valence and conduction bands, enhancing the optical absorption coefficient at specific wavelengths. This anisotropy in shape and size alters the spectral characteristics causes inhomogeneous broadening, and may impair the devices functionality [16]. As a result, research on different compositions, anisotropy, and size-dependent optical characteristics has significantly expanded in recent years [17]. The majority of the reported work on the optical investigation of QDs has concentrated on interband transitions involving energy levels between the conduction and valence bands, as well as bound-to-bound intersubband transitions within the conduction band [18, 19]. However, the optical response arising from intervalence sub-band transitions in p-type QDs has not been extensively studied. This is a significant gap because intervalence transitions provide access to multiple valence band levels, such as light-hole and heavy-hole states, which are spread over a wider energy range compared to conduction band transitions that involve only one type of electron. Understanding these effects is crucial for optimizing QD-based optoelectronic devices, particularly in applications requiring multi-band response [20]. The design and manufacturing of photonic devices like waveguides, laser structures, nonlinear photodetectors, and optical limiters all depend on the ability to regulate linear (LO) and nonlinear optical (NLO) properties. A refractive index that varies with light intensity is essential for ultrafast all-optical signal processing. Several significant physical phenomena, including self-trapping, ultra-short laser pulse filamentation, self-focusing, and optical phase conjugation can be attributed to intensity-dependent refractive index [21–23].

2 Theory

A thorough study of the semiconductor band configuration is needed to investigate the optoelectronic features of QDs. Since, the valence band (VB) in semiconductors is made up of degenerate heavy hole (HH), light hole (LH), and split-off hole sub-bands, it cannot be adequately modelled using an easy single parabolic band model when assessing the optoelectronic behaviour closely [24, 25]. A parabolic potential is particularly suitable for modeling QDs because it provides a simple, solvable, and physically meaningful approximation of the confining potential. Moreover, strain alters the electronic structure of QDs by modifying the band structure and effective masses of the carriers. Strain can shift the conduction band edge upward or downward, depending on the nature of the material (compressive or tensile strain). In many cases, strain lifts the degeneracy of the heavy-hole and light-hole bands, leading to a split in the valence band. The energy shifts induced by strain are superimposed on quantum confinement effects, creating strain-dependent changes in the quantized energy levels. This degeneracy is reduced by strain-induced effects and reducing the size in nanoscale dimensions facilitates substantial mixing and enables finite optical transitions among these sub-bands. To address the complicated VB structure, the 4 × 4 Luttinger Hamiltonian is taken into consideration. The 4 × 4 Luttinger Hamiltonian is a reduced form derived from the full 8 × 8 Kane model. It focuses on the heavy-hole (HH) and light-hole (LH) bands and the split-off bands. This approach facilitates the study of phenomena such as spin-orbit coupling, effective mass variations, and anisotropy. This method provides both the simplest calculation process as well as maintaining the spin-orbit interaction between HH and LH. The conduction band (CB) wave function is expressed as:

$$\psi^{c}_{m_s}(R) = f^{c}(R)u^{c}_{m_s}(R) \tag{1}$$

$u^{c}_{m_s}$ represent bulk Bloch functions at the lowest of the CB for spin-degenerate states ($s = 1/2$, $m_s = \pm 1/2$), $f^c(R)$ is the envelope function for conduction band which satisfies the equation:

$$\left[-\frac{\hbar^2}{2m_e{}^*}\nabla^2 + V_{x,y} + V(z)\right] f^c(r) = E_c f^c(r) \tag{2}$$

where $V_{x,\,y}$ is the parabolic natured anisotropic confinement potential. SKs-based QDs potential can be represented by:

$$V_{x,y} = \frac{1}{2}\mho_x x^2 + \frac{1}{2}\mho_y y^2 \tag{3}$$

The shape and size anisotropy of QDs determines the parameters $\mho_x$ and $\mho_y$, which are dependent on the strength of in-plane confinement as well as anisotropy. V(z) is the confinement potential in the direction z arises due to valence band offset ΔE^v between the barrier material and the dot represented by a finite well potential with width h^{QD}.

$$V(z) = \begin{cases} 0, & -h^{QD}/2 < z < h^{QD}/2 \\ & \Delta E^v, \ \textit{otherwise} \end{cases} \tag{4}$$

The shift in energy for the CB due to strain is [25, 26]

$$E^{c,st} = 2a_{cb}\left(\frac{\acute{c}_{11} - \acute{c}_{12}}{\acute{c}_{11}}\right)\epsilon \tag{5}$$

where a_{cb} is the hydrostatic deformation potential for CB, $\acute{c}_{11}$ are the elastic stiffness constants and ϵ denotes the elastic strain stems from lattice mismatch, which can be expressed as:

$$\varepsilon = \varepsilon^{xx} = \varepsilon^{yy} = \left(\frac{a_l(QD) - a_l(br)}{a_l(QD)}\right) \tag{6}$$

Here, $a_l(QD)$ denotes the lattice constant for QD and $a_l(br)$ is lattice constant for QD barrier respectively. The change in energy due to strain for the VB is given by:

$$E^{HH,st} = \left[-2a_V\left(\frac{\acute{c}_{11} - \acute{c}_{12}}{\acute{c}_{11}}\right) + \frac{b}{2}\left(\frac{\acute{c}_{11} + 2\acute{c}_{12}}{\acute{c}_{11}}\right)\right]\epsilon \tag{7}$$

$$E^{LH,st} = \left[-2a_V\left(\frac{\acute{c}_{11} - \acute{c}_{12}}{\acute{c}_{11}}\right) - \frac{b}{2}\left(\frac{\acute{c}_{11} + 2\acute{c}_{12}}{\acute{c}_{11}}\right)\right]\epsilon \tag{8}$$

where a_V and b are hydro-static and biaxial deformation potentials for VB. In semiconductors, tensile strain typically reduces the bandgap (redshift), while compressive strain increases the bandgap (blueshift). For conduction bands, the shift is dominated by hydrostatic strain. For valence bands, strain can cause both energy shifts and splitting due to shear components of the strain tensor. Deformation potentials quantify the coupling between strain (or lattice deformation) and the energy of electronic states in a material. They are material-dependent parameters that describe how the conduction band, valence band, or specific electronic states respond to lattice distortions. By influencing the band structure, deformation potentials affect the band gap, carrier mobility, recombination rates, and optical transitions. Considering the two spins degenerate LH and HH states in the VB and applying the envelope approximation method, the wave function for VB can be written as [27].

$$\Psi^v_{m_s}(R) = \sum_{m_j} f^v(R) u^v_{m_j}(R). \tag{9}$$

Here $u^v_{m_j}(R)$ represent the Bloch functions at the top of the VB with the quantum number for j = 3/2, HH and LH will take the following values: (HH : $m_j = \pm 3/2$; LH : $m_j = \pm 1/2$) and $f^v(R)$ is the envelope function for valence band. The envelope function can be determined by considering the finite potential well along the z-direction and parabolic confinement potential along the x-y plane. The energy eigenvalues for the CB are given by:

$$E^C = \left[\left(n_x + \frac{1}{2}\right)\hbar\omega_x + \left(n_y + \frac{1}{2}\right)\hbar\omega_y + E^z_e + E^{C,st}\right] \tag{10}$$

where, $\omega^{x,y} = \sqrt{(\mho_{x,y})/(m_e)}$, m_e is the effective mass of electron and E^z_e is the energy due to z confinement. Diagonalization of the block-diagonal Luttinger Hamiltonian gives a VB structure that consists of two degenerate HH bands and two degenerate LH bands. The 4 × 4 Luttinger Hamiltonian is given by [28–30].

$$H = \begin{bmatrix} H_{HH} & S & R & 0 \\ S^\dagger & H_{LH} & 0 & -R \\ R^\dagger & 0 & H_{LH} & S \\ 0 & -R^\dagger & S^\dagger & H_{HH} \end{bmatrix} \tag{11}$$

The harmonic oscillator wave function would be the result of the in-plane parabolic confinement potential without an off-diagonal term. With these wave functions serving as the basis of our system, the matrix components can be expressed as:

$$E^{HH} = \left[E^{z,HH} + h\omega_{HH}^{x}\left(n_{xx} + \frac{1}{2}\right) + h\omega_{HH}^{y}\left(n_{yy} + \frac{1}{2}\right) + E^{HH,st}\right]\delta_{n_{xx},n'_{xx}}\delta_{n_{yy},n'_{yy}} \tag{12}$$

$$E^{LH} = \left[E^{z,LH} + h\omega_{LH}^{x}\left(n_{xx} + \frac{1}{2}\right) + h\omega_{LH}^{y}\left(n_{yy} + \frac{1}{2}\right) + E^{LH,st}\right]\delta_{n_{xx},n'_{xx}}\delta_{n_{yy},n'_{yy}} \tag{13}$$

$$\omega_{HH(LH)}^{x} = (\omega_0 + \delta\omega)\sqrt{\gamma_1 \pm \gamma_2} \tag{14}$$

$$\omega_{HH(LH)}^{y} = (\omega_0 \mp \delta\omega)\sqrt{\gamma_1 \pm \gamma_2} \tag{15}$$

Solving above equation we get $\omega_0 = \left(\omega_{HH(LH)}^{x} + \omega_{HH(LH)}^{y}\right)/(2\sqrt{\gamma_1 \pm \gamma_2})$ and $\delta\omega = \left(\omega_{HH(LH)}^{x} - \omega_{HH(LH)}^{y}\right)/(2\sqrt{\gamma_1 \pm \gamma_2})$ and anisotropy is given by $A = \Delta\omega/\omega_0.\gamma_1, \gamma_2 and \gamma_3$ denotes Luttinger parameters. Diagonalization of the Luttinger Hamiltonian gives eigenvalues for energy and also calculations for wave functions. Further dipole matrix elements can be calculated by wave functions and energy eigenvalues based on which linear and nonlinear absorption as well as refractive index can be calculated. M_{fi} is the dipole matrix element and is defined as

$$M_{fi} = \langle \psi_{n_e}^{c} | E.P | \psi_{n}^{v} \rangle = \sum_{n'} a_{n,n'} I_{n',n_e} \left\langle u_{m_s}^{c} | E.P | u_{m_j}^{v} \right\rangle \tag{16}$$

here, **E** represents the direction of polarization, **P** represents momentum, $a_{n,n'}$ are eigenvectors, I_{n',n_e} is the overlap integral of CB and VB envelope functions. Due to the self-assembly growth method, size inhomogeneity exists in QDs, which leads them to absorb and emit light at slightly different wavelengths, leading to inhomogeneous broadening. The Gaussian function accurately describes this broadening because it effectively models the natural fluctuations in QD size within a given sample. Gaussian inhomogeneous broadening (IHB) is expressed as:

$$G(E) = \frac{1}{\sqrt{2\pi\sigma}} \exp\left\{\frac{f_v(E_f) - f_v(E_i) - \hbar\omega}{(2\sigma)^2}\right\}^2 \tag{17}$$

$f_v(E_f)$ is the Fermi-Dirac function for the valence band final state and $f_v(E_i)$ is the Fermi-Dirac function for the valence band initial state. Density of state is given as:

$$DOS = \frac{1}{\sqrt{2\pi\sigma}} \exp\left\{\frac{f_v(E_f) - f_v(E_i) - \hbar\omega}{(2\sigma)^2}\right\}^2 f_v(E_f) - f_v(E_i) \tag{18}$$

Inter-sub band absorption is given as:

$$\alpha = \frac{2\pi\, \hbar\omega M_{fi}^2\, \rho\{f_v(E_f) - f_v(E_i)\}}{n_r c\hbar\varepsilon 0}\left[\frac{1}{\sqrt{2\pi\sigma}} \exp\left\{\frac{f_v(E_f) - f_v(E_i) - \hbar\omega}{(2\sigma)^2}\right\}^2\right] \tag{19}$$

where σ is carrier density in QD, ϵ_0 is the dielectric permittivity of vacuum, n_r is the index of refraction and $\hbar\omega$ is photon energy.

3 Result

MATLAB codes are developed to calculate the density of states and dipole optical moments. Linear absorption is also calculated for different QD heights at the same strain value and for different strain values at the same dot length. The computed values are used to generate and analyze the plots. The parameters used for bulk CdSe are: n_r = 2.56 [30], $\varepsilon_r = 9.7$ [31], $E_p = 21$ eV [32], $m_e = 0.13\ m_0$ [33]. Luttinger parameters $\gamma_1 = 2.1, \gamma_2 = \gamma_3 = 0.55$ [34], $\acute{c}_{11} = 66.7$ GPa, $\acute{c}_{12} = 46.3$ GPa [35], $a_{QD} = 6.055$ Å, a_s = 5.815 Å [36]. Hydrostatic deformation potential for CB and VB is calculated as [37], $a_c = 2a/3, a_V = a/3$, where $a = -3.664$ eV, $b = -0.8$ eV [26], the VBO is 0.43 eV and CBO is 0.32 eV for the CdSe/ZnSe [38]. The variation in the absorption coefficient for CdSe/ZnSe strained QDs with heights of 5 nm, 6 nm, and 7 nm at a strain level of 1.42% are illustrated in Fig. 1. Reduction in the QD height results in the elevated absorption coefficient along with a blue shift of the absorption spectrum towards higher energies due to the quantum confinement effect. A blue shift in peak energy of approximately

Table 1. Comparison of peak linear absorption coefficients for various QDs

S. n	QD material	QD size (nm)/Transition type	Peak absorption coefficient (cm^{-1})	Method	Study
1.	CdSe	2 nm (diameter)/Interband	2×10^5	Experiment (absorption spectrophotometry)	[18]
2.	ZnSe	5 nm (diameter)/Interband	2.5×10^5	Experiment (UV–Visible spectrophotometer)	[38]
3.	CdSe	2.8 nm (radius)/Intraband (within conduction band)	8×10^4	Theory (effective mass approximation)	[19]
4.	CdSe	10 nm (radius)/Interband	4.5×10^5	Theory (effective mass approximation)	[39]
5.	CdS	6 nm (radius)/Intraband (within conduction band)	1.0×10^5	Theory (effective mass approximation)	[40]
6.	CdSe	5 nm (height)/Intraband (within valence band)	1.15×10^5	Theory (*k.p.* approximation)	Present work

7.3 meV is observed as the QD height is decreased from 7 nm to 6 nm. The value of the peak absorption coefficient is changed from 85 × 10^3 cm^{-1} to 115 × 10^3 cm^{-1} as the value for QD height changed from 7 nm to 5 nm. This peak corresponds to transitions between the heavy hole (HH) ground state and the light hole (LH) state in the valence band. The observed blue shift, which increases the energy separation between the HH and LH states as the QD height is reduced, leading to optical transitions at higher energies [23]. We also compared absorption coefficient values with those reported in the literature for similar QD structures (Table 1). This table highlights the consistency of our findings with previous studies.

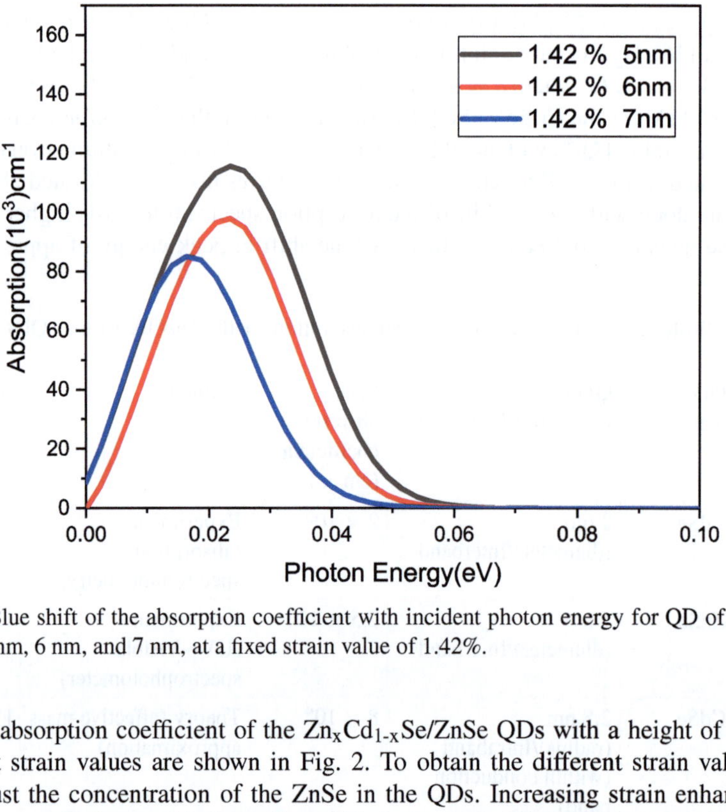

Fig. 1. Blue shift of the absorption coefficient with incident photon energy for QD of different height 5 nm, 6 nm, and 7 nm, at a fixed strain value of 1.42%.

The absorption coefficient of the $Zn_xCd_{1-x}Se/ZnSe$ QDs with a height of 9 nm at different strain values are shown in Fig. 2. To obtain the different strain values one can adjust the concentration of the ZnSe in the QDs. Increasing strain enhances the absorption coefficient and induces blue shifts in the peak. Blue shift in energy of 5 meV is observed on increasing strain value from 0.72% to 2.12%. The blue shift is caused by the increased energy separation between the HH and LH states, as they experience deformation potentials of opposite nature under strain [3, 15]. The peak values of the absorption coefficient for strain values of 0.72%, 1.42% and 2.12% are found to be 9 × 10^4 cm^{-1}, 15 × 10^4 cm^{-1} and 17 × 10^4 cm^{-1}, respectively. The mixing between the HH and LH enhances with increasing the strain, elevating the dipole matrix elements, and leading to the enhanced absorption coefficient. QDs with size-dependent emission and absorption spectra are used for tagging and imaging biological structures [41]. The blue shift in energy value can be used in third-generation photovoltaics, QDs enable absorption across a wider spectrum [42]. Increased absorption enables QD-based photodetectors to detect low-intensity light with greater efficiency, making them ideal for applications in imaging, remote sensing, and optical communication. By engineering QDs with specific absorption properties, photodetectors can be designed for targeted wavelengths, such as infrared or ultraviolet detection. QDs with increased absorption coefficients can exhibit strong nonlinear optical properties, useful in applications like frequency conversion, optical limiting, and the generation of entangled photons.

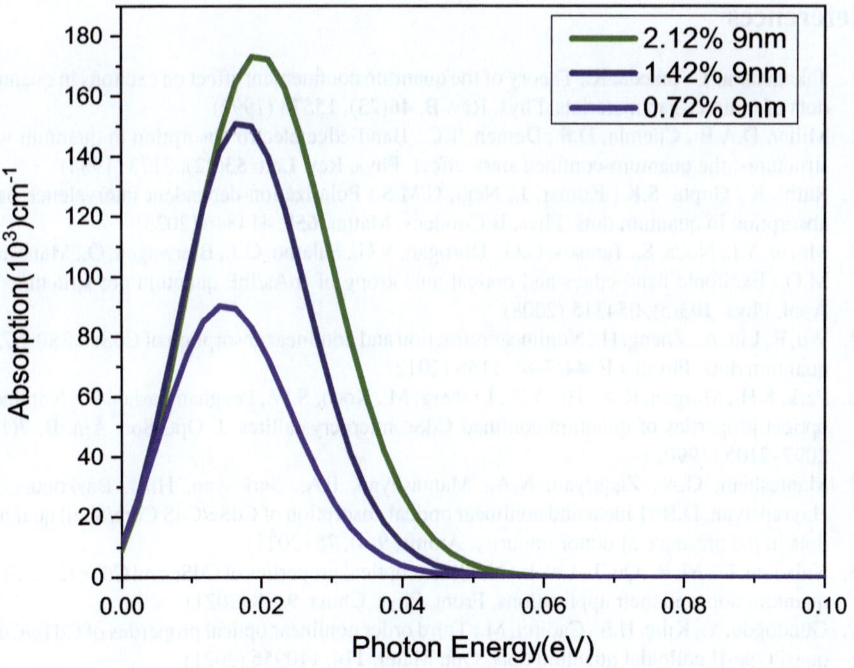

Fig. 2. Blue shift of absorption coefficient as a function of incident photon energy at different strain ($\varepsilon = 2.12$, 1.42 and 0.72%) at a fixed QD height (L_z) of 9 nm.

4 Conclusions

The linear absorption spectra of strained QDs have been theoretically investigated in this study. The analysis begins with the diagonalization of the Luttinger Hamiltonian, followed by an evaluation of optical properties through the calculation of energy eigenvalues and wave functions. The results indicate that reducing the QD height from 7 nm to 6 nm leads to a blue shift of approximately 7.3 meV in the linear absorption spectrum, accompanied by the enhancement of 30×10^3 cm^{-1} in the peak linear absorption coefficient. On the other hand, when strain was raised from 0.72 to 2.12%, we noticed a blue shift of 5 meV, along with a marked enhancement in linear absorption coefficient by 8×10^4 cm^{-1}, highlighting the influence of strain on the optical behavior of the

CdSe/ZnSe QDs. The present study contributes to calculate both linear (LO) and non-linear optical (NLO) properties of strained QDs, as well as other semiconductor-based material systems. By predicting these optical properties, this model facilitates the design and optimization of nano-optoelectronic devices, providing insights into performance improvements through tailored size, anisotropy, and material selection. This work paves the way for advancements in the understanding and application of QDs in cutting-edge optoelectronic technologies.

References

1. Takagahara, T., Takeda, K.: Theory of the quantum confinement effect on excitons in quantum dots of indirect-gap materials. Phys. Rev. B. **46**(23), 15578 (1992)
2. Miller, D.A.B., Chemla, D.S., Demen, T.C.: Band-edge electro absorption in quantum well structures: the quantum-confined stark effect. Phys. Rev. Lett. **53**(22), 2173 (1984)
3. Rathi, K., Gupta, S.K., Kumar, J., Negi, C.M.S.: Polarization-dependent intervalence band absorption in quantum dots. Phys. B Condens. Matter. **658**, 414846 (2023)
4. Mazur, Y.I., Noda, S., Tarasov, G.G., Dorogan, V.G., Salamo, G.J., Bierwagen, O., Manasreh, M.O.: Excitonic band edges and optical anisotropy of InAs/InP quantum dot structures. J. Appl. Phys. **103**(5), 054315 (2008)
5. Wu, F., Liu, A., Zheng, H.: Nonlinear refraction and nonlinear absorption of CdSe0.3S0.7/ZnS quantum dots. Physica E. **44**(7–8), 1158 (2012)
6. Park, S.H., Morgan, R.A., Hu, Y.Z., Linberg, M., Koch, S.W., Peyghambarian, N.: Nonlinear optical properties of quantum confined CdSe microcrystallites. J. Opt. Soc. Am. B. **7**(10), 2097–2105 (1990)
7. Mantashian, G.A., Zaqaryan, N.A., Mantashyan, P.A., Sarkisyan, H.A., Baskoutas, S., Hayrapetyan, D.B.: Linear and nonlinear optical absorption of CdSe/CdS Core/Shell quantum dots in the presence of donor impurity. Atoms. **9**(4), 75 (2021)
8. Kalsoom, U., Yi, R., Qu, J., Liu, L.: Nonlinear optical properties of CdSe and CdTe Core-Shell quantum dots and their applications. Front. Phys. China. **9**, 78 (2021)
9. Gündoğdu, Y., Kılıç, H.Ş., Çadırcı, M.: Third order nonlinear optical properties of CdTe/CdSe quasi type-II colloidal quantum dots. Opt. Mater. **114**, 110956 (2021)
10. Park, S.H., Kim, J.J., Kim, H.M.: Intraband relaxation time in short-wavelength II-VI CdZnSe/ZnSe quantum-well structures. J. Korean Phys. Soc. **55**(2), 671–674 (2009)
11. Rathi, K., Gupta, S.K., Kumar, J., Negi, C.M.S.: Impact of quantum dot parameters on the performance of p-type quantum dot infrared photodetectors. Superlattice. Microst. **156**, 106950 (2021)
12. Hong, H., Wu, C., Zhao, Z., Zuo, Y., Wang, J.: Giant enhancement of optical nonlinearity in two-dimensional materials by multiphoton-excitation resonance energy transfer from quantum dots. Nat. Photonics. **15**(7), 510–515 (2021)
13. Choubani, M., Maaref, H., Saidi, F.: Nonlinear optical properties of lens-shaped core/shell quantum dots coupled with a wetting layer: effects of transverse electric field, pressure, and temperature. J. Phys. Chem. Solids. **138**, 109226 (2020)
14. Ma, G., Shen, J., Zhang, Z., Hua, Z., Tang, S.H.: Ultrafast all-optical switching in one-dimensional photonic crystal with two defects. Opt. Express. **14**(2), 858–865 (2006)
15. Pacifici, D., Lezec, H.J., Atwater, H.A.: All-optical modulation by plasmonic excitation of CdSe quantum dots. Nat. Photonics. **1**(7), 402–406 (2007)
16. Negi, C.M.S., Kumar, D., Gupta, S.K., Kumar, J.: Theoretical analysis of resonant cavity p-type quantum dot infrared photodetector. IEEE J. Quantum Electron. **49**(10), 839–845 (2013)

17. Kumar, J., Kapoor, S., Gupta, S.K., Sen, P.K.: Theoretical investigation of the effect of asymmetry on optical anisotropy and electronic structure of Stranski–Krastanov quantum dots. Phys. Rev. B. Condens. Matter Mater. Phys. **74**(11), 115326 (2006)
18. Lim, S.J., Zahid, M.U., Le, P., Ma, L., Entenberg, D., Harney, A.S., Smith, A.M.: Brightness-equalized quantum dots. Nat. Commun. **6**(1), 8210 (2015)
19. Rodríguez-Magdaleno, K.A., Pérez-Álvarez, R., Ungan, F.A.T.İ.H., Martínez-Orozco, J.C.: Strain effect on the intraband absorption coefficient for spherical CdSe/CdS/ZnSe core–shell–shell quantum dots. Mater. Sci. Semicond. Process. **141**, 106400 (2022)
20. Sheng, W., Xu, S.J.: Optical characterization of structure for semiconductor quantum dots. Phys. Rev. B. Condens. Matter Mater. Phys. **77**(11), 113305 (2008)
21. Liao, Y.H., Liao, C.C., Ku, C.H., Chang, Y.C., Cheng, S.J., Jo, M., Kuroda, T., Mano, T., Abbarchi, M., Sakoda, K.: Geometrical impact on the optical polarization of droplet epitaxial quantum dots. Phys. Rev. B. Condens. Matter Mater. Phys. **86**(11), 115323 (2012)
22. Negi, C.M.S., Gupta, S.K., Kumar, D., Kumar, J.: Nonlinear optical absorption and refraction in a strained anisotropic multi-level quantum dot system. Superlattice. Microst. **60**, 462–474 (2013)
23. Boyd, R.W.: Nonlinear Optics, 3rd edn. Elsevier, Singapore (2008)
24. Kapoor, S., Kumar, J., Sen, P.K.: Magneto-optical analysis of anisotropic CdZnSe quantum dots. Physica E. **42**(9), 2380–2385 (2010)
25. Bir, G.L., Pikus, G.E.: Symmetry and Strain Induced Effect in Semiconductor. New York Willey, New York (1974)
26. Lozykowski, H.J., Shastri, V.K.: Excitonic and Raman properties of ZnSe/$Zn_{1-x}Cd_xSe$ strained-layer quantum wells. J. Appl. Phys. **69**(5), 3235 (1991)
27. Negi, C.M.S., Kumar, J.: Investigation of p-type multicolour-broadband quantum dot infrared photodetector. Superlattice. Microst. **82**, 336–348 (2015)
28. Luttinger, J.M., Kohn, W.: Motion of electrons and holes in perturbed periodic fields. Phys. Rev. **97**(4), 869–883 (1995)
29. Kumar, D., Negi, C.M.S., Gupta, S.K., Kumar, J.: Effect of shape anisotropy and size on the electronic structure of CdSe/ZnSe quantum dots. IEEE Trans. Nanotechnol. **12**(6), 925–930 (2013)
30. Bahae, M.S., Hagan, D.J., Strylandn, E.W.V.: Dispersion of bound electronic nonlinear refraction in solids. IEEE J. Quant. Electron. **27**(6), 1296–1309 (1991)
31. Krauss, T.D., Brus, L.E.: Electronic properties of single semiconductor nanocrystals: optical and electrostatic force microscopy measurements. Mater. Sci. Eng. B. **69**, 289–294 (2000)
32. Gerdova, I., Hache, A.: Third-order non-linear spectroscopy of CdSe and CdSe/ZnS core shell quantum dots. Opt. Commun. **246**(1–3), 205–212 (2005)
33. Adachi, S.: Properties of Semiconductor Alloys: Group-IV, III-V and II-VI Semiconductors, 1st. edn. John Wiley & Sons, West Sussex (2005)
34. Ekimov, A.I., Hache, F., Klein, M.C.S., Ricard, D., Flytzanis, C., Kudryavtsev, I.A., Yazeva, T.V., Rodina, A.V., Efros, A.L.: Absorption and intensity-dependent photoluminescence measurements on CdSe quantum dots: assignment of the first electronic transitions. J. Opt. Soc. Am. B. **10**(1), 100–107 (1993)
35. Martin, R.M.: Relation between elastic tensors of Wurtzite and zinc-blende structure materials. Phys. Rev. **6**(12), 4546–2553 (1972)
36. Benkabou, F., Aourag, H., Certier, M.: Atomistic study of zinc-blende CdS, CdSe, ZnS, and ZnSe from molecular dynamics. Mater. Chem. Phys. **66**(1), 10–16 (2000)
37. Cingolani, R., Prete, P., Greco, D., Giugno, P.V., Lomascolo, M., Rinaldi, R., Calcagnile, L., Vanzetti, L., Sorba, L., Franciosi, A.: Exciton spectroscopy in $Zn_{1-x}Cd_xSe$/ZnSe quantum wells. Phys. Rev. B. **51**(8), 5176–5183 (1995)
38. Toufanian, R., Zhong, X., Kays, J.C., Saeboe, A.M., Dennis, A.M.: Correlating ZnSe quantum dot absorption with particle size and concentration. Chem. Mater. **33**(18), 7527–7536 (2021)

39. Toscano-Negrette, R.G., León-González, J.C., Vinasco, J.A., Morales, A.L., Mora-Ramos, M.E., Duque, C.A.: Effect of external fields on the electronic and optical properties in ZnTe/CdSe and CdSe/ZnTe spherical quantum dots. Condens. Matter. **8**(3), 66 (2023)
40. Toscano-Negrette, R.G., León-González, J.C., Vinasco, J.A., Morales, A.L., Koc, F., Kavruk, A.E., Duque, C.A.: Optical properties in a ZnS/CdS/ZnS core/shell/shell spherical quantum dot: electric and magnetic field and donor impurity effects. Nanomaterials. **13**(3), 550 (2023)
41. Robert, E.B., Andrew, M.S., Shuming, N.: Quantum dots in biology and medicine. Physica E. **25**(1), 1–12 (2004)
42. Octavi, E.S., Joseph, M.L., Matthew, C.B.: Quantum dots for next-generation photovoltaics. Mater. Today. **15**(11), 508–515 (2012)

Performance Optimization of $CsBX_3$-Based Perovskite Solar Cells with Carbon Nanotube as a Back Electrode

Anurag Chandrakar and Ayush Khare(✉)

Department of Physics, National Institute of Technology Raipur, GE Road, Raipur 492010, India
akhare.phy@nitrr.ac.in

Abstract. Cesium-based inorganic perovskites are promising alternatives to silicon solar cells due to their stability, tunable bandgap, and efficient light absorption. Unlike organic-inorganic hybrids, $CsBX_3$ (B = Pb, Sn; X = I, Cl, Br) materials resist moisture and heat better. Within perovskite solar cells (PSCs), carbon-based electrodes such as carbon nanotubes (CNTs) and graphene, stand out for their low cost and long-term stability. In this study, SCAPS-1D software was applied to simulate and analyze the behaviour of $CsBX_3$-based PSCs incorporating CNT as the back electrode. The investigation focused on four different PSCs with Cs-based absorber layers viz. $CsPbI_3$, $CsPbBr_3$, $CsSnI_3$, and $CsSnBr_3$. The optimized device architecture consisted of FTO/SnO_2/$CsBX_3$/CNT layers with specific thicknesses. Based on our simulations, the best cell came out to be $CsSnI_3$, which had an impressive power conversion efficiency (PCE) of more than 25% and a fill factor of above 76%. This study shows that the performance of the device with CNT is similar to that of with gold electrode because the CNT layer serves multifunctional both as the hole transport layer (HTL) and back electrode. These results are expected to open new avenues for the development of incredibly stable and effective PSCs by offering researchers and industry professionals insightful theoretical information.

Keywords: Perovskite solar cells · Cs-based perovskites ($CsBX_3$) · Carbon nanotubes (CNTs) · Numerical simulation · Multifunctional layer

1 Introduction

Perovskite solar cells (PSCs), in particular, have emerged as a revolutionary technology in the solar energy sector due to their exceptional optoelectronic properties and rapid efficiency improvements. However, there are serious environmental concerns over the usage of lead (Pb) in PSCs, that limit their commercial viability. To address this issue, researchers have now diverted their focus on developing lead-free PSCs that retain high performance and cost-effectiveness while mitigating their ecological impact [1].

PSCs are anticipated to surpass the efficiency of c-Si-based SCs in the near future. But the conventional PSCs struggle with stability, leading researchers to explore Cs-based perovskites like $CsPbBr_3$ offer promising stability, $CsPbI_3$ shows better performance,

S. Kumari et al. (Eds.): ICMTP 2024, SPPHY 311, pp. 189–199, 2025.
https://doi.org/10.1007/978-981-96-6795-6_23

but their performance can be limited by factors like moisture sensitivity and lead content [2]. Lead-based perovskites, while efficient, produce hazardous by-products, such as PbI_2 upon decomposition. This environmental hazard underscores the importance of exploring alternative compositions. To mitigate these concerns, researchers are experimenting with lead-free alternatives, such as $CsSnX_3$ perovskites [3]. Researchers have experimentally shown that $CsPbBr_3$-based devices may achieve power conversion efficiencies (PCEs) of more than 11%. This can be attributed to improvements in film quality and device engineering [4].

$CsSnX_3$ materials, utilizing non-toxic Sn^{2+} ions, face challenges due to oxidation into the less stable Sn^{4+} state. However, their $CsBX_3$ structure enables intrinsic ion migration and excellent stability, making them promising for optoelectronic applications. Lead-free $CsSnX_3$ ($CsSnCl_3$, $CsSnI_3$) offers moisture stability, tunable bandgaps, and sustainability, positioning them as strong candidates for next-generation eco-friendly SCs [5]. $CsSnI_3$, with a 1.27 eV bandgap near the Shockley-Queisser limit, enables efficient visible light absorption. Its high hole mobility, low exciton binding energy, and effective charge separation enhance performance, making $CsSnX_3$ a promising lead-free perovskite with exceptional optoelectronic properties and a greener alternative to lead-based materials [6].

The longevity, manufacturing costs, and performance of SCs are all greatly impacted by the use of hole transport materials (HTMs). HTLs, such as PEDOT:PSS, Spiro-MeOTAD, PTAA, etc. offer excellent hole mobility but are costly and less stable under stress [5]. A promising alternative is HTL-free architectures using carbon-based electrodes like CNTs, which replace both HTLs and metal electrodes, reducing costs and complexity. Carbon electrodes offer high conductivity, mechanical strength, and stability, enhancing charge transport and device performance. CNT integration in HTL-free PSCs enables scalable, cost-effective solar technologies.

CNTs are cylindrical nanostructures with excellent conductivity, mechanical strength, and high surface area. In PSCs, they enhance charge transport, improve carrier extraction, and boost stability due to their high mobility and chemical durability. Their multifunctional properties make them an ideal replacement for HTLs and gold electrodes. Researchers in engineering and materials science have long studied CNTs for their exceptional charge carrier properties, stability, and conductivity, making them highly suitable for PV applications [7–9].

Moore et al. [10] highlighted the advantages of conductive carbon nanomaterials, including CNTs, which are abundant and cost-effective. CNTs are a promising HTL for PSCs, because of their work function (WF) falling between 4.8 eV and 5.1 eV and their intrinsic p-type nature at ambient temperature [11]. CNTs have also been effectively employed as p-type contacts in SCs, demonstrating their versatility. Using CNTs as HTMs offers several benefits over traditional organic or polymeric materials, which include higher stability and more affordability.

By optimizing several variables, including thickness, defect density, and acceptor concentration of perovskite absorber, this work aims to build and increase the efficiency of $CsBX_3$-based PSCs with CNT as back contact. This may demonstrate high PCE and improved device stability, enabling the development of more stable and effective PV devices for applications using renewable energy. This research uses Cs-based perovskite absorber layer (CALs) ($CsSnBr_3$, $CsSnI_3$, $CsPbBr_3$, $CsPbI_3$) to study all-inorganic PSCs, leveraging their potential for advancing third-generation PV cells.

The chosen device architecture for this numerical study is FTO/SnO_2/$CsBX_3$/CNT, as shown in Fig. 1(a), whereas Fig. 1(b) provides the bandgap values of used materials. Through simulation, we aim to determine which CAL offers the best performance with optimal band alignment for enhanced PV efficiency. To simulate the proposed device, we employed the one-dimensional solar cell capacitance simulator (SCAPS 1D) software, developed by Prof. Marc Burgelman [12].

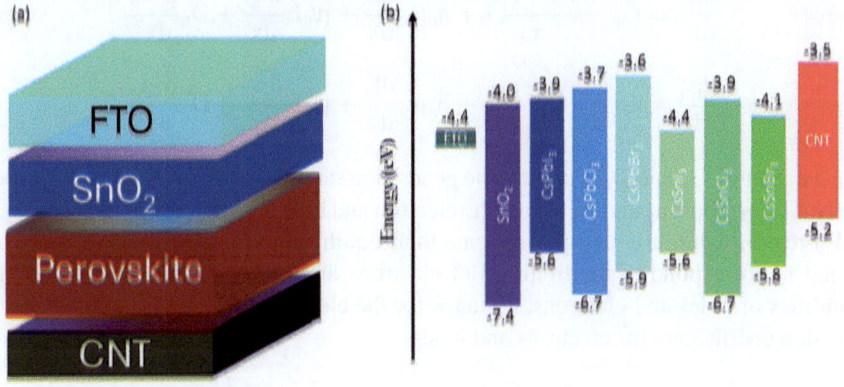

Fig. 1. (a) Device architecture and (b) bandgap values of used materials

This research is structured into three sections: "Methodology and Device Structure" covers the simulation tool and device setup, "Results and Discussion" analyzes the effects of CAL thickness, defect density, and acceptor concentration, and "Conclusion" summarizes device performance and key findings.

2 Numerical Methodology and Device Structures

Theoretical simulations provide a cost-effective and time-efficient approach to design and characterize devices. By virtually exploring various conditions and material properties, simulations help optimize device performance and reduce the need for extensive experimental trials.

2.1 Computational Simulation Details

In this study, SCAPS-1D software was employed to simulate the PV performance, including V_{OC}, short-circuit current density (J_{SC}), PCE and FF. SCAPS-1D is a widely used software among PV researchers owing to its accurate prediction of device behavior. This software solves the coupled one-dimensional continuity and Poisson's equations to simulate the transport of electrons and holes within the device [13]. Poisson's equation, which relates the charge distribution to the electrostatic potential, is given by:

$$-\frac{d^2\psi(x)}{dx^2} = \frac{q\left(p - n + N_D^+ - N_A^- + p_t(x) - n_t(x)\right)}{\varepsilon_0\varepsilon_r} \quad (1)$$

In this case, q stands for the electron charge, ψ for the electrostatic potential, ε_0 for the free space permittivity, and ε_r for the semiconductor material's relative permittivity, p/n is the hole/electron concentrations, N_D^+/N_A^- is the ionized donor/acceptor densities, n_t and p_t are the trapped electron and hole densities, and x is the position coordinate. The continuity equations, which provide about the change in n and p, are given by:

$$\frac{dn_p}{dt} = G_n - \frac{n_p - n_{p0}}{\tau_n} + n_p\mu_n\frac{dE}{dx} + \mu_n E\frac{dn_p}{dx} + D_n\frac{d^2n_p}{dx^2} \quad (2)$$

$$\frac{dp_n}{dt} = G_p - \frac{p_n - p_{n0}}{\tau_p} + p_n\mu_p\frac{dE}{dx} + \mu_p E\frac{dp_n}{dx} + D_p\frac{d^2p_n}{dx^2} \quad (3)$$

The parameters G_n and G_p represent the generation rates for electrons and holes, respectively. The symbols n_p and p_n denote the electron and hole concentrations in the p-region and n-region, while n_{p0} and p_{n0} indicate their equilibrium concentrations. The terms, τ_n and τ_p correspond to the lifetimes of electrons and holes, μ_p and μ_n represent the mobilities of holes and electrons, E stands for the electric field, and D_n and D_p are the diffusion coefficients for electrons and holes.

2.2 A Subsection Sample

In a typical PSC architecture, a perovskite absorber layer is sandwiched between an HTL and an ETL. These layers facilitate efficient charge carrier separation and transport, and prevent recombination and enable the extraction of holes and electrons to the respective electrodes. In this work, we replace the conventional HTL and metal electrode with a carbon-based CNT electrode, serving as both HTL and back electrode. This simplifies fabrication and lowers costs while ensuring efficient hole extraction and transport due to CNTs' excellent conductivity and mechanical properties. Device simulation was conducted using SCAPS-1D under AM 1.5G illumination at 300 K. The planar n-i-p PSC structure consists of a CAL between a glass substrate/FTO/SnO_2 and a CNT back electrode. The device configuration and energy levels are shown in Fig. 1(a, b). The necessary input parameters are extracted from the literature and are summarized in Table 1.

Table 1. Input parameters of different layers used for simulation.

Parameters	FTO [4]	SnO_2 [14]	$CsPbI_3$ [2]	$CsSnI_3$ [6]	$CsSnBr_3$ [15]	$CsPbBr_3$ [2]	CNT [16]
Thickness, d (nm)	200	100	700	700	700	700	400
Bandgap, Eg (eV)	3.5	3.5	1.73	1.27	1.75	2.3	1.55
Electron affinity, χ (eV)	4	4	3.95	4.3	4.17	3.4	3.64
Dielectric permittivity, ε_r	9	9	10	18	9.93	10	9
CB effective density of states, N_C (cm^{-3})	2.2×10^{18}	2.2×10^{17}	1.0×10^{19}	1.0×10^{17}	1.0×10^{18}	1.0×10^{19}	2.2×10^{18}
VB effective density of states, N_V (cm^{-3})	1.8×10^{19}	2.2×10^{17}	1.0×10^{17}	7.8×10^{18}	1.0×10^{18}	1.0×10^{18}	1.8×10^{19}
Electron/Hole thermal velocity (cm/s)	1.0×10^7	1.0×10^7	1.0×10^7	1.0×10^7	1.0×10^7	1.0×10^7	1.0×10^7
Electron/Hole mobility, ($cm^2V^{-1}s^{-1}$)	20/10	20/10	16/16	200/20	15/5.85	16/16	100/38.69
Donor density, N_D (cm^{-3})	2.0×10^{18}	1.0×10^{18}	–	–	–	–	–
Acceptor density, N_A (cm^{-3})	–	–	1.0×10^{15}	1.0×10^{17}	1.0×10^{18}	1.0×10^{15}	4.0×10^{18}

3 Result and Discussion

This section analyzes the PV performance of the designed $FTO/SnO_2/CsBX_3/CNT$ PSC architecture. To optimize the device performance, we first varied the different input parameters like acceptor concentration, thickness, and defect density of the CALs while keeping the thicknesses of FTO (200 nm), SnO_2 (100 nm), and CNT (400 nm) constant. These parameters are initially optimized using $CsSnI_3$ as the active layer. The acceptor concentration and defect density of all other layers are kept constant throughout the study as per Table 1. To maintain clarity, we focus on the optimized device configuration, i.e., $FTO/SnO_2/CsSnI_3/CNT$, in this discussion.

Figure 2(a-c) illustrates the impact of thickness, acceptor concentration, and defect density on the PV parameters. Figure 2(a) reveals that increasing the CAL's thickness from 0.2 μm to 1 μm leads to a significant rise in PCE, reaching a peak of over 25% and then saturating at around 0.6 μm. This trend can be attributed to two key factors [13]. Initially, thinner layers exhibit limited light absorption, especially in the extended wavelength range, resulting in fewer photogenerated carriers. As the thickness increases, light absorption improves, leading to increased carrier generation and consequently higher PCE [13]. However, excessively thick layers can hinder charge carrier collection due to increased recombination within the bulk of the perovskite. Furthermore, Fig. 2(a) shows that J_{SC} and V_{OC} generally increase with thickness, while the FF initially rises and then decreases.

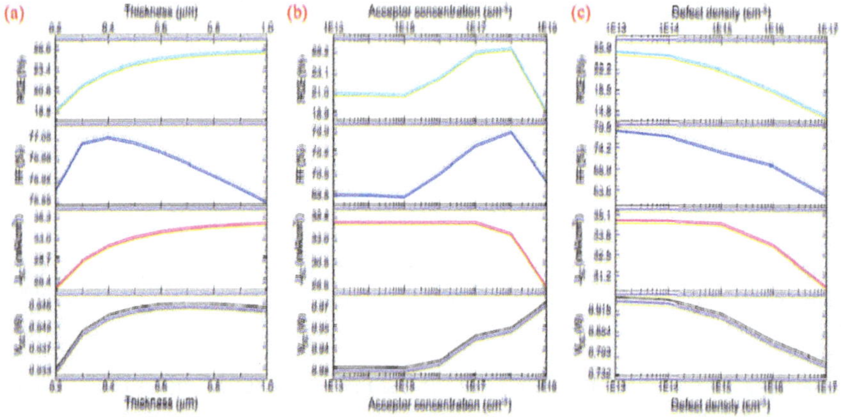

Fig. 2. Impact of varying (a) thickness, (b) acceptor concentration, and (c) defect density of absorbing layer on PV parameters

After optimizing the thickness of the absorber layer at 0.6 μm, we proceeded to optimize the acceptor concentration. Increasing the acceptor concentration initially boosts charge carrier generation, leading to higher PCE and V_{OC}. However, a further increase in acceptor concentration also increases the sheet resistance, as evident in Fig. 2(b). This results in a decline in PCE from 25.25% to 18.9% beyond the acceptor concentration of 1E18 cm^{-3}, accompanied by a decrease in J_{SC} from 34.1 mA/cm^2 to 28.6 mA/cm^2 after 1E17 cm^{-3} value of acceptor concentration. Therefore, an acceptor concentration of 1E17 cm^{-3} is deemed optimal.

As the next step, the defect density in the absorber layer, a critical parameter influencing device performance, is optimized. As we know the higher values of defect density accelerate carrier recombination [4], which reduces the number of electron-hole pairs travelling towards the electrodes, finally leading to a reduction in the PCE. In the present case also, as seen in Fig. 2(c), the PCE drops from 25.8% to 14.8% as the defect density rises from 1E13 cm^{-3} to 1E17 cm^{-3}. The value of V_{OC} also reduces from 0.945 V to 0.732 V in the same range while J_{SC} remains relatively constant at around 34.5 mA/cm^2 till 1E15 cm^{-3} defect density and then decreases to 30.6 mA/cm^2 at higher defect densities of 1E17 cm^{-3}. Although lower defect densities are desirable for higher performance; however, achieving extremely low defect densities in real devices can be challenging. Therefore, in the present work, a defect density of 1E14 cm^{-3} is chosen as a practical and suitable value.

The simulated devices are then subjected to a standard AM 1.5G solar spectrum at 300 K operating temperature. Upon irradiation, electron-hole pairs are generated in the perovskite absorber layer. These generated carriers diffuse through the perovskite layer, with electrons of the SnO_2 (ETL) and holes of the CNT (HTL and Electrode). Subsequently, the respective electrodes collect the charge carriers. Figure 3 illustrates the band alignment of different CALs in the proposed device with architecture FTO/SnO_2/CAL/CNT. The values of the optimized input parameters of these layers are already mentioned in Table 1. Figure 3 also demonstrates proper band alignment between the CNT and the perovskite layer. The varying bandgaps of different $CsBX_3$ materials result in different band alignments, which significantly impact charge carrier generation and recombination processes and, consequently, the overall device performance. The generation and recombination rates are discussed in the upcoming sections.

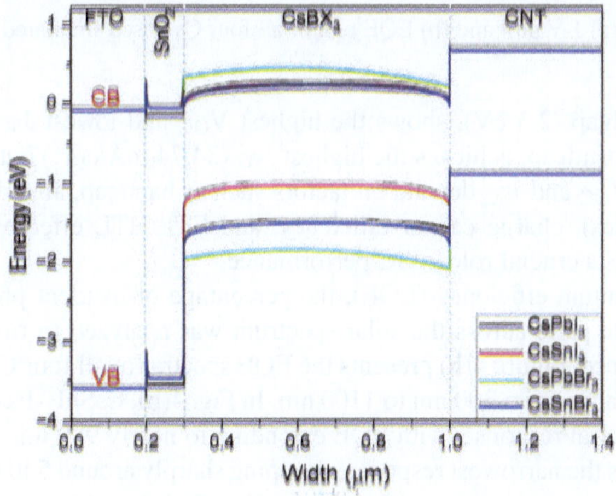

Fig. 3. Band alignment of simulated structure with different CALs

To evaluate the performance of the designed PSCs, the J-V characteristic curves, which determine crucial parameters like efficiency, FF, V_{OC}, and J_{SC} plotted and analyzed. These parameters were determined for the following four PSCs with different Cs-based perovskites: $CsPbI_3$, $CsPbBr_3$, $CsSnI_3$, and $CsSnBr_3$ (Fig. 4). As can be seen in Fig. 4(a), the $CsSnI_3$-based device exhibits the highest efficiency of 25.28%, with a J_{SC} of 34.74 mA/cm^2, V_{OC} of 0.94 V, and an FF of 76.97%. In contrast, the $CsPbBr_3$-based device demonstrates the lowest efficiency of 9.43%. The other two cells with $CsPbI_3$ and $CsSnBr_3$ as absorber layer achieved PCE of 20.4% and 15.51%, respectively. The corresponding FF values are 76.97% and 78.03%, respectively. Based on these results, it is inferred that PV parameters are significantly influenced by the bandgap of the perovskite material. It is important to mention that a wider bandgap limits the absorption of the solar spectrum, while a narrower bandgap increases thermalization losses.

The optimal SC bandgap ranges from 1.3 to 1.8 eV. The bandgap increases when transitioning from Sn to Pb and I to Br due to iodine's larger bond length, which weakens orbital overlap in $CsBX_3$ perovskites, lowering the energy difference between the valence band (VB) and conduction band (CB). Additionally, iodine's lower electronegativity (2.66) compared to bromine (2.96) further reduces the bandgap in $CsBI_3$. $CsPbBr_3$, with

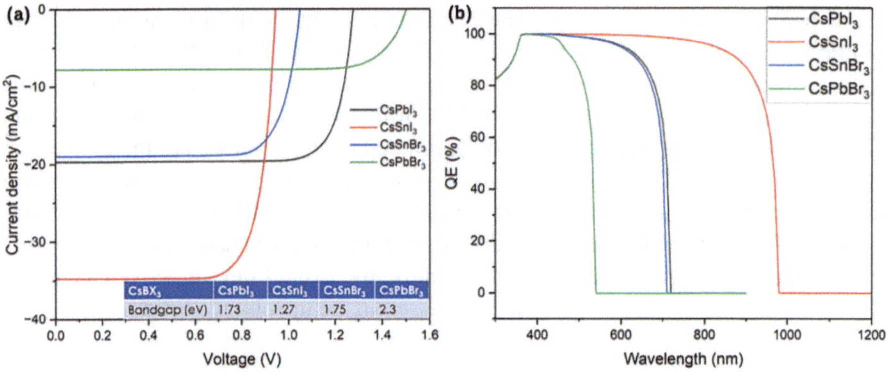

Fig. 4. (a) J-V plot and (b) EQE plot of all four Cs-based simulated devices

the highest bandgap (2.3 eV), shows the highest V_{OC} and lowest J_{SC}, while $CsSnI_3$, with the lowest bandgap, achieves the highest J_{SC} (34.74 mA/cm^2) but the lowest V_{OC} (0.94 V). Both V_{OC} and J_{SC} depend on factors such as bandgap, absorber layer quality (defects/impurities), charge carrier efficiency, and ETL/HTL effectiveness. Interface quality also plays a crucial role in SC performance.

External quantum efficiency (EQE), the percentage of incident photons converted into electron-hole pairs across the solar spectrum was analyzed to further understand device performance. Figure 4(b) presents the EQE spectra for all four Cs-based PSCs in the wavelength range from 300 nm to 1100 nm. In Fig. 4(b), $CsSnI_3$-based PSC exhibits the broadest spectral response, with EQE extending to nearly 980 nm, while $CsPbBr_3$-based PSC shows the narrowest response, dropping sharply around 540 nm. This finding is in accordance with the bandgap of each material. $CsSnI_3$, with its narrower bandgap (1.27 eV) can absorb a wider range of photons as compared to $CsPbBr_3$ with a wider bandgap.

Figure 5(a, b) illustrate the carrier generation and recombination rates within the perovskite layer for the devices. Upon illumination, photons generate excitons (bound electron-hole pairs) primarily within the perovskite layer. These excitons, possessing relatively long diffusion lengths, migrate toward the ETL/perovskite interface. At this interface, efficient exciton dissociation occurs, separating electrons and holes. Electrons and holes are then transported through the charge transport layer (CTL) towards the respective electrodes.

The results show that the peak generation rate occurs within the perovskite layer, typically between 0.3 µm and 1.0 µm, depending on the specific device parameters and material properties. SCAPS-1D determines the generation rate of electron–hole pairs. The above analysis provides valuable insights into the charge carrier dynamics within the PSC, highlighting the critical role of efficient exciton generation, separation, and transport in achieving high device performance.

The carrier generation rate varies significantly among the different CALs. The $CsSnI_3$-based PSC exhibits the highest generation rate, with a sharp peak exceeding 1E22 cm^{-3} s^{-1} at the ETL/perovskite interface, and decreases with the rising width of CAL, giving the lowest absorption rate of less than 1E21 cm^{-3} s^{-1} at perovskite/CNT

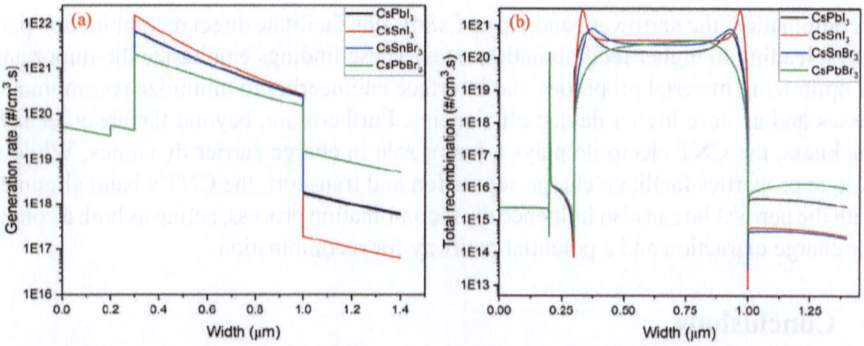

Fig. 5. (a) Generation rate and (b) recombination rate of all four simulated devices

interface. In contrast, the $CsPbBr_3$-based PSC demonstrates the lowest generation rate, with a value below 1E21 cm^{-3} s^{-1}. This variation in generation rates highlights the crucial role of the perovskite material in determining device performance. The high generation rate at the ETL/perovskite interface emphasizes CTLs' role in extracting carriers, reducing recombination, and enhancing transport efficiency.

Recombination, the process of electron-hole pair annihilation, significantly impacts the efficiency of PSCs. It is influenced by factors, such as charge carrier lifetime, density, etc. Defects within the perovskite layer, including vacancies and interstitials act as recombination centers, trapping charge carriers and reducing their availability for the current generation. High recombination rates were observed at the ETL/perovskite and perovskite/CNT interfaces for all devices, as evident from the band alignment diagram (Fig. 3). Energy barriers at these interfaces can hinder charge carrier transport, increasing the likelihood of recombination. A careful analysis reveals that the maximum recombination rates within the perovskite layer occurred within a thickness range of 0.3 μm to 1.0 μm for all the studied devices. Notably, the $CsSnI_3$-based PSC exhibited the highest recombination rate, with a pronounced peak exceeding 5E21 cm^{-3} s^{-1} at both the ETL/perovskite and perovskite/CNT interfaces. In contrast, the wider bandgap $CsPbBr_3$ (2.3 eV) shows significantly lower interfacial recombination rates. This observation can be attributed to several factors.

Maximizing device performance depends on interface band alignment, as energy barriers at ETL/perovskite and perovskite/CNT interfaces can hinder charge transport and increase recombination. In $CsSnI_3$, band alignment enables direct recombination, where electrons transition from the perovskite CB to the CNT VB, raising recombination rates. Conversely, $CsPbBr_3$'s wider bandgap creates a larger energy barrier, suppressing recombination. $CsPbBr_3$ shows no sharp peaks at interfaces, unlike others, due to CB and VB offsets. Additionally, defects in the perovskite layer act as recombination centers, trapping carriers and reducing current generation. The density and type of these defects vary across perovskite materials, significantly affecting recombination behavior.

Our analysis highlights the critical role of bandgap and its tuning, interface properties, and defect density in determining the recombination behavior within Cs-based PSCs. While wider bandgap materials like $CsPbBr_3$ generally exhibit lower interfacial recombination, the narrower bandgap of $CsSnI_3$ can facilitate direct recombination pathways, leading to higher recombination rates. These findings emphasize the importance of optimizing material properties and interface engineering to minimize recombination losses and achieve higher device efficiencies. Furthermore, beyond the absorber layer thickness, the CNT electrode plays a major role in charge carrier dynamics. While its unique properties facilitate charge separation and transport, the CNT's band alignment with the perovskite can also influence the recombination process, acting as both a conduit for charge extraction and a potential pathway for recombination.

4 Conclusions

This study comprehensively investigates the detailed numerical analysis of four Cs-based PSCs with the architecture of FTO (200 nm)/SnO_2 (100 nm)/$CsBX_3$ (700 nm)/CNT (400 nm), varying the acceptor concentration, thickness, and defect density of the CALs ($CsBX_3$), namely $CsSnI_3$, $CsSnBr_3$, $CsPbI_3$, and $CsPbBr_3$. By utilizing CALs with different bandgaps and WF, the study aims to determine the proper band alignment of CALs with FTO/SnO_2 and CNT. The results, after optimization, suggest that $CsSnI_3$ provides the best performance when paired with CNT as a replacement for the costly HTL and gold electrode, yielding a PCE of 25.28%, with a V_{OC} of 0.98 V, J_{SC} of 34.74 mA/cm^2, and an FF of 76.97%. The study underlines the importance of precise acceptor concentration, thickness, and defect density. Thin absorber layers absorb fewer photons, while thicker ones increase sheet resistance, reducing JSC and efficiency. Lower defect density improves performance, while higher acceptor concentration boosts absorption but risks recombination. These findings emphasize the need for optimal trade-offs between these parameters to achieve high efficiency. Also, they demonstrate the potential for efficient, stable, and cost-effective PSCs using multifunctional CNTs.

References

1. Singh, N., Agarwal, A., Agarwal, M.: Numerical investigation of the Pb-free titanium-based double-perovskite solar cell. J. Nanopart. Res. **26**, 1–17 (2024). https://doi.org/10.1007/s11051-024-06038-8
2. Shrivastav, N., Madan, J., Pandey, R.: Predicting photovoltaic efficiency in Cs- based perovskite solar cells: a comprehensive study integrating SCAPS simulation and machine learning models. Solid State Commun. **380**, 115437 (2024). https://doi.org/10.1016/j.ssc.2024.115437
3. Khalid Hossain, M., Pervez, M.F., Jalal Uddin, M., Tayyaba, S., Mia, M.N.H., Bashar, M.S., Jewel, M.K.H., Haque, M.A.S., Hakim, M.A., Khan, M.A.: Influence of natural dye adsorption on the structural, morphological and optical properties of TiO2 based photoanode of dye-sensitized solar cell. Mater. Sci. Pol. **36**, 93–101 (2018). https://doi.org/10.1515/msp-2017-0090
4. Rani, S., Kuar, D., Jeyachandran, V., Ghosh, D.S.: Opto-electrical modeling of CsPbBr3 based all-inorganic perovskite solar cell. Opt. Mater. (Amst). **131**, 112612 (2022). https://doi.org/10.1016/j.optmat.2022.112612
5. Saidani, O., Abderrahim, Y., Zitouni, M., Shankar, G., Rafik, S., Mohammad, Z.M.R.: Numerical study of high - performance lead - free CsSnCl 3 - based perovskite solar cells. (2024)

6. Seyed-Talebi, S.M., Mahmoudi, M., Lee, C.H.: A comprehensive study of CsSnI3-based perovskite solar cells with different hole transporting layers and back contacts. Micromachines. **14**, 1562 (2023). https://doi.org/10.3390/mi14081562
7. Advanced Science - 2021 - Yoon - Foldable Perovskite Solar Cells Using Carbon Nanotube-Embedded Ultrathin Polyimide.pdf
8. Jeon, I., Chiba, T., Delacou, C., Guo, Y., Kaskela, A., Reynaud, O., Kauppinen, E.I., Maruyama, S., Matsuo, Y.: Single-walled carbon nanotube film as electrode in indium-free planar heterojunction perovskite solar cells: investigation of electron-blocking layers and dopants. Nano Lett. **15**, 6665–6671 (2015). https://doi.org/10.1021/acs.nanolett.5b02490
9. Li, Z., Kulkarni, S.A., Boix, P.P., Shi, E., Cao, A., Fu, K., Sudip, K., Zhang, J., Xiong, Q., Wong, L.H., Mathews, N., Mhaisalkar, S.G.: Laminated carbon nanotube networks for metal electrode-free efficient perovskite solar cells. ACS Nano. **8**(7), 6797–6804 (2014)
10. Moore, K., Wei, W.: Applications of carbon nanomaterials in perovskite solar cells for solar energy conversion. Nano Mater. Sci. **3**, 276–290 (2021). https://doi.org/10.1016/j.nanoms.2021.03.005
11. Habisreutinger, S.N., Leijtens, T., Eperon, G.E., Stranks, S.D., Nicholas, R.J., Snaith, H.J.: Carbon nanotube/polymer composites as a highly stable hole collection layer in perovskite solar cells. Nano Lett. **14**, 5561–5568 (2014). https://doi.org/10.1021/nl501982b
12. Burgelman, M., Decock, K., Niemegeers, A., Verschraegen, J., Degrave, S.: SCAPS manual most recent. SCAPS Man. **155** (2021)
13. Ravidas, B.K., Roy, M.K., Samajdar, D.P.: Investigation of photovoltaic performance of lead-free CsSnI3-based perovskite solar cell with different hole transport layers: first principle calculations and SCAPS-1D analysis. Sol. Energy. **249**, 163–173 (2023). https://doi.org/10.1016/j.solener.2022.11.025
14. Yadav, N., Roy, P., Khare, A.: An investigation of defects, band-offset, and Schottky barrier height for boosting the performance of formamidinium mixed cation mixed halide based perovskite solar cell: theoretical approach. Mater. Sci. Eng. B. **293**, 116458 (2023). https://doi.org/10.1016/j.mseb.2023.116458
15. Nasrin, T., Mottakin, M., Selvanathan, V., Hossain, M.I., Shahiduzzaman, M., Islam, M.A., Ahamed, M.S., Alharbi, H.F., Akhtaruzzaman, M.: Performance optimization and defect studies of Pb-free CsSnBr3-based perovskite solar cells. Mater. Today Commun. **37**, 107000 (2023). https://doi.org/10.1016/j.mtcomm.2023.107000
16. Mohammed, M.K.A., Al-Mousoi, A.K., Singh, S., Kumar, A., Hossain, M.K., Salih, S.Q., Sasikumar, P., Pandey, R., Yadav, A.A., Yaseen, Z.M.: Improving the performance of perovskite solar cells with carbon nanotubes as a hole transport layer. Opt. Mater. (Amst). **138**, 113702 (2023). https://doi.org/10.1016/j.optmat.2023.113702

Temperature Dependent Structural and Optical Properties of $SnSe_2$ Nanosheets

Yogita Gendre(✉), K. S. Ojha(✉), and Chinmaya Mahapatra

Department of Physics, National Institute of Technology Raipur, Chhattisgarh 492010, India
{ygendre.phd2023.phy,ksojha.phy}@nitrr.ac.in

Abstract. In this paper, EDTA@$SnSe_2$ nanosheets have been prepared via cost effective and simple solvothermal method. The synthesized sample was heated up to 300 °C to examine the influence of annealing upon material's structural and optical characteristics. The structural analysis of sample annealed at 300 °C disclosed the formation of $SnSe_2$ nanosheets with hexagonal crystal structure. The field emission scanning electron microscope (FESEM) micrographs confirm the formation of plate-like morphology. Energy dispersive x-ray (EDX) analysis confirmed the formation of $SnSe_2$ and SnO_2 which was also verified by the X-ray diffraction (XRD) pattern. UV-Vis and photoluminescence spectroscopic techniques were used to investigate the absorption and emission characteristics of the samples. The energy bandgap observed for the sample annealed at 300 °C was found to be 2.75 eV whereas the PL spectra represented two emission bands lying at 380 nm and 472 nm. The vibrational properties of the $SnSe_2$ sample were examined by employing FTIR spectroscopic technique.

Keywords: Nanosheets · Solvothermal method · Capping agent · Annealing temperature · Light Emitting Diodes and Photodetectors

1 Introduction

Two-dimensional materials, specifically transition metal dichalcogenides (TMDs) composed of metals like molybdenum (Mo), tungsten (W), germanium (Ge) and tin (Sn) and chalcogen elements (S, Se and Te), have gained considerable research interest owing to their exceptional characteristics pertaining to both electrical and optical phenomena [1]. Tin diselenide ($SnSe_2$) is one of the IV–VI group metal dichalcogenides that has garnered significant attention because of their tunable band gap and found numerous applications in photovoltaics, infrared optoelectronic devices, energy storage devices, photocatalytic, gas sensors and materials used in lithium-ion battery electrodes [2–5]. Tin diselenide ($SnSe_2$), an n-type semiconductor, has a layered crystal structure that closely resembles the structure of cadmium iodide (CdI_2). In this structure, each Sn atom layer is embedded between two layers of Se atoms arranged in a hexagonal close-packed formation [6, 7]. The atomic configuration of $SnSe_2$ is typically characterized as a honeycomb-like dual layer made up of Se^{2-} and Sn^{4+} ions, with robust inter-layer bonding and exhibiting minimal van der Waals forces between layers, a phenomenon referred to as intercalation

The original version of the chapter has been revised. The chapter author Yogita Gendreis affiliation has been updated. A correction to this chapter can be found at
https://doi.org/10.1007/978-981-96-6795-6_50

S. Kumari et al. (Eds.): ICMTP 2024, SPPHY 311, pp. 200–213, 2025.
https://doi.org/10.1007/978-981-96-6795-6_24

[5]. Consequently $SnSe_2$ nanoparticles exhibit different morphological structures like sphere [2], rod [6], flower [8], flakes [9] etc. depends on controlled growth conditions during synthesis which is comparable to the various morphologies of tin selenide (SnSe) nanoparticles [7]. Tin diselenide ($SnSe_2$) exhibits a tunable bandgap, size-dependent optical characteristics. This material also displays anisotropic electrical conductivity, making it highly attractive for researchers in diverse optoelectronic applications [10].

Cheng et al. [3] produced $SnSe_2$ having various structures by modifying the reductant in the solvothermal synthesis process and reported that $SnSe_2$ in the form of interconnected nanosheets exhibit superior electrochemical properties as compared to the other structures. Liu et al. [9] synthesized single-phase $SnSe_2$ nanoflakes produced via hydrothermal method and reported that nanoflakes exhibit a regular shape and grow along the (0001) crystal plane. Ma et al. [6] used an organic solution-based method to produce two distinct nanostructures of $SnSe_2$—nanorods and nanoplates—simultaneously within a liquid paraffin medium and developed a plausible chemical reaction pathway to understand how these $SnSe_2$ nanostructures were formed. Liu et al. [11] successfully synthesized one-dimensional (1D) selenide materials, including Bi_3Se_4, $Cu_{2-x}Se$, $SnSe_2$, and MSe (where M represents divalent metal ions like Cadmium, Nickel, and Lead) using a solvothermal method and observed that the hydrazine hydrate significantly influenced the formation of these selenides due to its alkaline nature. They also investigated the absorption and emission properties of $SnSe_2$. Fang et al. [12] prepared $SnSe_2$ nanosheets using a simple method with benzyl alcohol and PVP and demonstrated their possible applications in photodetectors and photovoltaics via their unique band structures and photoelectrochemical properties. Kumar et al. [13] observed the improved crystallinity and reduced strain by increasing the substrate temperature during deposition of $SnSe_2$-SnSe alloy film and reported that the annealing temperatures modified the microstructure, leading to variations in particle size and composition. They also observed a strong optical absorption and emission which suggest potential applications in photonic devices and solar cells. Zhou et al. [14] successfully synthesized high-quality ultrathin $SnSe_2$ flakes using modified CVD method and demonstrated their use in photodetectors. Pawbake et al. [15] successfully synthesized ultrathin nanosheets of tin di-selenide ($SnSe_2$) using a solvothermal technique and investigated their humidity as well as photo sensing properties along with vibrational properties using temperature-dependent Raman spectroscopy. Das et al. [16] utilized a green synthesis approach to prepare non-toxic SnSe nanosheets using bovine serum albumin (BSA) as a stabilizer and suggested their potential for biomedical applications. Ren et al. [17] evaluated the photoelectrochemical (PEC) performance of SnSe nanosheets synthesized via a straight forward hydrothermal method and observed enhanced photo response activity by applying a positive potential. Tołłoczko et al. [18] studied the defects in SnX_2 (X = S, Se) crystals which strongly influence their optical behaviour, enabling bandgap tuning and affect the key properties like photocatalysis and thermoelectricity.

In this study, a straight forward and effective solvothermal route is used to synthesize $SnSe_2$ nanosheets controlled with ethylenediaminetetraacetic acid (EDTA). Traditional methods like coprecipitation method [19] and etc. for synthesizing $SnSe_2$ nanoparticles often face challenges in controlling their size, morphology, and surface properties [10, 20] but solvothermal technique offers advantages such as controlled reaction conditions, uniform particle size distribution, high temperature, high pressure, non-aqueous environment and the ability to tune the crystallinity of the resulting nanoparticles [20]. Moreover, the successful synthesis of $SnSe_2$ nanoparticles necessitates the use of suitable capping agents to stabilize the nanostructures and prevent aggregation [21]. One of the major challenges in working with $SnSe_2$ nanoparticles is their inherent tendency to agglomerate and oxidize, which can adversely affect their properties and performance [22]. To overcome these challenges, capping agents are commonly employed during the synthesis process to stabilize the nanoparticles and control their size and morphology [23]. The utilization of EDTA as a chelating agent behaves like a capping agent is particularly advantageous due to its strong binding affinity to metal ions, which helps to maintain a controlled particle size and enhances their stability as well as protection against oxidation. EDTA is a widely used chemical compound with remarkable chelating properties. Its structure with chemical formula $C_{10}H_{16}N_2O_8$ contains four carboxylic acid groups and two amine groups organized around a central ethylenediamine core. This structure enables EDTA to form stable complexes, known as chelates, with tin metal ions [24]. The crystal structure, particle size, band gap, high carrier mobility, surface area and morphology of $SnSe_2$ nanosheets significantly contribute in photodegradation, thermoelectric, imaging and medical applications [25]. This study investigates how different annealing temperatures influence the crystal structure and optical behaviour of 2D $SnSe_2$ nanoparticles and provides a scalable production route for $SnSe_2$ nanosheets [17]. Bandgap engineering has been a crucial aspect of optimizing optoelectronic devices. For instance, Singh et al. [26] demonstrated the influence of bandgap offsets on the performance of ZnO-based multi-quantum well light-emitting diodes, emphasizing the critical role of compositional adjustments in achieving high efficiency. By highlighting these connections, the work that has done on $SnSe_2$ underscores the significance of bandgap engineering in materials science while offering thermal annealing as an alternative method for achieving tunability, thus expanding the range of design strategies for advanced optoelectronic applications.

2 Experimental Approach

2.1 Required Reagents

Tin chloride dihydrate ($SnCl_2 \cdot 2H_2O$, 99.99%, Qualigens) and selenium metal powder (Se, 99.99%, Oxford) were employed as the respective tin and selenium sources. Ethylenediaminetetraacetic acid (EDTA, $C_{10}H_{16}N_2O_8$, 99.99%, Oxford) used as a surfactant. Ethylene glycol ($C_2H_6O_2$, 99.99%, Oxford) and deionized water were used as solvents.

2.2 Synthesis

Initially, Stannous chloride dihydrate ($SnCl_2 \cdot 2H_2O$) was solubilized in a 15 mL distilled water, while selenium fine particles (selenium powder) were dissolved separately in 15 mL of ethylene glycol, ensuring individual dissolution of each component. Following this, the solutions were continuously stirred using a magnetic stirrer for 40 min, maintaining a temperature within the range of 60–70 °C. Simultaneously, 10 mL of deionized water dissolved 0.01 M EDTA. The solution underwent high-speed stirring vigorously at 750 rpm for 20 min interval, while ensuring the temperature remained unchanged. The solution containing EDTA was added into transparent solution of stannous chloride during continuous stirring. After that this solution was mixed dropwise into the prepared black selenium solution, under stirring. Afterward, the prepared solution was cautiously transferred into a stainless-steel autoclave equipped with a Teflon liner for further processing. Subsequently, the autoclave was securely sealed and subsequently transferred to a digital oven and exposed to heat treatment at a consistent temperature of 180 °C for 20 h. Allow the autoclave to get cold gradually to surrounding temperature after the reaction is complete. Subsequently, isolate the precipitate containing the synthesized $SnSe_2$ nanostructures by employing centrifugation and filtration techniques. The collected precipitate is subjected to three consecutive washing cycles using deionized water. Subsequently, the precipitate is completely dried in a vacuum oven thermally treated at a consistent temperature of 130 °C for a period of 2 h. The prepared sample is safely enclosed in an airtight container to ensure its properties remain unchanged until further analysis can be conducted. A schematic diagram illustrating the experimental technique is presented in Fig. 1. The prepared sample was then heated from 100 to 300 °C to study their structural and optical properties with different annealing temperature.

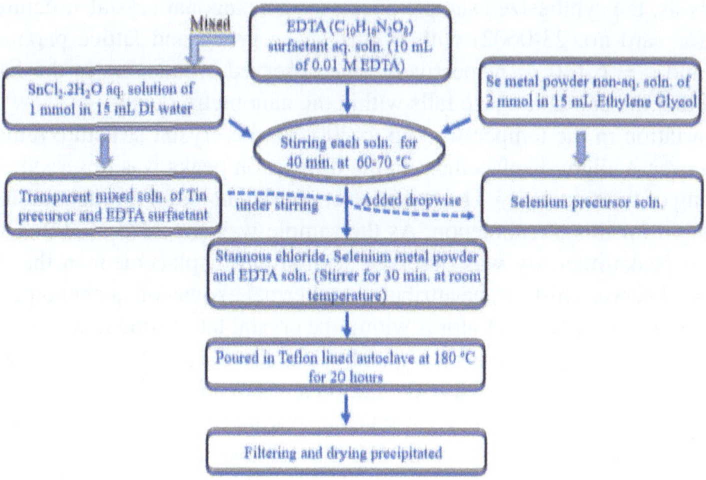

Fig. 1. EDTA@$SnSe_2$ Microstructure Preparation by Solvothermal Method.

2.3 Characterization Techniques

Structural characterization of the synthesized sample was carried out using powder X-ray diffraction (XRD) analysis with Cu-K_α radiation ($\lambda = 1.54$ Å) on a D8 Advance A25 diffractometer (BRUKER). Morphological investigations were conducted using a combination of imaging techniques, including Nova Nano-High resolution scanning electron microscopy (SEM 450 FEI, USA), Field Emission Scanning electron microscopy (SEM 500 KMAT, GEMINI), and High-resolution transmission electron microscopy (JEOL & JEM-F200 (CF-HR)). The elemental composition was examined using Energy-Dispersive X-ray Spectroscopy integrated with a High-resolution scanning electron microscope (HRSEM). Optical properties were assessed using a UV-Vis spectrophotometer (UV-1800, SHIMADZU) and a Photoluminescence spectrofluorometer (RF-5301PC, SHIMADZU). To conclude, Fourier transform infrared (FTIR) spectra were measured using a BRUKER FTIR spectrometer throughout the spectral region between 400 and 4000 cm^{-1}.

3 Results and Discussion

3.1 Structural Analysis

Figure 2 illustrates the X-ray diffraction (XRD) patterns of EDTA@$SnSe_2$ samples at various temperature 100 °C, 200 °C, and 300 °C including room temperature. The diffraction peaks detected in the XRD patterns were indexed as (100), (001), (101), (201) and (011) planes of EDTA@$SnSe_2$, corresponding to diffraction angles of 14.45°, 27.31°, 30.73°, 40.14°, and 47.67° respectively. The diffraction peaks found at 26.99° and 29.07° were ascribed as (011) and (200) crystallographic planes of SnO_2 [13]. According to XRD analysis, the synthesized sample possesses a hexagonal crystal structure (JCPDS standard for card no. 23-0602) with P-3m1 space group and lattice parameters a = 3.811 Å, and c = 6.140 Å, respectively. The observed broadening in the XRD peaks suggested that the crystallite size falls within the nanometre range.

The variation in the temperature up to 200 °C, the crystal structure remains unaltered, however, a slight displacement in the diffraction peaks was observed along with a weakening of their intensity. The presence of consistent strain throughout the material is responsible for this phenomenon. As the sample was annealed at 300 °C, a further decrease in peaks intensity was observed with slight displacement in the diffraction peaks. This observed shift can be attributed to thermal expansion, a phenomenon where the increased kinetic energy of atoms within the crystal lattice due to elevated temperatures causes the lattice parameters to expand, when a material is heated, its atoms vibrate more vigorously. This increased atomic motion leads to an expansion of the interatomic distances, effectively increasing the interplanar spacing (d) [27, 28]. The annealing temperature was kept within the range of 100-300 °C and not raised above 300 °C (i.e. ≤300 °C) to avoid a potential reduction in the crystallinity of $SnSe_2$. This reduction is likely due to the loss of selenium when it achieved 325 °C, a highly volatile element, as previously reported by D. et al. [29].

This peak broadening can also be attributed to factors such as strain, crystallographic imperfections, or distortions within the crystal structure, or potential limitations in the experimental setup [30]. The Debye-Scherrer equation, incorporating a shape factor (K) of 0.9, was utilized to determine average crystallite size of annealed samples. Williamson-Hall analysis was employed to study the effect of strain and other factors

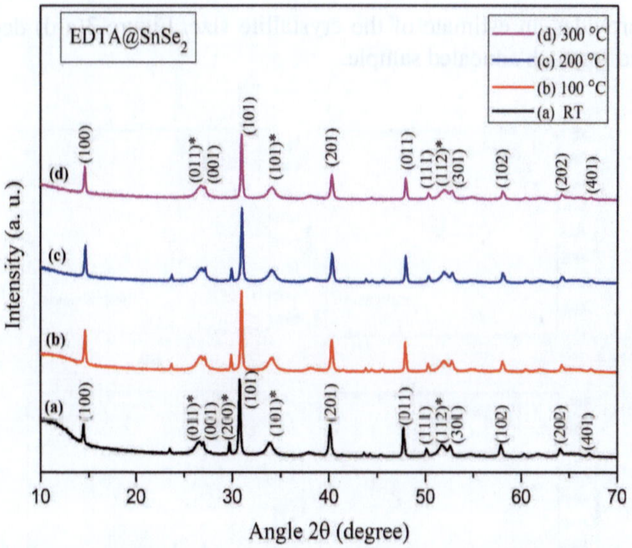

Fig. 2. XRD patterns of EDTA@$SnSe_2$ nanocrystals annealed at different temperatures: (a) RT (b) 100 °C (c) 200 °C (d) 300 °C

that broaden the XRD peak widths [30, 31].

$$\beta\cos\theta = \frac{K\lambda}{D} + 4\varepsilon\sin\theta \tag{1}$$

Here, D symbolizes the average dimension of the crystallites, λ represents wavelength of X-rays, θ denotes diffraction angle, and β signifies full width at half maximum of the XRD peak observed at angle θ, and ε characterizes the microstrain within the material. The average crystallite size and microstrain values for each annealed sample, calculated using the equation described above, are listed in Table 1:

Table 1. Analysis of crystallite size and micro strain in the prepared sample at varying annealing temperatures

Prepared sample at different annealing temperatures (°C)	Intercept	Slope	Crystalline size D using Scherrer Formula (in nm)	Microstrain $(\varepsilon)*10^{-3}$
31.5	0.00145	0.0025	25 ± 0.85	0.509
100	0.00184	0.00239	29 ± 0.59	0.559
200	0.00191	0.00229	28 ± 0.95	0.562
300	0.00204	0.00208	30 ± 0.75	0.502

A Williamson-Hall (W-H) plot, 4sin(θ) Vs. βcos(θ), exhibits a linear relationship. The slope of this line directly reflects the micro strain within the material, while the y-intercept provides an estimate of the crystallite size. Figure 3(a-d) depict the W-H plots obtained for each annealed sample.

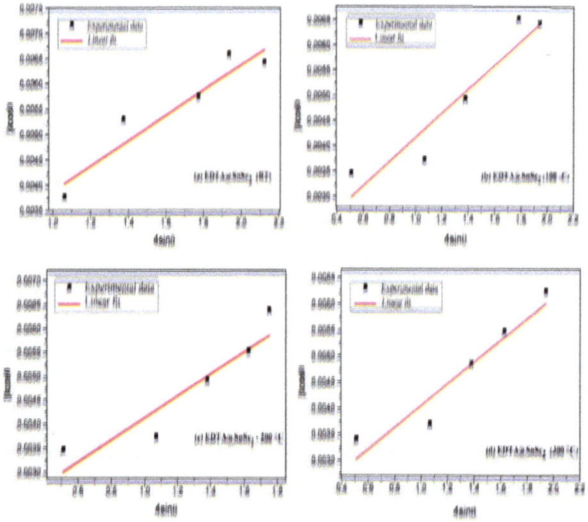

Fig. 3. Williamson-Hall plots for EDTA@$SnSe_2$ nanocrystals annealed at different temperature: (a) RT (b) 100 °C (c) 200 °C (d) 300 °C

3.2 Morphological Studies

Figure 4(a) revealed that the SEM micrograph of EDTA@$SnSe_2$ nanoparticles exhibit a predominantly plate-like morphology with some degree of agglomeration. It is interesting to note that the particle sizes measured by SEM micrographs appear to be exceeding the crystallite sizes calculated with the Debye-Scherrer formula. This discrepancy suggests that each particle observed in the SEM image is composed of multiple smaller crystallites, the results point towards a polycrystalline nature of the particles, as corroborated by XRD analysis [10, 15]. Figure 4(b) presents the results of transmission electron microscopy (TEM) analysis on $SnSe_2$ nanostructures. This clearly depicts the plate-like morphology of nanostructures, corroborating the findings of XRD analysis and aligning with the previously reported observations by Huang et al. [32].

3.3 EDX Analysis

The elemental composition of EDTA@$SnSe_2$ was investigated by means of Energy-dispersive X-ray spectroscopy (EDX), which are illustrated in Fig. 5. The atomic percentages of the constituent elements within the synthesized sample were determined and are tabulated below. The existence of carbon and oxygen was evident in the EDX spectrum, suggesting the possibility of impurities in prepared materials. Furthermore, the EDX data suggests the possible formation of a minor amount of SnO_2 alongside the

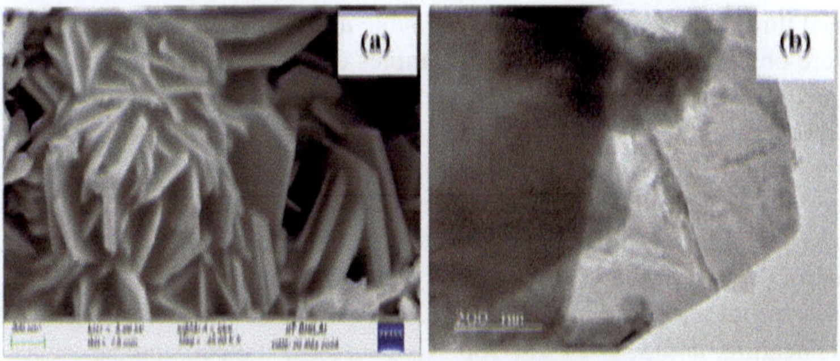

Fig. 4. (a) FESEM and (b) HRTEM micrograph of EDTA@$SnSe_2$ nanoparticles

primary EDTA@$SnSe_2$ phase [33]. The presence of SnO_2 into $SnSe_2$ can significantly influence its structural and optical properties. The incorporation of SnO_2 into $SnSe_2$ may result in lattice distortion due to differences in ionic radii and crystal structures between SnO_2 and $SnSe_2$. This distortion can cause strain within the crystal lattice, potentially leading to defects or dislocations. Such structural changes can alter the material's mechanical stability and influence its optical properties. The presence of SnO_2 can affect the band gap of $SnSe_2$. Since SnO_2 has a wider band gap compared to $SnSe_2$, its incorporation decreases the overall band gap of the prepared material. This alteration can shift the absorption edge towards longer wavelengths, affecting the material's optical absorption and photoluminescence properties [34].

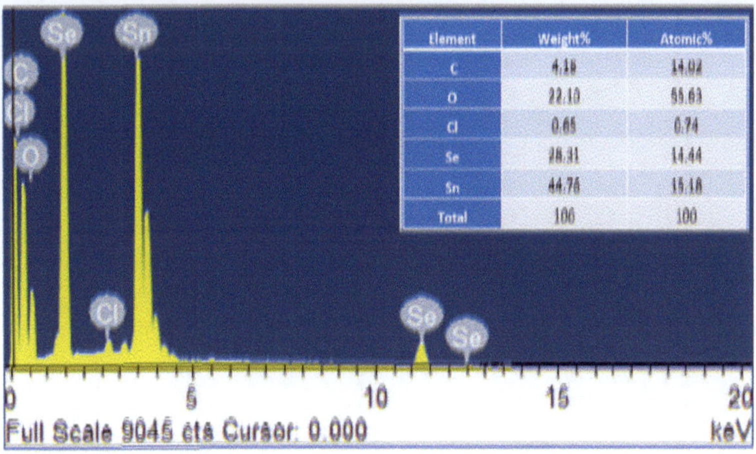

Fig. 5. Energy-dispersive X-ray spectrum of EDTA@SnSe2 nanocrystals.

3.4 UV-Vis Study

The energy bandgaps associated with all samples, encompassing those prepared at room temperature and those subjected to heat exposure at 100 °C, 200 °C, and 300 °C, were determined through analysis of their UV-Visible absorption spectra. Spectral measurements were obtained for these annealed samples over a wavelength range from 230 to 500 nm, and the obtained data are presented in Fig. 6. A prominent band edge emission was observed around 320 nm within the visible light spectrum, that the prepared samples could be effectively used for optoelectronic devices that operate within the visible light spectrum [35]. The optical energy bandgaps of the EDTA@$SnSe_2$ nanocrystals were derived from a Tauc's plot analysis, assuming a direct bandgap semiconductor, as illustrated in the insets of Fig. 6(a-d). The bandgap energies were found to be around 2.87 eV for room temperature annealing, 2.79 eV for 100 °C annealing, 2.89 eV for 200 °C annealing, and 2.75 eV for 300 °C. These calculated bandgap values suggest that the synthesized materials have strong potential for applications in optoelectronic devices [36]. The bandgap of $SnSe_2$ nanomaterials exhibits a dependence on the annealing temperature, influenced by factors such as increased grain size, quantum size effects, and increasing of charge carriers. These factors can lead to an initial increase followed by a decrease in the bandgap depending on the specific annealing temperature. Notably, all annealed samples demonstrate a red shift in their bandgaps compared to the bulk $SnSe_2$, which has a reported bandgap of 1.9 eV [28, 37]. The estimated energy bandgaps of the EDTA@$SnSe_2$ nanoparticles, calculated from the Tauc's plots, were observed to decrease from 2.87 eV (room temperature) to 2.75 eV (300 °C). This systematic reduction in the bandgap with increasing annealing temperature can be attributed to enhanced crystallinity and slight modifications in the particle size, as corroborated by XRD data. In comparison, bulk $SnSe_2$ typically exhibits a bandgap of ~1.2 eV to 1.4 eV, depending on its preparation method and purity, whereas previously reported [31] $SnSe_2$ nanosheet exhibit bandgaps in the range of 2.04 eV.

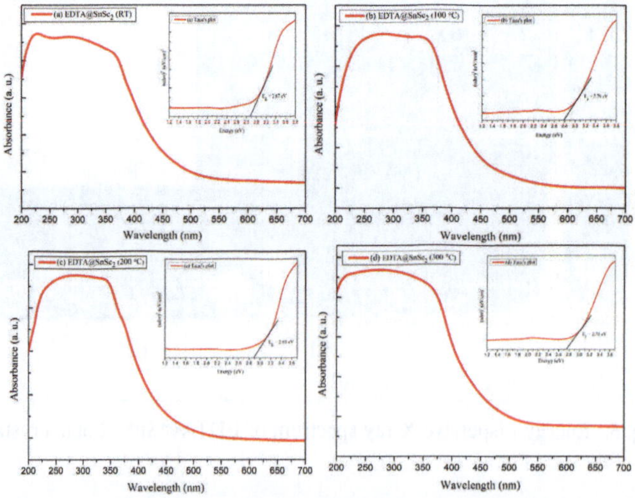

Fig. 6. UV-Vis absorption spectra and the associated Tauc's plots were obtained for all annealed EDTA@$SnSe_2$ nanocrystals: (a) RT (b) 100 °C (c) 200 °C and (d) 300 °C

3.5 PL Study

Figure 7(a-d) illustrates the room-temperature photoluminescence (PL) emission spectra of EDTA@$SnSe_2$ nanoparticles following annealing treatments at 100 °C, 200 °C, and 300 °C. A broad emission band is found to be over the range of wavelengths spanning 300-600 nm. The PL spectrum of the EDTA@$SnSe_2$ sample exhibits two emission bands. A violet emission band is observed with a peak centred at 380 nm, while another prominent emission peak is located at 472 nm, corresponding to blue emission. The intensity of these peaks increased with annealing temperature, indicating enhanced radiative recombination efficiency due to improved crystallinity and reduced defect states. The defects/vacancies are introduced due to the vertical orientation of nanosheets present in the morphology of prepared sample [39]. Consequently, the emission peaks are a result of radiative recombination. Specifically, electrons located at donor levels created by selenium vacancies recombine with holes localized at acceptor levels created by tin vacancies, leading to the observed emission as compared to bulk $SnSe_2$ typically shows weak or negligible PL emission, attributed to its indirect bandgap nature [38]. Reported $SnSe_2$ nanosheets, on the other hand, exhibit similar emission characteristics in the violet-to-blue region but with variations depending on synthesis methods and post-treatment processes [16, 40].

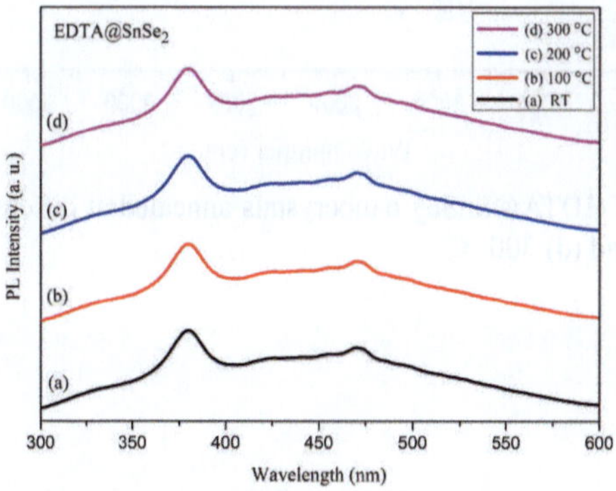

Fig. 7. Photoluminescence spectra of EDTA@$SnSe_2$ nanocrystals annealed at various temperatures: (a) RT (b) 100 °C (c) 200 °C and (d) 300 °C

3.6 FTIR Study

FTIR spectroscopic analysis was conducted to characterize the functional groups and chemical bonding within the EDTA@$SnSe_2$. Figure 8(a-d) illustrates the FTIR spectra of EDTA@$SnSe_2$ samples annealed at various temperatures including room temperature to 300 °C. A broad peak observed at 3422 cm^{-1} is indicative of -OH and -NH stretching vibrations. The narrow peaks located at 2957 cm^{-1} and 2347 cm^{-1} are characteristic of EDTA. The peak observed at 2957 cm^{-1} is attributed to the C-H bond stretching vibrations, despite the peak being located at 2347 cm^{-1} results from the C≡C bond stretching mode triple bond within the EDTA molecule. Peaks in the spectrum, located near 1692 cm^{-1} and 1614 cm^{-1}, are caused by the stretching of the carbonyl group (C=O) and the bending vibrational movements of the carbon-carbon double bond (C=C), respectively. Peaks are observed at 1450 cm^{-1} and 1378 cm^{-1} are related with the bending vibrations of the -CH groups. The spectral peaks located at 656 cm^{-1} and 567 cm^{-1} are assigned to Sn-Se stretching vibrations from $SnSe_2$ and Sn-O stretching vibrations from SnO_2 nanoparticles, respectively [21, 41].

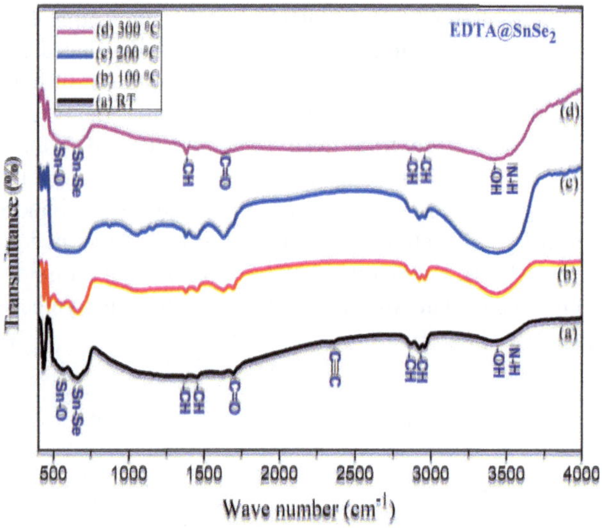

Fig. 8. FTIR spectra of EDTA@$SnSe_2$ nanocrystals annealed at different temperatures: (a) RT (b) 100 °C (c) 200 °C and (d) 300 °C

4 Conclusion

This study successfully synthesized EDTA@$SnSe_2$ nanoparticles utilizing the solvothermal method. The synthesized samples were subjected to annealing treatments at temperature of 100 °C, 200 °C, and 300 °C. Subsequently, their phase structure, morphological, and optical characteristics were comprehensively analysed. XRD analysis confirmed that the annealed EDTA@$SnSe_2$ nanoparticles possess a hexagonal crystal structure. Crystallite sizes, determined by the Scherrer equation, were 25.85 nm, 29.59 nm, 28.95 nm, and 30.75 nm for samples annealed sequentially at room temperature to 300 °C, respectively. EDX analysis confirmed material's composition as EDTA@$SnSe_2$. SEM imaging displayed a layered, plate-like morphology of the synthesized nanoparticles. The interactions and binding mechanisms between EDTA and $SnSe_2$ were examined using FTIR spectroscopy. The estimated energy bandgaps of the EDTA@$SnSe_2$ nanocrystals were found to range from 2.87 to 2.75 eV for samples annealed at room temperature to 300 °C, respectively, indicating slight variations in the bandgap energy with annealing temperature. A red shift in the bandgap was observed as compared to bulk $SnSe_2$, indicating a slight decrease in the bandgap indicating their potential suitability of the material for optoelectronic applications. Furthermore, the PL spectrum of each annealed EDTA@$SnSe_2$ sample exhibit violet and blue emission peaks centred at 380 nm and 472 nm, respectively. These findings suggest that EDTA@ $SnSe_2$ nanoparticles have the potential to be utilized as promising materials for the fabrication of photodetectors and light-emitting diodes.

References

1. Wang, J., Wei, Y., Li, H., Huang, X., Zhang, H.: Crystal phase control in two-dimensional material. Sci. China Chem. **61**, 1227–1242 (2018). https://doi.org/10.1007/s11426-018-9326-y
2. Yu, X., Zhu, J., Zhang, Y., Weng, J., Hu, L., Dai, S.: SnSe2 quantum dot sensitized solar cells prepared employing molecular metal chalcogenide as precursors. Chem. Commun. **48**, 3324–3326 (2012). https://doi.org/10.1039/c2cc17081g
3. Cheng, Y., Huang, J., Cao, L., Xie, H., Wang, Y., Wang, Y., Wu, H.: Structure-controlled SnSe2 nanosheets as high performance anode material for lithium ion batteries. Ionics. **26**, 2855–2862 (2020). https://doi.org/10.1007/s11581-019-03321-w
4. Rai, R.K., Islam, S., Roy, A., Agrawal, G., Singh, A.K., Ghosh, A., Ravishankar, N.: Morphology controlled synthesis of low bandgap SnSe 2 with high photodetectivity. Nanoscale. **11**, 870–877 (2019). https://doi.org/10.1039/c8nr08138g
5. Olimpio, G.D., Farias, D., Kuo, C., Ottaviano, L., Lue, C.S.: Tin diselenide (SnSe2) Van der Waals semiconductor: surface chemical reactivity, ambient stability. Chem. Opt. Sensors Mater. **15**, 1154 (2022). https://doi.org/10.3390/ma15031154
6. Ma, D.W., Cheng, C.: Synthesis of SnSe2 nanorods and nanoplates by an organic solution-phase route. J. Nanosci. Nanotechnol. **13**, 4433–4436 (2013). https://doi.org/10.1166/jnn.2013.7173
7. Shanmugam, G., Deshpande, U.P., Sharma, A., Shirage, P.M., Bhobe, P.A.: Impact of different morphological structures on physical properties of nanostructured SnSe. J. Phys. Chem. C. **122**, 13182–13192 (2018). https://doi.org/10.1021/acs.jpcc.8b03797
8. Lu, D., Yue, C., Luo, S., Li, Z., Xue, W., Qi, X., Zhong, J.: Phase controllable synthesis of SnSe and SnSe2 films with tunable photoresponse properties. Appl. Surf. Sci. **541**, 148615 (2020). https://doi.org/10.1016/j.apsusc.2020.148615
9. Liu, K., Liu, H., Wang, J., Feng, L.: Synthesis and characterization of SnSe2 hexagonal nano flakes. Mater. Lett. **63**, 512–514 (2009). https://doi.org/10.1016/j.matlet.2008.10.054
10. Yang, S., Liu, Y., Wu, M., Zhao, L., Lin, Z., Cheng, H.: Highly-anisotropic optical and electrical properties in layered SnSe. Nano Res. **11**, 554–564 (2018). https://doi.org/10.1007/s12274-017-1712-2
11. Liu, Y., Cao, J., Li, C., Zeng, J., Tang, K., Qian, Y., Zhang, W.: Hydrazine route to one-dimensional structural metal selenides crystals. J. Cryst. Growth. **261**, 508–513 (2004). https://doi.org/10.1016/j.jcrysgro.2003.08.079
12. Fang, Z., Hao, S., Long, L., Fang, H., Qiang, T., Song, Y.: The enhanced photoelectrochemical response of SnSe2 nanosheets. CrystEngComm. **16**, 2404–2410 (2014). https://doi.org/10.1039/C3CE42082E
13. Kumar, S., Banotra, A., Padha, N., Ahmed, S.: Effect of substrate and annealing temperature on the physical properties of the thin films of SnSe2-SnSe alloy. Opt. Mater. (Amst). **134**, 113078 (2022). https://doi.org/10.1016/j.optmat.2022.113078
14. Zhou, X., Gan, L., Tian, W., Zhang, Q., Jin, S., Li, H., Bando, Y., Golberg, D., Zhai, T.: Ultrathin SnSe2 flakes grown by chemical vapor deposition for high-performance photodetectors. Adv. Mater. **27**, 8035–8041 (2015). https://doi.org/10.1002/adma.201503873

15. Pawbake, A.S., Date, A., Jadkar, S.R., Late, D.J.: Temperature dependent Raman spectroscopy and sensing behavior of few layer SnSe2 nanosheets. Chem. Select. **1**, 5380–5387 (2016). https://doi.org/10.1002/slct.201601347
16. Das, L., Guleria, A., Adhikari, S.: Aqueous phase one-pot green synthesis of SnSe nanosheets in a protein matrix: negligible cytotoxicity and room temperature emission in the visible region. RCS Adv. **5**, 61390–61397 (2015). https://doi.org/10.1039/c5ra09448h
17. Ren, X., Qi, X., Shen, Y., Xu, G., Li, J., Li, Z., Huang, Z., Zhong, J.: Synthesis of SnSe nanosheets by hydrothermal intercalation and exfoliation route and their photoresponse properties. Mater. Sci. Eng. B. **214**, 46–50 (2016). https://doi.org/10.1016/j.mseb.2016.09.001
18. Tołłoczko, A.K., Peter, M., Rybak, M., Gorantla, S., Bachmatiuk, A., Kudrawiec, R.: Synthesis of SnSe nanosheets by hydrothermal intercalation and exfoliation route and their photoresponse properties. J. Phys. Chem. C. **2**, 1–32 (2024). https://doi.org/10.1021/acs.jpcc.4c03005
19. Dar, M.A., Govindarajan, D., Batoo, K.M., Hadi, M., Dar, G.N.: Photovoltaic and supercapacitor performance of SnSe nanoparticles prepared through co-precipitation method. Mater. Technol. **37**, 1396–1409 (2022). https://doi.org/10.1080/10667857.2021.1950887
20. Dar, M.A., Ahmed, S.R., Rather, M.A., Kalpana, S., Rather, A.A., Ahamed, S.R., Ahmad, Z.: The features of the solvothermal synthesis of SnSe nanosheets and nanoflowers for supercapacitor applications. Inorg. Chem. Commun. **153**, 110838 (2023). https://doi.org/10.1016/j.inoche.2023.110838
21. Mary, T.A., Christina, A., Sakthivel, F.P., Jesudurai, J.G.M.: A study on the role of surfactant on the layered growth of SnSe2 for electrical applications. J. Mater. Sci. Mater. Electron. **27**, 11041–11048 (2016). https://doi.org/10.1007/s10854-016-5220-z
22. Yamashita, A., Ogiso, O., Matsumoto, R., Tanaka, M., Hara, H., Tanaka, H., Takeya, H., Lee, C.: Influence of oxidation in starting material Sn on electric transport properties of SnSe single crystals. J. Phys. Soc. Jpn. **87**(065001), 1–2 (2018). https://doi.org/10.7566/JPSJ.87.065001
23. Baid, L., Ojha, K.S.: Effect of CTAB and PVP on morphological and optical properties of CdS microstructures. J. Mater. Sci. Mater. Electron. **35**, 1043 (2024). https://doi.org/10.1007/s10854-024-12797-w
24. Baid, L., Ojha, K.S.: Temporal evolution of optical and photocatalytic properties of hydrothermally synthesized EDTA-encapsulated CdS spherical nanostructures. Chem. Phys. Impact. **8**, 100623 (2024). https://doi.org/10.1016/j.chphi.2024.100623
25. Khan, I., Saeed, K., Khan, I.: Nanoparticles: properties, applications and toxicities. Arab. J. Chem. **12**, 908–931 (2019). https://doi.org/10.1016/j.arabjc.2017.05.011
26. Singh, A., Joshi, B.C., Kumar, D., Khanna, P.K., Kumar, M.: Effect of band gap offset on ZnO (Cd, Mg)/ZnO:P multi quantum well light emitting diodes: a simulation study. Energy Environ. Focus. **3**, 185–188 (2014). https://doi.org/10.1166/eef.2014.1095
27. Singh, K., Soni, Anwar, S., Dubey, P., Mishra, S.K.: Influence of temperatures on structure, thermoelectric, and mechanical properties of nanocrystalline SnSe thin films deposited by thermal evaporation. Mater. Today Commun. **32**, 103880 (2022). https://doi.org/10.1016/j.mtcomm.2022.103880
28. Orimi, R.L., Mirdeylami, S.J.: Effect of heat treatment on the phase transformation from SnS to pure SnS2. Adv. Mater. Res. **829**, 141–146 (2014). https://doi.org/10.4028/www.scientific.net/AMR.829.141
29. Ramachandran, M., Sergio, J., Rios, N.: Optical and electrical properties of SnSe2 and SnSe thin films prepared by spray pyrolysis. Thin Solid Films. **535**, 390–393 (2013). https://doi.org/10.1016/j.tsf.2012.12.081
30. Holder, C.F., Schaak, R.E.: Tutorial on powder X - ray diffraction for characterizing nanoscale materials. ACS Nano. **13**, 7359–7365 (2019). https://doi.org/10.1021/acsnano.9b05157

31. Mukhokosi, E.P., Krupanidhi, S.B., Nanda, K.K.: Band gap engineering of hexagonal SnSe2 nanostructured thin films for infra-red photodetection. Sci. Rep. **7**, 1–10 (2017). https://doi.org/10.1038/s41598-017-15519-x
32. Huang, Y., Xu, K., Wang, Z., Shifa, T.A., Wang, Q., Wang, F., Jiang, C., He, J.: Designing the shape evolution of SnSe2 nanosheets and their optoelectronic properties. Nanoscale. **7**, 17375–17380 (2015). https://doi.org/10.1039/c5nr05989e
33. Patel, K., Solanki, G., Patel, K., Pathak, V., Chauhan, P.: Investigation of optical, electrical and optoelectronic properties of SnSe crystals. Eur. Phys. J. B. **92**, 100306–100308 (2019). https://doi.org/10.1140/epjb/e2019-100306-8
34. Akram, M., Saleh, A.T., Ibrahim, W.A.W., Awan, A.S., Hussain, R.: Continuous microwave flow synthesis (CMFS) of nano-sized tin oxide: effect of precursor concentration. Ceram. Int. **42**, 8613–8619 (2016). https://doi.org/10.1016/j.ceramint.2016.02.092
35. Pallavolu, M.R., Minnam Reddy, V.R., Guddeti, P.R., Park, C.: Development of SnSe thin films through selenization of sputtered Sn-metal films. J. Mater. Sci. Mater. Electron. **30**, 15980–15988 (2019). https://doi.org/10.1007/s10854-019-01968-9
36. Siyar, M., Farid, M., Khan, H., Umar, M.A., Tanveer, W.H., Safdar, A.: Controlled morphology and its effects on the thermoelectric properties of SnSe2 thin films. Crystals. **11**, 1–8 (2021). https://doi.org/10.3390/cryst11080942
37. Pejjai, B., Minnam Reddy, V.R., Gedi, S., Park, C.: Status review on earth-abundant and environmentally green Sn-X (X = Se, S) nanoparticle synthesis by solution methods for photovoltaic applications. Int. J. Hydrog. Energy. **42**, 2790–2831 (2017). https://doi.org/10.1016/j.ijhydene.2016.11.084
38. An, B., Ma, Y., Zhang, G., You, C., Zhang, Y.: Controlled synthesis of few-layer SnSe2 by chemical vapor deposition. RSC Adv. **10**, 42157–42163 (2020). https://doi.org/10.1039/d0ra08360g
39. Sierra-Castillo, A., Haye, E., Acosta, S., Arenal, R., Bittencourt, C., Colomer, J.F.: Atmospheric pressure chemical vapor deposition growth of vertically aligned SnS2and SnSe2 nanosheets. RSC Adv. **11**, 36483–36493 (2021). https://doi.org/10.1039/d1ra05672g
40. Saray, A.M., Yousefi, R., Zare-Dehnavi, N., Jamali-Sheini, F.: Improvement visible-light photocatalytic performance of single-crystalline SnSe1±x NPs toward degradation of organic pollutants. Solid State Sci. **98**, 106044 (2019). https://doi.org/10.1016/j.solidstatesciences.2019.106044
41. Eagambaram, M., Kumar, K.: Design of an efficient tin selenide-based ternary nanocomposite electrode for simultaneous determination of paracetamol, tryptophan, and caffeine. ACS Omega. **7**, 35486–35495 (2022). https://doi.org/10.1021/acsomega.1c07306

Impact of Incorporating Trivalent Co-doping on Thermal and Structural Properties of $BaTiO_3$

Mitesh Shakya[1], Anju Dixit[2(✉)], Suresh Kumar[3], and Pramod S. Dobal[1]

[1] Department of Physics, VSSD College, Kanpur 208002, India
[2] Department of Physics, School of Basic Sciences, UIET, CSJM University, Kanpur 208024, India
anjudixit@csjmu.ac.in
[3] Ceramics and Ceramic Matrix Composites Division, DMSRDE, Kanpur 208013, India

Abstract. This study reports the successful synthesis of holmium (Ho) and scandium (Sc) co-doped barium titanate $Ba_{1-x}Ho_xTi_{1-y}Sc_yO_3$ using sol-gel method. Samples of compositions of $x = y = 0$ (BT) and $x = y = 0.05$ (BHTS) have been synthesized. The crystallization, phase formation, and micro-structural changes in these samples have been studied with the help of thermogravimetric and X-ray diffraction analyses. The X-ray diffraction pattern confirmed well-crystallized tetragonal phases for the samples, with minute variations in peak positions. The average particle sizes of 95 nm in BT were found to be 59 nm in BHTS samples. This size decrease was supported by FE-SEM. The peak shift in BHTS was better observed in some of the vibrational frequencies in FTIR spectra. The incorporations and elemental composition of Ho and Sc was confirmed from the EDX analysis of the samples.

Keywords: Ferroelectrics · Dielectric · Barium titanate · XRD · FTIR · SEM

1 Introduction

Over the last few decades, environmentally friendly ferroelectric ceramics composed of barium titanate $BaTiO_3$ (BT) have been highly significant to researchers as it serves as a suitable alternative to harmful $PbTiO_3$. It makes BT a part of the most preferred ferroelectric perovskite group. BT exhibits favorable dielectric, ferroelectric, and piezoelectric characteristics [1]. It possesses high dielectric properties and low dielectric loss, making it ideal for storage components like capacitors and multilayer capacitors [2]. When compared to non-ferroelectric piezoelectrics, the piezoelectric coefficients in BT compositions are significantly larger, leading to their common use in underwater sonar medical ultrasound devices and audible sound generating equipment. A high pyroelectric coefficient enables the use of BT in pyroelectric infrared sensors. The dielectric and electric properties of BT can be further improved or altered by doping it with trivalent and/or pentavalent impurities [3–8]. The incorporation of different ions at A and B sites of co-doped BT can lead to significant changes in the crystal structure, phase stability, dielectric, piezoelectric, and magnetic properties thus allowing better control

S. Kumari et al. (Eds.): ICMTP 2024, SPPHY 311, pp. 214–222, 2025.
https://doi.org/10.1007/978-981-96-6795-6_25

of the material's functional properties. This enables the optimization of $BaTiO_3$ for a wide variety of applications, including capacitors, sensors, actuators, energy harvesting devices, multiferroic materials, and memory devices. Thus, by controlling both doping sites, it is better possible to tailor the material towards the specific demands of advanced technologies in fields like electronics, energy, and sensing [3–6].

Based on the available literature holmium (Ho), a rare earth (4f) element is expected to occupy A-sites while the 3d-transition metal scandium (Sc) should occupy the B-sites of the BT lattice. Overall, the co-doping of Ho and Sc in BT could induce magnetic ordering and multiferroic properties required for multilayer ceramics, sensors, actuators, and transducers FeRAM and other applications [7–12]. Recently Zakeri et al. studied Al and Ce co-doped BT which showed that it has good mass-specific constant 224.18 Fg^{-1} and good water splitting properties, and has been utilized as electrochemical supercapacitors [13]. The primary goal of this study is to co-dope BT using Ho and Sc, and to synthesize the Ho and Sc co-doped ceramics of BT via the Sol-gel method. Also, their thermal, micro-structural, and vibrational characteristics have been analyzed. The electrical properties of these materials will be correlated with the structural changes arising due to the co-doping.

2 Synthesis and Experimental Details

Sol-gel synthesis was used to synthesize pure and Holmium (Ho) and Scandium (Sc) co-doped $Ba_{1-x}Ho_xTi_{1-y}Sc_yO_3$ compositions for $x = y = 0$ (BT) and $x = y = 0.05$ (BHTS). Precursor materials included holmium acetate hydrate (Sigma Aldrich 99.99%), scandium acetate hydrate (Sigma Aldrich 99.9%), titanium (IV) butoxide (Sigma Aldric regent grade 97%), and barium acetate (Thermo Scientific Chemicals 99.9%). Polyethylene glycol (PEG) (99%), acetic acid (99.5%), deionised (DI) water, and ethanol (99.9%) were employed as solvents. In the first step $Ba(CH_3COO)_2$, $Sc(CH_3COO)_3.xH_2O$ and $Ho(CH_3COO)_3.xH_2O$ was mixed with acetic acid and DI water and stirring at 60 °C for 30 min, a transparent white solution was obtained. In second step Titanium butoxide mixed with ethanol and PEG, stirred at 60 °C for 30 min resulted in a yellowish solution. Transparent white solution was added to this and continuous stirring for 1 h at 60 °C resulted in a transparent gel. After aging the transparent gel for 24 h and drying it at 100 °C for 12 h, Xerogel formed. Fine powder was obtained after calcining the grounded xerogel at 900 °C for 1 h [14].

Thermogravimetric analyzer from Perkin Elmer (STA 8000) was used for Thermogravimetric Analysis (TGA) of as grown xerogels after grinding. IR transmittance spectra were measured in 400–4000 cm^{-1} range using an FTIR spectrometer (Nicolet 5700, Thermo Fisher, USA). A D8 advance X-ray diffractometer (with Cu Kα radiation of wavelength $\lambda = 1.540598$ Å at 40 kV and 40 mA) in the 2 range of $10° < 2 < 80°$ with a step size of 0.02° was used to investigate the structure of the calcined samples. The surface morphology of the BHTS samples were observed by using the FESEM/EDX (Zeiss Gemini Sigma 360 VP) and particle size was calculated by image processing imageJ software.

3 Result and Discussion

3.1 Thermal Gravimetric Analysis

The thermogravimetric analysis (TGA) of xerogel samples of $Ba_{1-x}Ho_xTi_{1-y}Sc_yO_3$ for compositions $x = y = 0$ (BT) and $x = y = 0.05$ (BHTS) was done by heating them up to 1000 °C in air at a rate of 10 °C per minute. The resulting TGA curves for both BT and BHTS co-doped samples are displayed in Fig. 1 along with their derivative weight. These TGA curves clearly display three typical stages [14, 15]. A slight weight loss followed by the initial appearance of an endothermic peak was observed at about 78.8 °C and 182.6 °C. This is equivalent to the evaporation of the gel's volatile organic components, such as ethanol and acetic acid, as well as adsorbed moisture. The weight loss in the temperature range 100–200 °C is thus primarily reflects the evaporation of water content and other volatile organics from the sample. The second weight loss in between 300–600 °C is associated with the decomposition of organic compounds and species created during the gelation process. $BaCO_3$ and TiO_2 seem to react about 650 °C and the crystallization of $BaTiO_3$ phase above 750 °C. This observed crystallization behavior aligns with other studies reported on doped $BaTiO_3$ [15–18].

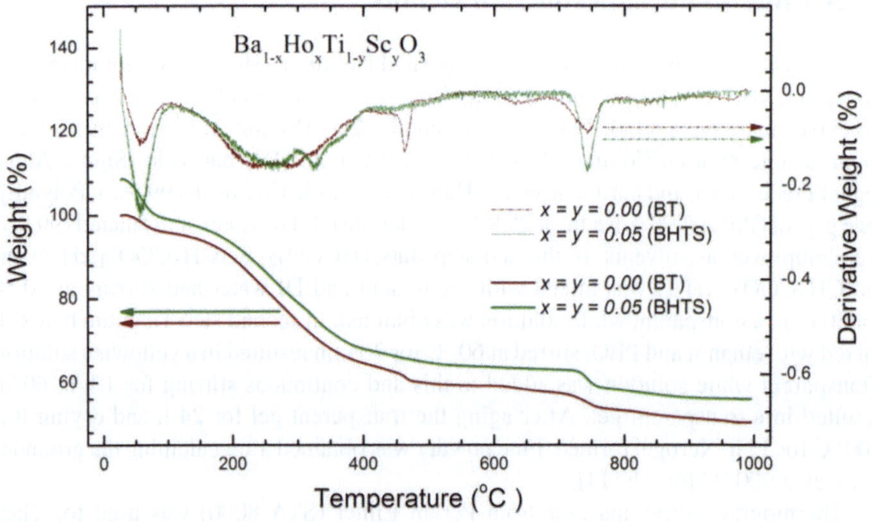

Fig. 1. TGA weight loss and derivative weight loss curves of $Ba_{1-x}Ho_xTi_{1-y}Sc_yO_3$ compositions for $x = y = 0$ (BT) and 0.05 (BHTS).

The derivative weight loss curve also reveals most prominent weight losses. As compared to undoped samples, relatively sharper dips reflecting more weight losses were observed from co-doped samples, especially for the moisture and volatile organics in the range 50–150 °C and at crystallization offset at about 750 °C. However, the effect of sharper weight loss dips in co-doped samples on material properties is still remains to be established.

3.2 Structural Analysis

XRD is a powerful method for examining crystalline properties of nano powders and other crystalline materials, such as the crystalline phases, associated planes, average crystalline diameters, and numerous other parameters [19]. Figure 2 shows the XRD data of undoped BT and $x = y = 0.05$ co-doped BHTS samples. Peaks located at about 22.1°, 23.9°, 31.5°, 38.8°, 45.1°, 50.9°, 56.1°, 65.8°, 74.9°, and 79.3°in undoped samples were attributed to (001), (100), (101), (111), (002), (102), (112), (202) (103), and (131) reflection peaks in that order [crystallographic open database card no COD1507756, and JCPDS card number 01-089-1428].

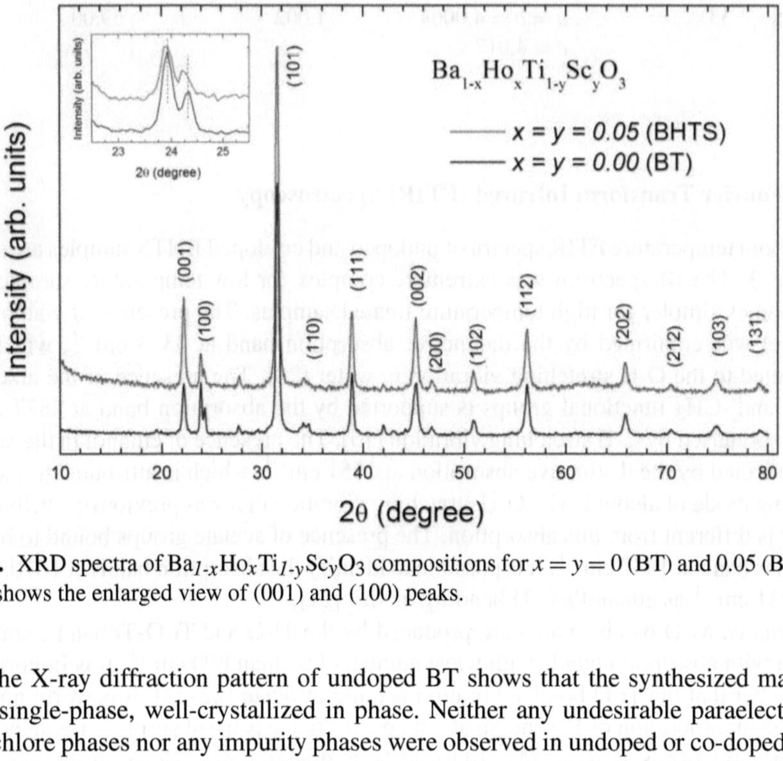

Fig. 2. XRD spectra of $Ba_{1-x}Ho_xTi_{1-y}Sc_yO_3$ compositions for $x = y = 0$ (BT) and 0.05 (BHTS). Inset shows the enlarged view of (001) and (100) peaks.

The X-ray diffraction pattern of undoped BT shows that the synthesized material is in single-phase, well-crystallized in phase. Neither any undesirable paraelectric or pyrochlore phases nor any impurity phases were observed in undoped or co-doped samples studied. The similarities of XRD peaks in undoped and co-doped samples clearly reveal that the Ho and Sc occupy lattice sites only. This was further confirmed by slight shift of diffraction peaks to lower angles as shown in the inset confirming the incorporation of co-doping in BT. The (001), (100) peak shifts caused by dopants ($\Delta 2\theta$ values) are in the same direction within 0.1–0.2°. The tolerance factor for these compositions of Ho and Sc (1.0575) is very near to those of undoped BT (1.0587) also revealing the tetragonal structure of the lattice. Therefore we do not expect a large variation in $\Delta 2\theta$ values for the compositions studied. This is the precise reason that the observed XRD patterns of undoped and co-doped samples reveal the tetragonal structures with almost similar tetragonality in the lattices. The calculated lattice parameters $a = b = 3.999$ Å and $c = 4.017$ Å and the resulting tetragonality factor $c/a = 1.004$ shows the tetragonal structure of the lattice. The calculated lattice parameters and tetragonality ratio along with the percentage weight loss and particle sizes for $Ba_{1-x}Ho_xTi_{1-y}Sc_yO_3$ compositions for $x = y = 0$ (BT) and 0.05 (BHTS) are given in Table 1.

Table 1. The percentage weight loss, lattice parameters, tetragonality ratio, and particle sizes for $Ba_{1-x}Ho_xTi_{1-y}Sc_yO_3$ compositions for $x = y = 0$ (BT) and 0.05 (BHTS).

Sample	Weight loss %	Lattice parameters (Å)	Tetragonality (*c/a*)	Particle size (nm)
BT	50	$a = b = 4.0065$ $c = 4.017$	1.002	94.85
BHTS	53	$a = b = 4.0064$ $c = 4.017$	1.002	59.00

3.3 Fourier Transform Infrared (FTIR) Spectroscopy

The room temperature FTIR spectra of undoped and co-doped BHTS samples are shown in Fig. 3. The IR spectrum was extremely complex for low temperature samples, but it becomes simpler for high temperature treated samples. The presence of water in the xerogel was confirmed by the distinctive absorption band at 3373 cm^{-1}, which was attributed to the O-H stretching vibration in water [20]. The presence of the alkoxides $-CH_2$ and $-CH_3$ functional groups is supported by the absorption band at 2877 cm^{-1}, which is caused by C-H stretching vibration [15]. The presence of ethanol in the xerogel is confirmed by the distinctive absorption at 1551 cm^{-1}, which is attributed to the -OH bending mode of alcohol. The O-H stretching vibration that was previously attributed to water is different from this absorption. The presence of acetate groups bound to barium atoms is band at 1440 cm^{-1}. It is possible to identify the absorption bands at 1070, 1019, and 931 cm^{-1} as ethanol's C-H bending modes [21].

Finally, M-O bands, which are produced by the Ti-O and Ti-O-Ti bands, are associated with absorption bands with wave numbers less than 800 cm^{-1}. It is important to remember that the Ti-O bond's vibration is quite intricate [22–24]. It is worth noticing that the incorporation of Ho and Sc at lattice sites reduces the absorption frequencies of undoped BT bands at about 851, 1440, 1571, and 3410 cm^{-1}. However, the compositional dependence of the frequency shift or bond strength of the functional groups is yet to be established for these co-doped samples.

3.4 Surface Analysis and EDAX

The surface morphology of the samples was studied by the field emission scanning electron microscopy (FE-SEM) coupled with EDAX. The SEM images of pure and co-doped barium titanate were taken at the magnification of 100.00 KX at 8 kV. Figure 4 shows the FESEM micrographs of undoped and co-doped barium titanate nano powders. The nanoscale crystallites in BT surface are clearly seen where the particles are clumped

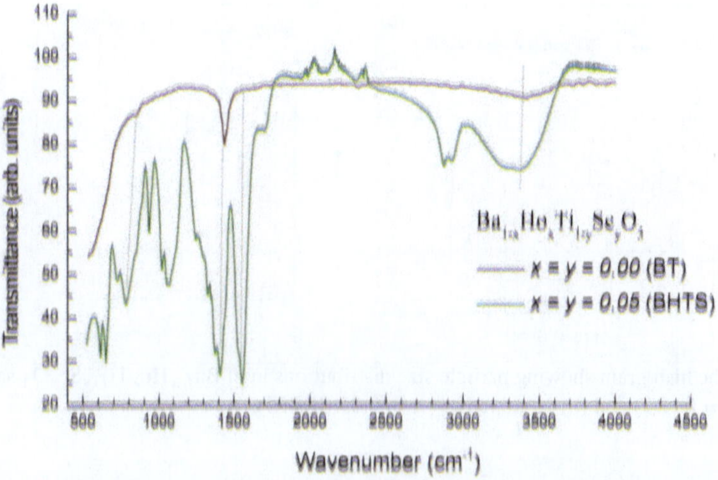

Fig. 3. FTIR spectra of $Ba_{1-x}Ho_xTi_{1-y}Sc_yO_3$ for $x = y = 0$ (BT) and 0.05 (BHTS).

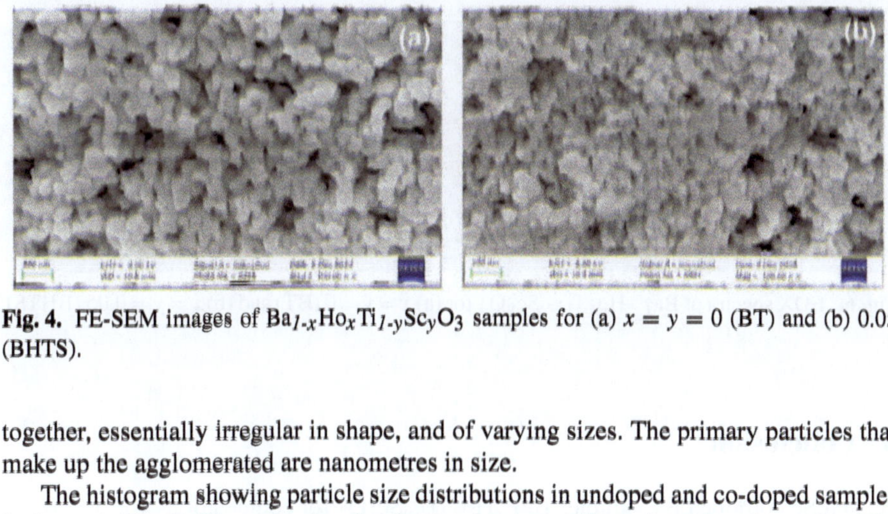

Fig. 4. FE-SEM images of $Ba_{1-x}Ho_xTi_{1-y}Sc_yO_3$ samples for (a) $x = y = 0$ (BT) and (b) 0.05 (BHTS).

together, essentially irregular in shape, and of varying sizes. The primary particles that make up the agglomerated are nanometres in size.

The histogram showing particle size distributions in undoped and co-doped samples is shown in Fig. 5. The grain size distribution is more uniform is undoped samples. The average grain size of pure barium titanate is around 95 nm (minimum size 57.59 nm and maximum size 148.57 nm). For co-doped samples, the average particle size is about 59 nm, minimum being 56.87 nm while maximum observed size was 87.22 nm (Table 1). This decrease in size in co-doped samples is primarily due to the lattice distortions caused by radius difference between the Ho/Sc and the replaced Ba/Ti elements [22–24]. With a decrease in grain size, we expect that the piezoelectric constant, maximum polarization, and remnant polarization should increase in co-doped samples.

Figure 6 displays the Elemental Dispersive X-ray (EDX) spectrum of undoped and co-doped samples. It shows that the barium titanate formed in this study has elemental composition Ba 56.8%, Ti 20.3%, and O 22.9%. A small quantity of Si impurity is

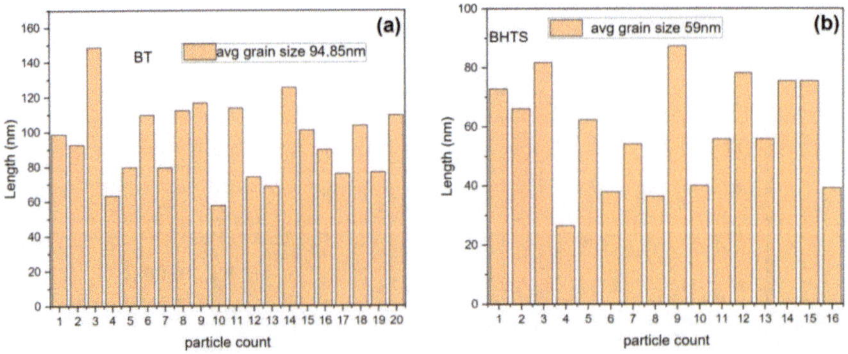

Fig. 5. The histogram showing particle size distributions in of $Ba_{1-x}Ho_xTi_{1-y}Sc_yO_3$ samples for (a) $x = y = 0$ (BT) and (b) co-doped $x = y = 0.05$ (BHTS).

detected because the images were taken on silicon wafer. In co-doped samples the Ho and Sc detected was found in line with the composition of those samples.

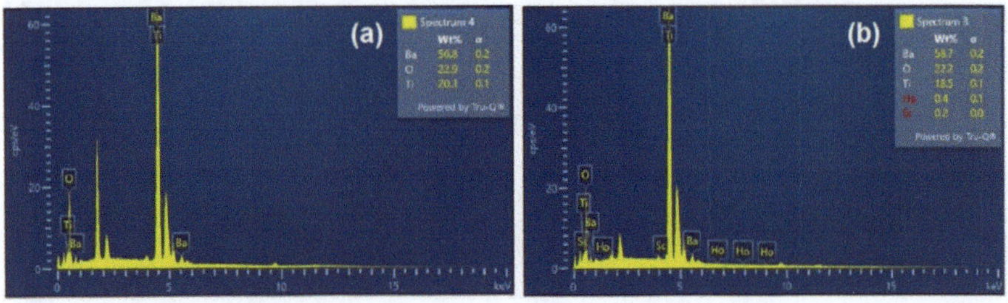

Fig. 6. EDX spectra of $Ba_{1-x}Ho_xTi_{1-y}Sc_yO_3$ for (a) $x = y = 0$ (BT) and (b) $x = y = 0.05$ (BHTS).

4 Conclusion

Ho and Sc co-doped BT samples $Ba_{1-x}Ho_xTi_{1-y}Sc_yO_3$ for compositions $x = y = 0$ and $x = y = .05$ were successfully synthesize by using the sol-gel method. From thermal and structural analysis it was revealed that the synthesized pure and co-doped BT crystallized in single phase and has tetragonal structure at room temperature. FTIR spectra show the presence of functional groups whose frequency shifts due to the mass differences of dopant and substituted ions. FESEM analysis shows almost uniform surface morphology of the samples. The BT particle size decreases with co-doping because of the lattice distortion caused by radius difference between the Ho/Sc and the replaced Ba/Ti elements. EDAX confirmed the elemental composition of the samples. Further work on electric and dielectric properties of these samples will be carried out.

References

1. Tsur, Y., Hitomi, A., Scrymgeour, I., Randall, C.A.: Site occupancy of rare-earth cations in $BaTiO_3$. Jpn. J. Appl. Phys. **40**, 255–258 (2001). https://doi.org/10.1143/JJAP.40.255
2. Whatmore, R.W., You, Y.-M., Xiong, R.-G., Eom, C.-B.: 100 years of ferroelectricity—A celebration. APL Mater. **9**, 070401 (2021). https://doi.org/10.1063/5.0059208
3. Lines, M.E., Glass, A.M.: Principles and Applications of Ferroelectrics and Related Materials. Oxford Academic, Oxford (2001). https://doi.org/10.1093/acprof:oso/9780198507789.001.0001
4. Singh, D., Dixit, A., Dobal, P.S.: Ferroelectricity and ferromagnetism in Fe-doped barium titanate ceramics. Ferroelectrics. **573**, 63–75 (2021). https://doi.org/10.1080/00150193.2021.1890464
5. Dobal, P.S., Dixit, A., Yu, Z., Guo, R., Bhalla, A.S., Katiyar, R.S.: Micro-Raman scattering and dielectric investigations of phase transition behaviour in $BaTiO_3$-$BaZrO_3$ system. J. Appl. Phys. **89**, 8085 (2001). https://doi.org/10.1063/1.1369399
6. Fan, T., Shi, Q., Xiao, W., Zhang, G., et al.: Colossal permittivity with ultra-wide temperature stability in Bi^+ Ca co-doped $BaTiO_3$. J. Mater. **11**, 100877 (2025). https://doi.org/10.1016/j.jmat.2024.03.019
7. Cernea, M., Galassi, C., Vasile, B.S., et al.: Electrical investigations of holmium-doped $BaTiO_3$ derived from sol-gel combustion. J. Mater. Res. **25**, 1057–1063 (2010). https://doi.org/10.1557/JMR.2010.0149
8. Aslla-Quispe, A.P., Miwa, R.H., Guerra, J.D.S.: Role of the rare-earth doping on the multiferroic properties of BaTiO3: first-principles calculation. Phys. B: Cond. Mat. **615**, 413107 (2021). https://doi.org/10.1016/j.physb.2021.413107
9. Sapkota, P., Fujii, I., Kim, S., Ueno, S., et al.: Mn–Nb co-doping in barium titanate ceramics by different solid-state reaction routes for temperature stable and DC-bias free dielectrics. Ceram. Int. **48**, 2154–2160 (2022). https://doi.org/10.1016/j.ceramint.2021.09.304
10. Vigneshwaran, B., Kuppusami, P., Ajithkumar, S., et al.: Temperature-dependent Raman and dielectric studies of Sm and Zr Co-doped $BaTiO_3$ ceramics. J. Mater. Sci. Mater. Electron. **35**, 353 (2024). https://doi.org/10.1007/s10854-024-12060-2
11. Khan, A.R., Goel, R., Gupta, A., et al.: Improved multiferroic and magnetodielectric properties in Ni and Fe co-doped barium strontium titanate ceramics. J. Mater. Sci. Mater. Electron. **35**, 1383 (2024). https://doi.org/10.1007/s10854-024-12951-4
12. Sun, Q., Gu, Q., Zhu, K., et al.: Crystalline structure, defect chemistry and room temperature colossal permittivity of Nd-doped barium titanate. Sci. Rep. **7**, 42274 (2017). https://doi.org/10.1038/srep42274
13. Zakeri, F., Javid, A., Orooji, Y., et al.: Al-Ce co-doped $BaTiO_3$ nanofibers as a high-performance bifunctional electrochemical supercapacitor and water-splitting electrocatalyst. Sci. Rep. **14**, 9833 (2024). https://doi.org/10.1038/s41598-024-54561-4
14. Xin, C.-R., Zhang, J., Liu, Y., et al.: Polymorphism and dielectric properties of Sc-doped $BaTiO_3$ nanopowders synthesized by sol–gel. Mater. Res. Bull. **48**, 2220–2226 (2013). https://doi.org/10.1016/j.materresbull.2013.02.044
15. Lu, D.-Y., Gao, X.-L., Wang, S., et al.: Abnormal Curie-temperature shift in Ho-doped $BaTiO_3$ ceramics with the self-compensation mode. Results Phys. **12**, 585–591 (2019). https://doi.org/10.1016/j.rinp.2018.11.094
16. Yu, D., Jin, A.-M., Zhang, Q.-L., et al.: Scandium and gadolinium co-doped BaTiO3 nanoparticles and ceramics prepared by sol–gel–hydrothermal method: facile synthesis, structural characterization and enhancement of electrical properties. Powder Technol. **283**, 433–439 (2015). https://doi.org/10.1016/j.powtec.2015.05.048

17. Ashiri, R.: Detailed FT-IR spectroscopy characterization and thermal analysis of synthesis of barium titanate nanoscale particles through a newly developed process. Vib. Spectrosc. **66**, 24–29 (2013). https://doi.org/10.1016/j.vibspec.2013.02.001
18. Nayak, S., Sahoo, B., Chaki, T.K., Khastgir, D.: Facile preparation of uniform barium titanate (BaTiO3) multipods with high permittivity: impedance and temperature dependent dielectric behavior. RSC Adv. **4**, 1212–1224 (2014). https://doi.org/10.1039/C3RA44815K
19. Elmahgary, M.G., Mahran, A.M., Ganoub, M., Abdellatif, S.O.: Optical investigation and computational modelling of $BaTiO_3$ for optoelectronic devices applications. Sci. Rep. **13**, 4761 (2023). https://doi.org/10.1038/s41598-023-31652-2
20. Cui, B., Yu, P., Tain, J., Chang, Z.: Preparation and characterization of Co-doped $BaTiO_3$ nanosized powders and ceramics. Mat. Sci. Eng. B. **133**, 205–208 (2006). https://doi.org/10.1016/j.mseb.2006.07.002
21. Burgos, M., Langlet, M.: The sol-gel transformation of TIPT coatings: a FTIR study. Thin Solid Films. **349**, 19–23 (1999). https://doi.org/10.1016/S0040-6090(99)00139-X
22. Ghamsari, M.S., Bahramian, A.R.: High transparent sol–gel derived nanostructured TiO_2 thin film. Mater. Lett. **62**, 361–364 (2008). https://doi.org/10.1016/j.matlet.2007.05.053
23. Jin, X., Sun, D., Zhang, M., et al.: Investigation on FTIR spectra of barium calcium titanate ceramics. J. Electroceram. **22**, 285–290 (2009). https://doi.org/10.1007/s10832-007-9402-1
24. Singh, M., Yadav, B.C., Ranjan, A., Kaur, M., Gupta, S.K.: Synthesis and characterization of perovskite barium titanate thin film and its application as LPG sensor. Sensors Actuators B Chem. **241**, 1170–1178 (2017). https://doi.org/10.1016/j.snb.2016.10.018

Design and Development of Rogers RT/Duroid 5880 Substrate-Based MIMO Antenna for Band-Rejected Ultrawideband Applications

Summaiyya Saleem(✉), Vinita Mirdha, Arti Meena, and Jitendra Kumar Verma

Department of Physics, University of Rajasthan, Jaipur, India
sumaiyya.jpr@gmail.com

Abstract. A WLAN band-notched multiple-input multiple-output (MIMO) ultra-wideband (UWB) antenna is designed and analyzed. The designed MIMO configuration utilizes 0.79 mm thick Rogers RT/duroid 5880 laminate (relative permittivity = 2.2, loss tangent = 0.0009), which is a ceramic-filled Polytetrafluoroethylene (PTFE) composite substrate. The multiple antenna system comprises of two orthogonally placed slotted circular radiators for achieving polarization diversity and high isolation performance. In order to eliminate WLAN band signals, the radiators are carved with split ring resonator (SRR) slots. The modelled diversity antenna having size of 54 × 30 mm^2 performs in the frequency range 3-13 GHz, with suppressed band ranging between 4.4 and 5.4 GHz. The isolation amid radiators is significantly higher than 25 dB and the highest gain of 6.5 dB is attained at 8.5 GHz. The proposed MIMO configuration exhibits good diversity performance. Additionally, the antenna performs well according to a number of examined metrics due to the use of low loss substrate with less relative permittivity. The recommended antenna is an appropriate choice for UWB MIMO communication systems, as demonstrated by the simulation outcomes.

Keywords: WLAN · UWB · band-notched · MIMO · PTFE · polarization diversity

1 Introduction

Modern wireless communication systems make extensive use of microstrip antennas because of their small size, low profile, affordability, simplicity of production, and adaptability. A microstrip antenna consists of a dielectric substrate material that separates the conducting patch from the conducting ground plane. They are especially popular in applications like mobile phones, Global Positioning System (GPS), satellite communication, and radar systems. For high radiation efficiency, common microstrip patch antennas operating at microwave frequencies employ substrates including quartz, PTFE, and honeycomb. Although these materials have exceptional electrical performance, their cost is quite large for commercial applications such as satellite and mobile communication. The radiation efficiency is altered by the material change, which is due to variation in the dielectric constant. When choosing a substrate, the loss tangent is a crucial factor. The

S. Kumari et al. (Eds.): ICMTP 2024, SPPHY 311, pp. 223–233, 2025.
https://doi.org/10.1007/978-981-96-6795-6_26

dielectric material will be better with lower losses if the loss tangent is less. Antenna performance is improved when the dielectric constant (ε_r) is low enough to create a greater aperture area, which raises the antenna's gain and bandwidth. For commercial use, the Federal Communications Commission (FCC) issued ultrawideband spectrum (3.1–10.6 GHz) [1]. UWB technology has garnered significant attention in the antenna sector because of its advantages, which include simplicity of production, affordable cost, low power consumption, high data rate, and huge frequency bandwidth [2]. This technology may be applied to a wide range of applications, including sensing and medical applications, Unmanned Aerial Vehicle (UAV) systems, wireless body area networks (WBANs), indoor localization systems, the Internet of Things (IoT) and weather radar applications [3–8]. Because UWB antennas have a low power spectrum density of −40 dBm, they typically interfere with narrowband signals. Band rejection characteristics in UWB antennas refer to the ability of an antenna to attenuate or reject signals within certain frequency bands while maintaining efficient operation over the UWB spectrum. These characteristics are important in UWB antennas to prevent interference from these narrowband or unwanted signals that might be present in the environment. Several ultrawideband monopole antennas implementing different band rejection strategies for use in a variety of wireless applications have been documented in the literatures [9–11]. Signal deterioration is another issue that UWB systems face due to multipath fading. Several antennas that serve as separate channels are used in the MIMO technology to send and receive data simultaneously. Without requiring more bandwidth or power, the MIMO approach can improve the antenna system's data throughput, channel capacity, and reliability. Additionally, it is more adept at using multipath to extend communication range, which makes it appropriate for the expanding number of modern wireless applications. A UWB-MIMO communication system that combines MIMO and UWB technologies can meet the need for improved overall system reliability and a high data transfer rate. Current research on diversity systems primarily focuses on band-notched, broadband, and high isolation features. Introducing conductor branches, slots, neutralization line [12], and decoupling element are ways to improve isolation between antenna elements in MIMO systems. A line slot placed in the ground plane and a T-shaped slot can assist increase in the impedance bandwidth and isolation [13]. By including slots, branches, parasitic components, metamaterial-based techniques, electromagnetic bandgap (EBG) structures, tuning stubs, and defected ground structures (DGS), the notched bands can potentially be incorporated to remove signal interference in UWB spectrum with present narrowband antenna systems. A compact MIMO antenna with wideband behaviour is described in [14]. To improve impedance bandwidth and isolation, a T-shaped stub was attached to ground. An UWB MIMO antenna with dual band elimination and polarization diversity features was devised for use with USB dongles [15]. Some other diversity antenna structures embedding the band stop filters for portable notched band ultrawideband applications have also been studied [16–20]. An ultrawideband MIMO antenna installed with a modified inter-digital capacitor (IDC) for realization of isolation of more than 16 dB is studied in [19]. It is incorporated with separate defected ground planes and the isolation structure is designed on the same side as that of the radiators. It eliminates two frequency bands by using ground stub and slot in patch, as well as by using a modified ground plane for improved isolation. A two-port MIMO antenna incorporated with stubs, chip resistors and DGS as decoupling structures for optimal isolation is investigated in [21].

A miniaturized MIMO antenna with polarization diversity and band suppression features for handheld ultrawideband devices is described in this research study. Polarization diversity technique due to orthogonal arrangement of antenna elements results into enhanced isolation with $|S_{12}| \leq -25$ dB. The achieved isolation is improved as compared to our earlier work based on spatial diversity feature [22]. The simulated reflection coefficient with $|S_{11}| \leq -10$ dB over the operating frequency spectrum 3-13 GHz (UWB) is achieved for the suggested MIMO antenna except at the suppressed WLAN band (4.4-5.4 GHz). The recommended antenna is well suited for UWB systems, according to simulation findings. The design and optimization have been carried out using 3D-electromagnetic (EM) simulation software Computer Simulation Technology (CST) Microwave Studio.

2 Design and Methodology of Single Antenna Element

Figure 1 displays the layout of a single UWB antenna, which is designed on an RT/Duroid 5880 substrate. This material was chosen because it is ideally suited for high frequency/broadband applications due to its low dielectric constant and low dielectric loss. Features of reinforced PTFE and RT/Duroid materials include superior chemical resistance, homogeneous electrical characteristics throughout frequency, reduced moisture absorption, and a low level of electrical loss. RT/Duroid materials are filled PTFE ceramic composite laminates, which are utilized in defence, aerospace, and high reliability applications. Space satellite transceivers, military radar systems, missile guidance systems, millimeter wave applications, airborne and ground-based radar systems would all make use of it. The modelled antenna is a slotted circular shaped radiator with radius a = 10.2 mm. For a 50-ohm impedance, the feedline's length and width are calculated as Lf + Lf1 = 8.85 mm and Wf1 = 3 mm. A partial ground structure with length Lg = 8.2 mm is used to produce a broad impedance bandwidth. The antenna has total dimensions of 30 × 24 mm^2. For a lower cut-off frequency of 3.05 GHz, the distance between the ground and the circular patch is considered to be g = 0.65 mm [23]. A split ring resonator (SRR) slot with outer and inner radii of 3.8 mm and 3.3 mm, acting as a band stop filter is added to the slotted patch in order to create a suppressed band at 5.2 GHz, thus to prevent WLAN band interference in the UWB spectrum. Table 1 lists the optimal parameters for the suggested diversity antenna.

Figure 2 displays the single antenna element's development phases and Fig. 3 depicts the corresponding return loss characteristics of these design steps. As depicted in Fig. 2, the design procedure started with a conventional circular patch antenna (Step 1), fed by a simple microstrip feedline. It resonates in the frequency range 3.34-9.05 GHz and exhibits wideband performance. The second design stage (Step 2) results after tapering the feedline from 2 mm to 3 mm, which results in the improvement of impedance bandwidth of the radiator. It thus shows ultrawideband (3.25-13.53 GHz) behavior.

The antenna resonates in the UWB band (3.2–13.49 GHz) and improved impedance matching is seen close to 6 GHz frequency when an open-ended slot is added to the

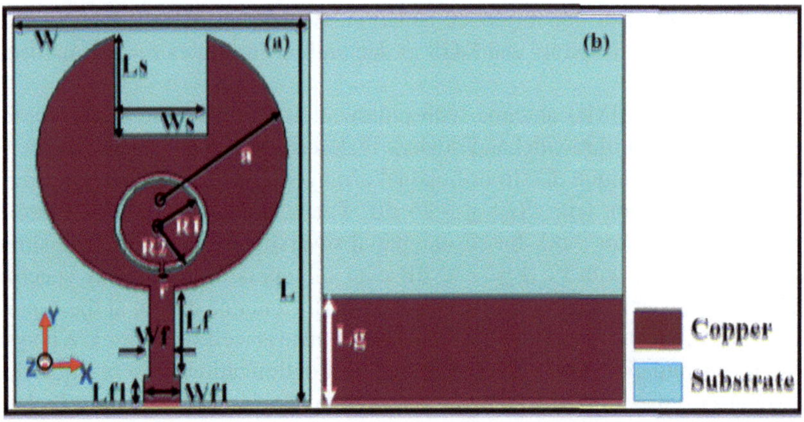

Fig. 1. Single UWB antenna layout (a) front side (b) back side.

Table 1. Optimized parameters for proposed ultrawideband diversity antenna.

Parameter	Value (mm)
W	24
L	30
a	10.2
Lg	8.2
Wf	2
R1	3.8
R2	3.3
Ls	8.58
Lf1	2
Wf1	3
r	0.3
Ws	7.5
Lf	6.85

top region of the antenna (Step 3). The last design phase (Step 4) is the outcome of the addition of SRR slot to the radiator's center in Step 3. The recommended monopole radiator has WLAN band (4.76 GHz–5.43 GHz) elimination feature and operates in the range 3.19-13.49 GHz. It is clearly indicated that the impedance bandwidth is not affected by introducing SRR, but only causes the notched band to occur. In Fig. 4, the current spread over antenna's surface (Step 2 and Step 3) at 5.2 GHz frequency is displayed. As illustrated in Fig. 4(a), the current is uniformly distributed over the slotted patch and feedline (in Step 3).

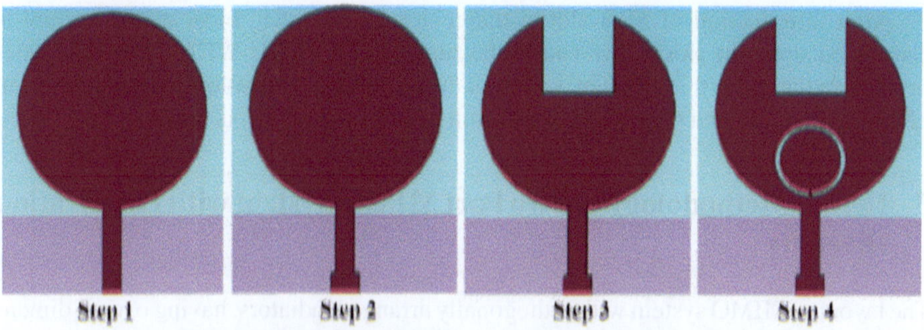

Fig. 2. Phases of development for a single antenna element.

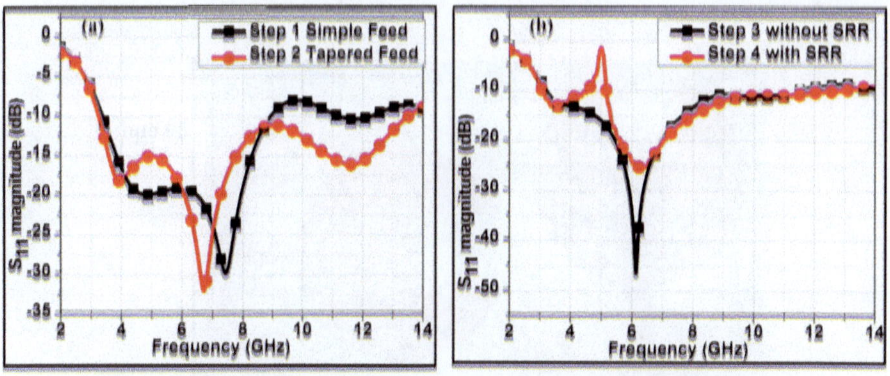

Fig. 3. Return loss profiles for the various phases of antenna development.

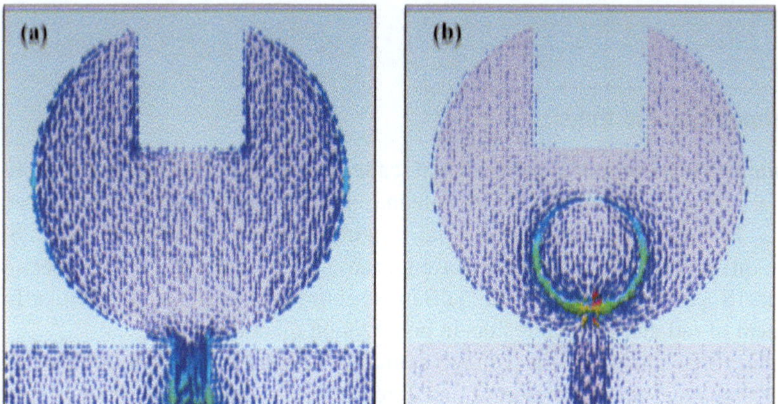

Fig. 4. Current spread on radiator's surface at 5.2 GHz notch frequency (a) Step 3 and (b) Step 4.

After introduction of SRR slot at the center of the patch, most of the current is distributed along the slot, which causes the suppressed 5.2 GHz WLAN band and deteriorated radiation at this frequency. To conduct all these simulation investigations, an EMF simulator named Computer Simulation Technology (CST) is utilized.

3 Design Methodology of Two Port MIMO System with Polarization Diversity

The layout of MIMO system with orthogonally arranged radiators, having overall dimension of $54 \times 30\ \text{mm}^2$ is illustrated in Fig. 5. Two identical slot-based antennas with different ground planes are combined to create the intended MIMO antenna. Polarization diversity improves the isolation between the antenna ports.

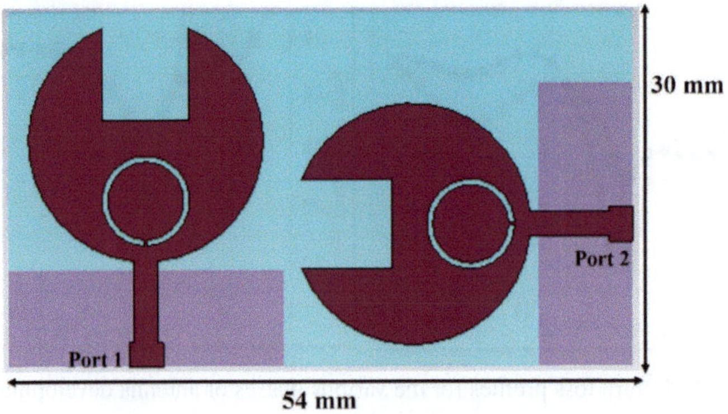

Fig. 5. Proposed ultrawideband diversity antenna.

3.1 Simulation Findings

Impedance Performance. The simulated scattering parameters of the proposed diversity antenna is plotted in Fig. 6(a). The antenna operates in the ultrawideband frequency range 3–13 GHz with suppressed band ranging between 4.4 and 5.4 GHz. S_{11} and S_{22} are different due to asymmetric orientation of radiators. The isolation between the radiators (S_{12} or S_{21}) is well above 25 dB in the entire operational band due to orthogonal alignment of radiators which results in polarization diversity.

Figure 6(b) illustrates the voltage standing wave ratio (VSWR) of the designed MIMO antenna at excitation ports 1 and 2, respectively. The VSWR value is less than 2 except at the rejected WLAN band, which denotes good impedance performance of the antenna.

Diversity Performance. The performance indicators like envelope correlation coefficient (ECC) and diversity gain (DG) are evaluated to investigate the efficacy of the designed diversity antenna in multi-path propagation environment. The ECC evaluates the field coupling amid closely spaced radiating elements and provides information

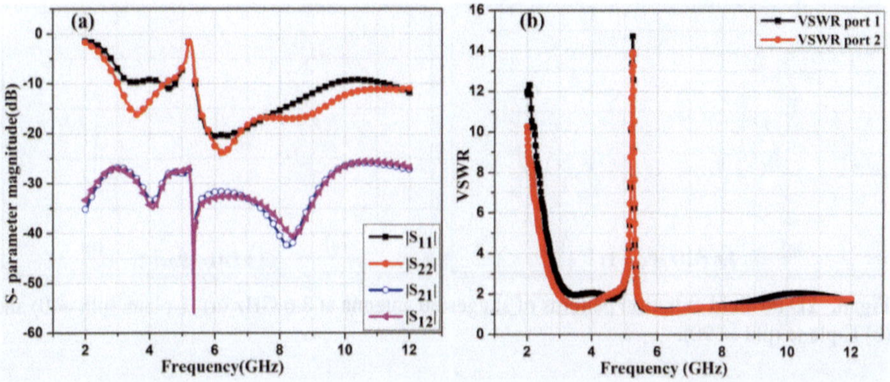

Fig. 6. (a) S-parameters and (b) VSWR versus frequency response for two ports of the suggested MIMO antenna.

on the correlation between various ports. The ECC value less than 0.5 is practically acceptable. DG measures the improvement in signal reliability achieved through some diversity combining schemes. The suggested antenna's simulated scattering characteristics are used to calculate its ECC and DG, so Fig. 7(a, b) shows the plots of these parameters as a function of frequency. With the exception of the notched band, DG is about 10 dB across the functional band, and the ECC value is less than 0.0025. Since the ECC and diversity gain values are within allowable limits, the antenna demonstrates good diversity behavior.

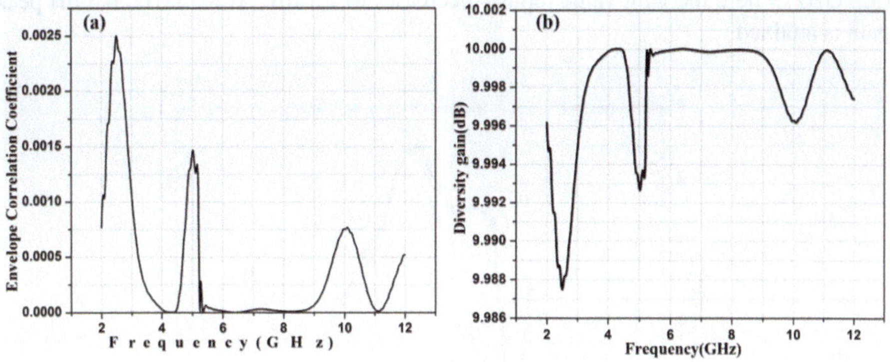

Fig. 7. (a) ECC versus frequency curve and (b) Diversity gain curve as a function of frequency.

Far-field Radiation Characteristics. The E-plane and H-plane radiation patterns are observed at two frequencies 3.6 GHz and 6.3 GHz for port 1 and port 2 for the designed MIMO antenna configuration as shown in Figs. 8 and 9. At two sample frequencies, bidirectional radiation patterns are seen in both major planes.

Figure 10 illustrates the simulated gain versus frequency curve. The simulated gain ranges from 2.9 to 6.5 dB in the operating ultrawideband, except at the notch frequency

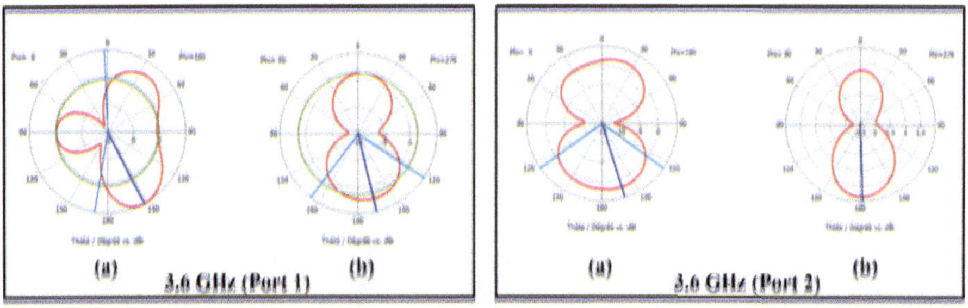

Fig. 8. 2D far-field radiation patterns of suggested antenna at 3.6 GHz (a) H-plane (phi = 0) and (b) E-plane (phi = 90).

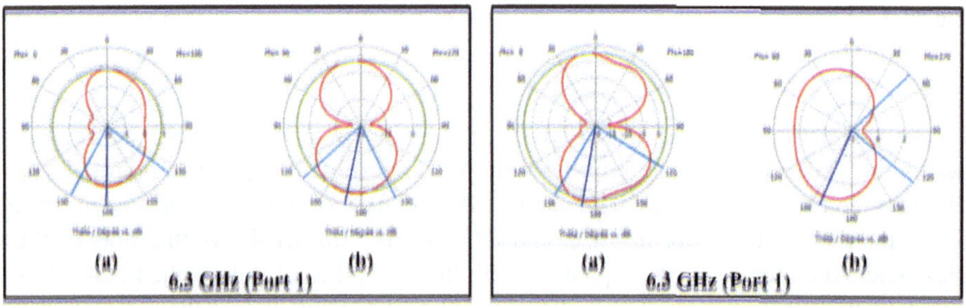

Fig. 9. 2D far-field radiation patterns at 6.3 GHz (a) H-plane (phi = 0) and (b) E-plane (phi = 90).

(5.2 GHz) where the gain value rapidly decreases to 2.1 dB. At 8.5 GHz, 6.5 dB peak gain is attained.

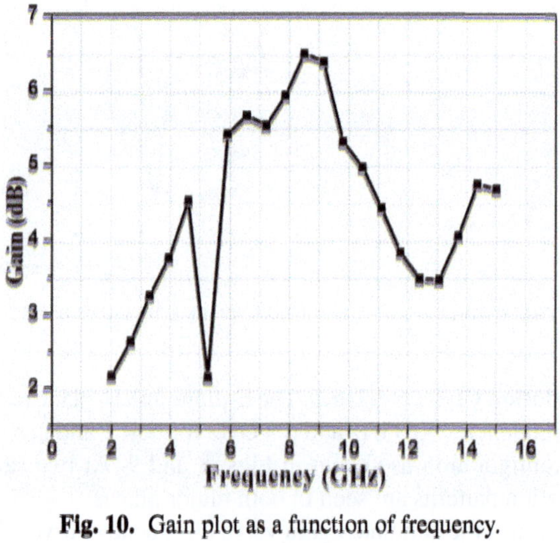

Fig. 10. Gain plot as a function of frequency.

Table 2. Comparison of performance with previous literatures.

Antenna	Size	Dielectric constant	Bandwidth	Isolation (dB)	Peak gain (dBi)	ECC	Rejected band
[18]	21 × 16 × 1	1.6	3.34–11.93	>20.3	3.5	<0.2	2.4–2.485 GHz, 5.15–5.825 GHz
[24]	25 × 45	4.4	3.1–12	>15	5.5	<0.2	NA
[25]	38.5 × 38.5	4.4	3.08–11.8	>15	3.6	<0.02	5.03–5.97 GHz
[26]	33 × 26	4.4	3.1–10.6	>15	6	<0.03	4.7–5.5 GHz
[27]	35 × 35	4.4	3–12	>20	–	<0.3	4.9–5.8 GHz
[28]	58 × 58	3.6	2.8–11	>14	2.2	<0.025	5.0–6.3 GHz
This work	54 × 30	2.2	3–13	>25	6.5	<0.0025	4.4–5.4 GHz

Table 2 provides a comparison of the designed antenna with previously reported comparable band notched diversity antennas. The suggested antenna has a lower ECC value, wider bandwidth, a greater gain, more isolation, and smaller size than the one described in [28]. Compared to all of the published literatures, the proposed antenna has low ECC, high gain, wide bandwidth, and significant isolation. As a result, the recommended antenna outperforms the previously reported antennas in a number of performance metrics.

4 Conclusion

This study proposes a miniaturized ultrawideband multiple antenna system based on a low loss dielectric substrate that may eliminate interference caused by WLAN band (4.5–5.4 GHz). Reduction in mutual coupling (S_{21} > 25 dB) is achieved by introducing polarization diversity to nearly placed antenna elements. The reflection coefficient, transmission coefficient, gain, far-field radiation patterns, surface current density, and diversity performance metrics are examined. The proposed antenna possesses large impedance bandwidth, large isolation, high gain, and excellent diversity behavior over the entire operating UWB range (3–13 GHz). The evaluated diversity performance parameters are also found to be within the acceptable limits. The suggested MIMO configuration is best suited for UWB diversity applications.

References

1. Federal Communications Commission: Federal Communications Commission revision of part 15 of the commission's rules regarding ultra-wideband transmission system from 3.1 to 10.6 GHz, ET-Docket, Washington, DC, USA, (2002)
2. Ahmed, S., Geok, T.K., Alias, M.Y., Hossain, F., Alsariera, H., Abdaziz, A., Soh, P.J.: A UWB antenna array integrated with multimode resonator bandpass filter. Electronics. **10**(5), 607 (2021)
3. Mahmud, M.Z., Islam, M.T., Misran, N., Almutairi, A.F., Cho, M.: Ultra-wideband (UWB) antenna sensor based microwave breast imaging: a review. Sensors. **18**(9), 2951 (2018)

4. Balderas, L.I., Reyna, A., Panduro, M.A., Del Rio, C., Gutierrez, A.R.: Low-profile conformal UWB antenna for UAV applications. IEEE Access. **7**, 127486–127494 (2019)
5. Yadav, A., Singh, V.K., Bhoi, A.K., Marques, G., Garcia-Zapirain, B., de la Torre Díez, I.: Wireless body area networks: UWB wearable textile antenna for telemedicine and mobile health systems. Micromachines. **11**(6), 558 (2020)
6. Karoui, M.S., Ghariani, N., Lahiani, M., Ghariani, H.: A compact UWB elliptic antenna for indoor localization system. IEICE Electron. Express. **16**, 1–5 (2019)
7. Saha, T.K., Knaus, T.N., Khosla, A., Sekhar, P.K.: A CPW-fed flexible UWB antenna for IoT applications. Microsyst. Technol. **28**, 5–11 (2022)
8. Slimani, A., Bennani, S.D., El Alami, A., Amellal, M.: Gain and bandwidth enhancement of new planar microstrip array antennas geometry for C band weather radar applications. Int. J. Microwave Wireless Technol. **9**(5), 1139–1146 (2017)
9. Lv, Y., Zhang, J., Hou, H.: A novel triple band-notched UWB printed monopole antenna. Progress Electromagn. Res. M. **81**, 85–95 (2019)
10. Garg, R.K., Nair, M.V.D., Singhal, S., Tomar, R.: Compact CPW-fed asymmetric UWB antenna with sufficient WLAN-band rejection. Progress Electromagn. Res. C. **108**, 201–210 (2021)
11. Mahfuz, M.M.H., Islam, M.R., Habaebi, M.H., Sakib, N., Hossain, A.K.M.Z.: A notched UWB microstrip patch antenna for 5G lower and FSS bands. Microwave Opt. Tech. Lett. **64**(4), 796–802 (2022)
12. Tang, T.-C., Lin, K.-H.: An ultrawideband MIMO antenna with dual band-notched function. IEEE Antennas Wireless Propagat. Lett. **13**, 1076–1079 (2014)
13. Luo, C.-M., Hong, J.-S., Zhong, L.-L.: Isolation enhancement of a very compact UWB-MIMO slot antenna with two defected ground structures. IEEE Antennas Wireless Propagat. Lett. **14**, 1766–1769 (2015)
14. Gurjar, R., Upadhyay, D.K., Kanaujia, B.K., Sharma, K.: A novel compact self-similar fractal UWB MIMO antenna. Int. J. RF Microwave Comp. Aided Eng., e21632 (2018)
15. Sipal, D., Abegaonkar, M.P., Koul, S.K.: Compact dual band notched UWB MIMO antenna for USB dongle application with pattern diversity characteristics. Progress Electromagn. Res. C. **87**, 87–96 (2018)
16. Yadav, D., Abegaonkar, M.P., Koul, S.K., Tiwari, V., Bhatnagar, D.: Two element band-notched UWB MIMO antenna with high and uniform isolation. Progress Electromagn. Res. M. **63**, 119–129 (2018)
17. Liu, L., Cheung, S.W., Yuk, T.I.: Compact MIMO antenna for portable UWB applications with band-notched characteristic. IEEE Trans. Antennas Propagat. **63**(5), 1917–1924 (2015)
18. Biswas, A.K., Basu, B., Nandi, A.: Compact wearable UWB MIMO antenna with reduced mutual coupling and notch characteristics of WLAN band. Arab. J. Sci. Eng. **47**, 14561–14569 (2022)
19. Kumar, A., Ansari, A.Q., Kanaujia, B.K., Kishor, J., Kumar, S.: An ultra-compact two-port UWB-MIMO antenna with dual band-notched characteristics. AEU-Int. J. Electron. Commun. **114**, 152997 (2020)
20. Khan, M.S., Naqvi, S.A., Iftikhar, A., Asif, S.M., Fida, A., Shubair, R.M.: A WLAN band-notched compact four element UWB MIMO antenna. Int. J. RF Microwave Comput. Aided Eng. **30**(9), e22282 (2020)
21. Li, Y., Bian, L., Xu, K., Liu, Y., Wang, Y., Chen, R., Xie, S.: Mutual coupling reduction for monopole MIMO antenna using L-shaped stubs, defective ground and chip resistors. AEU-Int. J. Electron. Commun. **160**, 154524 (2023)
22. Saleem, S., Kumari, S., Mirdha, V., Yadav, D., Bhatnagar, D.: Circular split ring resonator (SRR) slot and ground stub based slotted circular ultra-wideband MIMO antenna with WLAN band exclusion and high isolation performance. Sādhanā. **49**, 26 (2024)

23. Kumar, G., Ray, K. P.: Broadband Microstrip Antennas. Artech House (2003)
24. Mathur, R., Dwari, S.: Compact CPW fed ultrawideband MIMO antenna using hexagonal ring monopole antenna elements. AEU-Int. J. Electron. Commun. **93**, 1–6 (2018)
25. Kang, L., Li, H., Wang, X., Shi, X.: Compact offset microstrip-fed MIMO antenna for band notched UWB applications. IEEE Antennas Wireless Propagat. Lett. **14**, 1754–1757 (2015)
26. Liu, G., Liu, Y., Gong, S.: Compact uniplanar UWB MIMO antenna with band-notched characteristic. Microwave Opt. Technol. Lett. **59**(9), 2201–2207 (2017)
27. Zhu, J., Li, S., Feng, B., Deng, L., Yin, S.: Compact dual-polarized UWB quasi-self-complementary MIMO/diversity antenna with band-rejection capability. IEEE Trans. Antennas Propagat. Lett. **15**, 905–908 (2016)
28. Chacko, B.P., Augustin, G., Denidni, T.A.: Uniplanar slot antenna for ultrawideband polarization- diversity applications. IEEE Antenna Wireless Propagat. Lett. **12**, 88–91 (2013)

Fabrications and Characterizations of Ag/p-Si and Ag/PEO/p-Si Devices

Monu Singhal, Vipin Choudhary, Nairaj Jat, Kiran Githala, and Manisha Gupta(✉)

Department of Physics, University of Rajasthan, Jaipur 302004, Rajasthan, India
drguptamanisha@uniraj.ac.in

Abstract. The global electronics market's intense competition has driven increased academic interest in metal-semiconductor interactions, which are critical for enhancing electronic device efficiency. Researchers are exploring innovative methods to create reliable contacts using various metals, recognizing their importance in all modern electronics. This rapidly evolving field underscores the need to thoroughly understand the physical and electrical properties of these contacts. Schottky contacts, for instance, are pivotal in applications like radiation detection and radio frequency technology. High-quality ohmic contacts are essential for optimizing semiconductor device performance, as they ensure minimal resistance at the metal-semiconductor junction. This study specifically investigates the characteristics of aluminum ohmic contacts subjected to annealing at 550 °C for 15 min and examines the behaviour of Ag/p-Si and Ag/PEO/p-Si structures. The findings contribute to advancements in semiconductor technology, highlighting the significance of contact quality in the overall performance and reliability of electronic devices.

Keywords: Semiconductors · p-n Junction Diodes · MS Interface · Schottky Barrior Diodes · Ohmic Contacts

1 Introduction

Semiconductors are integral to modern electronics due to their tunable electrical properties. Defect engineering, particularly through controlled ion irradiation, plays a crucial role in modifying optical, chemical, and transport properties to enhance device performance [1–4].

Schottky barrier diodes (SBDs), formed by replacing a p-type semiconductor with a metal, enable fast switching by eliminating carrier recombination, making them ideal for high-frequency power applications [5–7]. Metal-semiconductor interfaces are categorized as ohmic contacts, which ensure minimal resistance, or Schottky barriers, which exhibit rectifying behavior similar to p-n junctions [8–10]. The demand for ultra-low resistance contacts has increased with VLSI advancements, necessitating stable electrical connections for efficient microwave power devices [11].

Polyethylene oxide (PEO), a linear polymer with polar ether groups, exhibits strong ionic complexation, making it suitable for polymer electrolytes. Its semi-crystalline

S. Kumari et al. (Eds.): ICMTP 2024, SPPHY 311, pp. 234–240, 2025.
https://doi.org/10.1007/978-981-96-6795-6_27

nature supports diverse applications in electronic devices. In this study, a PEO-based device was fabricated using high-molecular-weight PEO (>5 × 10^6 g/mol) sourced from Sigma-Aldrich, with films prepared via the solution-casting technique [12–15].

2 Experimental Procedure

The silicon wafers are cleaned using the RCA method, a standard process developed by Werner Kern (1960) to remove organic contaminants, heavy metals, and alkali residues. The cleaning process follows the RCA flowchart [17–19] (Fig. 1).

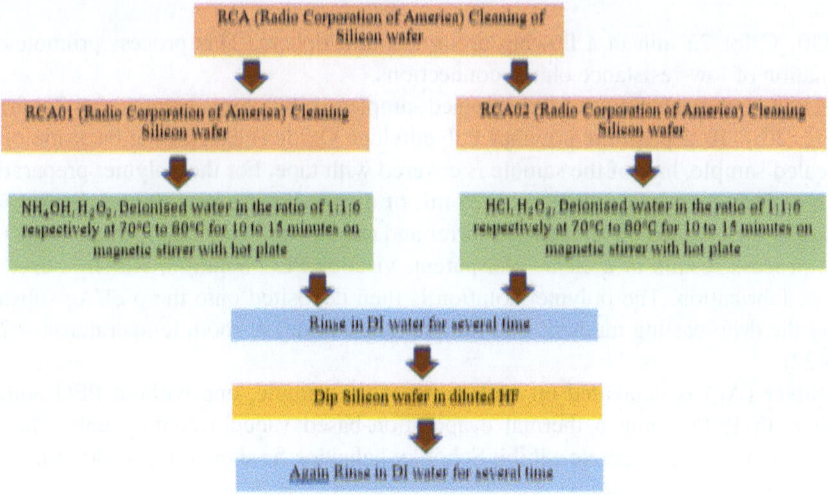

Fig. 1. Cleaning process through the RCA flowchart.

The cleaned silicon wafer is now ready for the fabrication of a metal-semiconductor Schottky diode. The single-side mirror-polished surface of the wafer is used to form ohmic contacts. To achieve this, a metal with a work function close to that of p-type silicon (p-Si) is selected, ensuring that the difference between the metal's work function and that of p-Si is less than 0.2 eV. The work function of p-Si is 4.85 eV, making silver (Ag), with a work function of 4.6 eV, a suitable choice. Therefore, Ag is an ideal material for the fabrication of ohmic contacts. To fabricate ohmic contacts, a silver (Ag) layer is deposited onto the mirror-polished surface of the p-Si wafer using a thermal evaporation-based vacuum coating unit. For metal deposition, the vacuum chamber pressure must be reduced to 10^{-6} mbar. Subsequently, a 100 nm thick Ag layer is deposited onto the p-Si wafer using the electron beam (e-beam) evaporation method [20, 21].

The current-voltage (I-V) characteristics of the coated sample are studied using the B2901A Source Meter. Multiple measurements are taken at varying voltage levels, and the corresponding current and voltage values are plotted, as illustrated in Fig. 2(a).

To convert the Schottky contacts into ohmic contacts, annealing of the sample is required. The coated sample is placed in a tube furnace and subjected to thermal treatment

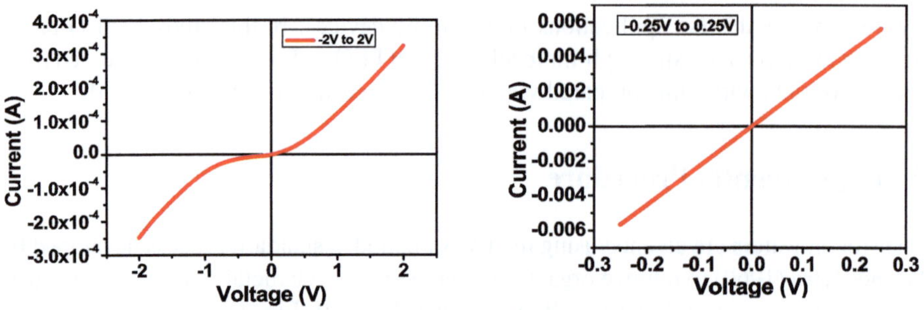

Fig. 2. (a) I-V Graph before annealing. (b) I-V Graph after annealing.

at 550 °C for 25 min in a flowing argon gas atmosphere. This process promotes the formation of low-resistance ohmic connections.

The I-V characteristics of the annealed sample exhibit ohmic behavior, as illustrated in Fig. 2(b). To deposit the polymer Polyethylene Oxide (PEO) on the backside of the annealed sample, half of the sample is covered with tape. For the polymer preparation, 0.5 g of PEO powder is mixed with 25 mL of distilled water in a 100 mL beaker. The beaker is then placed on a magnetic stirrer and stirred for 5 h at a temperature of 70 °C. This process results in a clear, transparent, viscous PEO solution, which is used for device fabrication. The polymer solution is then deposited onto the p-Si/Ag substrate using the drop-casting method. The sample is left to dry at room temperature for 24 h [19–22].

Silver (Ag) is deposited on both parts of the sample, one without PEO and the other with PEO, using a thermal evaporation-based vacuum coating unit. The I-V characteristics of the sample exhibit Schottky behavior, as shown in Fig. 3(a, b).

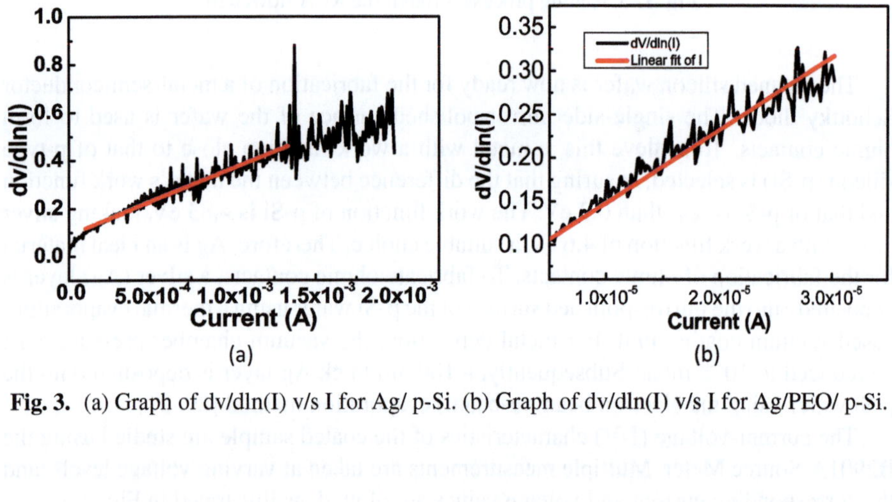

Fig. 3. (a) Graph of dv/dln(I) v/s I for Ag/ p-Si. (b) Graph of dv/dln(I) v/s I for Ag/PEO/ p-Si.

3 Experimental Results and Discussion

In Schottky contacts, the majority of charge carriers contribute to current transport, which is characterized by "thermionic emission" through the interface barrier. The traditional thermionic emission equation is applied to fit the experimental data. This equation is given by:

$$I = AA^*T^2\exp\left(\frac{-q\phi_B}{kT}\right)\left[\exp\left(\frac{q(V-IR_S)}{nkT}\right) - \exp\left\{\left(\frac{q(V-IR_S)}{kT}\right)\left(\frac{1}{n}-1\right)\right\}\right] \tag{1}$$

Neglecting second term of Eq. (1) and equation become

$$I = AA^*T^2\exp\left[\left(\frac{-q\phi_B}{kT}\right) + \left(\frac{q(V-IR_S)}{nkT}\right)\right] \tag{2}$$

Here A^* = Richardson constant, which corresponds to the effective mass m* of semiconductor and it is given by [16]

$$A^* = \frac{4\pi m^* qk}{h^3}$$

Taking the natural logarithm (*ln*) of both sides of Eq. (2) and differentiating with respect to $ln(I)$, the equation is then rearranged as follows:

$$\frac{dV}{d(ln(I))} = IR_S + n\left(\frac{kT}{q}\right) \tag{3}$$

This is known as the Cheung function [15, 18, 20]. It represents a straight-line relationship between $\frac{dV}{d(ln(I))}$ and I. The slope and intercept of this line correspond to series resistance of the Schottky barrier diode R_S and $n\left(\frac{kT}{q}\right)$ respectively. Here $k = 1.38 \times 10^{-23}$, $T = 300K$, $q = 1.6 \times 10^{-19}$ and Ideality factor (n) can be obtained using Eq. (4).

$$n = 38.65 \times (\text{Intercept}) \tag{4}$$

The series resistance (R_S) of a Schottky barrier diode (SBD) is a critical characteristic. Cheung and Cheung [16] proposed a method for calculating the series resistance using forward-bias I–V data. Cheung's functions can be applied to characterize the forward-bias I–V behavior of a Schottky structure with thermionic emission.

To calculate Schottky barrier height on the MS interface ϕ_B, a new function H(I) is defined using Eq. (2).

Taking the natural logarithm (ln) of both sides of Eq. (2) and rearranging it gives:

$$IR_S + n\phi_B = V - n\left(\frac{kT}{q}\right)ln\left(\frac{I}{AA^*T^2}\right) \tag{5}$$

The right-hand side (RHS) term of Eq. (4) can be defined as H(I). Therefore,

$$H(I) = IR_S + n\phi_B \tag{6}$$

The above equation takes the form of a straight line between H(I) and I. In this context, R_S and $n\phi_B$ represent the slope and intercept of the line, respectively, and IR_S corresponds to the voltage drop across the Schottky barrier diode.

$$\phi_B = \frac{\text{Intercept}}{n} \tag{7}$$

R_S can be calculated from the slope of the graphs in Figs. 3(a, b) and 4(a, b). The intercept value for the n is obtained from the graphs in Fig. 3(a, b). Similarly, the intercept value for the ϕ_B is derived from the graphs in Fig. 4(a, b).

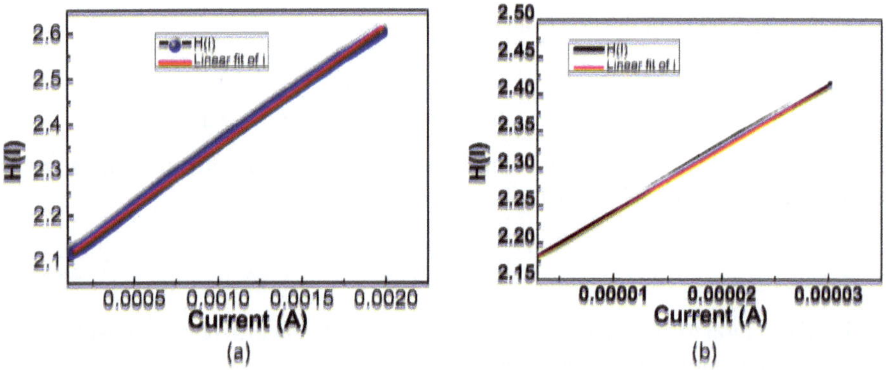

Fig. 4. (a) Graph of H(I) versus I for Ag/p-Si. (b) Graph of H(I) versus I for Ag/PEO/p-Si.

The Ag/p-Si material has a series resistance attained in Table 1, is significantly lower compared to the Ag/PEO/p-Si material. This suggests that the Ag/p-Si junction may have a lower energy barrier for charge flow than the Ag/PEO/p-Si junction. The series resistance can also be calculated using the Cheung equation, which employs a different methodology. The results obtained are consistent with those derived from the thermionic equation. This indicates that Ag/PEO/p-Si demonstrates significantly greater resistance, suggesting increased energy losses when current flows through the material. Ideality factor of the Ag/p-Si junction indicating greater deviation from ideal behavior. While Ag/PEO/p-Si has a higher potential height of 0.86 V. This implies that the Ag/PEO/p-Si junction has a slightly higher energy barrier for charge transport compared to Ag/p-Si.

Table 1. Electronic parameters of designed Ag/p-Si and Ag/PEO/p-Si.

Electronic parameter	Ag/p-Si	Ag/PEO/p-Si
Series resistance (Thermionic equation)	281.14441Ω	8375.48815 Ω
Series resistance (Cheung equation)	266.58766Ω	8508.5149 Ω
Ideality factor	2.779	2.506
Potential height	0.75 V	0.86 V

4 Conclusions

The comparative analysis of Ag/p-Si and Ag/PEO/p-Si materials highlights the distinct advantages and trade-offs associated with each. Ag/p-Si demonstrated lower resistance and a lower potential barrier, suggesting superior charge transport efficiency. However, its higher ideality factor indicates some non-ideal behavior, which may affect its overall performance in certain applications. On the other hand, Ag/PEO/p-Si exhibited higher resistance and a greater potential barrier but showed potential for better stability and unique characteristics that could be advantageous for specific applications. The synthesis of Ag/PEO/p-Si ohmic contacts through vapor deposition proved effective, with promising results including linear I-V characteristics and low contact resistance.

References

1. Chourasiya, H.K., Kulriya, P.K., Panwar, N., Kumar, S.: In-situ study of electrical transport in Pd/n-Si under high energy ion irradiation. Semicond. Sci. Technol. **35**(8), 085004 (2020)
2. Kanjilal, D.: Swift heavy ion-induced modification and track formation in materials. Curr. Sci. **81**, 1560–1566 (2001)
3. Avasthi, D.K., Pivin, J.C.: Ion beam for synthesis and modification of nanostructures. Curr. Sci. **98**(6), 780–792 (2010)
4. Rhoderick, E.H., Williams, R.H.: Metal-Semiconductor Contacts, 2nd edn. Clarendon, Oxford (1988)
5. Islam, N.E., Schamiloglu, E., Mar, A., Zutavern, F., Loubriel, G.M., Joshi, R.P.: Characteristics of trap-filled GaAs photoconductive switches used in high gain pulsed power applications. In: Ultra-Wideband, Short-Pulse Electromagnetics, vol. 5, pp. 461–466. Springer, Boston (2002)
6. Schroder, D.K.: Chapter 8. Semiconductor Material and Device Characterization. John Wiley & Sons, New York (1990)
7. Mugale, M.N., Shukla, S., Chourasia, M.K., Hanif, K., Nazir, A., Singh, S., Gayen, J.R., et al.: Regulatory safety pharmacology and toxicity assessments of a standardized stem extract of Cassia occidentalis Linn. in rodents. Regul. Toxicol. Pharmacol. **123**, 104960 (2021)
8. Muhammed, D.S., Brza, M.A., Nofal, M.M., Aziz, S.B., Hussen, S.A., Abdulwahid, R.T.: Optical dielectric loss as a novel approach to specify the types of electron transition: XRD and UV-vis as a non-destructive technique for structural and optical characterization of PEO-based nanocomposites. Materials. **13**(13), 2979 (2020)
9. Kumar, S., Katharria, Y.S., Kanjilal, D.: Swift heavy ion irradiation-induced defects and electrical characteristics of Au/n-Si Schottky structure. J. Phys. D. Appl. Phys. **41**(10), 105105 (2008)
10. Avasthi, D.K.: Modification and characterisation of materials by swift heavy ions. Def. Sci. J. **59**(4), (2009)
11. Kumar, S., Katharria, Y.S., Baranwal, V., Batra, Y., Kanjilal, D.: Inhomogeneities in 130 MeV Au12+ ion irradiated Au/n-Si (100) Schottky structure. Appl. Surf. Sci. **254**(11), 3277–3281 (2008)
12. Kumar, S., Katharria, Y.S., Kumar, S., Kanjilal, D.: Temperature-dependent barrier characteristics of swift heavy ion irradiated Au/n-Si Schottky structure. J. Appl. Phys. **100**(11), 113723 (2006)
13. Singh, R., Arora, S.K., Kanjilal, D.: Swift heavy ion irradiation induced modification of electrical characteristics of Au/n-Si Schottky barrier diode. Mater. Sci. Semicond. Process. **4**(5), 425–432 (2001)

14. Nava, F., Vittone, E., Vanni, P., Verzellesi, G., Fuochi, P.G., Lanzieri, C., Glaser, M.: Radiation tolerance of epitaxial silicon carbide detectors for electrons, protons and gamma-rays. Nucl. Instrum. Methods Phys. Res. A. **505**(3), 645–655 (2003)
15. Cheung, N.W., Liang, C.L., Liew, B.K., Mutikainen, R.H., Wong, H.: Buried dopant and defect layers for device structures with high-energy ion implantation. Nucl. Instrum. Methods Phys. Res. B. **37**, 941–950 (1989)
16. Elliman, R.G., Ridgway, M.C., Jagadish, C., Pearton, S.J., Ren, F., Lothian, J., Fullowan, T.R., et al.: Single-energy, MeV implant isolation of multilayer III-V device structures. J. Appl. Phys. **71**(2), 1010–1013 (1992)
17. Kumar, S., Katharria, Y.S., Batra, Y., Kanjilal, D.: Influence of swift heavy ion irradiation on electrical characteristics of Au/n-Si (100) Schottky barrier structure. J. Phys. D. Appl. Phys. **40**(22), 6892 (2007)
18. Cheng, H., Zhu, C., Huang, B., Lu, M., Yang, Y.: Synthesis and electrochemical characterization of PEO-based polymer electrolytes with room temperature ionic liquids. Electrochim. Acta. **52**(19), 5789–5794 (2007)
19. Crowell, C.R.: The Richardson constant for thermionic emission in Schottky barrier diodes. Solid State Electron. **8**(4), 395–399 (1965)
20. Cheung, S.K., Cheung, N.W.: Extraction of Schottky diode parameters from forward current-voltage characteristics. Appl. Phys. Lett. **49**(2), 85–87 (1986)
21. Brillson, L.J.: The structure and properties of metal-semiconductor interfaces. Surf. Sci. Rep. **2**(2), 123–326 (1982)
22. Lane, C.H.: Aluminum metallization and contacts for integrated circuits. Metall. Mater. Trans. B Process Metall. Mater. Process. Sci. **1**(3), 713–724 (1970)

Ab-Initio Investigation of Optical and Electronic-Structure Properties of Monoclinic Tungsten Trioxide (WO_3): A DFT Approach

Khushbu Dhaked[1](✉), Rimpy Shukla[1], Krishna S. Sharma[1], and Ramphal Sharma[1,2]

[1] Department of Physics, IIS (Deemed to be University), Jaipur 302020, Rajasthan, India
khushbudhakad@gmail.com

[2] Center of NanoScience and Technology, IIS (Deemed to be University), Jaipur 302020, Rajasthan, India

Abstract. In this study, a theoretical investigation of the electronic, structure and optical properties of monoclinic WO_3 is made using Density Functional Theory (DFT). The Full Potential (APW + lo) based code, WIEN2k is used along with Generalized Gradient Approximation (GGA) and Perdew-Burke-Ernzerhof (PBE) potential functional, which take proper account of exchange and correlation interactions, to analyze the electronic charge distribution in the unit cell of WO_3, band structure, density of states, and optical properties of WO_3. Crystal stability is checked through calculation of cohesive energy by using Quantum Espresso code, which uses plane wave basis sets and ultra soft pseudopotentials are used to calculate ground state energy. Our findings reveal occurrence of a DOS gap of about 2.0 eV just above the Fermi level in the DOS spectrum and direct band gap of about 2.0 eV occurs above the Fermi level at the symmetry point Γ (Gamma), which are in qualitative agreement with the reported optical band gap values of 2.6–3.0 eV. The optical properties such as optical conductivity, refractive index, absorption coefficient, extinction coefficient, dielectric function, and electron energy loss function, are also investigated, which provide valuable insight for the future application of WO_3 in areas like smart windows, sensors, and energy storage devices. This investigation offers a comprehensive theoretical framework that will meaningfully contribute to the understanding and development of WO_3-based technologies.

Keywords: Tungsten trioxide · PBE · DFT · band structure · DOS · optical and electronic structure properties

1 Introduction

Tungsten trioxide (WO_3) is a prominent transition metal oxide with a unique combination of electronic and optical properties [1], making it highly suitable for various industrial and technological applications [2]. The monoclinic phase of WO_3, stable at room temperature, has garnered significant interest due to its structural stability and the

S. Kumari et al. (Eds.): ICMTP 2024, SPPHY 311, pp. 241–254, 2025.
https://doi.org/10.1007/978-981-96-6795-6_28

possibility of phase transitions [3] which it can undergoes under different temperature conditions. These transitions contribute to its adaptability and functional capabilities in devices such as smart windows [4], gas sensors [5], and energy storage systems [6]. In recent years, the theoretical study of WO_3 has gained momentum, particularly through the application of Density Functional Theory (DFT), which provides a robust framework for investigating the material's electronic structure and optical behaviour [7]. In view of its potential for different types of applications, present study on WO_3 was undertaken to theoretically investigate its structural, electronic and optical properties by using DFT, as implemented in WIEN2k code [8]. Our focus has been on exploring the electronic charge distribution in the unit cell, band structure, density of states (DOS), and the comprehensive range of optical properties of monoclinic WO_3. The crystal stability is checked through computation of cohesive energy by using another DFT based code-Quantum Espresso, which is more suitable for this purpose. The insights derived from this study are expected to contribute significantly to the understanding and development of WO_3-based technologies, particularly in enhancing the performance and efficiency of smart windows, sensors, and energy storage devices.

2 Crystal Structure and Computational Details

WO_3 (tungsten trioxide), exhibits a monoclinic crystal structure at room temperature, which is the most stable phase among its various structural forms. In the present work we use the unit cell structure belonging to space group P21/c (International no. 14) with lattice parameters a = 5.31 Å, b = 5.21 Å, c = 7.69 Å [9]. The atomic positions in the primitive unit cell are given by: W (0.254165, 0.731671, 0.283241), O (0.540714, 0.965362, 0.783987), O (0.74832, 0.317102, 0.993935), and O(0.966751, 0.962717, 0.2162228) [9]. The crystal structure of Monoclinic WO_3 is shown in Fig. 1.

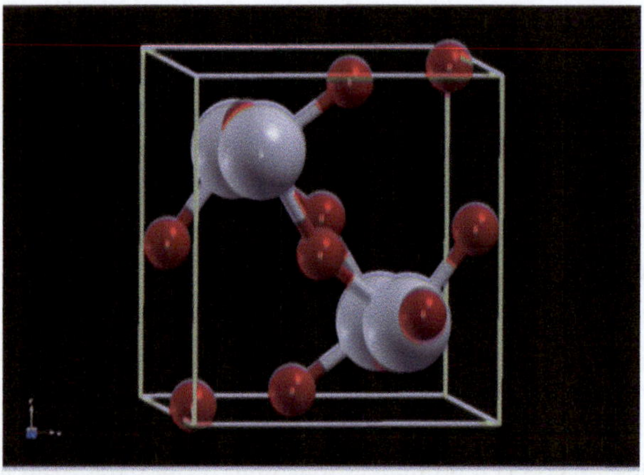

Fig. 1. Crystal structure of Monoclinic WO_3.

A first-principles calculation using Density Functional Theory (DFT) was performed to investigate the optical and electronic properties of WO_3, by using the WIEN2k code, which uses Linearized Augmented Plane Waves and local orbital as the basis sets and Generalized Gradient Approximation (GGA) [10] is used alongwith Perdew Burke-Ernzerhof (PBE) functional to take proper account of the exchange and correction interactions in the crystal. To check system stability, cohesive energy was calculated by using the Quantum Espresso software, which is another DFT based code, using a plane-wave basis set and ultra-soft atomic pseudopotentials are used, plane wave cutoff energy being kept 400 eV. The convergence threshold was set at 10^{-8} Ryd for self-consistent calculations. For self-consistent field (scf) calculation and non-self-consistent (nscf) calculations, the Monkhorst-Pack scheme was used for k-point grid sampling across the Brillouin zone, with a $3 \times 5 \times 5$ grid and $6 \times 10 \times 10$ grid respectively. As discussed above, The WIEN2k code was used to study the optoelectronic properties of WO_3. For this study, the RMT × Kmax was set to 7.0, and 1000 K-points were utilized, with energy and charge convergence criteria set at 0.0001 and 0.001, respectively. Additionally, G_{max} to 12 $(a.u.)^{-1}$. Calculations were performed on an 8-core machine operating in parallel computing mode. WIEN2k with APW + lo type of basis sets, can effectively account for the contributions of slowly converging d-orbitals in transition metals [11]. Due to its well-known opto-electronic properties, the optical characteristics of WO_3, such as dielectric function, absorbance, optical conductivity, refractive index, extinction coefficient, and energy loss function [12], were investigated using the WIEN2k code, within the DFT framework.

3 Results and Discussion

The crystal structure of Monoclinic WO_3 is shown in Fig. 1, which comprises a total of 16 atoms: 4 tungsten (W) atoms and 12 oxygen (O) atoms.

3.1 Volume Optimization

The total energy as a function of volume for this compound, as derived from WIEN2k by using the Murnaghan equation of state [13], is shown in Fig. 2. The volume at which the energy reaches its minimum value is 1445.29 au^3, which closely matches the primitive cell volume of 1435.7176 au^3 obtained from the cell parameters. Since these parameters also align well with the experimental results, we have adopted these parameters for the present study.

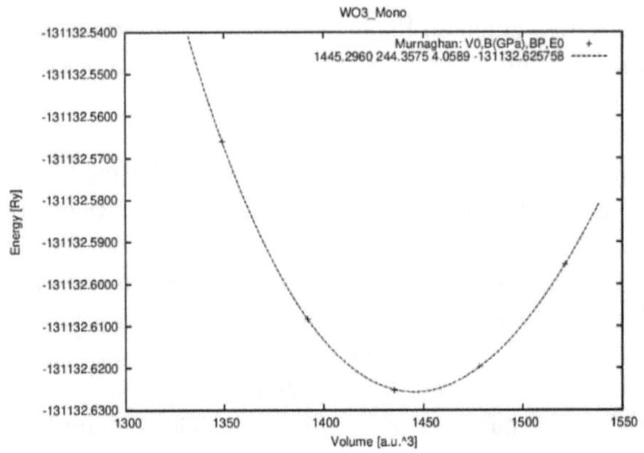

Fig. 2. Total energy V/s volume graph for Monoclinic phase of WO_3.

3.2 Structural Stability

To check the stability of this compound we calculated cohesive energy by using Quantum Espresso code. Cohesive energy is the amount of energy required to disassemble a solid into its individual atoms, indicating the strength of the bonds holding the atoms together in the solid. It is a measure of the stability of a solid and is calculated as the difference between the energy of the solid and the sum of the energies of its constituent isolated atoms. It was calculated by using the following equation:

$$E_{CE} = \frac{TE(A_m B_n) - mTE(A) - nTE(B)}{(A + B)} \tag{1}$$

In the given equation, A*m*B*n* represents a compound where *m* and *n* denote the number of A-type and B-type atoms, respectively. In the Eq. (1), TE ($A_m B_n$), TE (A), and TE (B) are the total energies of compound $A_m B_n$, isolated atom A, and isolated atom B, respectively and E_{CE} is the cohesive energy of the compound $A_m B_n$. In general, a structure is considered stable structure when its cohesive energy is negative. The cohesive energy of the monoclinic WO_3 structure calculated by following the procedure mentioned above, is found to be −8.583 eV/atom, indicating that it is a stable structure [14].

3.3 Electronic Properties of WO_3

Electronic Charge Distribution. To gain insight into the nature of bonding between the atoms in the Monoclinic WO_3 compound, we calculated the electronic charge density for the (100) and (110) planes. The electron density plots for the same are depicted in Figs. 3A and B, respectively.

The thermoscale indicates that the blue colour represents the highest value of charge density, while red colour signifies minimum value of charge concentration. To gain a deeper understanding of the bonding nature of the investigated compound, we also generated 3D plots and contour plots of the electronic charge distribution within the unit

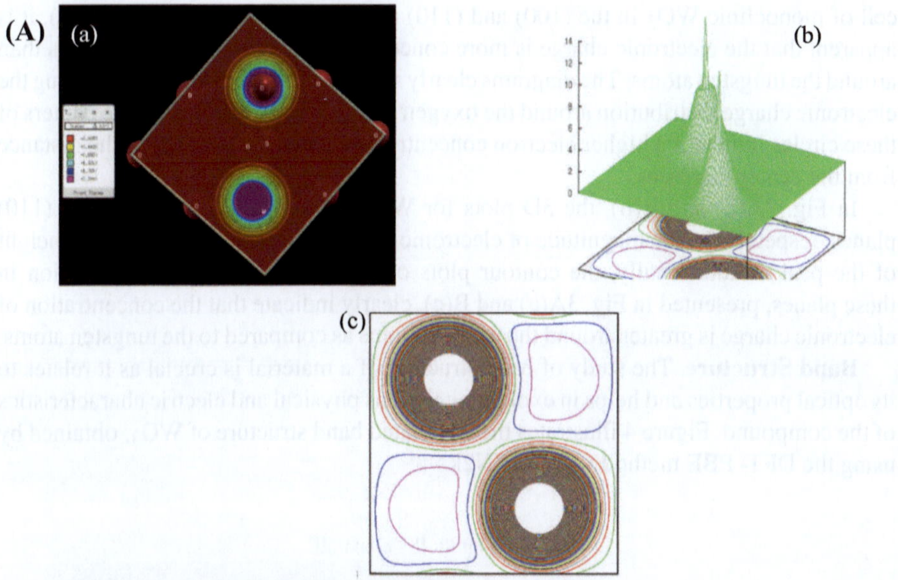

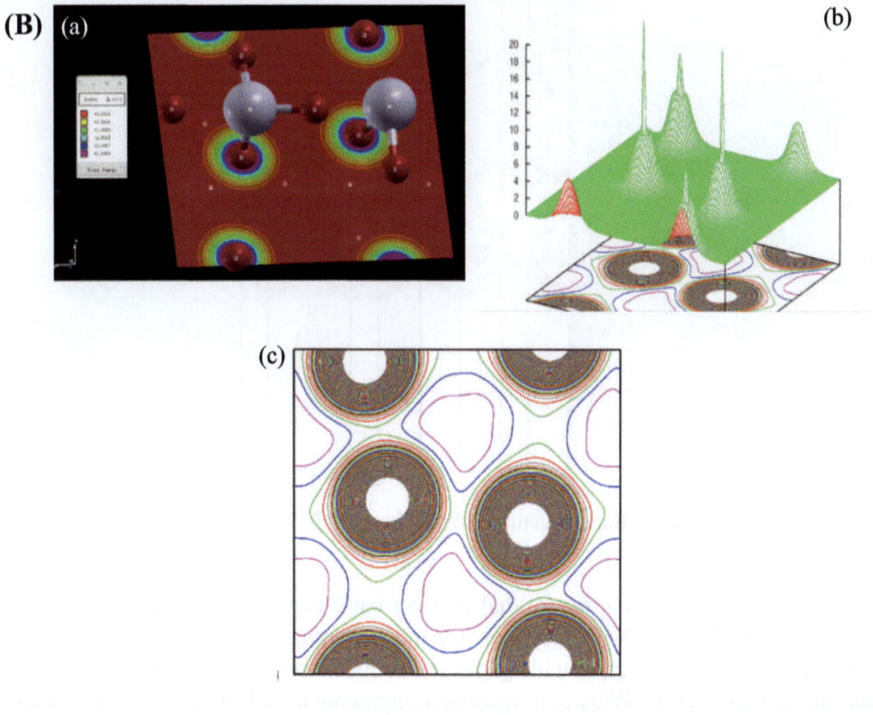

Fig. 3. (A) Electronic charge density distribution along (100) plane in the unit cell of Monoclinic WO_3: (a) 2D representation, (b) 3D representation; and (c) Contour plot. (B) Electronic charge density distribution in the unit cell of Monoclinic WO_3 along (110) plane: (a) 2D representation; (b) 3D representation; and (c) Contour plot.

cell of monoclinic WO_3 in the (100) and (110) planes. From Fig. 3A(a) and B(a), it is apparent that the electronic charge is more concentrated around the oxygen atoms than around the tungsten atoms. The diagrams clearly show concentric circles representing the electronic charge distribution around the oxygen atoms. The pink colour at the centers of these circles indicates a higher electron concentration, which diminishes as the distance from the center increases.

In Fig. 3A(b) and B(b), the 3D plots for WO_3 are shown in the (100) and (110) planes, respectively, the magnitude of electronic charge being represented by the height of the peaks. Additionally, the contour plots of the electronic charge distribution in these planes, presented in Fig. 3A(c) and B(c), clearly indicate that the concentration of electronic charge is greater around the oxygen atoms as compared to the tungsten atoms.

Band Structure. The study of band structure of a material is crucial as it relates to its optical properties and helps in explaining various physical and electric characteristics of the compound. Figure 4 illustrates the calculated band structure of WO_3, obtained by using the DFT- PBE method with WIEN2k.

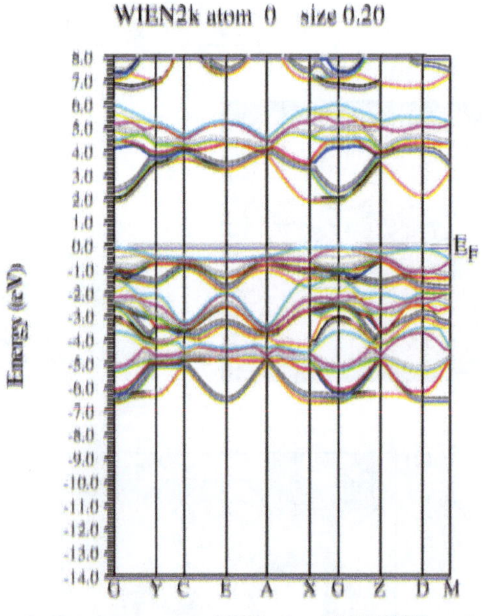

Fig. 4. Band structure of WO_3 from DFT-PBE method.

From this figure, it is apparent that a direct band gap of approximately 2.0 eV is observed at the Γ point, with the valence band maximum passing through the Fermi level at this point, indicating a wide gap semiconductor behaviour of WO_3. The calculated direct band gap of WO_3 is in qualitative agreement with its experimental values reported to be 2.6–3.0 eV from measurement of optical properties [15, 16]. These findings highlight WO_3's potential for applications in optoelectronics, photocatalysis, and semiconductor devices [17], offering valuable insights into its electronic and optical properties.

Density of States. The plots for calculated values of total DOS as well as (PDOS) partial density of states of Monoclinic WO_3 versus energy are shown in Fig. 5. Figure 5(a) displays the total DOS of WO_3 along with the atomic contributions to the total DOS, obtained from DFT calculations using WIEN2k. The diagram reveals a DOS gap of approximately 2.0 eV just above the Fermi level. It also shows that below the Fermi level, the O_1 atom contributes more to the total DOS than the O_2 and W atoms in the energy range of −0.5 to −2.0 eV. Outside this range, the contribution of O_2 atom dominates. Above the Fermi level, the W atom's contribution surpasses that of the O_1 and O_2 atoms. Significant peaks in the total DOS are observed at −6.5, −4.8, −2.7, −0.9, −0.4, 4.9, 5.2, and 5.9 eV, with corresponding DOS values of 12.5, 15.0, 15.8, 21.0, 18.6, 15.1, 12.9, and 9.2 states/eV, respectively.

Figures 5(b), 5(c), 5(d), and 5(e) depict the partial Density of States (PDOS) contributions of s, p, and d orbitals to the total DOS of the atom for W, O_1, O_2, and O_3 atoms, respectively. From Fig. 5(b), it is evident that the d-orbitals contribute more significantly to DOS of W atom than its s and p orbitals. Figures 5(c), 5(d), and 5(e) reveal that for O_1, O_2, and O_3 atoms, the p-orbitals are the dominant contributors as compared to their s-orbitals. Additionally, the unique characteristics of the DOS diagrams of three oxygen atoms (O_1, O_2, and O_3) can be attributed to their different distinct positions within the unit cell. Figure 5(b) also shows that the total DOS values for the W atom above the Fermi level are higher than its values below the Fermi level. Conversely, Fig. 5(c–e) indicate that the total DOS values for the O_1, O_2, and O_3 atoms are higher below the Fermi level as compared to their values above the Fermi level.

3.4 Optical Properties of WO_3

Dielectric Function. The dielectric function is crucial for understanding and designing materials with specific optical properties. The dielectric function, also known as the dielectric constant, is a complex function that describes how a material responds to an external electromagnetic field. The dielectric function $\varepsilon(\omega)$ is generally represented as:

$$\varepsilon(\omega) = \varepsilon_1(\omega) + i\varepsilon_2(\omega) \tag{2}$$

The imaginary part $\varepsilon_2(\omega)$ represents the absorption by the material corresponds to the electronic band structure of the material. The imaginary part $\varepsilon_2(\omega)$ is given by the following eq. [18]:

$$\varepsilon_2(\omega) = \frac{2e^2\pi}{\Omega\varepsilon_0}\sum_{k,v,c}\left|\langle\psi_k^c|\hat{u}.\vec{r}\,|\psi_k^v\rangle\right|^2 \times \delta\left(E_k^c - E_k^v - \hbar\omega\right) \tag{3}$$

Here e is the electronic charge, Ω represents the volume of the unit cell, $\varepsilon_0(\omega)$ is free space permittivity, $\hat{u}$ is the unit vector defining the polarization of the incident electric field and ω is the incident photon frequency. ψ_k^c and ψ_k^v represent the quasi-particle wave functions and E_k^c and E_k^v their corresponding energies, here v and c refer to valence band and conduction band respectively.

Dispersion of the material is represented by the real part $\varepsilon_1(\omega)$, which was extracted by using Kramer-Kronig relation [19] in the following form:

$$\varepsilon_1(\omega) = 1 + \frac{2}{\pi}P\int_0^\infty \frac{\omega'\varepsilon_2(\omega')}{(\omega')^2 - (\omega)^2}d\omega' \tag{4}$$

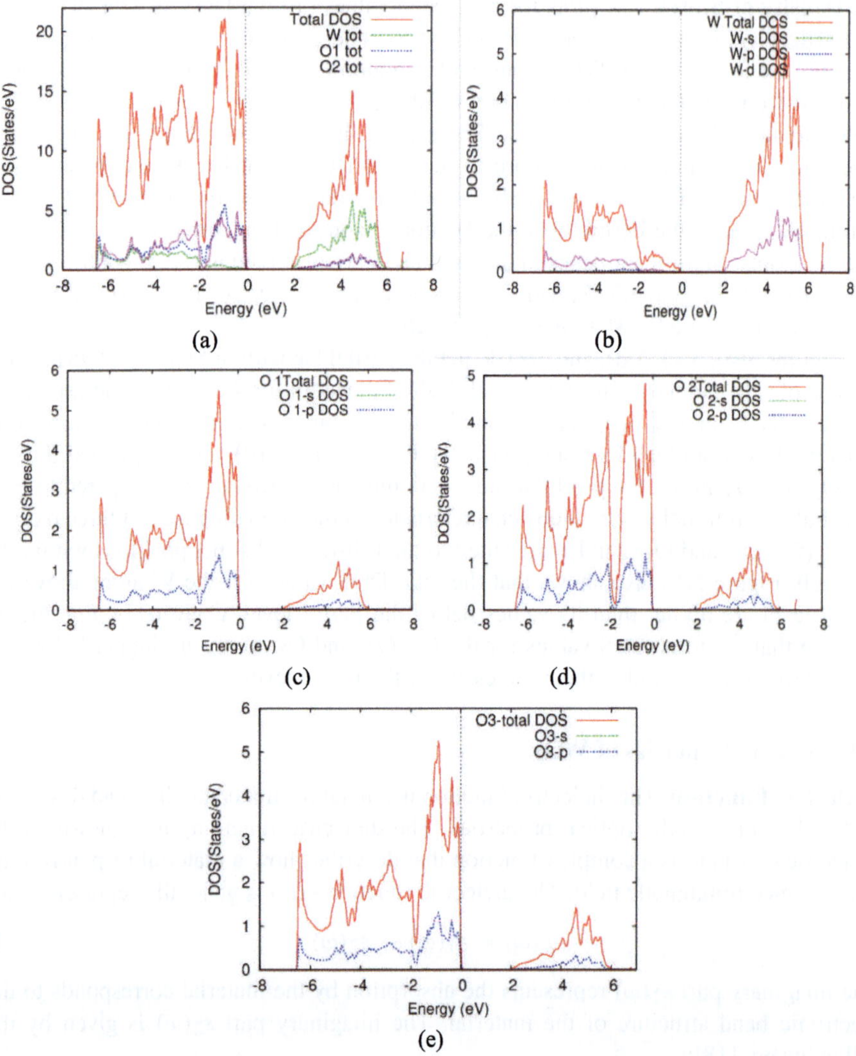

Fig. 5. (a) Total DOS of WO_3 along with atomic contributions; and Partial DOS due to s, p and d orbitals as compared to total DOS of the atom: (b) W atom; (c) O_1 atom;(d) O_2 atom; and (e) O_3 atom.

Here P represents principal value of the integral.

For WO_3, where the lattice parameters a, b and c are unequal, the incident photon can have three distinct polarization states: z (parallel to the z-axis), and x and y (perpendicular to the z-axis and orthogonal to each other). Consequently, this results in three separate components for both the imaginary part $\varepsilon_2(\omega)$ and real part $\varepsilon_1(\omega)$ of the dielectric function. The plots for calculated values of real and imaginary parts of dielectric function of WO_3 as funtion of the Photon energy are shown in Fig. 6, in the range of 0–15 eV.

From analysis of $\varepsilon_2(\omega)_{xx}$, $\varepsilon_2(\omega)_{yy}$, and $\varepsilon_2(\omega)_{zz}$ shown in Fig. 6(a), it can be seen that the dielectric loss is very low up to approximately 2 eV. This energy range aligns with the band gap of about 2.0 eV obtained from DFT calculations, indicating that electronic transitions from the valence band to the conduction band are not feasible below this energy, resulting in minimal energy absorption. Figure 6(a), indicate that there are strong absorption peaks for $\varepsilon_2(\omega)_{xx}$, $\varepsilon_2(\omega)_{yy}$, and $\varepsilon_2(\omega)_{zz}$ in the energy range 3–6 eV. So WO_3 can be used as a filter for varoius energies in the UV spectrum [17]. These curves also show secondary peaks in the energy range 9–11 eV. The maximum absorption peak in $\varepsilon_2(\omega)$xx is observed at 5.5 eV. For $\varepsilon_2(\omega)$yy, three peaks are found at approximately 6.9 eV, 6.2 eV, and 3.9 eV, while $\varepsilon_2(\omega)$zz shows a main peak at about 5.8 eV along with additional peaks at 3.0 eV, and 4.9 eV. These peaks are associated with various electronic transitions involving W 4s to 3d states and O 2p states.

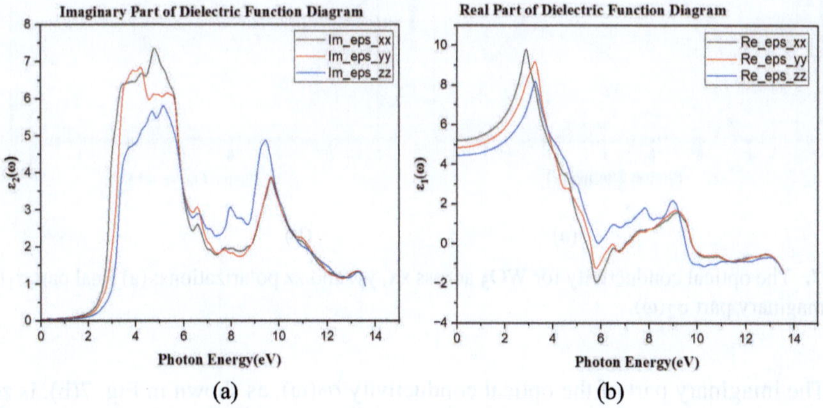

Fig. 6. Plots of di-electric function of WO_3 as a function of photon energy for three polarizations xx, yy and zz: (a) Imaginary part $\varepsilon_2(\omega)$; and (b) Real part $\varepsilon_2(\omega)$.

From Fig. 6(b), the values of $\varepsilon_1(\omega)$ at zero energy are found to be 5.2 eV,4.9 eV and 4.4 respectively for xx, yy and zz polarizations. The peak values of $\varepsilon_1(\omega)$ for three curves occur in the range 2.5–3.5 eV respectively and $\varepsilon_1(\omega)$ acquires negative values in the energy range 5.4–7.2 eV for three polarizations.

Optical Conductivity. The optical conductivity of a material is a complex quantity, expressed as:

$$\sigma_1(\omega) = \sigma_1(\omega) + i\sigma_2(\omega) \tag{5}$$

where $\sigma_1(\omega)$ and $\sigma_2(\omega)$ represents real and imaginary parts of the optical conductivity, respectively. They are directly proportional to the dielectric function components $\varepsilon_1(\omega)$ and $\varepsilon_2(\omega)$.

$$\sigma_1(\omega) = \frac{\omega}{4\pi}\,\varepsilon_2(\omega) \tag{6(a)}$$

$$\sigma_2(\omega) = \frac{\omega}{4\pi}\,\varepsilon_1(\omega) \tag{6(b)}$$

The real part of the optical conductivity $\sigma_1(\omega)$, shown in Fig. 7(a), begins around 2 eV, corresponding to the band gap of WO_3 from DFT calculations. For xx, yy and zz polarizations, peaks occur in the energy range 4–6 eV, indicating significant electronic transitions and strong absorption in this energy range. Sharp peaks occurring in the energy range 9–11 eV highlight strong optical absorption and electronic transitions within the material, with higher peaks suggesting stronger absorption at those specific photon energies.

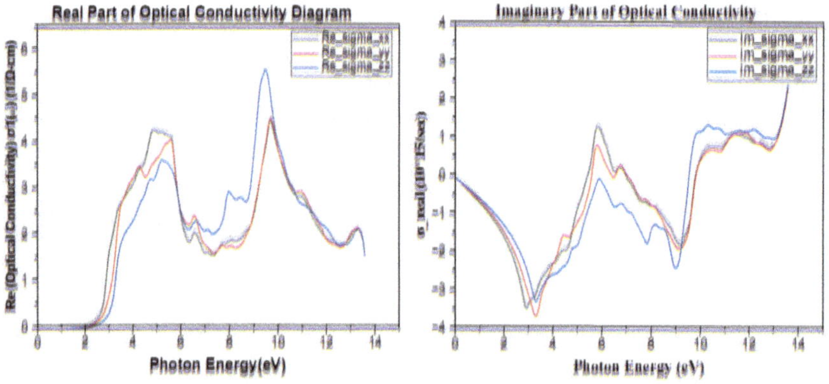

(a) (b)

Fig. 7. The optical conductivity for WO_3 across xx, yy, and zz polarizations: (a) Real part $\sigma_1(\omega)$; (b) Imaginary part $\sigma_2(\omega)$.

The imaginary part of the optical conductivity $\sigma_2(\omega)$, as shown in Fig. 7(b), is zero at zero energy for all polarizations and fluctuates between negative and positive values. Negative regions (0–5.5 eV) indicate inductive behaviour with a phase lag between the electric field and current density. Zero crossings mark the transition from inductive to capacitive behaviour. Positive peaks occurring in the energy range 5.5–7.2 eV suggest capacitive behaviour with current density leading the electric field, corresponding to resonant absorption and electronic transitions. $\sigma_2(\omega)$ again shows inductive behaviour in the energy range 7.2–10 eV. These insights are crucial for understanding the material's optical properties and potential optoelectronic applications.

Energy Loss Function. The energy loss function measures the energy lost by an electron moving through the material, indicating in which energy range the material strongly absorbs energy. Its value can be evaluated from the knowledge of $\varepsilon_1(\omega)$ and $\varepsilon_2(\omega)$ by using the following formula:

$$L(\omega) = \mathrm{Im}\left(\frac{-1}{\varepsilon(\omega)}\right) = \frac{\varepsilon_2(\omega)}{\varepsilon_1^2(\omega) + \varepsilon_2^2(\omega)} \tag{7}$$

The graph of the energy loss function for WO_3 across different photon energies, characterized by xx, yy, and zz polarizations, is shown in Fig. 8. The peaks in the loss function graph represent the photon energies at which the material exhibits significant energy loss due to electronic excitations or plasmons. The plasma frequency (ω_{pl}) is the frequency above which material shows dielectric behaviour [$\varepsilon_1(\omega) > 0$], while below which the material exhibits metallic properties [$\varepsilon_1(\omega) < 0$]. The peaks in L(ω) curve represent the transition points for change of material properties from metallic to dielectric and vice versa. The maximum energy loss occurs around 7 and 13 eV for all polarizations, indicating that these are the energy values where WO_3 absorbs energy most efficiently, likely due to the excitation of plasmons or high-energy electronic transitions. This is critical for applications like plasmonics [17], where controlling light at the nanoscale is essential.

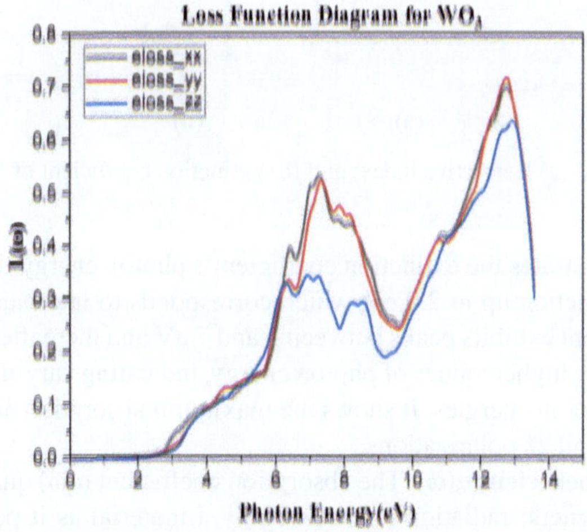

Fig. 8. Electron Loss Function v/s Photon Energy spectrum of WO_3 for xx, yy and zz polarizations.

Refractive Index n(ω) and Extinction Coefficient k(ω). The refractive index, a dimensionless number, describes how much light slows down and bends in a material, affecting the phase velocity of photons. The extinction coefficient measures how much light is absorbed by the material. Together, these properties offer a complete picture of light's interaction with a substance. The extinction coefficient and refractive index can be calculated by given equation:

$$k(\omega) = \frac{1}{\sqrt{2}}\left(\sqrt{\varepsilon_1^2(\omega) + \varepsilon_2^2(\omega)} - \varepsilon_1(\omega)\right)^{\frac{1}{2}} \tag{8a}$$

$$n(\omega) = \frac{1}{\sqrt{2}}\left(\sqrt{\varepsilon_1^2(\omega) + \varepsilon_2^2(\omega)} + \varepsilon_1(\omega)\right)^{\frac{1}{2}} \tag{8b}$$

Figure 9(a) shows the refractive index for WO_3. The refractive index remains relatively constant at lower energies, increases to peak values as the energy increases, and then decreases at higher energies shows showing dips and rising again to show peaks before falling again. The static refractive index n(0) values are 2.27 eV, 2.20 eV, and 2.11 eV for xx, yy, and zz polarizations, respectively, with maximum values of 3.15 eV, 3.09 eV, and 2.88 eV for these polarizations. Secondary peaks occur in n(ω) curves at about 9.12 eV.

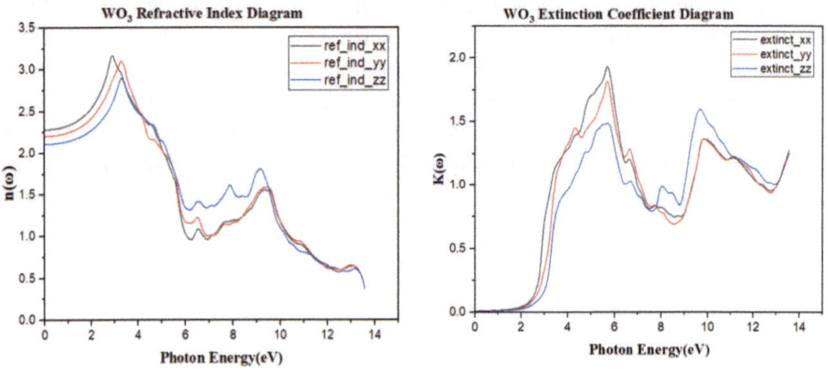

(a) (b)

Fig. 9. (a) Refractive index; and (b) extinction coefficient of WO_3.

Figure 9(b) illustrates the extinction coefficient's photon energy dependence, showing negligible extinction up to 2.0 eV, which corresponds to interband transitions. The extinction coefficient exhibits peaks between 4 and 7 eV and thereafter it gives dips, and secondary peaks for higher values of photon energy, indicating varying levels of absorption at different photon energies. It shows the maximum absorption at about 5.9 eV and 10 eV for xx, yy, and zz polarizations.

Absorption Coefficient α(ω). The absorption coefficient α(ω) quantifies how much light or electromagnetic radiation is absorbed by a material as it passes through that material. It measures the rate at which the intensity of light decreases with distance into the material. The absorption coefficient depends on the wavelength of the incident light and the material's properties. The Absorption coefficient α(ω) is given by following formula:

$$\alpha(\omega) = \sqrt{2}(\omega)\left(\sqrt{\varepsilon_1^2(\omega) + \varepsilon_2^2(\omega)} - \varepsilon_1(\omega)\right)^{\frac{1}{2}} \tag{9}$$

Figure 10 displays the dependence of absorption coefficient of incident electromagnetic radiation on the photon energy across the 0–15 eV range. Maximum absorption is observed at 9.7 eV for xx polarization and from 10 to 13 eV for both yy and zz polarizations. Absorption coefficient α(ω) also exhibits a peak value at about 5.7 eV for all the polarizations. Below 2 eV, the absorption coefficient α(ω) is almost zero, indicating that interband transitions do not occur for photon energies less than 2 eV.

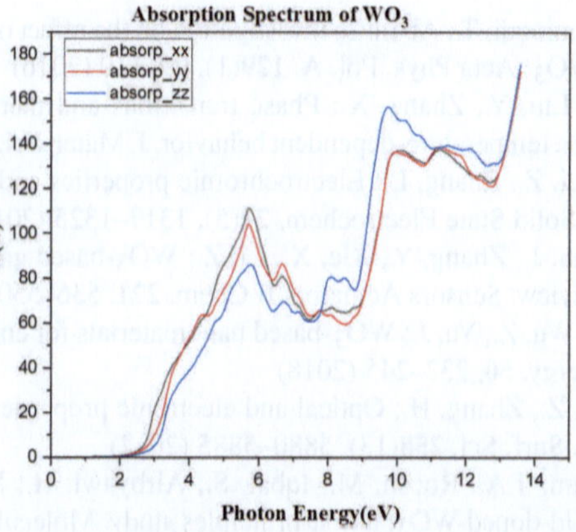

Fig. 10. Dependence of absorption coefficient α(ω) of WO_3 on energy of incident photon.

4 Conclusion

This study presents a comprehensive analysis of the electronic and optical properties of monoclinic WO_3 crystal using first-principles density functional theory (DFT) with the Generalized Gradient Approximation (GGA) and Perdew-Burke-Ernzerhof (PBE) potential functional to accurately account for exchange-correlation interactions. Calculations performed with WIEN2k effectively describe the electronic charge distribution, band structure, and density of states (DOS). The calculated band gap and DOS gap of about 2.0 eV are found to be inqualitative agreement with the reported optical bandgap values of 2.6–3.0 eV. Cohesive energy calculations using Quantum Espresso confirm the stability of the monoclinic WO_3 structure. Further, the optical properties of WO_3 have been examine by calculating the real and imaginary parts of the dielectric function, which are then used to derive the optical conductivity, energy loss function, absorption coefficient, refractive index, and extinction coefficient. The optical properties are well correlated with interband transitions and the band gap, providing valuable insights into the material's behaviour. The results presented in this paper give detailed information on the structural, electronic, and optical characteristics of WO_3, which could benefit experimental investigations and applications of this material for fabrication of various types of optical and electronic devices.

Acknowledgement. The author, Khushbu Dhaked, gratefully acknowledges the Department of Science and Technology (DST), Government of India, for the financial support provided under the 'WISE Fellowship for Ph.D.' programme (Ref. No. DST/WISE-PhD/PM/2023/105 (G)) to carry out this research work.

References

1. Babu, K.E., Murali, N., Vijaya Babu, K., Shibeshi, P.T., Veeraiah, V.: Structural, elastic, electronic, and optical properties of cubic perovskite $CsCaCl_3$ compound: an ab initio study. Acta Phys. Pol. A. **125**(5), 1039–1046 (2014)

2. Mansouri, M., Mahmoodi, T.: Ab initio investigation on the effect of transition metals doping and vacancies in WO_3. Acta Phys. Pol. A. **129**(1), 99–110 (2016)
3. Zhang, Y., Li, Z., Liu, Y., Zhang, X.: Phase transitions and thermochromic properties of tungsten oxide films: temperature-dependent behavior. J. Mater. Sci. **49**(3), 1229–1235 (2014)
4. Jiang, L., Li, M., Li, Z., Zhang, L.: Electrochromic properties and applications of WO_3 for smart windows. J. Solid State Electrochem. **21**(5), 1317–1325 (2017)
5. Gao, L., Li, Y., Zhu, J., Zhang, Y., Xie, X., Li, Z.: WO_3-based gas sensors for detection of harmful gases: a review. Sensors Actuators B Chem. **221**, 536–550 (2015)
6. Liu, Y., Zhang, L., Wu, Z., Yu, J.: WO_3-based nanomaterials for energy storage applications: a review. Nano Energy. **50**, 232–245 (2018)
7. Li, J., Yang, F., Li, Z., Zhang, H.: Optical and electronic properties of WO_3 for gas sensing applications. Appl. Surf. Sci. **258**(13), 5880–5885 (2012)
8. Bahadur, A., Anjum, T.A., Roosh, M., Iqbal, S., Alrbyawi, H.: Magnetic, electronic, and optical studies of Gd-doped WO_3: a first-principles study. Molecules. **27**(20), 6976 (2022)
9. Jain, A., Ong, S.P., Hautier, G., Chen, W., Richards, W.D., Dacek, S., Cholia, S., Gunter, D., Skinner, D., Ceder, G., Persson, K.A.: Commentary: the materials project: a materials genome approach to accelerating materials innovation. APL Mater. **1**(1), 011002 (2013)
10. Perdew, J.P., Burke, K., Ernzerhof, M.: Generalized gradient approximation made simple. Phys. Rev. Lett. **77**(18), 3865–3868 (1996)
11. Dhaked, K., Shukla, R., Sharma, K.S., Sharma, R.: Ab-initio investigation of electronic and optical properties of vanadium pentoxide (V_2O_5). Int. J. Mod. Phys. B. **39**(10), 2540032 (2025)
12. Ephraim Babu, K., Neeraja, K., Deenabandhu, D., Vijaya Ratna, A.M., Venkata Kumar, V., Bueala Kumari, K., Veeraiah, V.: First-principles study of structural, elastic, electronic and optical properties of cubic perovskite $LiMgF_3$ for novel applications. J. Phys.: Conf. Ser. **1495**, 012010. IOP Publishing, London (2020)
13. Jehan, A., Husain, M., Tirth, V., Algahtani, A., Uzair, M., Rahman, N., Khan, A., Khan, S.N.: Investigation of the structural, electronic, mechanical, and optical properties of $NaXCl_3$ (X = Be, Mg) using density functional theory. RSC Adv. **13**(41), 28395–28406 (2023)
14. Nash, P., O'Rourke, D.P.: First-principles study of the electronic and optical properties of WO_3. J. Appl. Phys. **111**(10), 103509 (2012)
15. Pérez, I., Martínez Faudoa, J.C., Abenuz Acuña, J.R., Elizalde Galindo, J.T.: Influence of lithium interstitial doping on the optoelectronic properties of NiO and WO_3. Comput. Mater. Sci. **190**, 110248 (2021)
16. Ramay, S.M., Shahabuddin, M., Alzayed, N.S.: Influence of Gd content on structural, electronic, thermoelectric, and optical properties of WO_3. Mater. Res. Express. **10**(10), 105901. IOP Publishing, London (2023)
17. Yao, Y., Sang, D., Zou, L., Wang, Q., Liu, C.: A review on the properties and applications of WO_3 nanostructure-based optical and electronic devices. Nanomaterials. **11**(8), 2136 (2021)
18. Yang, H., Sun, H., Li, Q., Li, P., Song, K., Song, B., Wang, L.: Structural, electronic, optical and lattice dynamic properties of the different WO_3 phases: first-principle calculation. Vacuum. **164**, 411–420 (2019)
19. Yu, P.: Fundamentals of Semiconductors, 2nd edn. Springer, Berlin (2005)

Synthesis and Characterization of Zirconia-Based Oxide (ZrO_2) Nanoparticles for Photocatalytic Degradation

Simranjeet Kaur, Nirmal Manyani, and S. K. Tripathi(✉)

Department of Physics, Center for Advanced Study in Physics, Panjab University, Chandigarh 160014, India
surya@pu.ac.in

Abstract. Recently, metal-oxide based nanoparticles have been extensively utilized as photocatalysts because of their exceptional physiochemical properties. In this study, ZrO_2 nanoparticles were prepared via a sol-gel approach to study their application as a photocatalyst. Structural analysis of the synthesized nanomaterial was done using X-Ray Diffraction (XRD). Using Debye-Scherrer's formula, the average crystallite size of ZrO_2 nanoparticles was determined to be approximately 23.85 nm. Optical properties of the fabricated nanomaterial were examined using UV-Visible spectroscopy and PL (photoluminescence) spectroscopy. On the basis of the absorption spectra, band gap of the nanomaterial was determined using the Tauc plot method. The band gap energy of ZrO_2 nanoparticles synthesized from the sol-gel method was found to be 4.03 eV. Further, ZrO_2 was employed to study the photocatalytic degradation of methyl red (MR) dye with a maximum duration of 120 min. Photodegradation of methyl red dye was studied with the help of UV-Visible spectroscopy. Degradation efficiency of ZrO_2 for methyl red (MR) dye was examined with different time periods and was found to be 41.07% for 120 min.

Keywords: ZrO_2 · Sol-Gel · Photocatalytic Degradation · Methyl Red

1 Introduction

Access to clean and safe drinking water, devoid of toxic substances, carcinogens and harmful bacteria is crucial for human health [1]. Industrial and technological advances have played a major role in the environmental release of toxic wastes, such as heavy metals and dyes. Dye contamination in wastewater threatens the environment, impacting both human health and ecosystems [2]. Over the years, different methods such as biological treatment, sedimentation and filtration have been utilized to tackle wastewater contaminants. However, due to their reliance on extensive infrastructure and specialized engineering expertise, these methods are generally ineffective at removing organic pollutants, making them complex and expensive [3].

Significant interest has been enticed by Advanced Oxidation Processes (AOPs) recently as a highly efficient method for eliminating contaminants from wastewater.

S. Kumari et al. (Eds.): ICMTP 2024, SPPHY 311, pp. 255–261, 2025.
https://doi.org/10.1007/978-981-96-6795-6_29

These processes employ powerful oxidizing agents such as ozone, hydrogen peroxide or ultraviolet (UV) radiation to degrade complex and resistant organic and inorganic contaminants in wastewater. Photocatalytic degradation is a promising AOP that has drawn significant attention for its ability to effectively degrade pollutants without producing harmful by-products. Moreover it is cost-effective and has the ability to operate using solar light as an energy source [4]. ZrO_2, a semiconductor with a wide bandgap, is currently being widely researched owing to its numerous oxygen vacancies on the surface, significant ion exchange capacity, excellent physicochemical stability and its environmentally friendly, non-toxic and cost-effective nature [5]. Methyl red is an anionic azo dye, a toxic and carcinogenic contaminant in aquatic environments. Its presence in water can lead to digestive problems and irritation of the eyes and the skin [6]. Ratnam et al. synthesized MgO NPs using the co-precipitation method and examined the photodegradation of methyl red dye under ultraviolet light [7]. A methyl red destruction rate of 97.08% was achieved by them under the optimal conditions determined through CCD. Aljaafari et al. synthesized ZnO nanosheets using a microwave-assisted hydrothermal method and investigated their photocatalytic activity on methyl red dye via UV- light irradiation [8]. Their study demonstrated the superior efficiency of ZnO nanosheets compared to commercial ZnO for the photodegradation of organic dyes. The sol-gel method is utilized in this study for the preparation of ZrO_2 nanoparticles as it produces products with high purity and uniformity, while also providing excellent control over the physiochemical properties of metal oxides at elevated temperatures [9].

2 Experimental Details

The sol-gel route was employed to prepare ZrO_2 nanoparticles (NPs). The precursor for ZrO_2 nanoparticles was prepared by dissolving 30 mg of $ZrOCl_2.8H_2O$ into 20 mL of D.I. water with continuous stirring. Separately, 30 mg of NaOH was dissolved in 20 mL of D.I. water to prepare the NaOH solution. The NaOH solution was cautiously added drop by drop to the $ZrOCl_2.8H_2O$ solution with continuous stirring using a magnetic stirrer. During the addition, the mixture initially turned white and gradually formed a white precipitate. The mixture was persistently stirred until it attained a gel-like state, confirming the successful formation of the target compound. The gel was dried at 80 °C to remove moisture and then subsequently calcined in a muffle furnace at 500 °C for 3 h to yield final product. Figure 1 shows the diagrammatic representation of ZrO_2 NPs preparation via the sol-gel approach. X-Ray Diffraction (XRD) analysis was employed to conclude the structure of the as prepared material. X-Ray Diffraction (XRD) was carried out by RigakuMiniflex 600 diffractometer. UV-Visible and PL spectroscopy were employed to study the optical characteristics of the synthesized ZrO_2 nanoparticles. UV-Visible spectroscopy was performed using Labindia UV-Visible spectrophotometer and PL by SHIMADZU RF-6000 Spectroflurophotometer. To evaluate the photocatalytic activity of ZrO_2 NPs over methyl red (MR), an aqueous solution of methyl red dye was formulated by dissolving 0.7 mg of dye in 70 mL of D.I. water. To this solution, 2-3 mg of ZrO_2 NPs were dissolved with continuous stirring to ensure uniformity of solution. A minute amount of hydrogen peroxide (H_2O_2) was added to the resulting mixture. The solution was then exposed to direct sunlight for a maximum duration of 120 min. The absorbance of methyl red (MR) dye was tracked with time via UV-Visible spectroscopy.

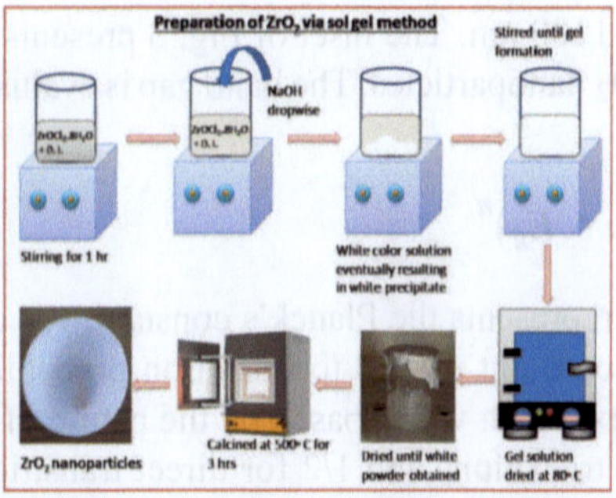

Fig. 1. Schematic illustration of preparation of ZrO_2 nanoparticles via sol-gel route

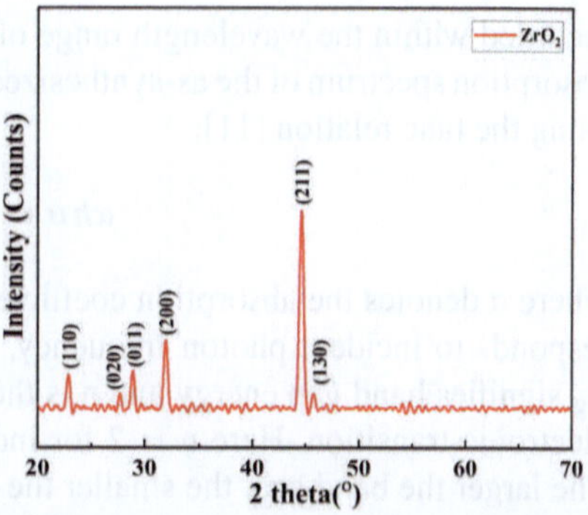

Fig. 2. XRD spectrum of ZrO_2 nanoparticles

3 Result and Discussion

Structural characterization of as-synthesized ZrO_2 NPs was done by X-Ray diffraction (XRD) technique at a scanning rate of 5° min^{-1} for a scanning angle range of $2\theta =$ 20°–70° as shown in Fig. 2. The crystallite size of ZrO_2 NPs was calculated for different planes using Debye-Scherrer's formula [10]:

$$D = \frac{K\lambda}{\beta cos\theta} \tag{1}$$

In this case, D, K, λ, β and θ correspond to the crystallite size, Scherrer's constant, X-ray wavelength, FWHM of diffraction peak and diffraction angle respectively.

Moreover, the d-spacing and strain for the significant planes were computed. Table 1 displays the calculated parameters for each plane. The average crystallite size of 23.85 nm confirms the nanostructure's presence. JCPDS card no. 01-083-0809 was used for indexing the X-Ray Diffraction (XRD) data.

Table 1. Micro structural parameters of ZrO_2 corresponding to prominent planes

2 θ (degree)	Lattice spacing 'd' (Å)	Lattice plane	Crystallite size 'D' (nm)	FWHM 'β' (°)	Lattice strain 'ε'
22.82	3.89	(110)	23.47	0.34	7.46
27.66	3.22	(020)	30.61	0.27	4.74
28.86	3.09	(011)	23.06	0.36	6.03
31.91	2.80	(200)	20.54	0.40	6.14
44.59	2.03	(211)	18.63	0.46	4.90
45.74	1.98	(130)	26.78	0.32	3.33

The optical properties of the as-prepared ZrO_2 NPs were studied using UV-Visible and PL spectroscopy. To find band gap of ZrO_2 NPs, the UV-Visible spectrum was recorded within the wavelength range of 200-1100 nm. The inset of Fig. 3 presents the absorption spectrum of the as-synthesized ZrO_2 nanoparticles. The band gap is evaluated using the tauc relation [11]:

$$\alpha h\upsilon = A(h\upsilon - E_g)^n \tag{2}$$

where α denotes the absorption coefficient, h represents the Planck's constant, ν is corresponds to incident photon frequency, A is constant related to transition probability, E_g signifies band gap energy and n is the index which varies based on the nature of the electronic transition. Here n = 2 for indirect transitions and 1/2 for direct transitions. The larger the band gap, the smaller the size of nanoparticles [10, 12].

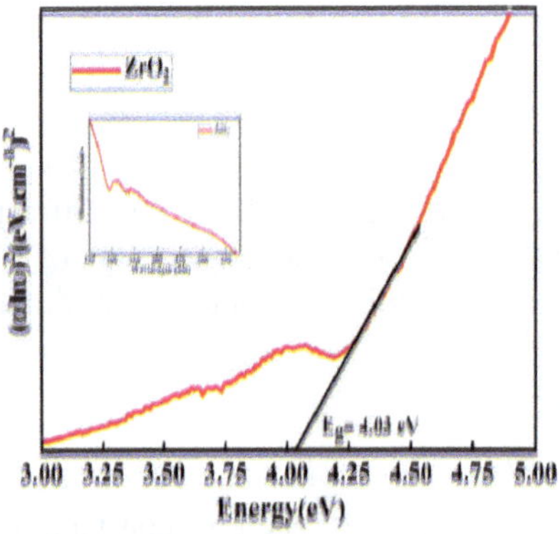

Fig. 3. Direct bandgap of ZrO_2 NPs

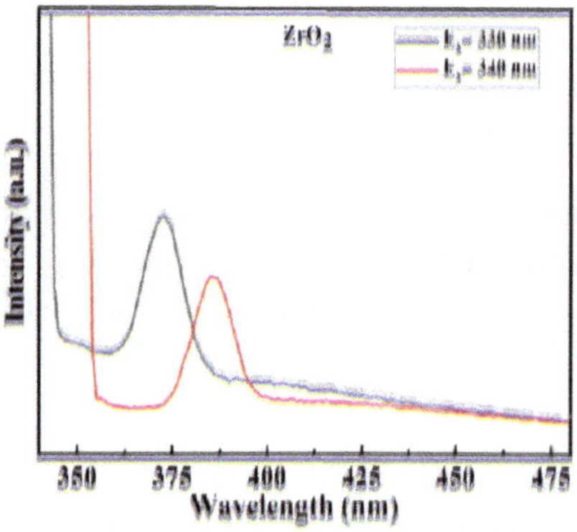

Fig. 4. PL spectrum of ZrO_2 NPs

ZrO_2 has a band gap that falls within the range of 3.25–5.1 eV. The high bandgap is associated with an enhanced quantum confinement effect and surface deformation observed in the sample. The band gap of the as-prepared nanoparticles was obtained to be 4.03 eV which indicates the formation of nanoparticles. The band gap energy value of the fabricated ZrO_2 nanoparticles is in close agreement with the optical band gap values of 4.8, 5.07, 5.50 eV as reported by Vinayagam et al. [13], Reddy et al. [14] and Horti et al. [15]. The PL spectrum of the as-synthesized ZrO_2 nanoparticles (NPs) was determined at room temperature using an excitation wavelength of 330 nm as depicted in Fig. 4. The band can be linked to the wide energy gap (4.03 eV) of ZrO_2 NPs from the valence to the conduction band. Results are aligning with those obtained from the UV-Visible spectroscopy [13].

The photocatalytic decomposition of methyl red dye using the prepared ZrO_2 nanoparticles under sunlight exposure was investigated at different time intervals. The UV–Visible spectral images of methyl red photodegradation reaction consisting of 10 ppm of methyl red dye and ZrO_2 nanoparticles at the time, t = 0 (initial) and 120 mins (completion of photocatalysis) is shown in Fig. 5. The photocatalytic degradation was confirmed by the UV–Visible spectrogram as the peak over 430 nm gradually diminished over time [13]. The degradation rate (or efficiency) of methyl red dye at various time intervals (Fig. 6) was then calculated using the formula [16, 17]:

$$\textbf{Degradation efficiency}\ (\%) = \frac{C_0 - C}{C_0} \times 100 \tag{3}$$

where C_0 is the absorbance at t = '0' min and C is the absorbance of methyl red dye at t = 't' min. The degradation efficiency was found to be 41.07% for 120 min. The moderate degradation efficiency in this case can be accounted to the fact that photocatalytic degradation is a highly selective process and the amount of degradation of different dyes will be different depending on the photocatalyst. Moreover, the degradation efficiency can be easily improved via doping or forming nanocomposites with suitable transition metals or their oxides.

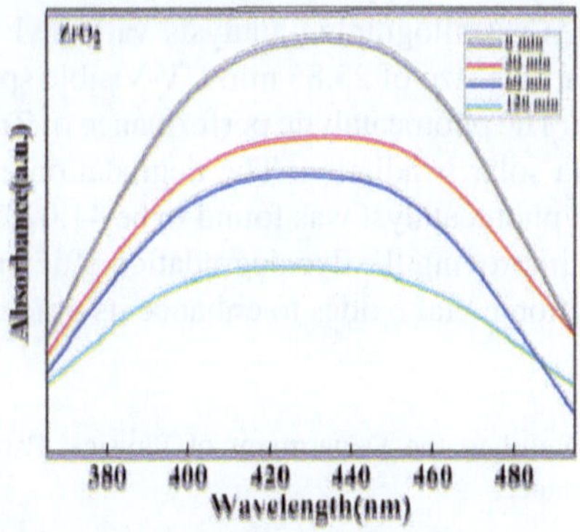

Fig. 5. Photocatalytic breakdown of methyl red by ZrO_2 NPs.

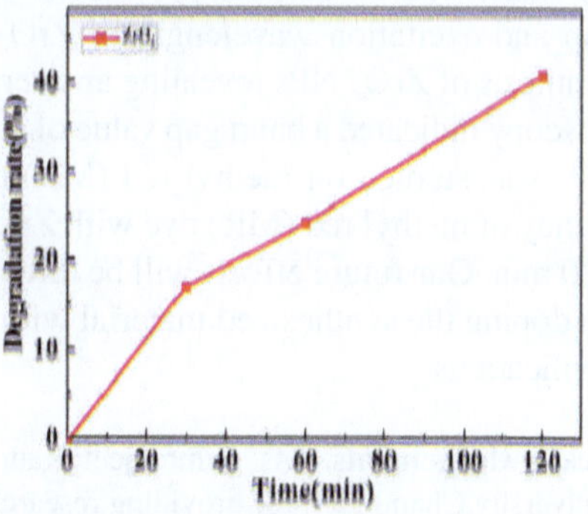

Fig. 6. Degradation rate (%) of methyl red at different time periods

The mechanistic principle behind the process of photocatalytic degradation of methyl red dye by ZrO_2 NPs is explained as follows. Upon exposure to sunlight, light energy causes an electron to transition from the valence band to the conduction band on the catalyst's surface, leading in the formation of e^-–h^+ pairs. These pairs interact with the pollutant, causing the generation of OH^- radicals, singlet oxygen radicals, or holes, all of which contribute to pollutant degradation. During their migration, the electron-hole pairs participate in redox reactions.

Holes combine with hydroxide ions to form hydroxyl radicals, while electrons react with oxygen in the system to produce superoxide and H_2O_2. The superoxide and hydrogen peroxide then subsequently combine to generate more hydroxyl radicals. In the final stage, these radicals interact with the pollutant (in this case, methyl red dye), forming intermediate compounds within a short period. Ultimately, these reactions break down the pollutant into harmless products such as CO_2 and H_2O [5, 12, 13].

$$ZrO_2\ NPs + h\vartheta \rightarrow ZrO_2\ NPs\left(h^+_{VB} + e^-_{CB}\right) \tag{4}$$

$$h^+ + OH^- \rightarrow OH\bullet \tag{5}$$

$$e^- + O_2 \rightarrow O_2^- \bullet + H_2O_2 \tag{6}$$

$$O_2^- \bullet + H_2O_2 \rightarrow OH\bullet \tag{7}$$

$$OH \bullet + O_2^- \bullet + MR \rightarrow Degradation\ Products\ (\ CO_2\ and\ H_2O) \tag{8}$$

4 Conclusion

ZrO_2 nanoparticles (NPs) were successfully synthesized via sol-gel route using $ZrOCl_2.8H_2O$. The X-Ray Diffraction (XRD) analysis verifies the successful synthesis of ZrO_2. The crystallographic analysis verifies the formation of ZrO_2 NPs with size in the nanometer range. The UV-Visible and PL spectroscopy was used to find the band gap and excitation wavelength of ZrO_2 NPs. Crystallographic analysis validated the synthesis of ZrO_2 NPs revealing an average particle size of 23.85 nm. UV-Visible spectroscopy indicated a band gap value of 4.03 eV. The photocatalytic performance of ZrO_2 NPs was studied on methyl red (MR) dye via solar irradiation. The degradation efficiency of methyl red (MR) dye with ZrO_2 as a photocatalyst was found to be 41.07% at 120 min. Our future efforts will be directed at improving the dye degradation efficiency by doping the synthesized material with transition metal oxides to enhance its practical applications.

Acknowledgements. Ms. Simranjeet Kaur is grateful to the Department of Physics, Panjab University Chandigarh for providing research assistance.

References

1. Ahmed, S.N., Haider, W.: Heterogeneous photocatalysis and its potential applications in water and wastewater treatment: a review. Nanotechnology. **29**(34), 342001 (2018)
2. Mustafa, G., Zahid, M.T., Kurade, M.B., Alvi, A., Ullah, F., Yadav, N., Jeon, B.H.: Microalgal and activated sludge processing for biodegradation of textile dyes. Environ. Pollut. **349**, 123902 (2024)
3. Mishra, S., Sundaram, B.: A review of the photocatalysis process used for wastewater treatment. Mater. Today: Proc. **102**, 393–409 (2024)
4. Zare, E.N., Iftekhar, S., Park, Y., Joseph, J., Srivastava, V., Khan, M.A., Varma, R.S.: An overview on non-spherical semiconductors for heterogeneous photocatalytic degradation of organic water contaminants. Chemosphere. **280**, 130907 (2021)
5. Gurushantha, K., Anantharaju, K.S., Kottam, N., Keshavamurthy, K., Ravikumar, C.R., Surendra, B.S., Murthy, H.A.: Synthesis of ZrO_2: Dy^{3+} nanoparticles: photoluminescent, photocatalytic, and electrochemical sensor studies. Adsorp. Sci. Technol. **2022**, 5664344 (2022)
6. Rajoriya, S., Saharan, V.K., Pundir, A.S., Nigam, M., Roy, K.: Adsorption of methyl red dye from aqueous solution onto eggshell waste material: kinetics, isotherms and thermodynamic studies. Curr. Res. Green Sustain. Chem. **4**, 100180 (2021)
7. Ratnam, M.V., Karthikeyan, C., Rao, K.N., Meena, V.: Magnesium oxide nanoparticles for effective photocatalytic degradation of methyl red dye in aqueous solutions: optimization studies using response surface methodology. Mater. Today: Proc. **26**, 2308–2313 (2020)

8. Aljaafari, A.: Size dependent photocatalytic activity of ZnO nanosheets for degradation of methyl red. Front. Mater. **7**, 562693 (2020)
9. Batool, T., Bukhari, B.S., Riaz, S., Batoo, K.M., Raslan, E.H., Hadi, M.: Microwave assisted sol-gel synthesis of bioactive zirconia nanoparticles–correlation of strength and structure. J. Mech. Behav. Biomed. Mater. **112**, 104012 (2020)
10. Manyani, N., Siwatch, P., Rana, S., Sharma, K., Tripathi, S.K.: Study of electrochemical behaviour of binder-free nickel metal-organic framework derived by benzene-1, 3, 5-tricarboxylic acid for supercapacitor electrode. Mater. Res. Bull. **165**, 112320 (2023)
11. Kumar, A., Singh, J., Tripathi, S.K.: High thermoelectric coefficients of β-Zn_4Sb_3 thin films at room temperature via chemical etching. Mater. Res. Bull. **184**, 113281 (2025)
12. Chu, M.N., Nguyen, L.T., Truong, M.X., Do, T.H., Duong, T.T.A., Nguyen, L.T., Pham, V.H.: Ce^{3+}/Ce^{4+}-doped ZrO_2/CuO nanocomposite for enhanced photocatalytic degradation of methylene blue under visible light. Toxics. **10**(8), 463 (2022)
13. Vinayagam, R., Singhania, B., Murugesan, G., Kumar, P.S., Bhole, R., Narasimhan, M.K., Selvaraj, R.: Photocatalytic degradation of methylene blue dye using newly synthesized zirconia nanoparticles. Environ. Res. **214**, 113785 (2022)
14. Reddy, C.V., Babu, B., Reddy, I.N., Shim, J.: Synthesis and characterization of pure tetragonal ZrO_2 nanoparticles with enhanced photocatalytic activity. Ceram. Int. **44**(6), 6940–6948 (2018)
15. Horti, N.C., Kamatagi, M.D., Nataraj, S.K., Wari, M.N., Inamdar, S.R.: Structural and optical properties of zirconium oxide (ZrO2) nanoparticles: effect of calcination temperature. Nano Express. **1**(1), 010022 (2020)
16. Kianfar, A.H., Arayesh, M.A., Momeni, M.M.: Degradation of MB and RhB by modified ZrO_2 nanoparticles via sunlight. Appl. Phys. A. **127**, 1–9 (2021)
17. Mahich, S., Saini, Y.K., Devra, V., Aggarwal, K., Kumar, A., Kumar, D., Arya, Y.: Metal-free adsorption and photodegradation methods for methylene blue dye removal using different reduction grades of graphene oxide. Heliyon. **10**(11), e31702 (2024)

Structural and Morphological Characterization of $Cr_{0.2}Ni_{0.8-x}Co_xFe_2O_4$ Nanocrystalline Samples Synthesized via Sol-Gel Auto-Combustion Method

Pooja Yadav[1], R. K. Beniwal[1], Anchal[1], Sarita[1], Ritu[1], S. N. Dolia[2], Parvez Ahmed Alvi[1], and B. L. Choudhary[1(✉)]

[1] Department of Physics, Banasthali Vidyapith, Banasthali 304022, Rajasthan, India
blcphysics@gmail.com

[2] Department of Physics, University of Rajasthan, Jaipur 304022, India

Abstract. The study investigated the structural and morphological characteristics of nanocrystalline $Cr_{0.2}Ni_{0.8-x}Co_xFe_2O_4$ samples with xx values of 0.0, 0.2, 0.4, and 0.6. The samples were synthesized via the sol-gel auto-combustion method, utilizing nitrates as the primary precursors dissolved in distilled water to form a homogeneous solution. The solution's pH was adjusted incrementally by adding ammonia (NH_3) dropwise. Following gel formation, the mixture was stirred continuously on a hot plate at approximately 120 °C to ensure uniformity. The gel underwent auto-combustion, resulting in fine nanoparticles.

X-ray diffraction (XRD) analysis confirmed the single-phase cubic crystal structure of the $Cr_{0.2}Ni_{0.8-x}Co_xFe_2O_4$ samples. Field emission scanning electron microscopy (FESEM) revealed the homogeneity of the nanoparticles, while energy-dispersive X-ray spectroscopy (EDAX) verified the presence of the intended elemental composition. Fourier-transform infrared (FTIR) spectroscopy identified the characteristic bonds associated with the doped elements within the nanocrystalline matrix.

Keywords: Nanocrystalline · XRD · Ferrite

1 Introduction

Spinel ferrite nanoparticles exhibit a high surface area and find applications across diverse fields, including high-density data storage, catalysis, and drug delivery. These materials follow the general formula AB_2O_4, with a structure based on a close-packed oxygen lattice. The cations occupy either tetrahedral (A sites) or octahedral (B sites) interstitial positions [1–9]. Spinel ferrites are classified as normal when divalent ions exclusively occupy the tetrahedral sites and as inverse when these ions are located in the octahedral sites. The unique properties of spinel ferrites stem from the interactions between cations at the A and B sites. In this study, we sought to enhance the intrinsic properties of Co-ferrite by progressively substituting Cr^{3+} ions using the sol-gel method.

S. Kumari et al. (Eds.): ICMTP 2024, SPPHY 311, pp. 262–266, 2025.
https://doi.org/10.1007/978-981-96-6795-6_30

Introducing different metal ions into the spinel ferrite lattice enables significant tuning of magnetic and electrical properties. Due to the antiferromagnetic nature of Cr ions, such substitution offers a practical way to control key magnetic parameters, such as remanence and coercivity, essential for developing advanced functional materials.

Various chemical techniques, including hydrothermal co-precipitation and sol-gel synthesis, have been employed to produce ultrafine ferrite powders. Among these, the sol-gel method stands out for providing highly homogeneous compositions within a relatively short processing time and at low temperatures. Understanding the interplay between structural, particle size, and magnetic properties is crucial for designing new magnetic materials. This work focuses on investigating the effects of Cr^{3+} ion concentration on the structural and magnetic properties of Co-ferrite nanoparticles synthesized via the sol-gel method.

2 Synthesis

Cobalt-nickel ferrite nanoparticles substituted with Cr^{3+}, having the chemical formula $Cr_{0.2}Ni_{0.8-x}Co_xFe_2O_4$ where x = 0.2, 0.4 and 0.6) were synthesized using the sol-gel method. Analytical-grade $Fe(NO_3)_3{\bullet}9H_2O$, $Co(NO_3)_2{\bullet}6H_2O$, $Cr(NO_3)_3{\bullet}6H_2O$, $Ni(NO_3)_2{\bullet}6H_2O$, and citric acid ($C_6H_8O_7$) were employed as precursor materials [2, 3, 9–12]. The chemicals were accurately weighed based on the desired stoichiometric ratios, and deionized (DI) water was added to achieve complete dissolution. Ammonia was introduced dropwise to adjust the solution's pH to 8, while continuous stirring was maintained using a magnetic stirrer. The solution was subsequently subjected to evaporation through vigorous stirring and heating at 100 °C for 6–7 hours, ensuring a consistent temperature until the sol transitioned into a gel. The resultant gel was dried and mechanically ground using a pestle and mortar to produce the nanoferrite particles.

3 XRD Measurements

To determine the crystal structure and size, the XRD technique was used. Structural analysis of the synthesized $Cr_{0.2}Ni_xCo_{0.8-x}Fe_2O_4$ (where x = 0.2, 0.4 and 0.6) were performed using an XRD (Bruker D8 Advance) used with a CuKα source (λ = 1.5406 Å). X-ray data were measured at 300 K temperature over a 2θ range of 20°–70° [13–18]. The diffraction peaks established the construction of a cubic spinel single-phase structure. The average crystallite size of all samples was determined measuring the FWHM values of the highest intensity peak (311) from the XRD pattern and using the following Debye-Scherer's formula (Fig. 1b)

$$D = k\lambda/\beta\cos\theta \tag{1}$$

where D represents the crystallite size, k is the shape factor, λ denotes the wavelength of the incident X-rays, β is the full width at half maximum (FWHM) of the diffraction peak, and θ corresponds to the Bragg diffraction angle. The lattice constant, have been found 8.38 Å, 8.35 Å, and 8.30 Å (x = 0.2, 0.4 and 0.6 respectively). Investigation of the XRD pattern also established the nanocrystalline nature of the sample, with no additional or impurity peaks observed. the Debye-Scherrer method was used for the crystallite size, based on the FWHM of XRD peaks, and was found to be approximately 37 nm, 18 nm and12 nm (x = 0.2, 0.4 and 0.6 respectively). The broad peaks in the diffraction patterns further validate the nanoscale dimensions of the particles.

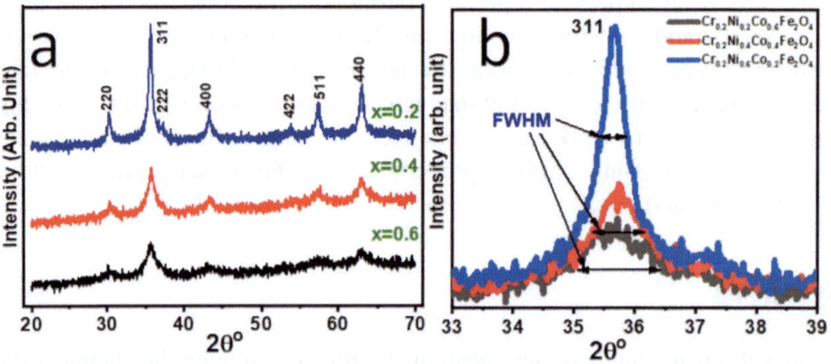

Fig. 1. (a) XRD pattern of $Cr_{0.2}Ni_xCo_{0.8-x}Fe_2O_4$ (where x = 0.2, 0.4 and 0.6) (b) FWHM of the all the samples.

4 FTIR Measurements

The FTIR spectrum presented in Fig. 2 displays a distinct peak at 3747 cm^{-1}, attributed to the O–H vibrational mode. A subtle shoulder in the 3200–3270 cm^{-1} range corresponds to O–H vibrations of adsorbed water molecules. Vibrational modes related to adsorbed and free anions, as well as twisting vibrations of interlayer and adsorbed H_2O molecules, are identified within the 2500–1000 cm^{-1} range. A twisting vibration of H_2O molecules is observed at 1692 cm^{-1}, while a weak peak at 2354 cm^{-1} is associated with nitrate

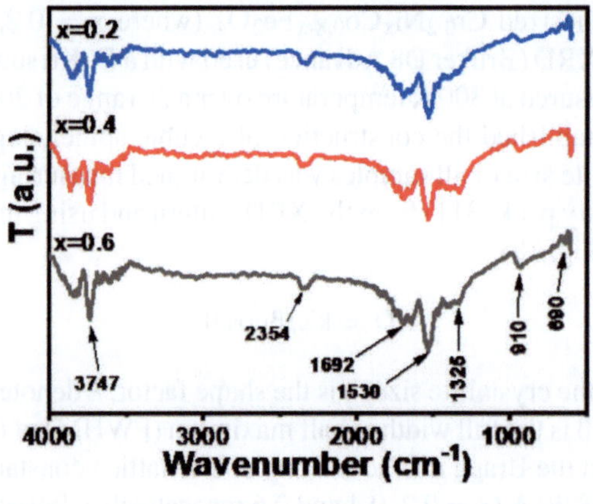

Fig. 2. FTIR Spectrum of $Cr_{0.2}Ni_xCo_{0.8-x}Fe_2O_4$ (where x = 0.2, 0.4 and 0.6) Ferrite samples.

groups. Furthermore, a broad peak at 1325 cm^{-1} (ν_3 irregular stretching) and faint peaks at 910 cm^{-1} (ν_1 proportional stretching), 830 cm^{-1} (ν_2 out-of-plane bending), and 690–696 cm^{-1} (ν_4 scissoring distortion) correspond to the fundamental vibrational modes of NO_3^- ions originating from the initial nitrate salts.

5 Conclusion

Cr^{3+}-substituted cobalt-nickel ferrite nanoparticles with the formula $Cr_{0.2}Ni_{0.8-x}Co_xFe_2O_4$ (x = 0.2, 0.4, 0.6) were synthesized via the sol-gel method. XRD investigation confirmed the creation of a single-phase cubic spinel structure, with no detectable impurity peaks. The crystallite size, calculated using the Debye-Scherrer formula, decreased from 37 nm to 12 nm as the Co content increased. This decrease in crystallite size, along with the broadening of XRD peaks, validated the nanocrystalline nature of the samples. The lattice parameters showed a systematic decrease with increasing Co content, reflecting structural adjustments due to ion substitution. FTIR spectroscopy provided complementary insights, revealing characteristic vibrational modes associated with the spinel structure and confirming the presence of hydroxyl and nitrate groups. Peaks corresponding to O–H stretching and nitrate ion vibrations were observed, further supporting the purity and chemical composition of the synthesized nanoferrites. These findings highlight the effectiveness of the sol-gel method in controlling particle size and structural properties, making Cr-substituted cobalt-nickel ferrites promising candidates for applications requiring nanostructured materials with tailored magnetic and structural properties.

References

1. Wei, Q.-M., Li, J.-B., Chen, Y.-J.: Cation distribution and infrared properties of NixMn1 − xFe2O4 ferrites. J. Mater. Sci. **36**, 5115–5118 (2001). https://doi.org/10.1023/A:1012473207424
2. Sarita, Anchal, Priya, Kumari, N., Palsaniya, K.K., Rulaniya, M.S., Alvi, P.A., Dolia, S.N., Choudhary, B.L.: Synthesis and characterization of nano-crystalline Ni-doped CoFe2O4 ferrite for biomedical applications. AIP Conf. Proc. **3149**, 020007 (2024). https://doi.org/10.1063/5.0224545
3. Anchal, Sarita, Priya, Kumari, N., Palsaniya, K.K., Rulaniya, M.S., Alvi, P.A., Dolia, S.N., Choudhary, B.L.: Characterization and synthesis of nanocrystalline CoFe2O4 ferrites prepared by sol-gel method with citric acid variation. AIP Conf. Proc. **3149**, 020006 (2024). https://doi.org/10.1063/5.0224532
4. Beniwal, R.K., Rulaniya, M.S., Saini, P.M., Yadav, P., Kumar, U., Alvi, P.A., Choudhary, B.L.: Development and characterization of superparamagnetic Zn-Doped Nickel ferrite nanoparticles. J. Magn. Magn. Mater. **610**, 172547 (2024)
5. Choudhary, B.L., Kumar, U., Kumar, S., Chander, S., Kumar, S., Dalela, S., Dolia, S.N., Alvi, P.A.: Irreversible magnetic behavior with temperature variation of Ni0. 5Co0. 5Fe2O4 nanoparticles. J. Magn. Magn. Mater. **507**, 166861 (2020)
6. Sarita, Anchal, Priya, Beniwal, R.K., Rulaniya, M.S., Saini, P.M., Yadav, P., Kumar, U., Aakansha, Alvi, P.A., Alvi, P.A., Choudhary, B.L.: Development and characterization of superparamagnetic Zn-Doped Nickel ferrite nanoparticles. J. Magn. Magn. Mater. **610** (2024). https://doi.org/10.1016/j.jmmm.2024.172547

7. Choudhary, B.L., Krishnamurthy, A., Srivastava, B.K.: Size dependence of magneto-resistive behaviour of nano-particle magnetite. Indian J. Pure Appl. Phys. **49**, 190–194 (2011)
8. Kumari, N., Sarita, Anchal, Palsaniya, K.K., Choudhary, S.R., Beniwal, R., Priya, Saini, P.M., Dolia, S.N., Alvi, P.A., Choudhary, B.L.: Structural and magnetic behavior in nanocrystalline Ni0.5Co0.5Fe2O4. Mater. Today Proc. **74**, 289–292 (2022). https://doi.org/10.1016/j.matpr.2022.08.241
9. Kumari, N., Sarita, Anchal, Priya, Palsaniya, K.K., Beniwal, R.K., Choudhary, S.R., Rulaniya, M.S., Saini, P.M., Dolia, S.N.: The role of citric acid for formation of nanocrystalline MnFe2O4 ferrite. Appl. Phys. A. **130**, 266 (2024)
10. Anchal, Sarita, Palsaniya, K.K., Priya, Kumari, N., Dolia, S.N., Alvi, P.A., Choudhary, B.L.: Tailoring quantum dots through citric acid modulation of CoFe2O4 ferrite. Mater. Chem. Phys. **313**, 128820 (2024). https://doi.org/10.1016/j.matchemphys.2023.128820
11. Choudhary, B.L., Kumari, N., Kumar, A., Kumari, J., Dolia, S.N.: Relaxation mechanism in Ni0. 5Zn0. 5Fe2O4 nanocrystalline ferrite at a lower temperature. Mater. Lett. **304**, 130731 (2021)
12. Choudhary, B.L., Chander, S., Dolia, S.N., Alvi, P.A.: Synthesis of Ni-and Co-doped ferrite nanoparticles and study of magnetic behaviour with temperature variation. In: AIP Conference Proceedings. AIP Publishing (2020)
13. Lal, G., Punia, K., Dolia, S.N., Alvi, P.A., Choudhary, B.L., Kumar, S.: Structural, cation distribution, optical and magnetic properties of quaternary Co0. 4+ xZn0. 6-xFe2O4 (x= 0.0, 0.1 and 0.2) and Li doped quinary Co0. 4+ xZn0. 5-xLi0. 1Fe2O4 (x= 0.0, 0.05 and 0.1) nanoferrites. J. Alloys Compd. **828**, 154388 (2020)
14. Sarita, Anchal, Priya, Choudhary, S.R., Rulaniya, M.S., Kumar, A., Dolia, S.N., Alvi, P.A., Choudhary, B.L.: Studies of structural, optical, and Raman analysis of Ni-substituted CoFe2O4. Phys. Status Solidi A. **221**, 2400083 (2024). https://doi.org/10.1002/pssa.202400083
15. Swami, S.K., Khan, J.I., Dutta, V., Lee, J., Laquai, F., Chaturvedi, N.: Spray-deposited aluminum-doped zinc oxide as an efficient electron transport layer for inverted organic solar cells. ACS Appl. Energy Mater. **6**, 2906–2913 (2023). https://doi.org/10.1021/acsaem.2c03858
16. Kumar, A., Swami, S.K., Dutta, V.: The role of electric field during spray deposition on fluorine doped tin oxide film. J. Alloys Compd. **588**, 546–550 (2014). https://doi.org/10.1016/j.jallcom.2013.11.130
17. Swami, S.K., Chaturvedi, N., Kumar, A., Kumar, V., Garg, A., Dutta, V.: Spray deposited gallium doped zinc oxide (GZO) thin film as the electron transport layer in inverted organic solar cells. Sol. Energy. **231**, 458–463 (2022). https://doi.org/10.1016/j.solener.2021.12.002
18. Swami, S.K., Chaturvedi, N., Kumar, A., Dutta, V.: Effects of electric field during deposition on spray deposited indium-doped zinc oxide films. Prog. Photovolt. Res. Appl. **24**, 74–82 (2016). https://doi.org/10.1002/pip.2649

Performance Evaluation of MEMS Flexural Resonators for Different Types of Electromechanical Materials

Anshu Sharma[1](✉) and Shaminder Kaur[2]

[1] Chitkara University School of Engineering and Technology, Chitkara University, Baddi, Himachal Pradesh, India
anshu.sharma@chitkarauniversity.edu.in

[2] Department of Electronics and Communication Engineering, Chitkara University Institute of Engineering and Technology, Chitkara University, Rajpura, Punjab, India

Abstract. Micro-resonators have gained popularity owing to the advantages of frequency output, small Size, batch fabrication and high sensitivity. Due to their inherent advantages, they provide a suitable substitute for macro-resonators. This paper presents a numerical study for obtaining material-based performance evaluation of certain flexural MEMS (micro-electro-mechanical systems) resonators such as fixed-free beam, fixed-fixed beam and double ended tuning fork. A COMSOL based simulative approach is adopted to study the vibrational response of resonators depending upon the type of material used. The materials referred for analysis are semiconductors, compound semiconductors and piezo-electric materials used in micromachining technology. These correspond to commonly used MEMS materials such as Silicon, Silicon carbide, Polysilicon, Aluminium nitride, Gallium nitride, Quartz etc. The resonant parameters analyzed are frequency, modal difference, order of operational mode and stress at the fixed ends of the resonator. Based upon the study conducted it can be inferred that amongst the materials being studied, Silicon is the material with suitable values for the analyzed parameters. Thus, utility of Silicon in micro-devices is being emphasized upon.

Keywords: MEMS · resonators · materials · resonant frequency · Simulation · modes

1 Introduction

A resonator is a mechanical structure that vibrates at a particular frequency in a specific shape with its maximum amplitude. Mainly fundamental modes are used as resonance modes but sometimes higher modes are also desirable [1]. With the advent of MEMS sensors, resonators hold a crucial place in micro-sensor technology. Micro-resonators are amongst the fundamental constituents in sensors, filters, accelerometers, oscillators etc. [2]. A resonant device comprises of a vibrating resonator as the basic element along with associated electronics. The sensor operation is based on the detection of frequency change in resonator vibration caused by the measurand [3]. Thus, these are benefitted

S. Kumari et al. (Eds.): ICMTP 2024, SPPHY 311, pp. 267–277, 2025.
https://doi.org/10.1007/978-981-96-6795-6_31

with digital output. Further, the transduction sensitivity of these is much higher than that of other sensors [4]. MEMS resonators offer high quality factor, need less power and have more stable operational characteristics [5].

The resonators are broadly of three types: bulk, flexural and torsional [6]. These resonator categories are based on the type of vibration they exhibit during oscillatory motion. The movement of resonator for bulk mode is expansion and contraction, for flexural resonators is to and fro motion and for torsional resonators is paddled motion [5]. Shape of a resonator is the most crucial aspect for understanding its operational behavior. The material chosen for the respective shape is another parameter which is detrimental for the resonator behavior. In this paper three micro-flexural resonator designs: fixed-free beam, fixed-fixed beam and double ended tuning fork are discussed and analyzed. First of all, a brief description of the three resonator shapes and various materials used in MEMS sensors is presented. A finite element approach is then used for analysis of the three resonant structures chosen for study. FEM simulations are performed using Multiphysics Software COMSOL 5.6. The vibrational behavior of the resonator is analyzed for various materials used in MEMS technology. The resonators are studied for the impact of material selection on the position of operational mode, difference between frequency of operational mode and adjacent modes, resonant frequency and stress at the resonator fixed ends.

2 Resonator Features

Nathanson reported the phenomenon of resonance in mechanical micro-structures in his early work [7]. The shape of resonator is important in determining the vibration shape, vibration balance, energy losses, quality factor and stability of resonator [4, 8]. The resonator's geometry is the factor that can be modulated as per the requirements to achieve the required device characteristics. For achieving flexural motion, the beam-shaped resonators are the most basic resonator shapes used in literature. Mainly fundamental modes are used as resonance modes but sometimes higher modes are also desirable [9, 10]. This section discusses three basic flexural resonator geometries:

- Fixed-free beam
- Fixed-fixed beam
- Double ended tuning fork

2.1 Fixed-Free Beam

A fixed-free (FF_R) beam resonator as the name suggests has one end of the beam fixed at its root while the other is free to move [11]. The basic shape of a FF_R beam is shown in Fig. 1(a). These have simple design and provide great avenues for sensing at the micro-scale [12]. These are also known as cantilever and are used in variety of sensors such as pressure sensor [13], mass sensor [14], chemical sensor [15] etc. These are used for both lateral [9] and transverse modes [16]. Schematics for the two vibrational mode shapes are shown in Fig. 1(b) and (c) respectively. Expression for resonant frequency f_{fr}, for FF_R beam resonator is given below in Eq. (1) [16].

$$f_{fr} = \frac{\alpha h}{2\pi l^2}\sqrt{\frac{E}{12\rho}} \tag{1}$$

where α is the mode constant depending on mode number, l and h are the length and height of the resonator while E and ρ are the Young's modulus and density of the material used.

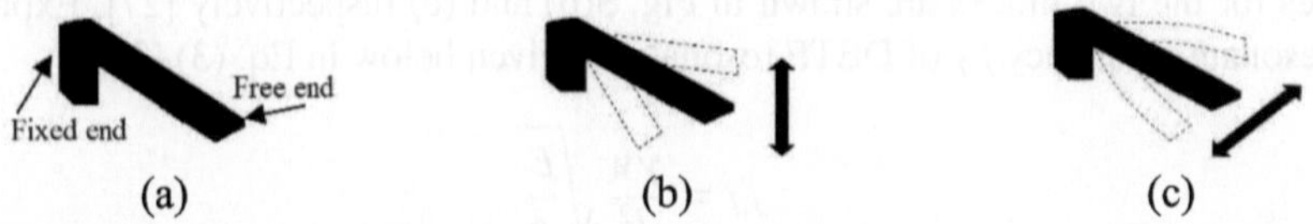

Fig. 1. Schematic for (a) Fixed-free beam and its (b) Transverse vibration mode and (c) Lateral vibration mode (Dotted lines depict the motion of resonator)

2.2 Fixed-Fixed Beam

In a fixed-fixed (FF) beam resonator, both the ends of a beam are fixed and the structure acts like a bridge structure [17]. The basic shape of FF beam is shown in Fig. 2(a). These are also known as clamped-clamped (CC) beams and are used for pressure sensing [18], magnetometer [19], mass sensing [20] etc. FF beam resonators are vividly used owing to their simple geometry and can be excited in both lateral mode [21] as well as transverse mode [22] as shown in Fig. 2(b) and (c) respectively. Expression for resonant frequency f_{ff} for FF beam resonator is given below in Eq. (2) [23].

$$f_{fr} = \frac{\beta h}{2\pi l^2}\sqrt{\frac{E}{12\rho}} \tag{2}$$

where β is the mode constant depending on mode number, l and h are the length and height of the resonator while E and ρ are the Young's modulus and density of the material used.

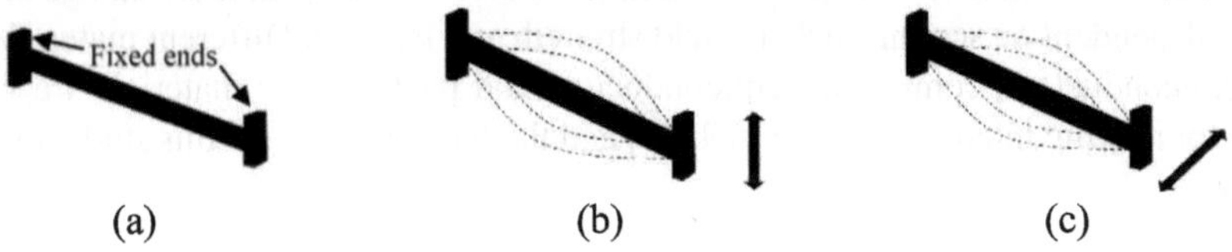

Fig. 2. Schematic for (a) Fixed-fixed beam and its (b) Transverse vibration mode and (c) Lateral vibration mode (Dotted lines depicts the motion of resonator)

2.3 Double Ended Tuning Fork

When two parallel FF beams are joined at their respective ends, then it forms a double ended tuning fork (DETF) structure. The basic shape of a DETF is shown in Fig. 3(a) and this structure is dynamically balanced [24, 25]. The desired shape of vibration for a DETF resonator is lateral anti-symmetric vibration mode where the two beams vibrate in opposite directions [26]. The two beams can vibrate in the same plane (lateral) or can vibrate in opposite planes (transverse) in the anti-symmetric mode. The basic vibration shapes for the two modes are shown in Fig. 3(b) and (c) respectively [27]. Expression for resonant frequency f_d of DETF resonator is given below in Eq. (3) [28].

$$f_d = \frac{\gamma w}{l^2}\sqrt{\frac{E}{\rho}} \tag{3}$$

where γ is the mode constant depending on mode number, l and w are the length and width of the resonator while E and ρ are the Young's modulus and density of the material used.

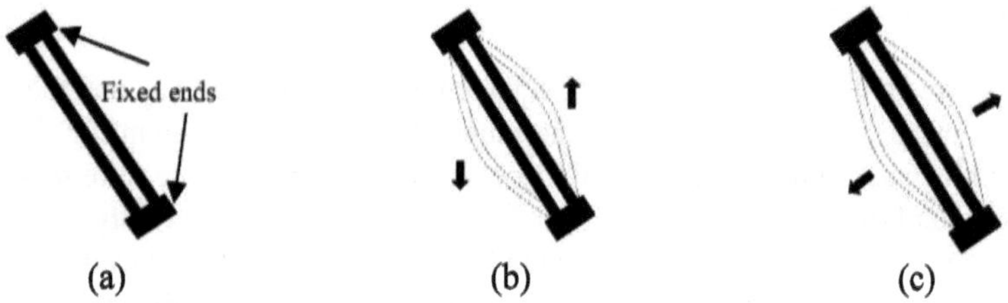

Fig. 3. Schematic for (a) Double ended tuning fork and its (b) Transverse vibration mode and (c) Lateral vibration mode (Dotted lines depicts the motion of resonator)

3 Materials and Methods

A vital factor in the popularity of the MEMS technology is the choice of materials available for fabricating complex MEMS sensors. The micro-sized dimensions of the devices enable choice from a wide range of materials. Also, it is easier to take advantage of properties dependent on scaling such as yield strength and fracture. Different materials such as semiconductors, compound semiconductors and piezoelectric materials are vividly used for making micro-resonators [29, 30] and the ones selected for this study are given below:

- *Semiconductors*: Silicon, Polycrystalline Silicon (PolySi), Silicon carbide (SiC), Gallium nitride (GaN) and Gallium arsenide (GaAs)
- *Piezoelectrics*: Quartz, Aluminium nitride (AlN), Zinc oxide (ZnO), Lead zirconate titanate (PZT) and Polyvinylidene fluoride (PVDF)

Silicon is the most versatile material in MEMS industry. It is available in pure form and has great mechanical properties. For applications that need high strength such as turbo machinery and pressure sensor, Silicon is a very promising material. The versatility also stems from the fact that fabrication techniques are well-established for Silicon [29, 31]. Silicon and Polysilicon, both exhibit piezoresistive properties. Single-crystal

Silicon is used for making thermal piezoresistive resonators and Polysilicon is very advantageous for mixed-mode micromachining fabrication techniques. SiC is another very attractive material for microsystems. It exhibits many useful properties such as: chemically inert, high operational temperature, higher thermal conductivity and great mechanical strength. It has significant utility for sensors in which specific stiffness is a crucial parameter [32–35].

Some compound semiconductors like GaN and GaAs are also widely used for making micro-resonators. They are highly benefitted by their capability to form high quality single crystalline multilayer heterostructures. Further their wide band gap and high breakdown properties enable integration with optical and electronic devices [36–38]. Quartz is a piezoelectric material popularly used for resonators and also exhibits insulative properties. AlN, PZT, ZnO are generally used as contour mode resonators owing to piezoelectric properties though such contour mode resonators can vibrate as flexural resonators also. In surface micromachining ZnO is also used as sacrificial layer [7, 39–42]. A polymer PVDF that exhibits piezoelectric properties is another material explored for micro-systems. Though it has lesser Young's modulus but its electromechanical properties are comparable to others and it gives high energy density [43].

3.1 FEM Simulation

The vibrational parameters of the three resonator shapes considered in this paper are analyzed using different materials for lateral vibration mode only using finite element analysis tool COMSOL 5.6. The different materials used in simulations along with their properties are tabulated in Table 1. A beam of $(150 \times 4 \times 8)$ μm is used for all the three resonators. For the DETF resonator the two beams are joined with the gap of 3 μm. Solid mechanics physics is used in COMSOL to analyze the mechanical vibrational behavior for the three resonators. Boundary conditions were set by fixing the respective ends of the beams. Then the eigen frequency analysis is performed to evaluate the performance of the resonators. The XY plane is which resonators are placed is shown in Fig. 4(a). The simulated mode shapes of lateral mode for the FF_R beam and FF beam resonators are shown in Fig. 4(b) and (c) while mode shape of the anti-symmetric lateral mode for DETF resonator is shown in Fig. 4(d).

4 Results and Discussion

The simulation-based analysis helped to observe the impact of material selection on vibrational performance of three resonators namely FF_R beam, FF beam and DETF. The simulations performed for micro-resonators were done to study the impact of material selection on the resonant frequency, frequency difference between operational mode and adjacent modes, operational modes position and stress at fixed ends of resonator. All these parameters were studied for the lateral mode of vibration for the resonators depicted above in Fig. 4(b–d) and the results for conducted parametric study are given below in Fig. 5(a–d) and Fig. 6(a–d).

From Fig. 5(a) it can be seen that maximum value for resonant frequency is obtained for SiC and minimum value is obtained for ZnO for all the three resonators. This can be

Table 1. Material properties used for COMSOL Simulation [39, 43–47]

Material	Young's Modulus (GPa)	Density (Kg/m^3)	Poisson's Ratio
Silicon	170	2329	0.28
PolySi	160	2320	0.22
SiC	748	3216	0.45
GaAs	85.9	5316	0.31
GaN	200	6070	0.23
AlN	300	3300	0.22
PZT	56	7500	0.36
ZnO	6.83	5680	0.20
PVDF	6.9	1780	0.35
Quartz	90	2651	0.08

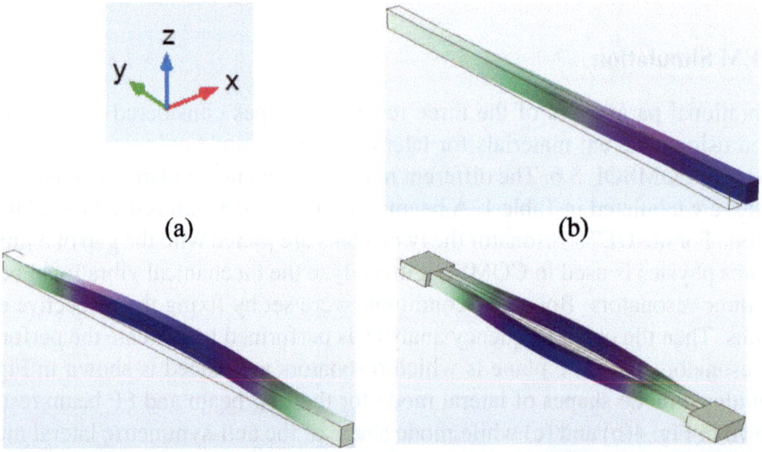

Fig. 4. Schematic of (a) XY Plane (b) FF_R beam lateral mode (c) FF beam lateral mode (d) DETF lateral anti-symmetric mode

justified from Eqs. (1), (2) and (3) and from Table 1 which shows that maximum value of Young's modulus is for SiC and minimum value is for PVDF. Stress at the ends of a resonator is desired to be less since this reduces the loss of vibrational energy [25]. From Fig. 5(b–d) it can be seen that the pattern followed for variation in stress with change in material is almost same for all the three resonator shapes being studied. It shows that SiC resonators have the maximum stress at their ends while it is least for PVDF resonators. Figure 5(b–d) also shows that piezoelectric materials on the whole, show less stress at resonator ends except for AlN which has second highest stress. Amongst

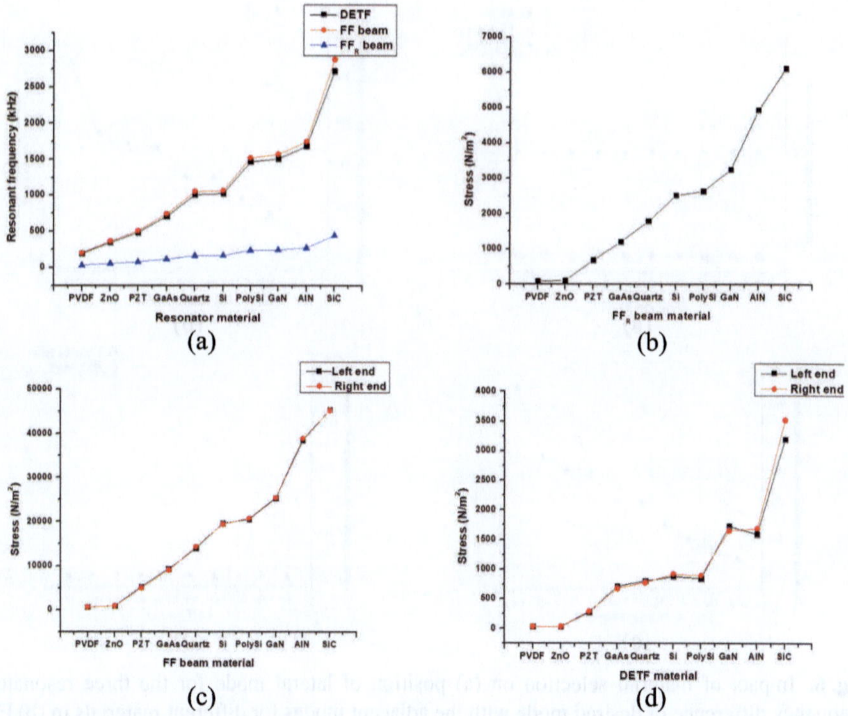

Fig. 5. Impact of material selection on (a) resonant frequency for lateral vibration mode of the three resonators; Stress at the fixed ends of (b) FF_R beam (c) FF beam (d) DETF

semiconductors, GaAs resonators exhibit least stress followed by Silicon, GaN and Polysilicon resonators.

It is desirable to have operational mode at the lower order position amongst different vibrational modes of the resonator [48]. Mode position is mainly dependent on resonator geometry. Simulations have also shown that material selection has not much impact on the position of operational mode. Figure 6(a) shows that for the chosen dimensions FF_R and FF beams have fundamental mode as lateral mode while for DETF, lateral mode is at second position for all the materials. Further, it is desirable to have a large frequency difference between the mode frequencies to prevent spurious modes [49]. Thus, frequency difference of operational lateral mode will be assessed for previous mode and next mode in DETF and for next mode only in the other two shapes: FF_R and FF beams. The impact of material selection on the frequency difference of the lateral mode with adjacent modes for the three resonators is shown in Fig. 6(b–d). The order here is also more or less same with ZnO resonators being the one with the least value. All the piezoelectric resonators have less difference as compared to others except AlN resonator. SiC resonator again has the highest frequency difference from the adjacent mode. Amongst semiconductors PolySi resonator exhibits highest frequency difference followed by Silicon, GaAs and GaN resonators.

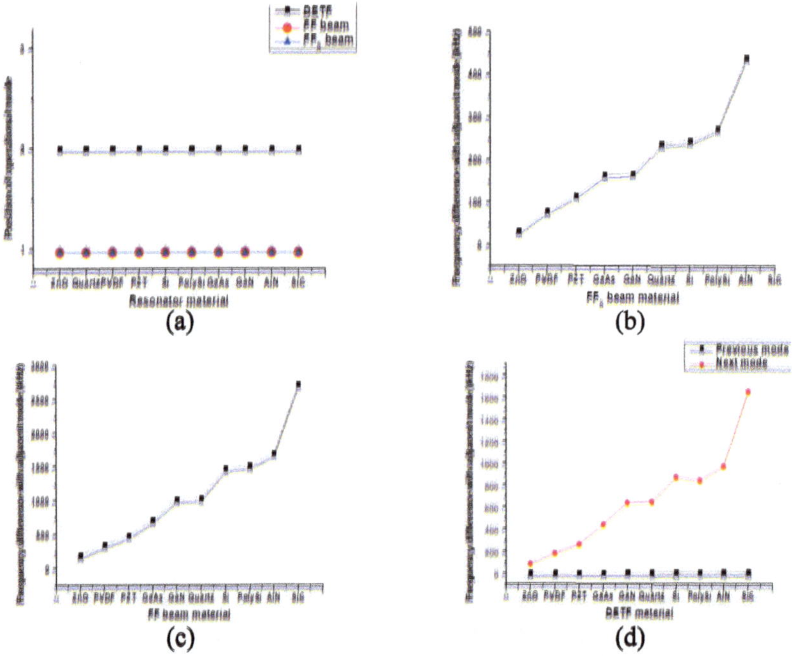

Fig. 6. Impact of material selection on (a) position of lateral mode for the three resonators; Frequency difference of desired mode with the adjacent modes for different materials in (b) FF_R beam (c) FF beam (d) DETF

Overall, from Fig. 5 and Fig. 6 it can be observed that for the piezoelectric materials, generally low values were obtained for all the evaluated parameters. Thus, they do not satisfy high frequency difference requirement for operational and adjacent modes. SiC has higher values for all the parameters. Though high resonant frequency can be useful and higher frequency difference is desirable but higher stress values at the ends of resonator is not at all desirable. Thus, SiC failed in the category of low stress values. The values for all the parameters were optimum for Silicon with neither being too low nor being too high. The results for Silicon resonators are better than that of compound semiconductor resonators and are in proximity to Quartz and Polysilicon resonators. Stress at the fixed ends in Silicon resonator lies between that of Quarts and Polysilicon resonators. However, frequency difference with adjacent mode is higher for Silicon in comparison to Quartz but is less than that of Polysilicon. Silicon thus has required less stress value and high frequency difference value in the studied resonator shapes. The results obtained from the simulative analysis presented in this paper thereby contribute to the fact that Silicon has appreciable properties and is the most widely used micromachining material.

5 Conclusion

This paper presents an overview of material selection-based performance evaluation for certain flexural resonators. In this paper three micro-resonator geometries namely FF_R beam, FF beam and DETF are analyzed and reviewed for different materials. The materials chosen for study were semiconductors, compound semiconductors and piezoelectrics. This paper highlights the parameters required to evaluate resonator vibrational performance and presents the impact of material selection on these parameters. Different important parameters such as position of operational mode, its frequency difference with adjacent modes, its resonant frequency and stress at resonator ends during its vibration are assessed. This is accomplished by using finite element method approach of COMSOL software. It was observed that material selection has not much impact on the position of the operational mode in a particular resonator shape. However, other parameters were impacted based on the selection of material for studied resonator shapes. Amongst different materials that were studied, it was observed that Silicon has the optimum performance parameters in comparison to other materials. Also, from the studied literature [4, 16, 21–23, 29] it can be inferred that Silicon is a widely used material in the MEMS technology. Hence, the study presented here supports this very fact that Silicon is a vividly preferred material in the micromechanical devices due to its suitable properties.

References

1. Tilmans, H.A., Elwenspoek, M., Fluitman, J.H.: Micro resonant force gauges. Sensors Actuators A Phys. **30**(1–2), 35–53 (1992)
2. Paci, D., Mastrangeli, M., Nannini, A., Pieri, F.: Modeling and characterization of three kinds of MEMS resonators fabricated with a thick polysilicon technology. Analog Integr. Circ. Sig. Process. **48**, 41–47 (2006)
3. Putty, M.W., Chang, S.C., Howe, R.T., Robinson, A.L., Wise, K.D.: One-port active polysilicon resonant microstructures. In: IEEE Micro Electro Mechanical Systems, Proceedings, 'An Investigation of Micro Structures, Sensors, Actuators, Machines and Robots', pp. 60–65. IEEE, Salt Lake City (1989)
4. Stemme, G.: Resonant silicon sensors. J. Micromech. Microeng. **1**(2), 113 (1991)
5. Basu, J., Bhattacharyya, T.K.: Microelectromechanical resonators for radio frequency communication applications. Microsyst. Technol. **17**, 1557 (2011)
6. Chandorkar, S.A., Agarwal, M., Melamud, R., Candler, R.N., Goodson, K.E., Kenny, T.W.: Limits of quality factor in bulk-mode micromechanical resonators. In: 21st International Conference on Micro Electro Mechanical Systems, pp. 74–77. IEEE, Tucson (2008)
7. Lee, J.E., Yan, J., Seshia, A.A.: Study of lateral mode SOI-MEMS resonators for reduced anchor loss. J. Micromech. Microeng. **21**(4), 045010 (2011)
8. Judge, J.A., Photiadis, D.M., Vignola, J.F., Houston, B.H., Jarzynski, J.: Attachment loss of micromechanical and nanomechanical resonators in the limits of thick and thin support structures. J. Appl. Phys. **101**(1), 013521 (2007)
9. Kolesar, E.S., Allen, P.B., Howard, J.T., Wilken, J.M., Boydston, N.: Thermally-actuated cantilever beam for achieving large in-plane mechanical deflections. Thin Solid Films. **355**, 295–302 (1999)
10. Nayfeh, A.H., Younis, M.I., Abdel-Rahman, E.M.: Dynamic pull-in phenomenon in MEMS resonators. Nonlinear Dyn. **48**, 153–163 (2007)
11. Ababneh, A., Al-Omari, A.N., Dagamseh, A.M.K., Qiu, H.C., Feili, D., Ruiz-Diez, V., Manzaneque, T., Hernando, J., Sanchez-Rojas, J.L., Bittner, A., Schmid, U.: Electrical characterization of micromachined AlN resonators at various back pressures. Microsyst. Technol. **20**, 663–670 (2014)

12. Patel, P.N., Yadav, R., Adhvaryu, M.: Design and analysis of diversified micro-cantilever structure for sensor applications. In: 2nd International Conference on Emerging Technology Trends in Electronics, Communication and Networking, pp. 1–5. IEEE, Surat (2014)
13. Krakover, N., Ilic, R., Krylov, S.: Resonant pressure sensing using a micromechanical cantilever actuated by fringing electrostatic fields. In: Micro Electro Mechanical Systems (MEMS), pp. 846–849. IEEE, Belfast (2018)
14. Rajak, A.K., Dash, R., Kumari, A., Bhattacharyya, A.S.: Sensitivity in nanomechanical pedestal MEMS cantilever. Mater. Today Commun. **38**, 107891 (2024)
15. Abdul, B., Shibly, M.A.H., Asary, A.R.: Development of a novel design and modeling of MEMS piezoelectric cantilever-based chemical sensors. Eng. Proc. **56**(1), 105 (2023)
16. Waggoner, P.S., Craighead, H.G.: Micro-and nanomechanical sensors for environmental, chemical, and biological detection. Lab Chip. **7**(10), 1238–1255 (2007)
17. Ryou, J.H., Gorman, J.J.: Mode selection for electrostatic beam resonators based on motional resistance and quality factor. J. Appl. Phys. **120**(21) (2016)
18. Li, P., Luo, Y., Guo, L., Zhang, Y.: Design and simulation of an aluminum nitride piezoelectric resonant pressure sensor. AIP Adv. **12**(6) (2022)
19. Kanoun, O., Derbel, F., Derbel, N. (eds.): Sensors, Circuits and Instrumentation Systems: Extended papers, vol. 6. Walter De Gruyter GmbH & Co KG (2017)
20. Li, L., Zhang, Y., Ma, C., Liu, C., Peng, B.: Anti-symmetric mode vibration of electrostatically actuated clamped–clamped microbeams for mass sensing. Micromachines. **11**(1), 12 (2019)
21. Xie, B., Jiao, H., Wang, J., Chen, D., Chen, J.: An electrostatically-driven and capacitively-sensed differential lateral resonant pressure microsensor. In: The 8th Annual International Conference on Nano/Micro Engineered and Molecular Systems, pp. 1271–1274. IEEE, Suzhou (2013)
22. Petersen, K., Pourahmadi, F., Brown, J., Parsons, P., Skinner, M., Tudor, J.: Resonant beam pressure sensor fabricated with silicon fusion bonding. In: TRANSDUCERS'91: International Conference on Solid-State Sensors and Actuators. Digest of Technical Papers, pp. 664–667. IEEE, San Francisco (1991)
23. Tang, Z., Fan, S., Cai, C.: A silicon micromachined resonant pressure sensor. J. Phys. Conf. Ser. **188**(1), 012042 (2009)
24. Cheshmehdoost, A., Jones, B.E., O'Connor, B.: Characteristics of a force transducer incorporating a mechanical DETF resonator. Sensors Actuators A Phys. **26**(1–3), 307–312 (1991)
25. Sharma, A., Kaur, H.J., Kumar, V.: Dimensional optimization and modelling of a novel double-ended-tuning-fork micro-resonator for high frequency applications. Sādhanā. **48**(1), 17 (2023)
26. Ueda, T., Kohsaka, F., Ogita, E.: Precision force transducers using mechanical resonators. Measurement. **3**(2), 89–94 (1985)
27. Huang, B.-S., Chang-Chien, W.-T., Hsieh, F.-H., Chou, Y.-F., Chang, C.-O.: Analysis of the natural frequency of a quartz double-end tuning fork with a new deformation model. J. Micromech. Microeng. **26**(6), 065006 (2016)
28. Cheng, R., Li, C., Zhao, Y., Li, B., Tian, B.: A high performance micro-pressure sensor based on a double-ended quartz tuning fork and silicon diaphragm in atmospheric packaging. Meas. Sci. Technol. **26**(6), 065101 (2015)
29. Spearing, S.M.: Materials issues in microelectromechanical systems (MEMS). Acta Mater. **48**(1), 179–196 (2000)
30. Pratap, R., Arunkumar, A.: Material selection for MEMS devices. Indian J. Pure Appl. Phys. (2007)
31. Greenwood, J.C.: Silicon in mechanical sensors. J. Phys. E Sci. Instrum. **21**(12), 1114 (1988)
32. Howe, R.T.: Applications of silicon micromachining to resonator fabrication. In: Proceedings of 48th Annual Symposium on Frequency Control, pp. 2–7. IEEE, Boston (1994)

33. Watkins, C.A., Lee, J., McCandless, J.P., Hall, H.J., Feng, P.X.L.: Single-crystal silicon thermal-piezoresistive resonators as high-stability frequency references. J. Microelectromech. Syst. (2024)
34. Abdolvand, R., Johari, H., Ho, G.K., Erbil, A., Ayazi, F.: Quality factor in trench-refilled polysilicon beam resonators. J. Microelectromech. Syst. **15**(3), 471–478 (2006)
35. Sarro, P.M.: Silicon carbide as a new MEMS technology. Sensors Actuators A Phys. **82**(1–3), 210–218 (2000)
36. Yamaguchi, H.: GaAs-based micro/nanomechanical resonators. Semicond. Sci. Technol. **32**(10), 103003 (2017)
37. Rais-Zadeh, M., Gokhale, V.J., Ansari, A., Faucher, M., Théron, D., Cordier, Y., Buchaillot, L.: Gallium nitride as an electromechanical material. J. Microelectromech. Syst. **23**(6), 1252–1271 (2014)
38. Popa, L.C.: Gallium nitride MEMS resonators. PhD Thesis, Massachusetts Institute of Technology (2015)
39. Danel, J.S., Delapierre, G.: Quartz: a material for microdevices. J. Micromech. Microeng. **1**(14), 187 (1991)
40. Piazza, G., Stephanou, P.J., Pisano, A.P.: Piezoelectric aluminum nitride vibrating contour-mode MEMS resonators. J. Microelectromech. Syst. **15**(6), 1406–1418 (2006)
41. Kasambe, P.V., Asgaonkar, V.V., Bangera, A.D., Lokre, A.S., Rathod, S.S., Bhoir, D.V.: Piezoelectric lead zirconate titanate (PZT) ring shaped contour-mode MEMS resonators. IOP Conf. Ser. Mater. Sci. Eng. **310**(1), 012069 (2018)
42. Vellekoop, M.J., Visser, C.C.O., Sarro, P.M., Venema, A.: Compatibility of zinc oxide with silicon IC processing. Sensors Actuators A Phys. **23**(1–3), 1027–1030 (1990)
43. Jiang, Y., Shiono, S., Hamada, H., Fujita, T., Higuchi, K., Maenaka, K.: Low-frequency energy harvesting using a laminated PVDF cantilever with a magnetic mass. Power MEMS. **2010**, 375378 (2010)
44. Kisielowski, C., Krüger, J., Ruvimov, S., Suski, T., Ager, J.W., Jones, E., Liliental-Weber, Z., Rubin, M., Weber, E.R., Bremser, M.D., Davis, R.F.: Strain-related phenomena in GaN thin films. Phys. Rev. B. **54**(24), 17745–17753 (1996)
45. Akashah, N.A., Rohani, M., Rosmi, A.S., Isa, M., Ismail, B., Khalid, K.N., Mustafa, W.A., Wooi, C.L.: Modelling of piezoelectric sensor with different materials approach for partial discharge detection on power transformer: PZT-5H, ZnO and AlN. IOP Conf. Ser. Mater. Sci. Eng. **864**(1), 012203 (2020)
46. Ji, S., Li, L., Motra, H.B., Wuttke, F., Sun, S., Michibayashi, K., Salisbury, M.H.: Poisson's ratio and auxetic properties of natural rocks. JGR Solid Earth. **123**(2), 1161–1185 (2018)
47. Cheng, Y., Li, D.: Numerical analysis of piezoelectric signal of PVDF membrane flapping wing in flight. IOP Conf. Ser. Mater. Sci. Eng. **774**(1), 012090 (2020)
48. Sharma, A., Kumar, V., Kaur, H.J.: Finite element analysis for design optimization of MEMS double ended tuning fork resonator. ECS Trans. **107**(1), 4697 (2022)
49. Vigevani, G., Przybyla, R.J., Yen, T.-T., Lin, C.-M., Guedes, A., Kuypers, J.H., Boser, B.E., Horsley, D.A., Pisano, A.P.: Characterization of a single port aluminum nitride tuning fork. In: International Ultrasonics Symposium, pp. 1281–1285. IEEE, San Diego (2010)

Gallium Nitride Channel-Based Junctionless Transistors for Futuristics Power Electronics Energy Applications

Pooja Srivastava[1(✉)], Aditi Upadhyaya[1], Shekhar Yadav[1], C. M. S. Negi[1], and Arvind Kumar Singh[2]

[1] Department of Physical Sciences, Banasthali Vidyapith, Radha Kishnpura 304022, Rajasthan, India
poojasrivastava@banasthali.in

[2] Department of Mathematics, Institute of Science, Banaras Hindu University, Varanasi 221005, Uttar Pradesh, India

Abstract. An investigation on cylindrical gate-all-around Junctionless Field Effect Transistors (JLFETs) with no gradients in doping concentration and non-availability of junctions has been conducted. The findings indicate that the suggested device provides a unique architecture that significantly enhances transistor performance while being less vulnerable to short-channel effects (SCEs). Specifically designed for power electronics applications, the Atlas 3D device simulator has been utilized to analyze the performance of the proposed JLFET. Gallium Nitride (GaN) III-V compound semiconductors are used in this study to examine the electrical performance of a JLFET. Comparing the III-V JLFET to other circuit topologies, the simulations showed a favorable subthreshold slope and a lower drain-induced barrier lowering (DIBL). The subthreshold slope is approximately 63 mV/dec, which is extremely near the ideal 60 mV/dec and exhibits a low DIBL of approximately 30 mV/V. According to simulations and experimental evidence, GaN-based JLFETs have a higher power efficiency than silicon-based devices. The potential use of GaN-based JLFET as a cutting-edge substitute for advanced energy applications is covered in this research.

Keywords: Drain-induced barrier lowering · Subthreshold Slope · Gate-all-around Junction-less FET · III-V Compound Materials

1 Introduction

The last 40 years have steadily reduced the physical size of metal-oxide-semiconductor field-effect transistor (MOSFET) devices. New technologies are being introduced more quickly to expand MOS scaling beyond the 100-nm node due to the high cycle of MOSFET downscaling. This acceleration necessitates a thorough investigation of the detrimental short-channel effects (SCEs) and their solutions to enhance performance and maintain the historical pace of downsizing. This work aims to identify the least allowable channel length and avoid SCEs, which are considered the most difficult

S. Kumari et al. (Eds.): ICMTP 2024, SPPHY 311, pp. 278–287, 2025.
https://doi.org/10.1007/978-981-96-6795-6_32

obstacle to sub-100 nm MOSFET scaling. For this, the focus is detailing the latest developments in unconventional MOS device topologies. When modifying the SCEs of deep-submicron MOSFETs and overcoming the carrier transport efficiency bottleneck, the source/drain engineering, substrate engineering, and gate engineering approaches show great promise. Due to the ongoing development of semiconductors, devices have become smaller, increasing their operational speed and requiring less space in VLSI circuits. SCEs beyond 45 nm technology deteriorate the performance of MOSFETs. To improve the performance of these devices, designers have proposed different architectures, including bulk and silicon-on-insulator (SOI) FET. The downscaling process is still being used to meet the need for electronics with higher performance. Compared to CMOS, transistors with smaller dimensions have superior characteristics, such as stronger current drive and faster processing speed [1].

Many digital devices' capabilities are entirely reliant on Moore's law. The legislation is named after Gordon E. Moore, a co-founder of Intel. From 1958 until 1965, when the integrated circuit was invented, the number of components in integrated circuits increased annually, and he projected that this trend would last for at least 10 years. The semiconductor industry uses this law, proving his forecast was correct [2]. This law updates the semiconductor industry's International Technology Road Map (ITRS). Energy scarcity is a problem that the entire world is dealing with. As a result, everyone is looking for new technologies to create energy-efficient devices. Other MOSFET structures are required because conventional transistors (CTs) are getting so thin that they make it harder to overcome SCEs. The age of nanowires or nano-engineering has arrived, and current research is centered on a novel technology called a Junctionless Field Effect Transistor (JLFET) [3].

The increasing demand for high efficiency and high power density in contemporary power electronics has fueled the search for new transistor designs and alternative materials. Gallium Nitride (GaN) is becoming a serious contender for the upcoming generation of power devices due to its advantageous material characteristics, which include broad energy bandgap, extremely high electron mobility, remarkably low resistance, high saturation velocity, outstanding thermal conductivity, and exceptionally high frequencies Compared to the traditional MOSFET, JLFETs have a simpler structure and fewer SCEs, which facilitates performance scaling for high-power devices. This study describes the design, construction, and performance assessment of GaN channel-based JLFET, highlighting their superiority over traditional silicon-based transistors in terms of power management, efficiency, and dependability. GaN transistors demonstrate superior power efficiency when facing silicon transistors due to their technical material properties combined with modern device design. The wide GaN bandgap that reaches 3.4 eV exceeds that of Si′s 1.1 eV, thus enabling material breakdown resistance at elevated voltage and temperature conditions while lowering leakage current during stress points, which benefits high-power applications. The wide bandgap structure allows high electron mobility and faster efficient switching, which decreases the emergence of transition state energy losses. The outstanding thermal conductivity of GaN technology enables devices to dissipate heat efficiently; thus, they can function optimally during high-temperature usage without requiring advanced cooling technologies. Transistors built from GaN technology use a junctionless structure because it removes resistance typical of traditional Si junctions, maximizing electron mobility and minimizing voltage losses. The manufacturing process benefits from simplified methods, and high-frequency power applications achieve better operational efficiency because of lower parasitic capacitance and resistance. GaN transistors have become fundamental for high-power and energy-efficient electronics because of their superior properties, which have established advanced performance requirements [4].

Our findings establish GaN JLFETs as a crucial technology for sustainable energy solutions by showcasing their potential for power conversion, renewable energy systems, and electric cars. This research is a foundation for the potential integration of GaN-based JLFET, which promises energy efficiency in newly built power electronics. The requirement for power devices that can manage large power densities and deliver energy efficiency while maintaining electrical and thermal stability has increased dramatically due to the quick advancements in modern electronic technology. Traditional silicon-based transistors have been at the core of power electronics for a long time. They present serious obstacles to meeting the high-efficiency standards for power conversion, electric vehicles (EVs), and energy from renewable energy sources. Poor heat conductivity, large switching losses, and performance degradation at high voltages are the main factors restricting them. GaN has become a viable substitute material for power electronics in recent years. Because of this, GaN-based power devices are quickly being used in various applications, such as power converters, inverters, and power management systems, where efficiency and power handling are crucial [5].

2 Theory and Simulation Framework

JLFET is one of the most important recent developments in transistor design. The JLFET has a simpler topography with an uninterrupted channel of semiconductor material, negating the need for a junction, in contrast to the conventional MOSFET, whose current modulation is carried out solely by a well-defined junction. This design approach is becoming increasingly common for high-end use in power electronics applications because it reduces SCEs and improves scalability [6]. This study focuses on a novel development and performance assessment of GaN channel-based JLFET for more sophisticated power electronics applications. In addition to having structural advantages, these GaN JLFETs assist in closing several gaps left by traditional power devices. The potential existence of GaN JLFETs and their performance advantages over conventional silicon-based power transistors have been confirmed in this study using efficiency, power density, and thermal behavior as criteria. In actuality, the development of JLFET based on GaN suggested in this work greatly advances more general objectives, such as designing power electronics that are more sustainable, dependable, and efficient to serve the energy industry better.

GaN can withstand higher voltages than silicon before experiencing an electrical breakdown because of its higher breakdown voltage. It implies that GaN-based devices can use higher power levels without affecting functionality. GaN-based JLFET is being developed for advanced power electronics applications in energy. These transistors hold promise for future energy technologies. GaN semiconductors provide exceptional speed and efficiency while significantly reducing carbon emissions. It makes them a revolutionary development in the field of electronics. The application of GaN technology is observed in numerous industries, such as automotive, data centers, and optoelectronics. GaN material is the optimal solution for implementing smart power integration. A vital component of modern energy systems, power electronics handles and distributes electrical power efficiently and effectively. All too frequently, constraints in power devices based on traditional silicon power devices are encountered in renewable energy, electric car technology, and power grid infrastructure, where high-efficiency power systems have exponential demand for such devices. Silicon transistors have high switching losses, decreased efficiency at high voltages, and thermal problems that restrict their use in high-demand applications as power devices get smaller and operating frequencies rise [7].

As energy demands have grown and system efficiency has become more crucial, the electronics industry has shown interest in wide-bandgap materials like gallium nitride (GaN), which offer better properties than silicon. Because of their superior thermal conductivity, high breakdown voltage, and increased electron mobility, these materials can be used in applications that call for high power, high temperatures, and high frequencies [8]. Because of its great efficiency, small size, and ability to function at higher voltages and temperatures than silicon devices, GaN has emerged as the most advanced material for power electronics. Furthermore, GaN's wide bandgap of 3.4 eV makes it a good choice for demanding applications like power converters, motor drives, and electric cars because it can function effectively at high voltages and temperatures. GaN is especially well-suited for high-power and high-frequency applications due to its high saturation velocity and higher electron mobility at around 2000 $cm^2/V\ s$ [9].

3 Device Structure and Methodology

The suggested architecture has gate lengths of 40 nm with a constant radius of 14 nm, as illustrated in Fig. 1. Figure 1 illustrates how this suggested device shape was created using GaN as the channel material, and Fig. 2 illustrates the 40 nm structure of silicon channel-based JLFET. The 40 nm gate length has been selected in this work to achieve minimal SCEs alongside practical manufacturing capabilities for robust device performance. A 40 nm gate length configuration enables sufficient SCE constriction because SCE intensity increases while the gate length decreases even more. A gate length of 40 nm is established as an ideal basic unit that delivers proper electrical functioning yet avoids operational complexities and reliability issues. Decreasing gate lengths below 40 nm produces additional source/drain-induced barrier-lowering effects and threshold voltage roll-off phenomena. A transistor operates ineffectively when SCEs occur because they limit the gate's control of the channel and create unwanted leakage currents, which decreases the device's total efficiency. The control and management of SCEs become more challenging when gate lengths drop below 40 nm because present manufacturing methods approach their physical minimums.

The gate-all-around (GAA) structure enables this device to succeed in SCE reduction by providing extensive gate control across all boundaries of the channel area. The geometrical shape improves gate control over the channel through electrostatic means to minimize SCE effects, including DIBL and subthreshold slope. Uniform electric field distribution becomes possible because the cylindrical gate structure prevents hot spots

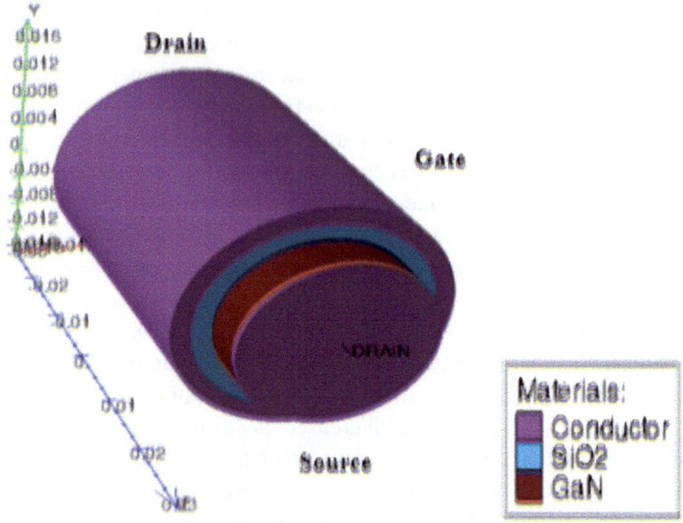

Fig. 1. 40 nm structure of GaN channel-based JLFET.

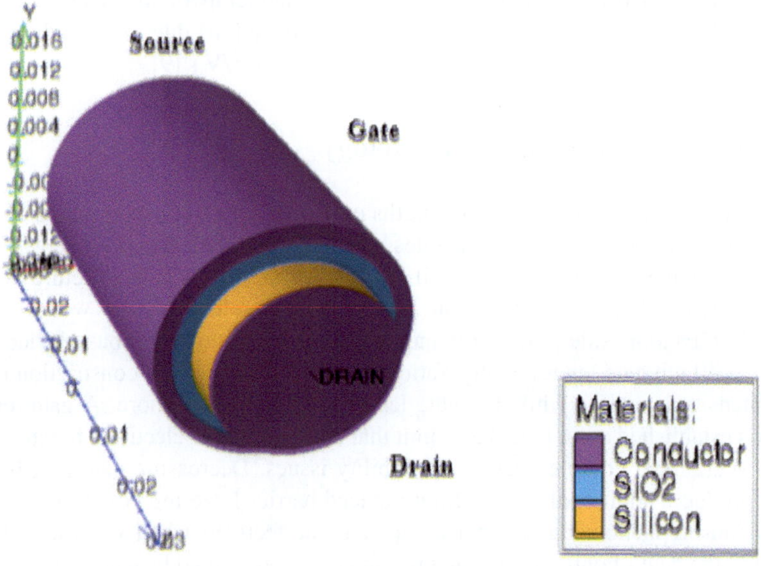

Fig. 2. 40 nm structure of silicon channel-based JLFET.

and uneven doping effects producing SCEs. Through advanced lithography and fabrication techniques, scientists have proven experimental fabrication works for producing GaN-based junctionless transistors with 40 nm gates. Self-aligned junctionless design enables improved manufacturing processes since it removes the requirement of creating precise nanoscale junctions with high dopants. Complete experimental success requires accurate etching and deposition techniques coupled with high-quality GaN material and repeatability of the cylindrical gate structure fabrication. Precision and material quality would require high-resolution electron-beam lithography and atomic layer deposition.

The ITRS road map has led to mesh design and geometrical settings, and this important effort has ultimately produced the required technology node and feature size. The channel, source, and drain region concentration is N-type and 10^{19} cm^{-3}. Recombination effects at high current density were examined using the Shockley-Read-Hall and AUGER recombination models. Equations with linked and decoupled forms were modified to include the NEWTON and GUMMEL models [10]. This study compares the electrical characteristics of a JLFET design using GaN as a III-V compound material and silicon channel. The simulations revealed that GaN has a favorable subthreshold slope and reduced DIBL compared to other circuit topologies. The study also found a substantial association between the subthreshold slope (SS) values and DIBL in GaN. The subthreshold slope is around 63 mV/dec, close to the optimum value of 60 mV/dec, and demonstrates a DIBL of approximately 30 mV/V.

An evenly doped GaN channel sandwiched between two source and drain areas makes up a GaN JLFET structure. The transistor's geometry is tailored for device performance and manufacturing-related practical restrictions. TCAD technologies like Silvaco aid simulation model development. Several operating circumstances, including threshold voltage, subthreshold swing, drain-source current, and on/off current ratio, are used to replicate the electrical behavior of GaN JLFET. Doping profiles of different concentrations were examined to determine the ideal concentration for channel doping to avoid SCEs while maximizing conductivity in the on-state. For the device to function well at high voltages and temperatures necessary in power electronics, GaN material features, including electron mobility and bandgap, were simulated. GaN-based JLFET can, therefore, be applied to power applications that require constant operation in extremely hot environments. Silicon-based devices are only used in energy-intensive applications like industrial power converters and electric vehicles despite having high leakage currents and poor performance at high temperatures [11].

4 Result and Discussion

The transfer characteristics of the GaN-based JLFET, as shown in Fig. 3, demonstrate the impact of GaN's wide bandgap (~3.4 eV) on device performance. The graph highlights a clear transition from the subthreshold region to the active conduction region, characterized by a low subthreshold current due to the material's low intrinsic carrier concentration and strong thermal stability. Beyond threshold voltage, the drain current rises sharply, indicating effective channel modulation through gate bias, and eventually saturates at higher gate-to-source voltage, showcasing the device's ability to maintain stable current under strong electric fields. The wide bandgap of GaN contributes to reduced leakage currents, low subthreshold slope, and enhanced breakdown voltage [12].

As a result, GaN JLTs have substantially less subthreshold swing than traditional devices. The measured subthreshold swing for GaN JLT was around 63 mV/dec, significantly less than the reported subthreshold slope for silicon-based MOSFETs (~70–80 mV/dec). The subthreshold slope determines transistor state switching efficiency because lower values indicate improved performance between on and off states. The gate-all-around junctionless transistor design offers improved electrostatic control, enabling

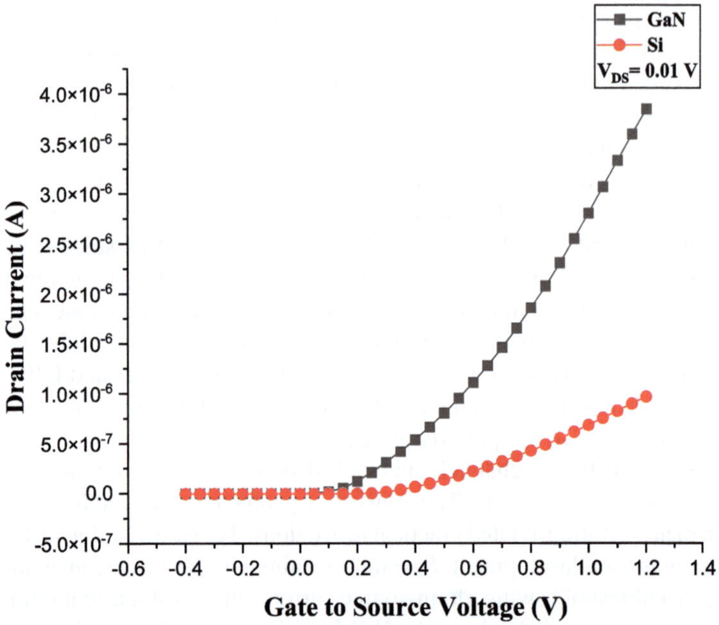

Fig. 3. Transfer Characteristics of GaN and Si-based 40 nm JLFET

the transistor to reach close to optimal 60 mV/dec subthreshold slope operations. The device architecture effectively reduces leakage and SCEs since these characteristics are essential in scaled technologies. The turn-on characteristics at the threshold become sharp because of GaN's wide bandgap and high electron mobility, which enable a low subthreshold slope. The documented DIBL value of 30 mV/V shows that the device maintains stability against changes in drain voltage because it achieves strong channel control and minimizes short-channel impact, a common scalability issue in the transistor field.

The output characteristics of GaN JLFET have been shown in Fig. 4. Reduced leakage during off-state situations and enhanced channel current regulation at low gate voltages with greater energy efficiency result from reduced subthreshold slope. It is also highly desirable for low-power operation and decreased static power dissipation in power devices. GaN JLT had an on/off current ratio of roughly 10^7, significantly higher than traditional silicon-based MOSFETs (~10^5). A high on/off ratio like this raises the prospect of superior switching with a low off-state leakage current. Because high on-state current also raises the prospect of reduced conduction losses, GaN JLTs are especially well-suited for high-power applications with high efficiency.

Because of the GaN's better electron mobility and decreased gate capacitance, switching losses were between 30 and 40 percent lower than those of silicon MOSFETs. The transconductance of GaN JLFET has been shown in Fig. 5. A smaller subthreshold slope is desirable because a higher on/off current ratio causes a quicker transition from on-state to off-state [13]. The present research analyzed the electrical behavior of

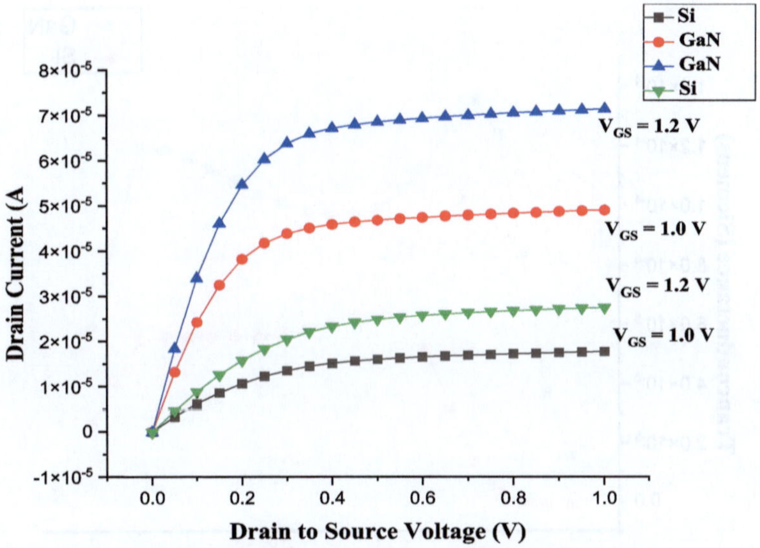

Fig. 4. Output Characteristics of GaN and Si-based 40 nm JLFET

GaN-based junctionless transistors at a 40 nm gate length with great detail. The subthreshold slope measurement for these devices produced almost theoretical perfection at 63 mV/dec, while DIBL showed a low threshold of 30 mV/V. The research shows that gate-all-around technology optimizes electrostatic control operations inside the channel. GaN's superior semiconductor properties emerged clearly from transconductance evaluations because of its high sensitivity to gate voltage modifications.

5 Experimental Fabrication Feasibility of GaN Junctionless Transistors

The semiconductor technology advancements explored in this research initiate proper inquiry regarding the real-world manufacturing potential of GaN junctionless transistors with GAA designs. Nanofabrication technology has progressed to offer new ways of fabricating these structures. The methods of electron-beam lithography (EBL) and atomic layer deposition (ALD) now meet the technical demands needed to create gate structures relevant to GAA devices. High-resolution patterning from electron-beam lithography allows users to create precise nanometer-scale dimensioned gates that align with the 40 nm gate sizes discussed in this work. Using atomic layer deposition, researchers can generate precise and thin insulation layers that GAA transistors require for effective function. The fabrication of these devices relies on reactive ion etching (RIE) because this technique provides clean and precise material removal, which is essential for creating the junctionless channel.

The high-volume manufacturing capability of GaN material through MOCVD corrosion supports the large-scale production of GaN-based devices. MOCVD is an established technique for manufacturing high-grade GaN films essential for junctionless transistor semiconductor layers. Contemporary technology can make these devices, but

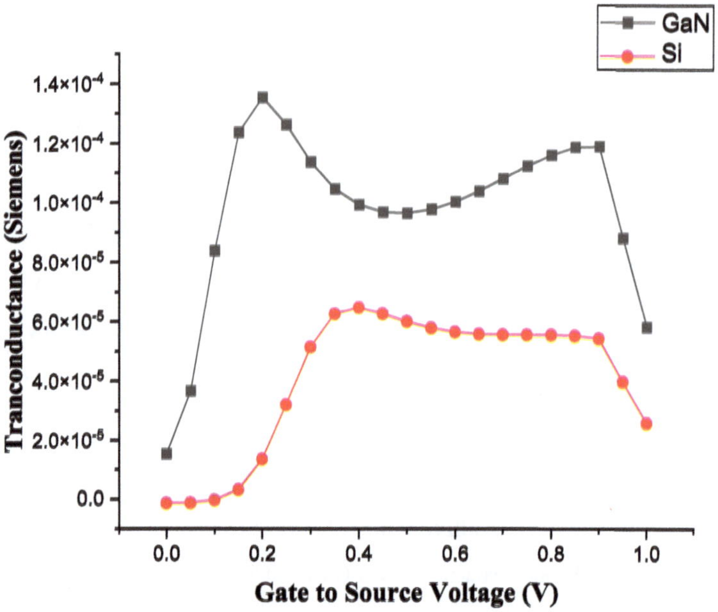

Fig. 5. Transconductance of GaN and Si-based 40 nm JLFET

researchers face multiple substantial obstacles. The control of dopant concentration and uniformity across semiconductor channels must be improved because these factors determine device performance consistency. Additional study is needed to assess the production cost-effectiveness of developing these complex devices beyond the laboratory stage. High-precision lithography and deposition methods demonstrate costs that restrain these transistors' expansion into low-cost commercial applications. The experimental production of junctionless GaN transistors with GAA structure remains possible through existing technological means, although these technologies require sophisticated process optimization and precision. Future investigations should direct their efforts toward making these production methods more efficient and cost-efficient, enabling widespread deployment of these sophisticated semiconductor elements.

6 Conclusion

JLFET is the possible alternative and promising emerging device to the conventional junction-based transistor. The fabrication process of JLFET eliminates the need for costly ultrafast annealing techniques. Even on a small scale, these structures are simple to construct, representing a significant advancement in possible cost savings. The analysis of JLFET has been provided to support the researchers for product development in ULSI design. Future studies clarify how non-silicon materials, such as polysilicon germanium, GaAs, and indium tin oxide, can be used with silicon materials for superior outcomes. It is especially well-suited for next-generation power systems in electric vehicles, aircraft, telecommunications, and renewable energy applications. Despite its promise, the GaN

JLT is still a long way off because there are many obstacles to realizing its full potential. There are serious issues with GaN device fabrication techniques, particularly for junctionless-structured devices: material flaws, doping homogeneity, and interface quality are the main issues. Moreover, certain obstacles can be overcome when scaling GaN JLTs to high power levels to enhance gate control and device reliability. Researchers are putting much effort into developing new doping strategies, enhanced gate architectures, and sophisticated fabrication procedures to overcome these obstacles.

References

1. Dhiman, R., Chandel, R.: Nanoscale VLSI: Devices, Circuits and Applications, p. 319. Springer Nature (2020)
2. Moore, G.E.: Progress in digital integrated electronics. IEDM Tech. Digest. **21**, 11–13 (1975)
3. Colinge, J.P., Lee, C.W., Afzalian, A., Akhavan, N.D., Yan, R., Ferain, I., Murphy, R.: Nanowire transistors without junctions. Nat. Nanotechnol. **5**(3), 225–229 (2010)
4. Smith, J., et al.: Power electronics: challenges and future directions. IEEE Trans. Power Electron. **35**(7), 2180–2194 (2020)
5. Kumar, A., et al.: The Role of GaN in Power Electronics. IEEE J. Electron. Mater. **40**, 1052–1065 (2019)
6. Colinge, J.P.: Junctionless transistors. In: 2012 IEEE International Meeting for Future of Electron Devices, Kansai. IEEE (2012)
7. Choi, S.I., et al.: GaN power devices for high-efficiency energy conversion. IEEE Trans. Electron. Devices. **65**(5), 2041–2049 (2018)
8. McCluskey, M.R., et al.: Wide-bandgap semiconductors for power electronics. Mater. Sci. Eng. R. Rep. **128** (2021)
9. Lee, H., et al.: Gallium nitride: a revolutionary material for power electronics. Nat. Electron. **2**, 315–320 (2021)
10. Ghosh, S., Jana, A., Agnihotri, A.K., Kundu, S., Das, D., Sarkar, S.K.: DC and analog/RF performance comparison of renovated GAA JLFET structures. In: IEEE VLSI Device Circuit and System (VLSI DCS), pp. 80–84. IEEE (2022)
11. Wang, K., et al.: High efficiency GaN-based power electronics for energy systems. IEEE Power Electron. Mag. **9**(4), 42–49 (2019)
12. Lee, S.W., et al.: Challenges and advancements in GaN power transistors. IEEE Trans. Power Electron. **66**(8), 6662–6675 (2020)
13. Yoder, P.D., et al.: Junctionless transistors for high-performance electronics. IEEE Trans. Electron. Devices. **64**, 4592–4597 (2017)

Comprehensive Analysis of Physical Properties of Barium Based Inorganic Metal Halide [$BaNaX_3$ (X = Cl, Br, I)] Perovskites

Devender Kumar[1], Dheeraj Sharma[1], Amarjeet Singh[1], Naincy Pandit[2], Anusha Dubey[3], and Ajay Singh Verma[4](✉)

[1] Department of Physics, Himachal Pradesh University, Shimla 171005, Himachal Pradesh, India

[2] Department Physics, School of Allied Sciences, Dev Bhoomi Uttarakhand University, Dehradun 248007, India

[3] Department of Physics, Noida Institute of Engineering and Technology, Greater Noida 201306, India

[4] Division of Research & Innovation, School of Applied and Life Sciences, Uttaranchal University, Dehradun 248007, India

Abstract. Here in, present study is about the investigation of the structural, electronic and optical properties of $BaNaX_3$ (X = Cl, Br, I) perovskites by using DFT (Density Functional Theory) based WIEN2k code. Density of States (DOS) reveals the distribution of electronic states in a material and availability of energy level for electron occupation. $BaNaCl_3$ materials have the direct band gap of 4.04 eV and the other two compounds $BaNaBr_3$ and $BaNaI_3$ indirect band gap of 3.01 eV and 2.24. Also this study contains the detailed examination of the density of states for the $BaNaX_3$ (X = Cl, Br, I) compounds to unravel the origins of its band structure. Further in this context various optical constant, such as dielectric function, absorption coefficient and refractive index are studied. Finally the results are compared with available prior results, shows a good agreement.

Keywords: Density Functional Theory · WIEN2k · Density of States · Band Structure · Absorption coefficient

1 Introduction

Recently, Perovskite materials with diverse composites and structures have gained significant attention for their potential applications across various industrial and technological fields [1]. Since last two decades, perovskites are attaining much popularity in the photovoltaic and thermoelectric industries due to their attractive properties [2–5]. Perovskites are basically the material having chemical formula ABX_3, where A and B are two cations and X is anion [6–8]. In addition perovskites can be found in various other stable structure like ABX_3, $A2BX_6$, $A_2BB'X_6$ and A_3B2X_9 etc [9]. Many researchers are studying the perovskites structure and its properties to make them more efficient and suitable for various optoelectronic properties. Further, it is possible to change or modify

S. Kumari et al. (Eds.): ICMTP 2024, SPPHY 311, pp. 288–296, 2025.
https://doi.org/10.1007/978-981-96-6795-6_33

the physical or chemical properties of any perovskite compound by adding some doping of monovalent or trivalent atoms to make it more suitable for photovoltaic applications [10]. With this motivation, this study is carried out and contains the study of the structural, electronic and optical properties of Barium based non-metalic halide $BaNaX_3$ (X = Cl, Br, I) perovskite. Here in, the full potential linear augmented plane wave (FP—LAPW) with local density approximation (LDA) and wu-cohen (WC) exchange correlation are employed. First, the structural stability of the compound $BaNaX_3$ (X = Cl, Br, I) by using WC potential are studied. Energy bandgap and density of states are also computed for the same compound. In optical properties dielectric constant, optical conductivity, absorption coefficient and refractive index are also calculated. Its tunable optical properties allow its to use in various optoelectronic applications.

2 Methods and Computational Details

Here all the results are obtained computationally by using WIEN2k software where density functional theory (DFT) is implemented. It uses the full potential linear augmented plane wave (FP-LAPW) [11] method to solve the quantum mechanical calculations. Inside LAPW method, there is Kohn-Sham eq. [12] which have been solved for a unit cell. The value of the $R_{mt} \times K_{max} = 7$, here R_{mt} known as muffin tin radii of atom and K_{max} is the wave vector. Number of K-point are taken 1200K for study of both electronic and optical properties. First optimization (energy equilibrium) is done by using Wu-Cohen (WC) exchange correlation potential, which gives the equilibrium energy, lattice parameter and bulk modulus of these compound. All other calculations electronics and optical properties are calculated by using local density approximation (LDA) potential [13].

3 Result and Discussion

3.1 Structural Properties

The barium based non metallic halide perovskite $BaNaX_3$ (X = Cl, Br, I) have the simple cubic crystal structure. First, optimization curve are obtained for all three compound which is the plot between energy and unit cell volume. This curve show the variation of energy of material with volume of unit cell which is obtained by using Wu-Cohen exchange correlation potential, hence gives the equilibrium energy, lattice parameter and bulk modulus. Variation of energy with the volume of unit cell is shown in Fig. 1. The lattice constant increase as one move towards the higher atomic number halide (Table 1).

This happen due to increase in the ionic radii of higher atomic number halide. There is no experimental result to compare with these results, but theoretical prior findings of some parameter of compound $BaNaCl_3$ are available. If these, present value of lattice parameter are compared with previous existing value, only the difference of 0.17% we got, which confirms the accuracy of our research. On the basis of this closely findings, it confirms the accuracy of our results for other two compounds because, there is no previous existing data for other two compounds.

Bulk modulus and its pressure derivative is also calculated which are also very close to previously reported values. Here the material is not able to oppose the external mechanical pressure for long time because the low value of its bulk modulus.

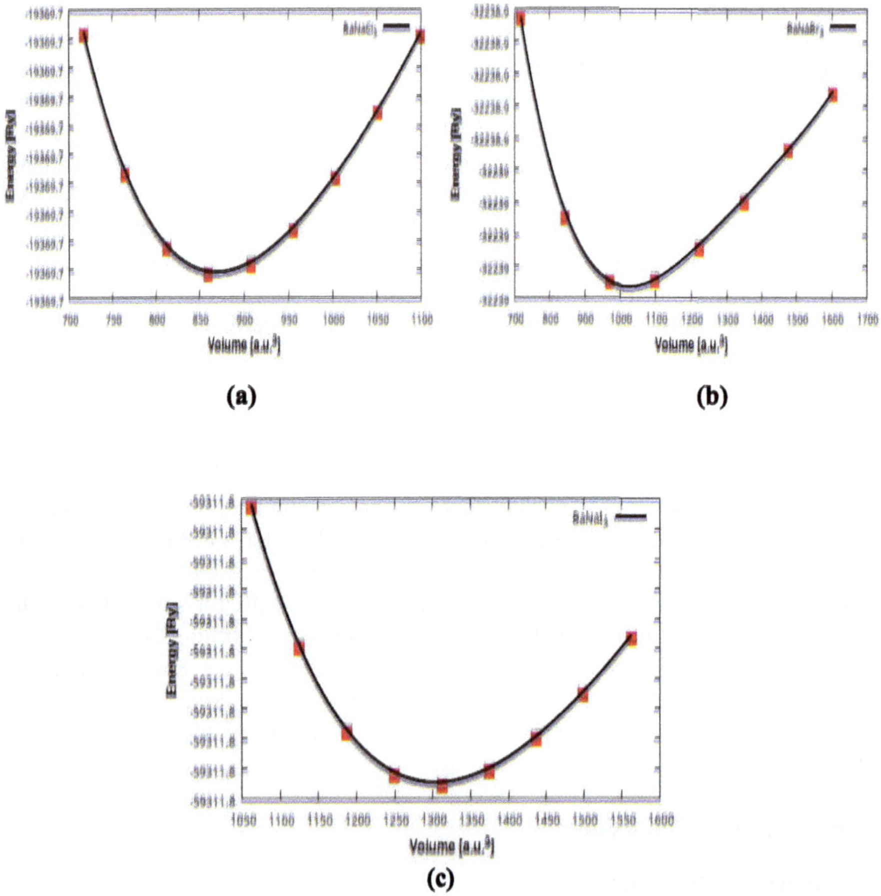

Fig. 1. Optimization curve for (a) $BaNaCl_3$ (b) $BaNaBr_3$ (c) $BaNaI_3$.

Table 1. Lattice constant, unit cell volume, bulk modulus, pressure derivative of bulk modulus and equilibrium energy for all material are listed below in Table 1.

Property	$BaNaCl_3$		$BaNaBr_3$		$BaNaI_3$	
	Present value	others	Present	others	Present	others
a (Å)	5.0504	5.0492[a]	5.3409		5.7828	
V(a.u)^3	869.3913		1028.6449		1304.4131	
B (GPa)	31.7743	31.9[a]	26.7108		20.4272	
B′	4.7799	4.77[a]	4.7544		4.9767	
E0(Ry)	−19369.7304		−32239.0317		−59311.8124	

[a]Ref. [14]

3.2 Electronic Properties

This section contains the electronic properties of perovskite material $BaNaX_3$ (X = Cl, Br, I) including band gap and density of states. Here LDA potential is used to study the band gap and DOS properties of all the compound. Table 2 reveals that band gap of the compound decrease towards the lower value as we move towards the higher atomic number halide. It can be seen that band gap values obtained here agree with the prior theoretical findings which are results of GGA potential.

Table 2. Calculated band gap Eg (in eV) and its comparison with prior findings as per available data.

	Present value (LDA potential)	GGA-PBEsol (others)
$BaNaCl_3$	4.03(Direct)	4.14[a]
$BaNaBr_3$	3.01(Indirect)	
$BaNaI_3$	2.24(Indirect)	

[a]Ref. [14]

Comparison of $BaNaCl_3$ with previous theoretically reported data, got the difference of only 2.65%, which assure our findings to be much correct. Band structure of all three compound is shown in Fig. 2. We found that the $BaNaCl_3$ material have the direct band gap of type (Γ→Γ) which is 4.03 eV while other two compound $BaNaBr_3$ and $BaNaI_3$ have the indirect band gap of type (R→Γ) with value having 3.01 eV and 2.24 eV. One thing that can notice here is that band gap value obtained by using LDA approximation is not much deviate from the value reported by using GGA exchange correlation potential. It is not able to compare the values of other two compound because of not availability of prior findings.

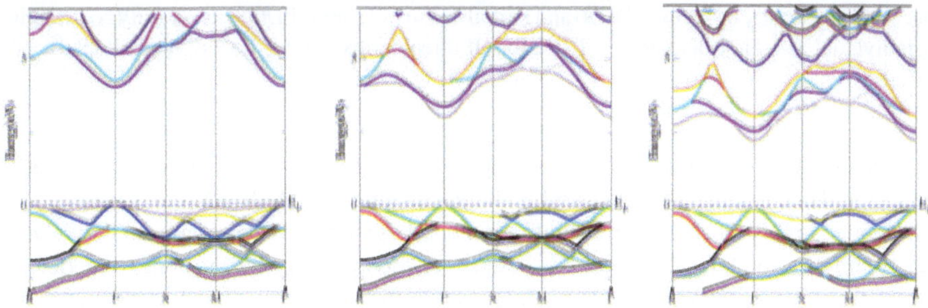

(a) (b) (c)

Fig. 2. The energy band curves for all three compound (a) BaNaCl3 (b) BaNaBr3 (c) BaNaI3 reported by using LDA approximation.

Total density of the states occupied by these atoms or contribution by individual atom is also calculated by using same exchange correlation approximation i.e. LDA is shown in Fig. 3.

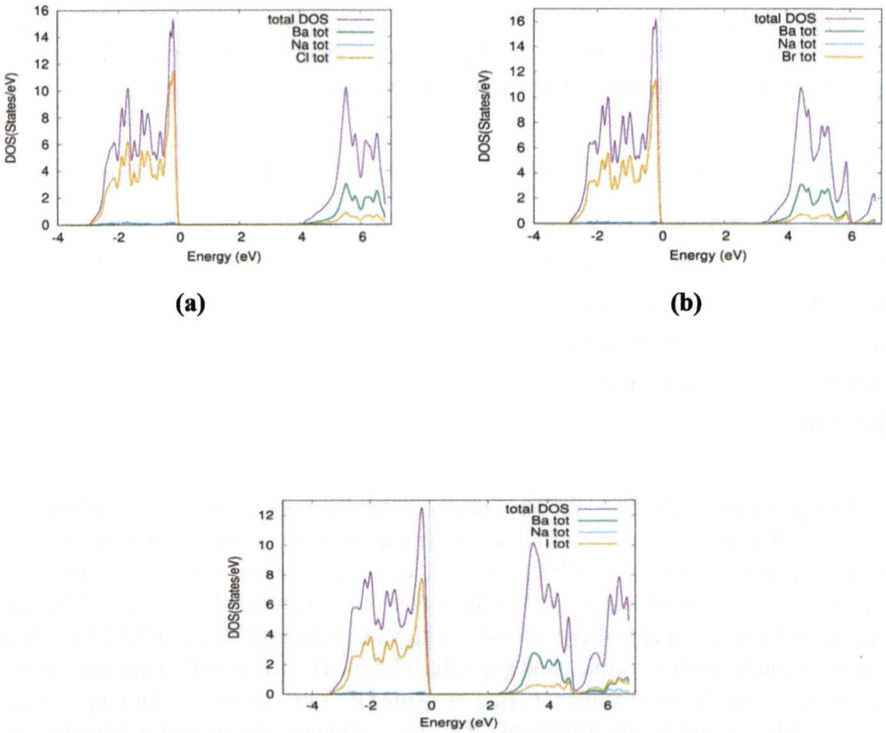

(a) (b)

(c)

Fig. 3. Total density of states (TDOS) and Partial density of states (PDOS) i.e. density contribution by individual atom is calculated by LDA for all three compound (a) $BaNaCl_3$ (b) $BaNaBr_3$ (c) $BaNaI_3$.

Density of occupied states is maximum at the fermi level and almost zero near in the region between the valence band and conduction band. Number of occupied states in valence band is more than the conduction band as shown by the graph. In the valence band of all three material $BaNaX_3$ (X = Cl, Br, I), most of contribution to the total density comes from halide atoms (X = Cl, Br, I) but in the conduction band of all materials, Barium (Ba) atoms contribute the most.

3.3 Optical Properties

Optical properties of $BaNaX_3$ (X = Cl, Br, I) perovskites were successfully calculated by using LDA exchange correlation function. Figure 3(a) show the real part of the dielectric function for all three compounds $BaNaX_3$ (X = Cl, Br, I) respectively. As the replacement of the Cl with Br and I, the peak value of the real part of dielectric constant move towards the lower value of energy.

In this study, the peak values of real part of dielectric constant for all three compound $BaNaX_3$ (X = Cl, Br, I) are 6.3, 6.9 and 8.4 respectively, which occurs at particular energy values which are 5.3 eV, 4.4 eV and 3.5 eV. Also the static dielectric function, $\varepsilon_1(0)$ values comes out from this study are 3.1, 3.75 and 4.7. It is the important point that should note here, static value of dielectric function $\varepsilon_1(0)$ increases with decreases in band gap of these perovskite compound as we move from Cl to I, which is the statement of Penn's Model [15]. Moreover, for better optoelectronic performance large value of static dielectric should be there.

Imaginary part of dielectric constant for all perovskites compound are shown in Fig. 3(b). Imaginary part show the absorption capability of the material [16]. As we replace the Cl with Br and I, the threshold value of energy shifted toward the lower side of x-axis. Threshold value is the value of energy from where material starts to absorbs the light incident on it. Prominent absorption peaks for all the perovskites materials $BaNaX_3$ (X = Cl, Br, I) lies at energy values 5.9 eV, 4.7 eV and 3.8 eV. Hence we can say that material $BaNaI_3$ starts to absorb light earlier as compare to other two compound due its low band gap value. Figure 3(c) predicts the absorption behavior of all the three compounds $BaNaX_3$ (X = Cl, Br, I). Here again $BaNaI_3$ show the wide range of absorption as compared to other two compounds.

Optical conductivity of all three compound are also calculated as shown in Fig. 4(a), which shows that perovskites with Iodine halide starts to conduct earlier than other two material. After the prominent conductive peak there are ups and down peaks for all three compound $BaNaX_3$ (X = Cl, Br, I) upto energy 10–12 eV.

Loss function is also calculated for all materials is shown in Fig. 4(b). Energy loss is closely related to the absorption and reflection of incident radiation [17]. In this study we can say that the material with good absorption coefficient have much tendency of energy loss in the form of scattering or reflection when light incident on it. Refractive index of all material as a function of incident energy is calculated and shown in Fig. 4(c). We can see here that refractive index curve are similar to that of real part of dielectric function. Here calculated static refractive index i.e. $\eta(0)$ are 1.75, 1.88 and 2.2 for all three material $BaNaX_3$ (X = Cl, Br, I) perovskites. Where peak values of refractive index for all three materials are 2.6, 2.7 and 3 which lies with the energy scale at 5.75 eV, 4.4 eV and 3.8 eV respectively. After exhibit these peak value the refractive index starts to decreases gradually for all material.

4 Conclusion

This study investigated the structural, electrical and optical properties of cubic perovskites materials $BaNaX_3$ (X = Cl, Br, I) by using DFT based WIEN2k package along with FP-LAPW (full potential—linear augmented plane wave) method implemented in it. In this study we use the WC and LDA exchange correlation function to study various characteristics of perovskites material. The main conclusion of this study are:

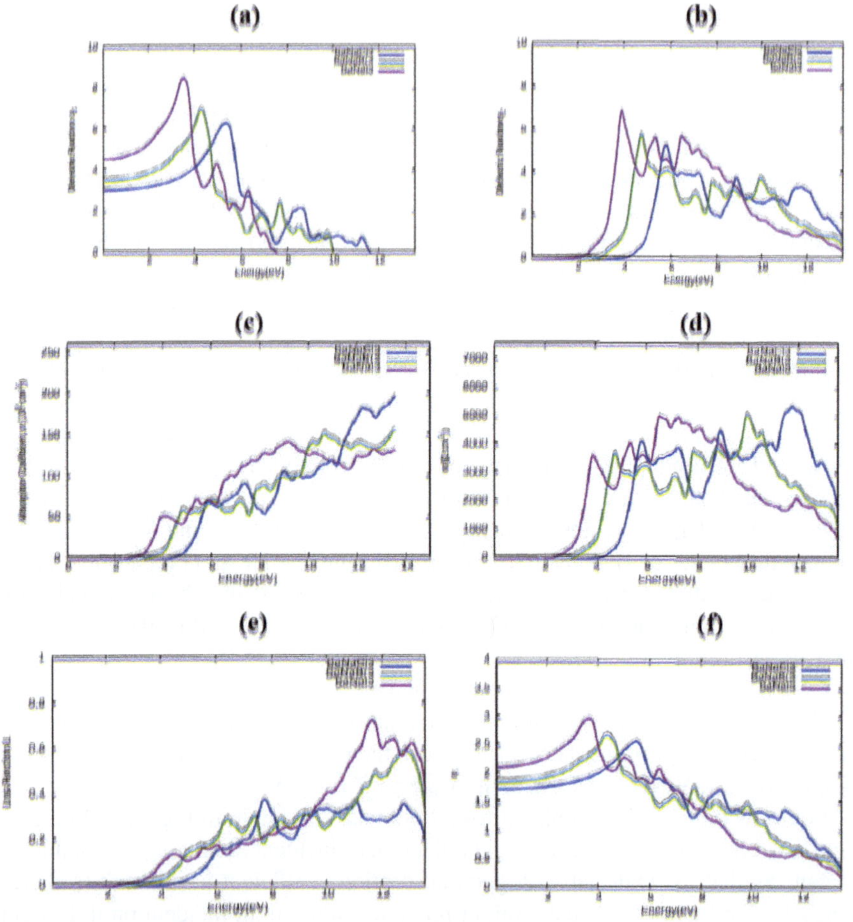

Fig. 4. Real and imaginary part of Dielectric constant spectra shown in (a) and (b). Absorption coefficient, α for all material is shown in (c) Optical conductivity, loss function and refractive index spectra of all the materials in shown in (d), (e), and (f).

- The compounds $BaNaX_3$ (X = Cl, Br, I) exhibit the stable cubic structure. In present study the calculated lattice parameter highly coincide with the prior findings.
- The band gap calculated in this study shows that two compound out of three are of insulator category. We got band gap values 4.04 eV, 3.01 eV and 2.24 eV for $BaNaX_3$ (X = Cl, Br, I) material respectively. Here $BaNaCl_3$ and $BaNaBr_3$ shows insulating properties while $BaNaI_3$ move towards semiconductor category.
- In study of optical properties, as we move from Cl to I, the static dielectric function $\varepsilon_1(0)$ increases but band gap values decreases in accordance with Penn Law. Also absorption spectra of all the compounds $BaNaX_3$ (X = Cl, Br, I) shows the absorption in photovoltaic range.

Hence the optical properties of these materials predict that they have the strong potential in photo-electronics area.

References

1. Bakar, A., Afaq, A., Ahmed, M., Ahmad, Z., Ramay, S.M.: Revealing the remarkably improved structural, phonon, elastic, optoelectronic and thermoelectric properties of $TlXF_3$(X Ca, Cd) for energy applications using first-principles approach with hybrid and range-separated hybrid functionals. Solid State Commun. **394**, 115699 (2024). https://doi.org/10.1016/j.ssc.2024.115699
2. Chaturvedi, N., Swami, S.K., Dutta, V.: Spray deposition of poly(3-hexylthiophene) and [6,6]-phenyl-C_{61}-butyric acid methyl ester blend under electric field for improved interface and organic solar cell characteristics. Thin Solid Films. **598**, 82–87 (2016). https://doi.org/10.1016/j.tsf.2015.12.017
3. Chaturvedi, N., Dutta, V.: Effect of electric field on the spray deposited PEDOT. PSS Film. Energy Procedia. **33**, 228–232 (2013). https://doi.org/10.1016/j.egypro.2013.05.062
4. Nagarjuna, P., Chaturvedi, N., Bagui, A., Runjhun, R., Garg, A., Singh, S.P.: Solution-processed organic solar cells using new electron acceptor derived from naphthalene and fluorene unit. Chem. Select. **2**, 7913–7917 (2017). https://doi.org/10.1002/slct.201700910
5. Balawi, A.H., Kan, Z., Gorenflot, J., Guarracino, P., Chaturvedi, N., Privitera, A., Liu, S., Gao, Y., Franco, L., Beaujuge, P., Laquai, F.: Quantification of photophysical processes in all- polymer bulk heterojunction solar cells. RRL Solar. **4**, 2000181 (2020). https://doi.org/10.1002/solr.202000181
6. Sadia Islam Ria, M., Ghosh, A., Azizur Rahman, M., Al-Humaidi, J.Y., Awwad, N.S., Ibrahium, H.A., Rahman, M.M.: Examining anion influence on the physical properties and performance analysis of lead-free calcium-based Ca_3NX_3 (X=F, Cl, Br and I) perovskite. Mater. Sci. Eng. B. **310**, 117674 (2024). https://doi.org/10.1016/j.mseb.2024.117674
7. Abdullah, D., Gupta, D.C.: Investigation of structural, electronic, optical, and elastic properties of Ru-based inorganic halide perovskites $ARuBr_3$ (A = K, Rb, Cs) via DFT computations. Arab. J. Sci. Eng. (2024). https://doi.org/10.1007/s13369-024-09524-2
8. Pingak, R., Bouhmaidi, S., Setti, L., Pasangka, B., Bernandus, B., Sutaji, H., Nitti, F., Ndii, M.: Structural, electronic, elastic, and optical properties of cubic $BaLiX_3$ (X = F, Cl, Br, or I) perovskites: an *Ab-initio* DFT study. Indones. J. Chem. **23**(3), 843–862 (2023). https://doi.org/10.22146/ijc.83261
9. Nazir, A., Ahmad, K.E., Althubeiti, K., Al Otaibi, S., Soliyeva, M., Sharma, R., Manzoor, M.: Impact of halogens modifications on physical characteristics of lead-free hybrid double perovskites compounds Cs_2YCuX_6 (X=Cl, Br, and I) for energy storage applications: first principles investigations. Chem. Phys. **588**, 112449 (2025). https://doi.org/10.1016/j.chemphys.2024.112449
10. Zhang, H., Ou, T., Jiang, W., Wang, X., Zhuang, Q., Feng, S., Sun, Y., Li, P., Ma, X.: The impact of halogens on the structural, electronic, and optical properties of vacancy-ordered double perovskites Rb_2SeX_6 (X=I, Br, Cl). J. Solid State Chem. **340**, 125002 (2024). https://doi.org/10.1016/j.jssc.2024.125002
11. Schwarz, K.: DFT calculations of solids with LAPW and WIEN2k. J. Solid State Chem. **176**, 319–328 (2003). https://doi.org/10.1016/S0022-4596(03)00213-5
12. Schwarz, K., Blaha, P.: Solid state calculations using WIEN2k. Comput. Mater. Sci. **28**, 259–273 (2003). https://doi.org/10.1016/S0927-0256(03)00112-5
13. Boran, Y., Kara, H.: A comprehensive density functional theory analysis on structural, electronic, and optical properties of BiOF. Braz. J. Phys. **54**, 159 (2024). https://doi.org/10.1007/s13538-024-01523-w
14. Chaba, M.S., Radjai, M., Bouhemadou, A., Rahman, M.A., AU-Kara, H., Houatis, D., Allali, D., Sâad Essaoud, S., Allaf, H.: Structural, electronic, and optical characteristics of $BaXCl_3$ (X = Li, Na) perovskites. Mater. Sci. Eng. B. **308**, 117578 (2024). https://doi.org/10.1016/j.mseb.2024.117578

15. Cirilo-Lombardo, D.J.: Semiconductor dielectric function, excitons and the Penn model. Philos. Mag. **95**(9) (2024). https://doi.org/10.1080/14786435.2015.1015468
16. Talukder, M.R., Alam, M.S., Hasan, M.M., Sarker, M.A., Islam, M.R., Ahmad, S.: Indirect to direct band gap transition of ABI_3 (A = Rb, Cs; B = Ca, Sr) perovskites under hydrostatic pressure for photovoltaic and optoelectronic applications: a DFT study. Phys. B Condens. Matter. **695**, 416538 (2024). https://doi.org/10.1016/j.physb.2024.416538
17. Manzoor, M., Dixit, A., Waqas, I.M., Sadique, I., Alotaibi, N.H., Mohammad, S., Sharma, R., Ur Rehman, I., Ismayilova, N.A.: Explored the potential of Rb_2SeX_6 (X=Cl, Br) double perovskites for energy harvesting devices: a DFT study. Mater. Sci. Eng. B. **310**, 117710 (2024). https://doi.org/10.1016/j.mseb.2024.117710

Amendments in Sensing Response of Multi-Walled Carbon Nanotube (MWCNT) with Integration of Metal Oxide

Komal Kaliraman[1], R. Ridhi[2(✉)], and S. K. Tripathi[1(✉)]

[1] Department of Physics, Panjab University, Chandigarh, India
surya@pu.ac.in

[2] Mehr Chand Mahajan DAV College for Women, Sector-36A, Chandigarh, India
r.ridhi@mcmdavcwchd.in

Abstract. The present work demonstrates the role of organic and inorganic composites in amending the sensing efficiency towards methanol, ethanol and propanol. The hazardous effects of alcohol inhalation, such as blindness, organ failure and other harmful effects, make it indispensable to fabricate suitable sensing devices with higher sensing responses. The higher operating temperatures (>200 °C) of metal oxides act as bottlenecks for their utilisation as gas sensors under ambient conditions. Therefore, adding suitable fillers like Multi-Walled Carbon Nanotubes to metal oxides enhances their sensitivity and selectivity under ambient conditions. For this purpose, metal oxide Titanium dioxide is integrated into a Multi-Walled Carbon Nanotube to form thin film chemical sensors for methanol, ethanol and propanol vapour detection. The Titanium dioxide (TiO_2)-Multi-walled carbon nanotube (MWCNT) film has been deposited using a drop-casting method, and the solution of the nanocomposite film has been synthesised using the chemical mixing and dispersion route. The sensing properties of Titanium Dioxide-Multi-Walled Carbon Nanotube (TiO_2-MWCNT) films were evaluated through electrical conductivity measurements and compared with optical (UV-visible spectroscopy) and spectroscopic analyses (Fourier Transform Infrared (FTIR) spectroscopy) under ambient conditions. Among the tested alcohol vapours, the TiO_2-MWCNT film exhibited the highest sensitivity to methanol (12.29 ppm^{-1}) and the lowest sensitivity to propanol (4.9 ppm^{-1}). The electrical sensing results were in strong agreement with the optical and spectroscopic findings, as methanol exposure led to the most significant deviations in band shifts and peak positions of the sensing film.

Keywords: Gas sensor · TiO_2 · MWCNT

1 Introduction

Gas sensors play a vital role in maintaining efficiency, safety, and health in a variety of industries [1]. By identifying hazardous gases, preventing mishaps, and maintaining worker safety, these sensors are essential to industrial settings [2–4]. With the widespread

S. Kumari et al. (Eds.): ICMTP 2024, SPPHY 311, pp. 297–307, 2025.
https://doi.org/10.1007/978-981-96-6795-6_34

use and related hazards in a variety of industries, the detection of methanol, ethanol, and propanol is essential. Methanol must be closely monitored to ensure safety in commercial and industrial settings because it is extremely poisonous, even in low quantities, and can cause blindness or death [5]. Propanol and ethanol are frequently employed as solvents, disinfectants, and medications where precise concentration control is necessary for both safety and efficacy [6]. When these alcohols are released into the air, water, or soil as a result of industrial operations or accidents, pollution may result. On account of their unique special structural and functional characteristics of multi-walled carbon nanotube (MWCNT)-titanium dioxide (TiO_2) nanocomposites have become very successful chemiresistive sensors for the detection of alcohol vapours like methanol, ethanol, and propanol [7]. The highly conductive network of multi-walled carbon nanotubes (MWCNT) enables effective electron mobility and increases the composite's sensitivity to resistance changes when exposed to target analytes [8]. Because of high surface activity and chemical reactivity, titanium dioxide (TiO_2) reacts successively with alcohol molecules, causing notable changes in electrical resistance [9]. Better adsorption of alcohol molecules is made possible by the nanocomposite's large surface area, and a quick and reversible reaction is enabled by its hybrid structure. With applications in industrial safety, environmental monitoring, and breath analysis in medical settings, these characteristics make MWCNT-TiO_2 nanocomposites ideal for monitoring methanol, ethanol, and propanol [10, 11].

2 Experimental

2.1 Reagents and Materials

Glass substrates were thoroughly cleaned using de-ionised water and acetone, both supplied by Central Drug House (CDH). TiO_2 powder was obtained from Sigma Aldrich and was used as received. For the sensing experiments, specific amounts of ethanol, propanol, and methanol having a purity of 99.0% were sourced from Fisher Scientific. All chemicals were employed in their original form.

2.2 Fabrication of MWCNT-TiO_2 Thin Film

A chemical mixing and dispersion process was used to create the MWCNT-TiO_2 nanocomposite. 2 mg of MWCNTs were first ultrasonically dispersed in 2 mL of Deionised water to create a uniform dispersion. After adding 20 mg of TiO_2 nanopowder to a homogenous solution under constant stirring, the mixture was ultrasonically agitated for 6 hours at room temperature (RT). To get the nanocomposite powder, the mixture was heated to 80 °C under constant stirring to completely evaporate the water. It was then dried in a muffle furnace for an hour at 110 °C [12].

Fabrication of Device and Chemical Sensing Set-up. The device construction depicted in Fig. 1 has been employed to investigate the sensing mechanism of MWCNT-TiO_2 film. We applied a silver metal contact to the thin film of MWCNT-TiO_2. The Keithley 6517A electrometer and the pico-ammeter were used in a two-probe approach to evaluate electrical current variations under the applied voltage. When exposed to alcohol vapours, specifically methanol, ethanol, and propanol, the changes in electrical current are examined at room temperature with a fixed applied voltage of 50 V. During conductivity measurements, the gap between the electrodes placed on the film was maintained at approximately 1 mm [2, 3, 16].

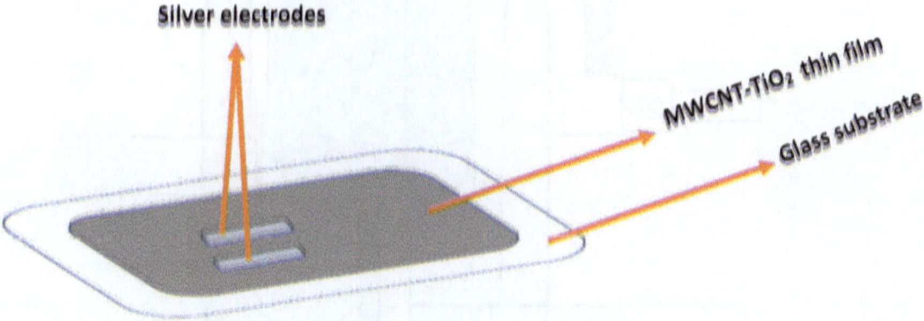

Fig. 1. Device design and fabrication.

The same sensing setup described in our previous work [5] (Fig. 2) was used, featuring a 0.5-L stainless steel test chamber to evaluate the vapour sensing capabilities of MWCNT-TiO_2 thin films. The sensing characteristics were examined under standard atmospheric pressure (760 mmHg) and at room temperature (25 °C). To eliminate any water and organic molecules that have already been adsorbed, the MWCNT-TiO_2 thin film was pre-conditioned under a vacuum (10^{-3} mbar) at room temperature for about an hour before conducting the sensing tests. The baseline current of the MWCNT-TiO_2 thin film under dry air, ambient temperature, and vacuum conditions was recorded as 5.0×10^{-12} A. Before introducing alcohol vapours, the initial current of the thin film was measured under a vacuum pressure of approximately 10^{-3} mbar. To ensure a vapour concentration of 2 parts per million (ppm), calibrated volumes of alcohol were added to the test chamber for sensing investigations [5].

For an effective comparative analysis of the sensing mechanism, this concentration has been kept constant for all the vapours. Alcohol vapours were introduced into the sensing chamber and individually exposed to MWCNT-TiO_2 thin films to investigate their interaction mechanism and overall sensing behaviour. Over time, the changes in the sensing current were observed while keeping the applied voltage at 50 V.

Instrumentation. The Bruker Alpha II Spectrometer was used to record Fourier Transform Infrared (FTIR) spectra to examine alterations in the molecular structure of MWCNT-TiO_2 following exposure to alcohol vapours. UV-visible spectra in the MWCNT-TiO_2 thin films brought on by vapour exposure, UV-visible spectra in the 200–900 nm range were recorded using a PerkinElmer UV WinLab Spectroscan 30 spectrometer to investigate changes in the electronic transitions. The UV-visible study was performed using a pristine glass slide as the reference.

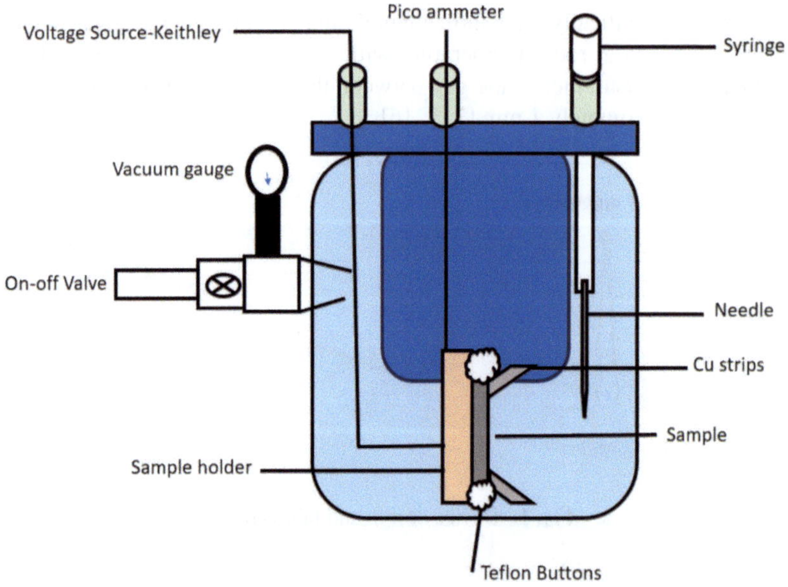

Fig. 2. Sensing set-up.

3 Analysis and Interpretation of Results

3.1 Response and Recovery Behaviour of MWCNT-TiO_2 Thin Film

Figure 3 displays the MWCNT-TiO_2 thin film's response-recovery characteristic after exposure to three alcohol vapours at 2 ppm: methanol, ethanol, and propanol. We observe that the MWCNT-TiO_2 thin film shows a change in current when exposed to vapours and returns to its baseline state as the vapours are sucked out. MWCNT-TiO_2 has shown the maximum current with methanol exposure, followed by ethanol and propanol vapours at similar vapour concentrations.

3.2 Comparative Study of MWCNT-TiO_2 Thin Film Sensitivity to Alcohol Vapours

The following formula [5, 18] was used to determine the MWCNT-TiO_2 thin film's sensitivity values to the propanol, ethanol, and methanol vapours (Fig. 4)

$$S = \frac{1}{c}\left(\frac{\Delta I}{I}\right) \tag{1}$$

where I is the thin film's initial baseline current before exposure, and ΔI is the change in current of the MWCNT-TiO_2 thin film when exposed to organic vapours. The organic vapour concentration is represented by the parameter c, and the sensitivity, S, is given in ppm^{-1}. At a concentration of 2 ppm, methanol has the maximum sensitivity (12.29 ppm^{-1}), followed by ethanol (6.4 ppm^{-1}) and propanol (4.9 ppm^{-1}).

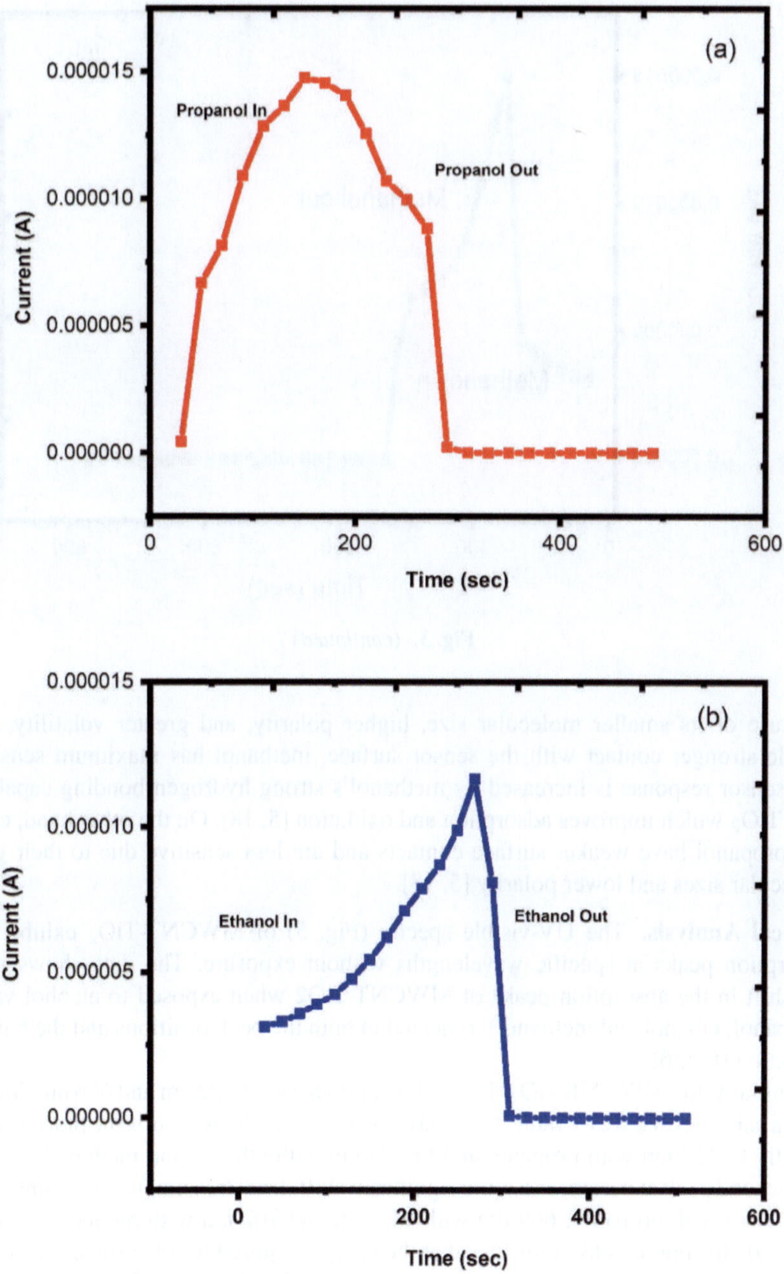

Fig. 3. Response-Recovery of MWCNT-TiO_2 with (a) propanol, (b) ethanol, (c) methanol.

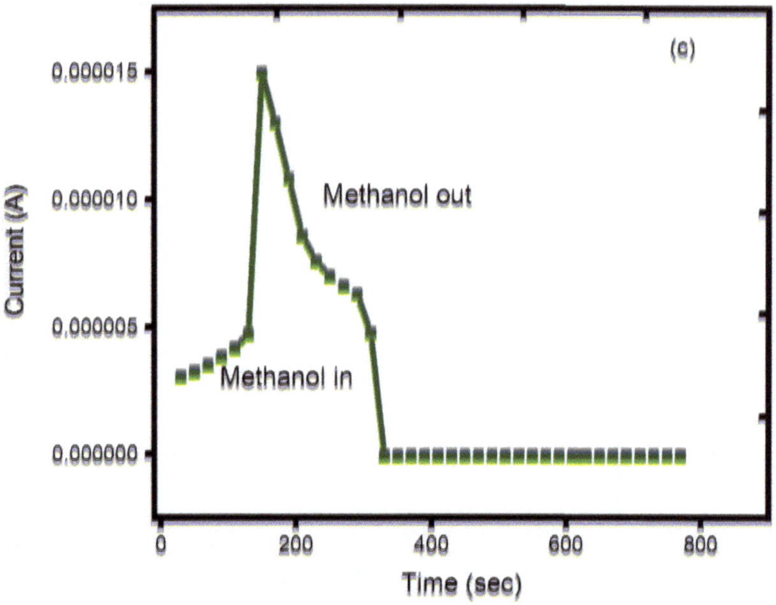

Fig. 3. *(continued)*

Because of its smaller molecular size, higher polarity, and greater volatility, which enable stronger contact with the sensor surface, methanol has maximum sensitivity. The sensor response is increased by methanol's strong hydrogen bonding capabilities with TiO_2 which improves adsorption and oxidation [5, 18]. On the other hand, ethanol and propanol have weaker surface contacts and are less sensitive due to their greater molecular sizes and lower polarity [5, 18].

Optical Analysis. The UV-visible spectra (Fig. 5) of MWCNT-TiO_2 exhibit stable absorption peaks at specific wavelengths without exposure. The data shows a clear red shift in the absorption peaks of MWCNT-TiO2 when exposed to alcohol vapours (propanol, ethanol and methanol), reflected in both the peak positions and the band gap reduction (Fig. 6).

Initially, the MWCNT-TiO_2 shows absorption peaks at 422 nm and 606 nm (Table 1), with a band gap of 2.2 eV (Table 2). Upon exposure to alcohol vapours, the first peak shifts slightly to 423 nm with propanol and to 424 nm with ethanol and methanol (Table 1). The second peak experiences a more significant shift: from 606 nm in the pristine sample to 626 nm with propanol, 669 nm with ethanol, and 702 nm with methanol (Table 1). This red-shifting of peaks indicates that the energy required for electronic transitions is decreasing, which corresponds to a narrowing of the band gap [19, 20]. The intensity of the absorption peaks decreases as the organic vapour is exposed, suggesting enhanced charge transfer or increased electronic interactions between the alcohol vapours and the MWCNT-TiO_2 composite. The shifts in absorbance, particularly in the second peak, suggest that methanol induces the greatest effect on the electronic structure of the composite due to stronger electron-donating properties [19, 20]. This leads to a further lowering of

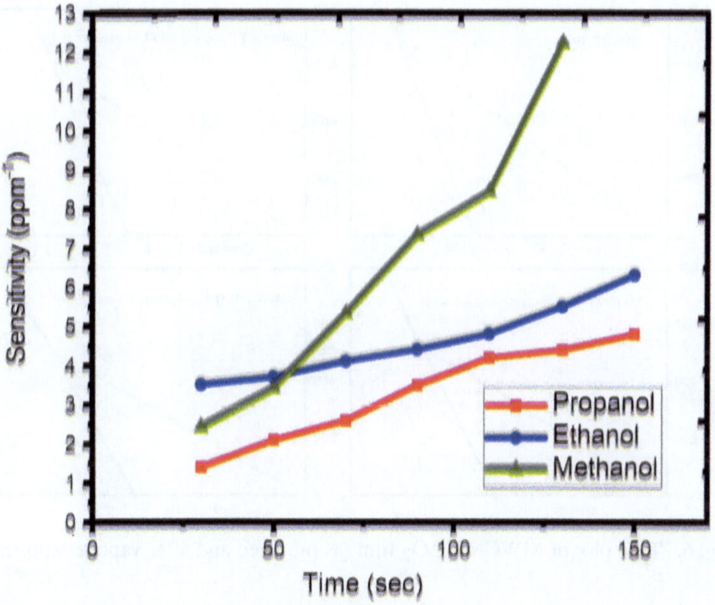

Fig. 4. Sensitivity curve of MWCNT-TiO_2 with propanol, ethanol and methanol.

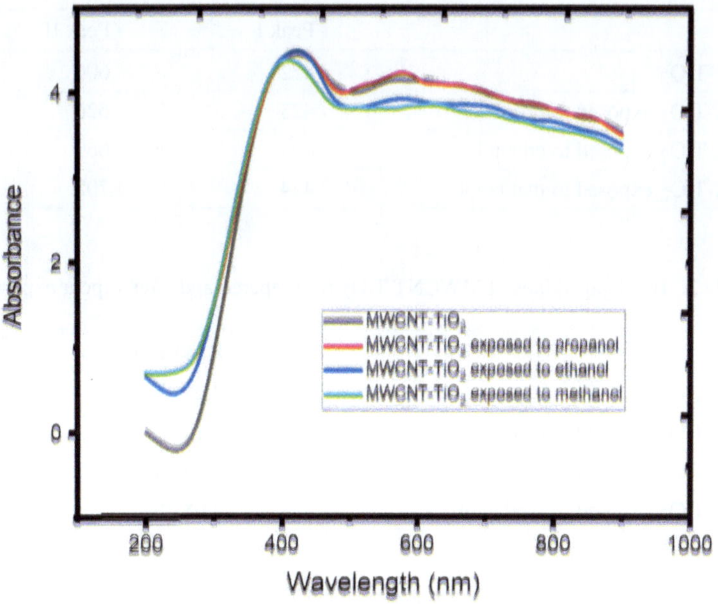

Fig. 5. UV-vis spectra of MWCNT-TiO_3 film (as prepared and after vapour exposure).

the band gap from 2.2 eV in MWCNT-TiO_2 to 2.08 eV (Table 2) in the methanol-exposed sample.

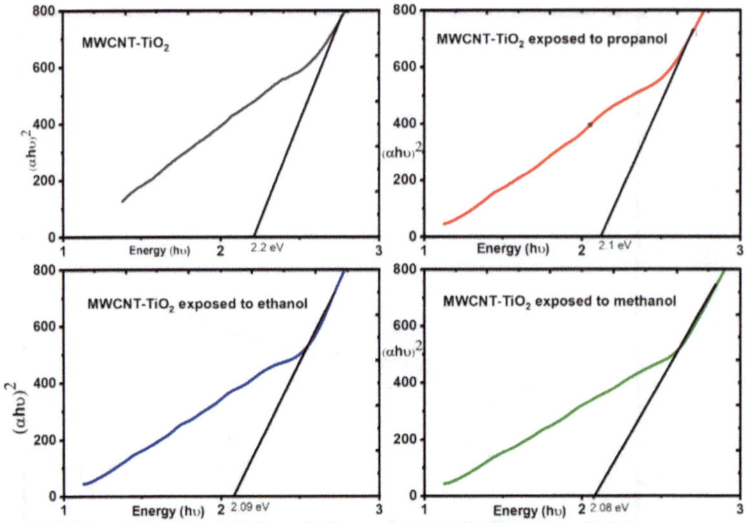

Fig. 6. Tauc plot of MWCNT-TiO_2 film (as prepared and after vapour exposure).

Table 1. Peak shifts in UV-vis spectroscopy of MWCNT-TiO_2 (as prepared and after vapour exposure).

Sample	Peak I	Peak II
MWCNT-TiO_2	422	606
MWCNT-TiO_2 exposed to propanol	423	626
MWCNT-TiO_2 exposed to ethanol	424	669
MWCNT-TiO_2 exposed to methanol	424	702

Table 2. Band gap values of MWCNT-TiO_2 (as prepared and after vapour exposure).

Sample	Band gap
MWCNT-TiO_2	2.2 eV
MWCNT-TiO_2 exposed to propanol	2.1 eV
MWCNT-TiO_2 exposed to ethanol	2.09 eV
MWCNT-TiO_2 exposed to methanol	2.08 eV

Such behaviour implies that methanol is more effective at altering the electronic states of the MWCNT-TiO_2 component, enhancing its ability to absorb light at longer wavelengths and potentially improving its sensing capabilities [19, 20].

Spectroscopic Analysis. The Fourier Transform Infrared spectroscopy data for MWCNT-TiO_2 (Fig. 7) shows distinct changes in the absorption peaks after exposure to different alcohol vapours, propanol, ethanol and methanol, indicating that the alcohols interact with the film and change its surface properties. In the pristine MWCNT-TiO_2 sample, the peak around 3746 cm^{-1} (Table 3) is associated with O-H stretching vibrations, which is typically due to hydroxyl groups on the surface of TiO_2. After exposure to alcohol vapours, the peak shifts to higher wavenumbers, such as 3808 cm^{-1} with propanol, 3837 cm^{-1} with ethanol, and 3740 cm^{-1} with methanol (Table 3). These shifts indicate stronger hydrogen bonding interactions between the hydroxyl groups, with ethanol and methanol showing a more significant shift due to stronger interaction with the film [13–17]. The peak around 1758 cm^{-1} in the unexposed sample is attributed to C=O stretching, possibly from surface carbonyl groups or adsorbed organic species. After exposure to the alcohol vapours, this peak shifts slightly to lower wavenumbers (1746 cm^{-1} with propanol, 1756 cm^{-1} with ethanol and methanol) (Table 3), indicating that the alcohol vapours modify the surface structure or induce some electronic changes around the carbonyl groups [13–17].

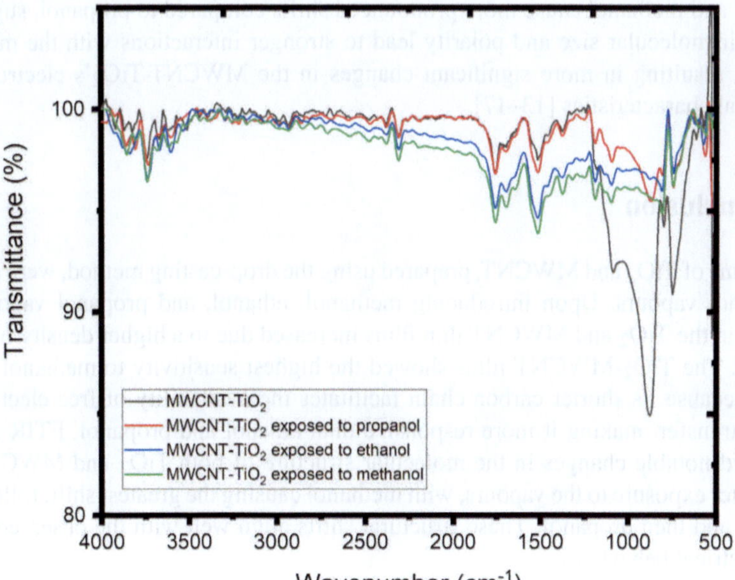

Fig. 7. Spectroscopic spectra of MWCNT-TiO_2 film (as prepared and after vapour exposure).

The peak at 874 cm^{-1} corresponds to C-H bending vibrations from the MWCNT in the composite. After exposure to alcohol vapours, this peak shifts slightly (869 cm^{-1} with propanol, 807 cm^{-1} with ethanol) (Table 3), which is due to the interaction between the alcohol vapours and the MWCNT, potentially affecting its electronic properties [13–17]. The Ti-O stretching peaks, located around 748 cm^{-1} and 523 cm^{-1} in the unexposed MWCNT-TiO_2 (Table 3), correspond to the TiO_2 component. After alcohol vapour exposure, these peaks show minimal shifts but do change in intensity. The peak at 523 cm^{-1} shifts to 561 cm^{-1} with propanol, 572 cm^{-1} with ethanol, and 556 cm^{-1} with methanol (Table 3), indicating a change in the Ti-O bonding environment due to

Table 3. Observed peaks in spectroscopic spectra of MWCNT-TiO_2 film (as prepared and after vapour exposure).

Sample	Peaks observed
MWCNT-TiO_2	3746,1758,1085,874,748,523
MWCNT-TiO_2 exposed to propanol	3808,3743,1756,1515,869,742,561
MWCNT-TiO_2 exposed to ethanol	3837,3736,3605,1756,1516,807,714,572
MWCNT-TiO_2 exposed to methanol	3740,1756,1515,805,742,556

the solvent interactions. These shifts are due to the formation of new surface complexes [13–17].

Overall, the changes in FTIR peaks reflect the influence of the alcohol vapours on the surface properties and chemical interactions within the MWCNT-TiO_2 composite. Ethanol and methanol cause more pronounced shifts compared to propanol, suggesting that their molecular size and polarity lead to stronger interactions with the material's surface, resulting in more significant changes in the MWCNT-TiO_2's electronic and structural characteristics [13–17].

4 Conclusion

Thin films of TiO_2 and MWCNT, prepared using the drop-casting method, were exposed to alcohol vapours. Upon introducing methanol, ethanol, and propanol vapours, the current in the TiO_2 and MWCNT thin films increased due to a higher density of charge carriers. The TiO_2-MWCNT films showed the highest sensitivity to methanol vapour, likely because its shorter carbon chain facilitates the availability of free electrons for charge transfer, making it more responsive than ethanol and propanol. FTIR analysis indicated notable changes in the molecular structure of both TiO_2 and MWCNT thin films after exposure to the vapours, with methanol causing the greatest shift, followed by ethanol and then propanol. These structural shifts align well with the observed optical and electrical behaviour.

Acknowledgement. Ms. Komal would like to thank the UGC (University Grants Commission) for financing this project and the Instrumentation Lab at D.A.V College, Sec. 10, for technical help.

References

1. Ridhi, R., Saini, G.S.S., Tripathi, S.K.: Nanotechnology as a sustainable solution for the proliferating agriculture sector. Mater. Sci. Eng. B. **304**, 117383 (2024)
2. Kumar, A., Kumar, M., Singh, A., Kumar, S., Kumar, D.: Nickel silicide formation by electroless technique for ULSI technology. Microelectron. Eng. **87**(3), 286–289 (2010)

3. Kumar, A., Suhag, A.K., Singh, A., Sharma, S.K., Kumar, M., Kumar, D.: Deposition and characterisation of amorphous electroless Ni-Co-P alloy thin film for ULSI application. Mater. Res. Express. **1**(3), 035001 (2014)
4. Ridhi, R., Saini, G.S.S., Tripathi, S.K.: Study of sensing mechanism of heterocyclic hazardous vapours with metal phthalocyanines. Dyes Pigments. **216**(3), 111328 (2023)
5. Ridhi, R., Saini, G.S.S., Tripathi, S.K.: Sensing of volatile organic compounds by copper-phthalocyanine thin films. Mater. Res. Express. **4**(2), 025102 (2017)
6. Ridhi, R., Saini, G.S.S., Tripathi, S.K.: Sensing of organic vapours by sulfonated copper-phthalocyanine salt thin films. Mater. Focus. **6**(4), 386–393 (2017)
7. Mhalsekar, M., Kole, P., Borker, V.: Synthesis, characterization and application of MWCNT/TiO_2 composite for photocatalytic reduction of benzophenone to benzopinacol. J. Environ. Chem. Eng. **12**(6), 114681 (2024)
8. Wong, W.P., Walvekar, R., Vaka, M., Khalid, M.: Hybrid MWCNT/TiO_2 nanoparticles based high temperature quinary nitrate salt mixture for thermal energy storage applications. J. Energy Storage. **73**, 108792 (2023)
9. Gawri, I., Ridhi, R., Tripathi, S.K.: Chemically synthesized TiO_2 and PANI/TiO_2 thin films for ethanol sensing applications. Mater. Res. Express. **5**(2) (2018). https://doi.org/10.1088/2053-1591/aaa9f1
10. Ridhi, R., Saini, G.S.S., Tripathi, S.K.: Study of interaction mechanism of metal phthalocyanines dispersed sol-gel glasses with chemical vapors. Sens. Actuators B Chem. **255**(1) (2017)
11. Oliveira, R.H., Gonçalves, D.A., Reis, D.D.: TiO_2/MWCNT/Nafion-modified glassy carbon electrode as a sensitive voltammetric sensor for the determination of hydrogen peroxide. Sensors. **23**(18), 7732 (2023)
12. Kumar, A., Kumari, P., Kumar, M.S., Gupta, G., Shivagana, D.D., Bapna, K.: A high-performance flexible humidity sensor based on a TiO_2–MWCNT nanocomposite for human healthcare applications. Phys. Chem. Chem. Phys. **26**, 21186–21196 (2024)
13. Ridhi, R., Chouksey, A., Gautam, S., Rawat, J.S., Jha, P.: Study of the effect of flow rate and decomposition temperature on sensing of ammonium nitrate by carbon nanotubes. Sens. Actuators B Chem. **2021**, 123456 (2021)
14. Ridhi, R., Neeru, Gautam, S., Saini, G.S.S., Tripathi, S.K., Rawat, J.S., Jha, P.: Study of the effect of orbital on interaction behaviour of SWCNT-metal phthalocyanines composites with ammonia gas. J. Mol. Struct. **1234**, 56789–56796 (2021)
15. Askari, M.B., Tavakoli Banizi, Z., Seifi, M., Bagheri Dehaghi, S., Veisi, P.: Synthesis of TiO_2 nanoparticles and decorated multi-wall carbon nanotube (MWCNT) with anatase TiO_2 nanoparticles and study of optical properties and structural characterization of TiO_2/MWCNT nanocomposite. J. Nanostructure Chem. **7**, 567–578 (2017)
16. Božič, M., Ban, I., Hribernik, S., Fakin, D., Stana Kleinschek, K.: Surface engineering of TiO_2-MWCNT nanocomposites towards tuning of functionalities and minimizing toxicity. J. Mater. Sci. **52**, 6789–6798 (2017)
17. Shafique, S., Khalid, Z., Arif, S., Noreen, Z., Tabassam, L., Waseem, M.: ZnO, TiO_2 and Ag_2O decorated MWCNTs for enhanced antibacterial and antidiabetic activities. J. Nanomater. (2024)
18. Ridhi, R., Saini, G.S.S., Tripathi, S.K.: Effect of central metal ions on interactions of metal phthalocyanines composites with chemical analytes. Appl. Phys. **130**(5), 359 (2024)
19. Wang, Q., Wang, Y., Duan, B., Zhang, M.: Modified sol-gel synthesis of carbon nanotubes supported titania composites with enhanced visible light induced photocatalytic activity. J. Nanomater. **2016**(41), 1–6 (2016)
20. Ghasemi Koozekonan, A., Monazzam Esmaeilpour, M.R., Kalantary, S., Karimi, A., Azam, K., Golbabaei, F.: Fabrication and characterization of TiO2 and MWCNT coated electrospinning nanofibers for UV protection properties. J. Ind. Text. (2021)

Design and Analysis of Plasmonic Antenna for 6G-Enabled WBANs in Personalized Healthcare

Simran, Mohd. Gulman Siddiqui(✉), and Shuvabrata Bandophyay

Department of Physical Sciences, Banasthali Vidyapith, Radha Kishnpura, Rajasthan, India
mohdgulmansiddiqui@banasthali.in

Abstract. This paper has proposed a plasmonic multi-ring structure for an antenna that aims to be deployed in 6G-enabled wireless body area networks (WBANs) for personalized healthcare. Due to their nanoscale dimensions, plasmonic antennas are suitable for integration into ultra-compact wearable sensors and implantable devices used in WBANs, allowing data transmission and continuous health monitoring without sacrificing convenience or portability. The multi-ring type antenna design is a prominent wideband antenna design technique that consists of multiple concentric ring elements made of conductive material, arranged in a planar configuration. The proposed concentric multi-ring antenna structure for terahertz applications is designed with the help of a CST tool, and its features are analyzed. The proposed antenna structure resonates at a frequency of 5.75 THz, with a minimum S11 of −18 dB. It also achieved a bandwidth of 20.27% between 5.23 THz and 6.41 THz. Its dimensions in the micrometer range make it suitable for use in wearable or implanted WBAN nodes.

Keywords: Wireless body area networks (WBANs) · 6G communication · terahertz communication · multi-ring structure

1 Introduction

The Wireless Body Area Network (WBAN) is a network of wearable and implantable devices that is considered the key enabler for next-generation personalized healthcare solutions [1]. The WBAN facilitates early diagnosis with seamless integration of real-time health monitoring with advanced AI technologies, leading to personalized treatment plans. The successful implementation of WBAN lies in real-time data transmission between wearable or implantable devices and healthcare systems [2]. The quality of service (QoS) requirements for the communication network to support applications related to WBAN are ultra-low latency, reliability and high reliability features to ensure immediate and accurate data transmission, strong security protocols to ensure protection of sensitive health information, and high energy efficiency of the network to maintain the longevity of devices [3, 4].

After the successful rollout of 5G technology, the International Telecommunication Union (ITU) has released the next version of its International Mobile Telecommunications (IMT) series, IMT 2030, defining the framework and vision for the next generation

S. Kumari et al. (Eds.): ICMTP 2024, SPPHY 311, pp. 308–315, 2025.
https://doi.org/10.1007/978-981-96-6795-6_35

of mobile networks, commonly known as 6G. The 6G is to be considered a milestone for future digital ecosystems with near-terabyte data rates, ultra-low latency, enhanced connectivity, and AI-powered network intelligence [5–7]. With features like ultra-high data rates and near-zero latency, along with AI-driven network intelligence, the 6G network is best suited for WBAN applications like real-time health monitoring, predictive healthcare, and personalized treatment [8, 9].

This paper is aims to design and analysis of plasmonic antennas for 6G-enabled WBANs in personalized healthcare. The Plasmonic antenna is a leading candidate for 6G application due to their ability to operate at high frequencies, enabling ultra-high-speed communication [10]. Due to their nanoscale dimensions, plasmonic antennas are suitable for integration into ultra-compact wearable sensors and implantable devices used in WBANs, allowing data transmission and continuous health monitoring without sacrificing convenience or portability [11, 12]. Plasmonic antennas, with its ability to enhance and localize terahertz waves at the nanoscale considered to be suitable in terahertz communication enabling high data transfer rates and compact device integration [13, 14]. Due to its exceptional electrical conductivity and flexibility, and its support for surface plasmon polaritons, graphene is the key element in plasmonic antenna design.

The critical feature for high-speed terahertz communication systems is wideband antenna design. The multi-ring type antenna design is a prominent wideband antenna design technique that consists of multiple concentric ring elements made of conductive material, arranged in a planar configuration [15]. Due to its high gain, compact design, and circular polarization features, along with wideband operation, the concentric multi-ring type antenna is suitable for space-constrained communication systems like WBAN nodes. Confinement of electromagnetic field at the nanoscale level can be achieved with a plasmonic multi-ring antenna that combines concentric ring elements with plasmonic materials [16].

In this paper, a plasmonic concentric multi-ring antenna structure for terahertz applications is designed with the help of a CST tool, and its features are analyzed. The proposed antenna structure resonates at a frequency of 5.75 THz, with a minimum S11 of −18 dB. It also achieved a bandwidth of 20.27% between 5.23 THz and 6.41 THz. Its dimensions in the micrometer range make it suitable for use in wearable or implanted WBAN nodes. The organization of the rest of the paper is given as follows: the design methodology and analysis in Section II, results and their discussion in Section III, and paper conclusions in Section IV.

2 Plasmonic Antenna Design and Methodology

The proposed antenna's patch (Fig. 1) is made up of several concentric rings. The size of the proposed antenna (L × W × h) is 25 μm × 16 μm **x** 5 μm. The antenna is constructed on substrate of polyamide with an electric permittivity of 3.5. The base layer consists of copper as a ground while top layer consists of frequency-sensitive graphene as a patch. The parametric dimensions for the antenna designed are given in Table 1.

This work presents the implementation of plasmonic antenna design iterations for improved parametric analysis. The Fig. 1 represent different design iterations.

The calculations of proposed antenna sizeusing substrate material and desired frequency [17, 18] are as follows:

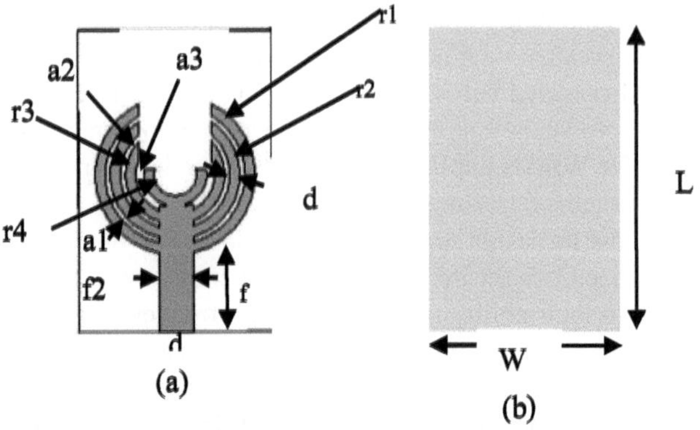

Fig. 1. Design of Top and Bottom view of antenna (a) and (b).

Table 1. Parametric dimensions (in μm) of the antenna.

Parameters of antenna	Dimensions
W	16
L	25
h	5
f1	7.5
f2	2.8
r1	5.75
r2	4.5
r3	3.25
r4	1.75
p	6
a1	0.5
a2	0.5
a3	0.75
d	0.86

Step I. Effective dielectric constant (ε_{eff})

$$\varepsilon_{eff} = \frac{\varepsilon r + 1}{2} + \frac{\varepsilon r - 1}{2}\left[1 + 12\frac{h}{W}\right]^{-\frac{1}{2}} \quad (1)$$

where, ε_{eff} represents effective dielectric constant, ε_r to Dielectric constant, h to height of dielectric, and W to Width of patch.
Step II. Effective length ($Leff$):

$$Leff = \frac{c}{2fo\sqrt{\varepsilon_{eff}}} \tag{2}$$

Step III. Extension length (ΔL):

$$\frac{\Delta L}{h} = 0.412\frac{(\varepsilon_{eff} + 0.3)}{(\varepsilon_{eff} - 0.258)}\frac{(\frac{W}{h} + 0.264)}{(\frac{W}{h} + 0.8)} \tag{3}$$

Step IV. The length of the micro strip patches (L):

$$L = L_{eff} - 2\Delta L \tag{4}$$

The size of the radiator patch and the height of substrate depend upon the fringing field. Also, the effective length of the THz antenna has more than its actual size due to fringing field and the performance of the antenna can be altered.

3 Results and Discussion

Various results of THz antenna with simulations are given here.

(A) **Reflection coefficient versus Frequency Plot**

Figure 2 shows that the antenna resonates at the frequency of 5.75 THz with the minimum S11 of −18 dB. It also depicts that the bandwidth of 20.27% is achieved between 5.23 THz and 6.41 THz frequencies.

(B) **Voltage Standing Wave Ratio versus Frequency Plot**

The VSWR is less than 2 in the relevant frequency bands and the impedance matching between the load and the field as shown in Fig. 3. Stated otherwise, the antenna's optimal value should be more than 1 but less than 2 but at the resonating frequencies, this value is less than 2 that is desired standing wave ratio attained by this antenna.

(C) **Radiation pattern**

The Fig. 4 illustrates the directivity of far-field radiation pattern at 90° for different Theta (or angles) variation with respect to isotropic antenna at 5.75 THz. The major lobe indicates maximum power radiation while side lobes suggestminor radiated power. Thus, a more stable pattern with Omni-directional capabilities exhibited in azimuth plane and in the elevation plane. The red circular line in the Fig. 4 shows the Phi variation in XY-plane which is azimuthal plane, while the blue circular line shows the Theta variation in XZ or YZ-plane which is the elevation plane.

These results have an edge over existing literature as shown in Table 2. The designed antenna has a clear edge over other designs because due to its novel combination of fractional bandwidth, manufacturing ease, and application-specific tuning for 6G-enabled WBANs, even though it is not the one with the minimum

S11 or maximum gain. In contrast to its graphene-based counterparts, which require complicated manufacturing, the utilization of polyamides provides low-cost manufacturing and flexibility, rendering it ideal for wearable and bio-integrated devices. Moreover, despite a small reduced gain, the antenna retains uniform plasmonic behavior and effective impedance matching in the THz range for reliable operation

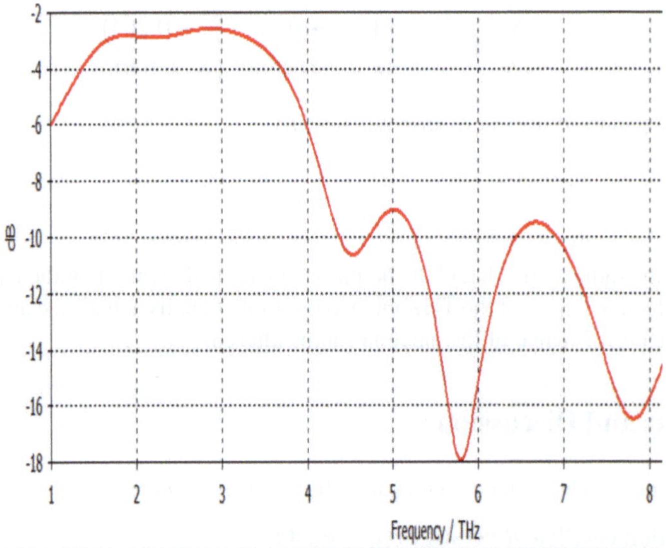

Fig. 2. Reflection coefficient versus Frequency Plot.

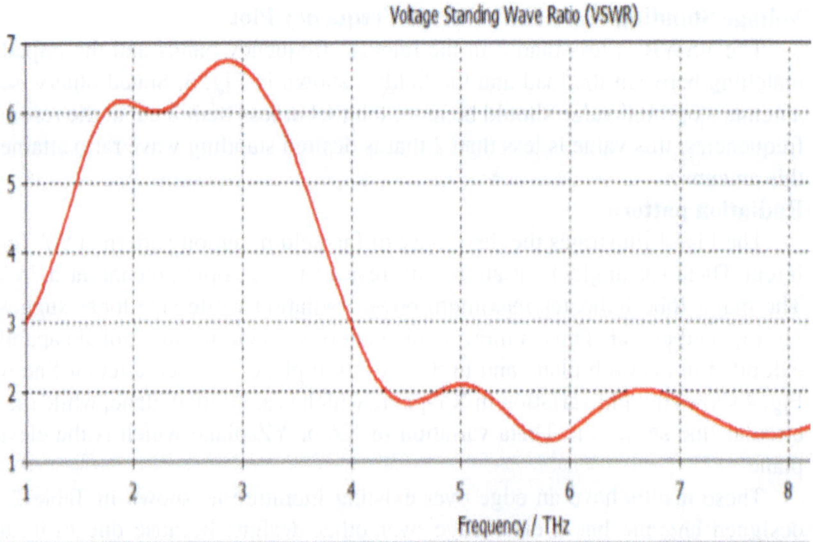

Fig. 3. VSWR versus frequency plot.

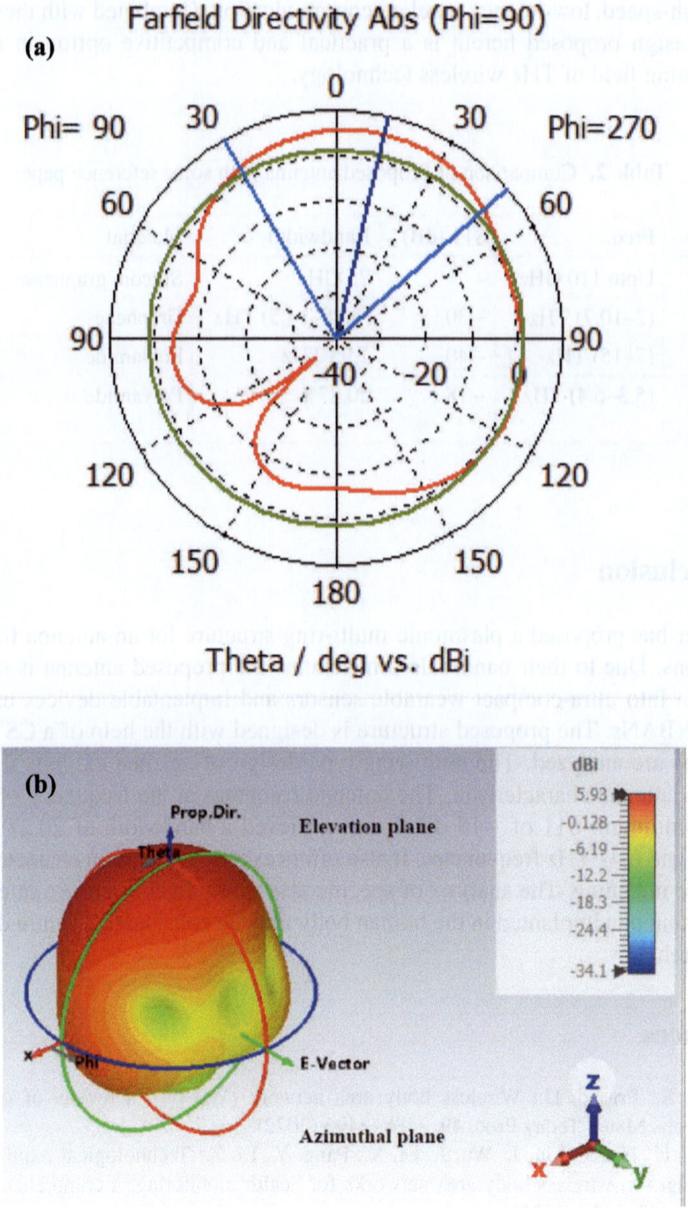

Fig. 4. (**a**) 2D and (**b**) 3D Radiation Pattern of Proposed Antenna at 5.75 THz.

in high-speed, low-latency wireless communication. Combined with these features, the design proposed herein is a practical and competitive option in the rapidly emerging field of THz wireless technology.

Table 2. Comparison of Proposed antenna with some reference paper.

Reference	Freq.	S11 (dB)	Bandwidth	Material	Gain
[12]	Upto 110 GHz	–	75 GHz	Silicon, graphene	N.A.
[13]	(2–10.7) THz	−30	(3.98–14.5) THz	Graphene	7.7 dBi
[14]	(7–15) THz	−40	1.95 THz	Polyamide	20.4 dB
Proposed antenna	(5.3–6.4) THz	−18	20.27%	Polyamide	5.93 dBi

4 Conclusion

This paper has proposed a plasmonic multi-ring structure for an antenna for terahertz applications. Due to their nanoscale dimensions, the proposed antenna is suitable for integration into ultra-compact wearable sensors and implantable devices used in 6G-enabled WBANs. The proposed structure is designed with the help of a CST tool, and its features are analyzed. The multi-ring type design of antenna exhibits a prominent wideband antenna characteristic. The antenna resonates at the frequency of 5.75 THz with the minimum S11 of −18 dB. It has achieved a bandwidth of 20.27% between 5.23 THz and 6.41 THz frequencies. It also offers excellent radiation characteristics and impedance matching. The analysis of specific absorption rates to ensure safe use of the proposed antenna implanted in the human body may be considered a future direction of this research.

References

1. Singh, S., Prasad, D.: Wireless body area network (WBAN): a review of schemes and protocols. Mater. Today Proc. **49**, 3488–3496 (2022)
2. Zhong, L., He, S., Lin, J., Wu, J., Li, X., Pang, Y., Li, Z.: Technological requirements and challenges in wireless body area networks for health monitoring: a comprehensive survey. Sensors. **22**, 3539 (2022)
3. Salayma, M., Al-Dubai, A.Y., Romdhani, I., Nasser, Y.: Reliability and energy efficiency enhancement for emergency-aware wireless body area networks (WBANs). IEEE Trans. Green Commun. Netw. **2**, 804–816 (2018)
4. Sun, G., Luo, L., Wang, K., Yu, H.: Toward improving QoS and energy efficiency in wireless body area networks. IEEE Syst. J. **15**(1), 865–876 (2020)
5. Bisoyi, S., Pasupuleti, M., Kuchi, K., Rao, K., Kumar, H., Kumari, S., Amuru, S.: IMT 2030 candidate technologies: a white paper by IIT Hyderabad & Wisig networks. arXiv preprint arXiv:2307.14978. (2023)

6. Jiang, W., Han, B., Habibi, M., Schotten, H.: The road towards 6G: a comprehensive survey. IEEE Open J. Commun. Soc. **2**, 334–366 (2021)
7. Pattnaik, S., Samal, S., Bandopadhaya, S., Swain, K., Choudhury, S., Das, J., Poulkov, V.: Future wireless communication technology towards 6G IoT: an application-based analysis of IoT in real-time location monitoring of employees inside underground mines by using BLE. Sensors. **22**(9), 3438 (2022)
8. Dao, N.: Internet of wearable things: advancements and benefits from 6G technologies. Futur. Gener. Comput. Syst. **138**, 172–184 (2023)
9. Cornet, B., Fang, H., Ngo, H., Boyer, E., Wang, H.: An overview of wireless body area networks for mobile health applications. IEEE Netw. **36**(1), 76–82 (2022)
10. Cubukcu, E., Kort, E., Crozier, K., Capasso, F.: Plasmonic laser antenna. Appl. Phys. Lett. **89**(9) (2006)
11. Ebadi, S., Khani, S., Örtegren, J.: Ultra-compact multifunctional surface plasmon device with tailored optical responses. Results Phys. **61** (2024)
12. Ding, Y., Cheng, Z., Zhu, X., Yvind, K., Dong, J., Galili, M., Hu, H., Mortensen, N.A., Xiao, S., Oxenløwe, L.K.: Ultra-compact integrated graphene plasmonic photodetector with bandwidth above 110 GHz. Nanophotonics. **9**(2), 317–325 (2020)
13. Kumari, R., Agarwal, R., Sharma, M.K., Singh, V.: Metamaterial-inspired multi-port tunable THz antenna with self-multiplexing and MIMO capability for 6G wireless and sensing applications. Braz. J. Phys. **55**(85) (2025)
14. Patel, S.K., Jansari, D., Jamaesha, S.S., Shah, R.P.: THz plasmonic metamaterial antenna with high isolation and geometrical performance optimization for 6G/TWPAN applications. Plasmonics. **19**(3), 2257–2271 (2024)
15. Jayant, S., Birwal, A., Patel, S., Patel, K.: Four-port multi-ring-based MIMO antenna for wideband applications. In: IEEE 2nd International Conference on Device Intelligence, Computing and Communication Technologies (DICCT), pp. 1–5. IEEE (2024)
16. Thiruppathi, P., Balasubrahmaniam, L., Manoharan, K., Balasubramanian, K.G.A.: Multiband plasmonic MIMO antenna array for 6G communications. Opt. Quant. Electron. **56**(6), 969 (2024)
17. Ansari, J.A., Mishra, A., Yadav, N.P., Singh, P., Vishvakarma, B.R.: Analysis of W-slot loaded patch antenna for dual-band operation. AEU-Int. J. Electron. Commun. **66**, 32–38 (2012)
18. Bahl, I.J., Bhartia, P.: Microstrip Antennas. Artech House, Dedham (MA) (1980)

Morphological and Optical Investigation of CTAB and PVP Capped ZnS Microspheres

Priya Chandra(✉) and K. S. Ojha

Department of Physics, National Institute of Technology Raipur, Raipur 492010, Chhattisgarh, India
pchandra.phd2022.phy@nitrr.ac.in

Abstract. N–cetyl-N, N, N–tri–methyl-ammonium-bromide (CTAB) and Polyvinylpyrrolidone (PVP) capped ZnS nanocrystals were synthesized in parallel employing a one-step hydrothermal technique to explore their enhanced stability and optical properties for photonic applications. Structural and morphological analysis authenticated the formation of the zinc-blende structure of ZnS with spherical morphology. The crystalline size estimated by Scherer's equation was around 5.8 nm. FTIR spectra confirmed the bonding of ZnS with capping agents and ensured effective capping. UV-visible absorption spectra indicate a direct bandgap of 3.7 eV for each sample. Photoluminescence analysis revealed the emissions in bluish-green region with enhanced intensity for CTAB-capped samples. These findings offer valuable role of capping agents in nanocrystal synthesis and their potential applications in photonic devices.

Keywords: Zinc sulfide · hydrothermal technique · capping agent · CTAB · PVP

1 Introduction

Over the past several decades, extensive research has been conducted on semiconductor nanomaterials due to their prospective significance in electronics and photonics [1, 2]. Researchers have shown great interest in developing novel nanomaterials to enhance their optical characteristics, for instance, broader absorption ranges, narrower emission bands, and higher quantum yields. Zinc sulfide (ZnS) is a II-VI group semiconductor material that attracts considerable research attention due to its distinctive characteristics such as wide bandgap, exceptional luminescent properties, and minimal toxicity as compared to cadmium- or lead-based semiconductors which make it a key material for applications in photonics and optoelectronics [3, 4]. Various approaches have been employed to synthesize ZnS nanomaterials such as the co-precipitation method [5], microwave irradiation [6], sol-gel [7, 8], microwave assist synthesis [9, 10], liquid-crystal template [11], spray deposition technique [12–14], solvothermal and hydrothermal methods [15, 16]. Among these, the hydrothermal approach allows the use of aqueous solutions, making it environment-friendly and suitable for laboratory synthesis as well as industrial synthesis. It offers advantages such as restrained size distribution, enhanced crystallinity, reduced agglomeration, and highly stable particles. It promotes nucleation and

S. Kumari et al. (Eds.): ICMTP 2024, SPPHY 311, pp. 316–327, 2025.
https://doi.org/10.1007/978-981-96-6795-6_36

growth of nanomaterials, resulting in well-defined structures with improved properties [17, 18]. Bhushan et al. [19] prepared n-type ZnS nanomaterials having a cubic structure with spherical morphology via co-precipitation method for absorption of the solar spectrum in UV-visible region. Cholan et al. [20] synthesized PEG-capped $ZnS:Ce^{3+}$ nanomaterials, having a cubic structure with a crystallite size of 2.8 nm, by chemical precipitation method and observed two emission peaks in the ultra-violet and visible regions respectively. Devi et al. [21] examined the consequences of starch capping on the photoluminescence (PL) characteristics of ZnS nanomaterials prepared by the co-precipitation technique and reported that PL intensity decreases by increasing the starch concentration. Hathwara et al. [22] synthesized ZnS nanomaterials using a microwave-assisted wet chemical method in presence of a copolymer capping agent and observed a defect-related emission at 440 nm as well as reduced bandgap for copolymer-capped ZnS. Liu et al. [23] employed ZnS precursors prepared via microemulsion technique to synthesize uniformly distributed spherical ZnS nanomaterials through a hydrothermal process and confirmed that the hydrothermal temperature had a more significant impact on ZnS crystallite formation compared to reaction time. The synthesized nanomaterials exhibit higher PL intensity and a noticeable blue shift as compared to bulk ZnS particles. Wang et al. [24] successfully synthesized the ZnS nano-spheres without surfactants via a hydrothermal method and established that the sulfur concentration, reaction time, and temperature significantly influenced the preparation of these nano-spheres. The prepared ZnS nano-spheres exhibit a strong green emission suggesting their potential applications in novel luminescent devices. In addition, they also proposed an innovative mechanism for the formation of these nano-spheres based on experimental results. Kumar et al. [25] synthesized silica-capped Mn-doped ZnS QDs, having cubic structure with crystallite size of 2 nm, using a chemical precipitation technique and reported a tunable emission and variable intensity with silica capping. The literature reviewed reveals that earlier workers have examined various synthesis parameters and investigated their effects on the properties of ZnS nanomaterials. Several studies demonstrate that the choice of capping agents significantly influences the properties of nanomaterials, offering an additional approach to optimize their performance. Capping agents applied to ZnS nanomaterials act as a stabilizer and prevent the nanomaterials from agglomeration by providing a protective coating on their surface and ensuring that the nanomaterials remain dispersed and stable [20]. Additionally, the surfactant can modify the surface chemistry of the synthesized nanomaterials which improves their compatibility with various solvents and enables their functionalization with other molecules for specific applications. PVP and CTAB are commonly used as capping agents, they provide stabilization, prevent agglomeration, and enhance the dispersibility of various nanomaterials. It can also influence the shape and size of the nanomaterials, affecting their catalytic and optical properties. This study aims to compare the impact of CTAB and PVP on the structural, optical, and morphological properties of ZnS nanocrystals (NCs) under constant reaction parameters.

2 Materials and Methodology

2.1 Chemicals

Zinc acetate dihydrate ($Zn(CH_3COO)_2{\cdot}2H_2O$), thiourea ($CS(NH_2)_2$), acetone, and deionized water were purchased from Qualigens. Polyvinylpyrrolidone (PVP, $(C_6H_9NO)_n$, MW = 10,000) was purchased from Alfa Aesar and N-cetyl-N, N, N-tri-methyl-ammonium bromide (CTAB, $C_{19}H_{42}BrN$) was purchased from Oxford. All the chemicals were employed without any further alteration.

2.2 Synthesis of ZnS NCs

The samples were prepared using a hydrothermal approach, which involved specific temperature and pressure conditions. To maintain these conditions, an autoclave with a thick-walled steel cylinder was used in which a Teflon vessel was inserted inside to prevent corrosion [26]. To prepare the CTAB-capped ZnS sample (CTAB@ ZnS), 20 mL of 1 M zinc acetate solution, 20 mL of 1 M thiourea solution, and 10 mL of 0.01 M CTAB solution were prepared separately. The solutions were mixed dropwise under continuous stirring until a clear solution was formed. This solution was poured into a 100 mL Teflon-lined autoclave and heated at 200 °C for 24 h. After cooling to room temperature, the precipitates were washed and filtered 3–4 times with distilled water, and dried in an oven at 100 °C for 3 h [27]. The resulting powder was stored in an airtight container for further characterization. For the PVP-capped ZnS sample (PVP@ZnS), the same procedure was repeated.

2.3 Experimental Techniques

A Pan-Analytical 3 kW X'pert powder multifunctional diffractometer was used to record diffraction pattern. All samples were scanned from 10° to 90° in the 2θ scale with an angular step of 0.02°. Using this technique, crystallographic features such as crystal structures, phase identification, and other associated properties were examined by interpreting the diffraction pattern. A Zeiss manufactured (EVO-18) scanning electron microscope (SEM) equipped with energy dispersive X-ray spectroscopy (EDS) was employed to evaluate the surface morphology and compositional information. A Bruker Fourier transform infrared (FTIR) spectrometer, Shimadzu UV spectrophotometer (UV-1800), and Shimadzu spectro-fluoro-photo-meter (RF-5301 PC) were used to conduct functional group study, optical absorption investigations and analyze the sample's room-temperature PL behavior, respectively.

3 Result and Discussion

3.1 Structural Investigation

The XRD data of synthesized samples were examined via Rietveld refinement method shown in Fig. 1(a). It validates that CTAB@ZnS and PVP@ZnS have cubic zinc blende structures (F-43m space group). The diffraction patterns closely matched with JCPDS card 01-080-0020 [28]. Both CTAB@ZnS and PVP@ZnS samples showed three broad peaks analogous to (111), (220), and (311) planes, indicating that the crystal structure of the NCs remained unchanged after stabilization with different capping agents. The lattice parameters and unit cell volume for synthesized samples were determined using Rietveld refinement and are presented in Table 1. The average crystallite size (D) of the synthesized samples was estimated using the Debye-Scherrer equation (Eq. 1),

$$D = \frac{k\lambda}{\beta \cos\theta} \tag{1}$$

where λ is the wavelength of the X-ray, β is full width at half maximum (FWHM), θ is Bragg angle, and k is the shape factor (considered as 0.9 for spherical particle) [29]. Scherrer's equation assumes that the broadening in XRD peaks is only due to the crystallite size. It does not consider instrumental broadening and microstrain. The Williamson-Hall (W-H) plot is a graphical method used to separate, and quantify crystallite size (β_D) and microstrain (β_ε) contributions to the broadening of diffraction peaks. The W-H plot uses the following equations,

$$\beta = \beta_D + \beta_\varepsilon \tag{2}$$

$$\beta = \frac{k\lambda}{D\,cos\theta} + 4\varepsilon\, tan\theta \tag{3}$$

Rearranging Eq. 3 gives

$$\beta \cos\theta = 4\varepsilon\, sin\theta + \frac{k\lambda}{D} \tag{4}$$

Equation 4 represents a straight-line equation, where the linearly fitted slope and intercept provide the value of lattice strain (ϵ) and crystallite size respectively. The W-H plots exhibit a negative strain, signifying the existence of compressive strain. Table 1 provides detailed information regarding the crystallite size and microstrain found from the W-H plot (Fig. 1(b, c)). The broad peaks suggest that the crystallites are in nano size, and it further indicates that no significant change has been observed in crystallite size with the use of CTAB and PVP surfactants.

3.2 Morphological Study

The SEM micrograph of CTAB and PVP capped ZnS NCs is shown in Fig. 2(a, b, c) and Fig. 2(e, f, g) respectively at different magnifications. The shape of individual particles is spherical and individually dispersed with irregular size distribution. With the aid of ImageJ software, the microsphere's average size has been calculated. The average sizes found were approximately 4 μm for both CTAB and PVP-capped ZnS shown in Fig. 2(d, h) [30].

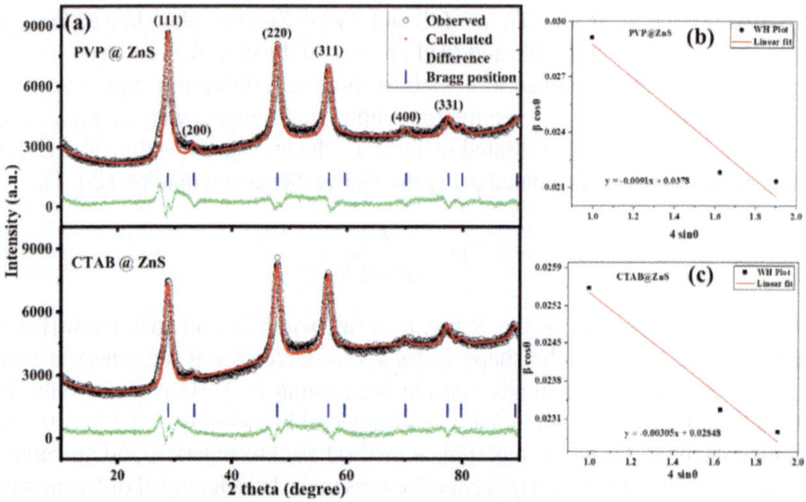

Fig. 1. (a) Rietveld refined XRD profile of synthesized ZnS NCs, (b) W-H plot for PVP@ZnS, and (c) W-H plot for CTAB@ZnS.

Table 1. Crystallographic parameter of synthesized ZnS samples.

S.no.	Sample name	Peak position, 2theta (degree)	FWHM (degree)	hkl	from Scherrer's equation		From Rietveld refinement		From W-H plot		
					Crystallite size (nm)	Average Crystallite size (nm)	Lattice constant $a = b = c$ (Å)	Volume of unit cell (Å^3)	Crystallite size (nm)	Strain ($\varepsilon * 10^{-3}$)	Dislocation density ($1/D^2$), (cm$^2 * 10^{12}$)
1	CTAB @ZnS	28.88	1.51	111	5.43	5.82	5.38	155.4	4.87	−3.05	4.2
		47.98	1.46	220	5.96						
		56.82	1.49	311	6.06						
2	PVP @ZnS	28.76	1.72	111	4.76	5.87	5.36	154.5	3.67	−9.10	7.4
		47.88	1.37	220	6.35						
		56.74	1.39	311	6.50						

3.3 Compositional Study

The prepared sample will be examined for composition by EDS. The EDS spectra represented in Fig. 3(a, b) along with the table for elements vs atomic weight percentage values showed that all the elements were present in the sample in stoichiometric proportions, and no additional elements were found. The presence of S and Zn resultant of ZnS, and others N, Br, O, and C are related to the capping agents [22]. The prepared sample's purity and the proper material formation were both confirmed by the EDS spectra.

Fig. 2. SEM micrographs (a, b, c) CTAB@ZnS and (e, f, g) PVP@ZnS at different magnifications, Histograms for particle size distributions (d) CTAB@ZnS, and (h) PVP@ZnS.

4 FTIR Study

FTIR measurements were made in the 4000–400 cm^{-1} region to analyze the bonding interactions between the stabilizer molecule and zinc sulfide. The investigations have been made by preparing a disc of ZnS powders mixed with KBr. KBr is used because it is transparent from near UV to long IR wavelength regions of light. The spectrum exhibits IR absorption due to several vibrational and stretching modes. Both samples have peaks near 490 cm^{-1} and 670 cm^{-1} for Zn-S stretching vibration [20, 31]. A broad peak near 3500 cm^{-1} may be accredited to the –OH stretching vibrations of water molecules adsorbed by sample's surfaces. FTIR spectrum (Fig. 4) of CTAB@ZnS shows

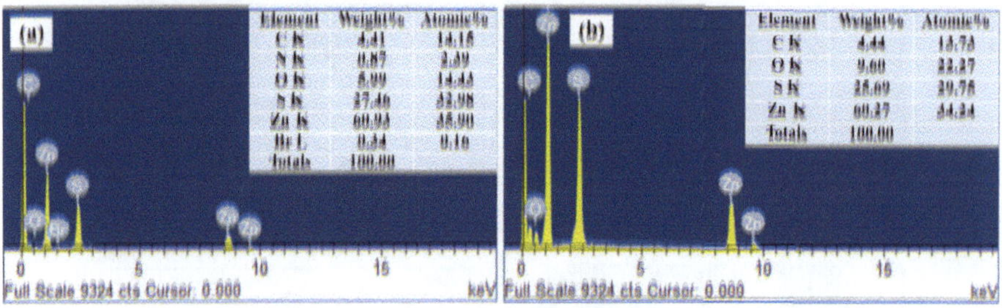

Fig. 3. EDS plot for synthesized (a) CTAB@ZnS and (b) PVP@ZnS.

several characteristic peaks that confirm the capping of ZnS with CTAB. The vibrational bands appearing at 2924 cm^{-1} and 2853 cm^{-1} are may be accredited to symmetrical and asymmetrical stretching of the $-CH_2$ groups present in the alkyl chain of CTAB [32]. The peaks at 1553 cm^{-1} and 1449 cm^{-1} represent the C-H scissoring vibrations of the $-N-CH_3$ group present in CTAB, indicating the presence of the quaternary amine group. The peak at 950 cm^{-1} is accredited to the C-O functional groups, further confirming the presence of a stabilizer. Also, the FTIR spectrum of PVP@ZnS has some characteristic peaks that confirm the capping of ZnS with PVP. The peak at 2913 cm^{-1} is attributed to C-H stretching of the methylene groups while the peak at 1640 cm^{-1} is associated with the C=O stretching of the amide group present in PVP. A peak at 1289 cm^{-1} accredits to the C-N stretching while the band around 1420–1460 cm^{-1} is related to the C-H bending of methylene groups in PVP [33]. These distinctive peaks confirm the capping of PVP over ZnS.

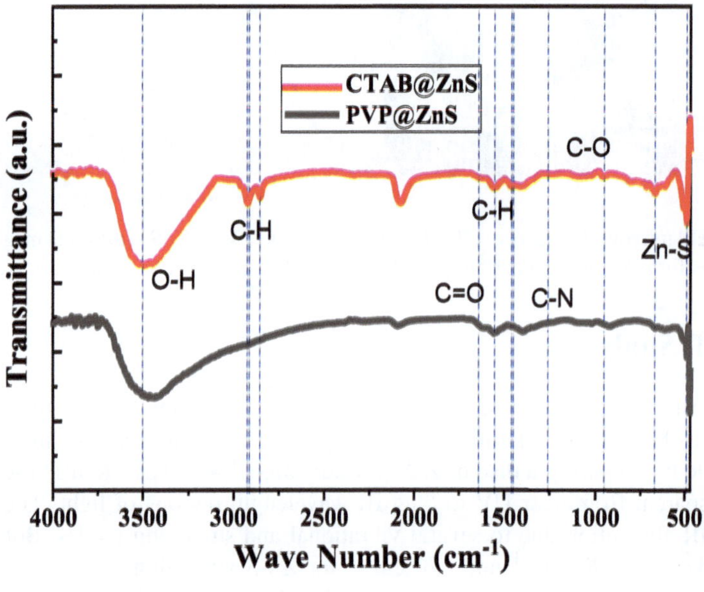

Fig. 4. FTIR spectra for synthesized ZnS NCs.

4.1 UV Vis Study

The UV-visible spectra, illustrated in Fig. 5(a, b), were recorded by dissolving the synthesized ZnS powder in ethanol as a reference. All samples display strong absorption in the UV region [34]. Also, the absorption peaks are not sharp, indicating the presence of small particle size distribution, which is also confirmed by the SEM images [21]. The Tauc plot method (Eq. 5) was employed to estimate the optical energy gap (E_g) of the ZnS NCs,

$$\alpha h\nu = c\left(E_{h\nu} - E_g\right)^n \tag{5}$$

where $E_{h\nu}$ is the energy of the incident photon, α represents absorption's coefficient and C represents a constant. Since ZnS is a direct band-gap semiconductor, an index $n = 1/2$ was used [35]. The absorption coefficient α can be evaluated through the Beer-Lambert relation (Eq. 6),

$$\alpha = \frac{2.303\,A}{d} \tag{6}$$

where A is absorbance obtained from the UV-Vis spectrum and d is path length (1 cm in this case). Figure 5(a, b) shows the Tauc plot (insets) used to calculate the optical energy gap, which is approximately 3.7 eV for both CTAB and PVP-capped ZnS [36].

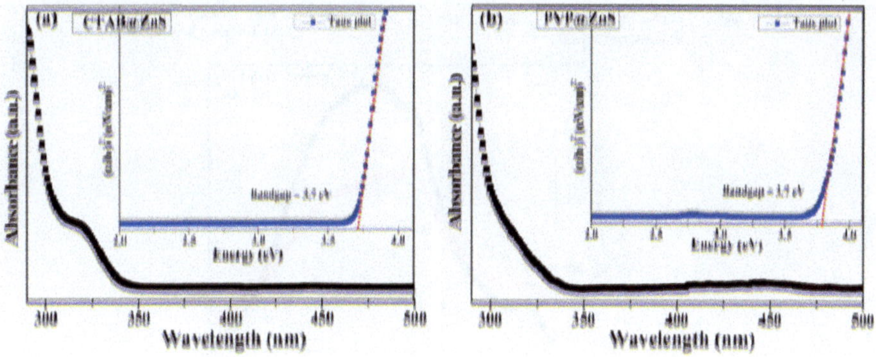

Fig. 5. Tauc plot inset with UV-Vis absorption spectra of (a) CTAB@ZnS and (b) PVP@ZnS.

4.2 Photoluminescence Study

The PL spectra exhibit emission peaks at specific wavelengths, which correspond to various energy levels and electronic transitions within the material. The precise shape and position of these PL peaks can vary with crystal structure, the presence of dopants or impurities, defects, and measurement conditions [37]. However, certain general characteristics of ZnS NCs can be identified. To examine the emission behaviors of the prepared ZnS samples, PL spectra were recorded using an excitation wavelength (300 nm) at room temperature. The PL spectra showed a broad emission band in the visible range, with peak positions at 380, 444, 469, 497, and 536 nm for CTAB@ZnS and 380, 469, 497, and 555 nm for PVP@ZnS, as determined by Gaussian fitting in Fig. 6(a, b). The peak is centered around the blue-green region, indicating that the sample can emit light in the blue-to-green region. This emission is commonly attributed to holes-electrons recombination in the material. The broad 497 nm peak suggests the presence of multiple peaks within its envelope, which points to the existence of several defect states in ZnS. These defects, often involving sulfur and zinc, can act as either electron acceptors or donors, which contributes to the observed emission behavior [38]. The peak lying at 380 nm is possibly due to near band edge emission of ZnS (3.7 eV). Figure 6(c) illustrates the variations in the intensity of the PL spectra as CTAB@ZnS and PVP@ZnS samples [33]. It shows that CTAB@ZnS shows higher PL intensity compared to PVP@ZnS NCs.

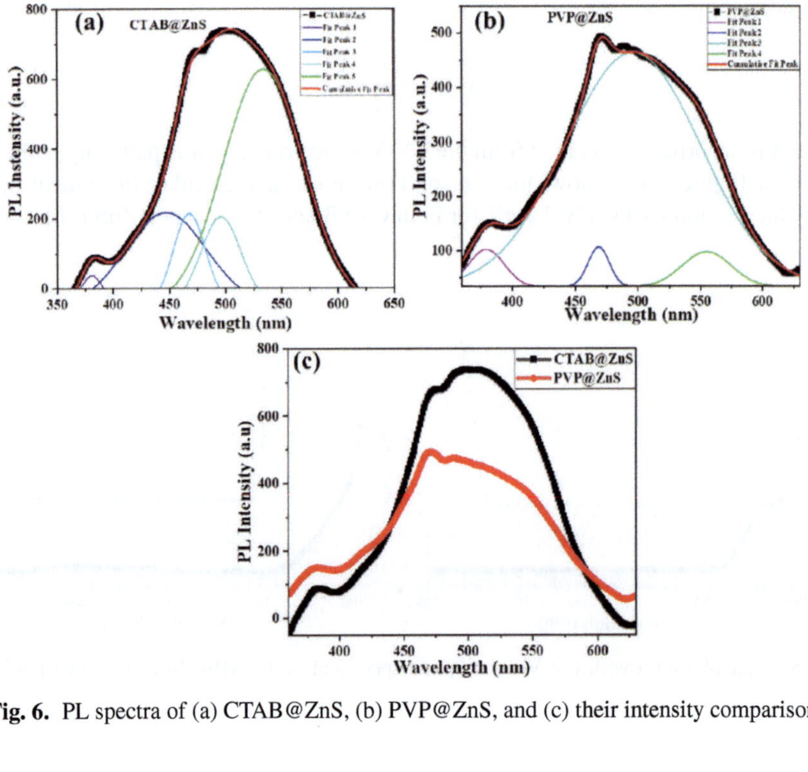

Fig. 6. PL spectra of (a) CTAB@ZnS, (b) PVP@ZnS, and (c) their intensity comparison.

5 Conclusion

The one-step hydrothermal method was employed to synthesize ZnS NCs, utilizing two organic compounds CTAB and PVP as stabilizing and capping agents separately. The comprehensive investigations of the synthesized ZnS NCs through XRD, SEM, EDS, FTIR, UV-visible absorption spectroscopy, and PL spectroscopy techniques provided valuable insights into the structural, optical, and morphological characteristics of the prepared material. XRD analysis confirmed that the prepared ZnS samples exhibited a cubic zinc blende structure with average crystallite sizes of approximately 5.8 nm. The SEM image shows the spherical morphology of all samples with irregular size distribution. The sample's purity and proper material formation were confirmed by the EDS spectra. The formation of CTAB@ZnS and PVP@ZnS NCs was confirmed by FTIR spectra. The UV-Vis study demonstrated ZnS has excellent absorption of UV radiation, allowing for efficient detection and measurement of ultraviolet light. PL spectra demonstrate tunable PL intensity related to specific capping. CTAB@ZnS NCs show higher PL intensity as compared to PVP@ZnS. Overall, the results highlight that the prepared CTAB@ZnS and PVP@ZnS NCs exhibited better structural, morphological, and optical properties. These findings highlight the promising nature of ZnS as a UV light sensing material, paving the way for further advancements in UV sensor technology and its diverse applications in photonic industries such as environmental monitoring, aerospace, phototherapy, and optoelectronic.

References

1. Samadi, M., Sarikhani, N., Zirak, M., Zhang, H., Zhang, H.-L., Moshfegh, A.Z.: Group 6 transition metal dichalcogenide nanomaterials: synthesis, applications and future perspectives. Nanoscale Horizons. **3**, 90–204 (2018). https://doi.org/10.1039/C7NH00137A
2. Dong, B., Cao, L., Su, G., Liu, W.: Synthesis and characterization of Mn doped ZnS d-dots with controllable dual-color emissions. J. Colloid Interface Sci. **367**, 178–182 (2012). https://doi.org/10.1016/j.jcis.2011.10.010
3. Alexander-Dann, B., Pruteanu, L.L., Oerton, E., Sharma, N., Berindan-Neagoe, I., Módos, D., Bender, A.: Developments in toxicogenomics: understanding and predicting compound-induced toxicity from gene expression data. Mol. Omics. **14**, 218–236 (2018). https://doi.org/10.1039/C8MO00042E
4. Wu, P., Yan, X.-P.: Doped quantum dots for chemo/biosensing and bioimaging. Chem. Soc. Rev. **42**, 5489 (2013). https://doi.org/10.1039/c3cs60017c
5. Sharma, H.K., Shukla, P.K., Agrawal, S.L.: Effect of sulphur concentration on the structural and electronic properties of ZnS nanoparticles synthesized using chemical precipitation method. J. Mater. Sci. Mater. Electron. **28**, 6226–6232 (2017). https://doi.org/10.1007/s10854-016-6302-7
6. Wang, H., Xu, J.-Z., Zhu, J.-J., Chen, H.-Y.: Preparation of CuO nanoparticles by microwave irradiation. J. Cryst. Growth. **244**, 88–94 (2002). https://doi.org/10.1016/S0022-0248(02)01571-3
7. Mandal, S., Ali, S.I., Mandal, A.C.: Investigation of structural, optical and photoluminescence properties of the sol–gel synthesized powder ZnS nanoparticles. Appl. Phys. A Mater. Sci. Process. **129**, 219 (2023). https://doi.org/10.1007/s00339-023-06499-z
8. Srivastava, M., Chaubey, S., Ojha, A.K.: Investigation on size dependent structural and magnetic behavior of nickel ferrite nanoparticles prepared by sol-gel and hydrothermal methods. Mater. Chem. Phys. **118**, 174–180 (2009). https://doi.org/10.1016/j.matchemphys.2009.07.023
9. Cadiş, A.I., Muresan, L.E., Perhaiţa, I., Popovici, E.J., Barbu-Tudoran, L., Indrea, E.: Studies regarding ZnS:Mn nanopowders prepared from single source molecular precursor using microwave-assisted decomposition. Mater. Res. Bull. **84**, 57–64 (2016). https://doi.org/10.1016/j.materresbull.2016.07.023

10. Kuzmin, A., Dile, M., Laganovska, K., Zolotarjovs, A.: Microwave-assisted synthesis and characterization of undoped and manganese doped zinc sulfide nanoparticles. Mater. Chem. Phys. **290** (2022). https://doi.org/10.1016/j.matchemphys.2022.126583
11. Zhang, D., Qi, L., Cheng, H., Ma, J.: Preparation of ZnS Nanorods by a Liquid Crystal Template. J. Colloid Interface Sci. **246**, 413–416 (2002). https://doi.org/10.1006/jcis.2001.8081
12. Barkhouse, D.A.R., Gunawan, O., Gokmen, T., Todorov, T.K., Mitzi, D.B.: Yield predictions for photovoltaic power plants: empirical validation, recent advances and remaining uncertainties. Prog. Photovolt. Res. Appl. **20**, 6–11 (2015). https://doi.org/10.1002/pip
13. Swami, S.K., Chaturvedi, N., Kumar, A., Dutta, V.: Effect of electric field on spray deposited zinc sulphide films. AIP Conf. Proc. **1665**, 1–4 (2015). https://doi.org/10.1063/1.4917979
14. Kumar, A., Swami, S.K., Dutta, V.: The role of electric field during spray deposition on fluorine doped tin oxide film. J. Alloys Compd. **588**, 546–550 (2014). https://doi.org/10.1016/j.jallcom.2013.11.130
15. Yoon, S.Y., Kim, Y.H., Jo, D.Y., Jo, J.H., Lee, S.H., Kim, H.M., Kim, Y., Kim, S.K., Yang, H.: Efficient synthesis of multinary Zn-Cu-Ga-Se1−xSx quantum dots as full visible-covering emitters and their tricolored white electroluminescence. Chem. Eng. J. **410** (2021). https://doi.org/10.1016/j.cej.2021.128426
16. Rose, A., Shunmugapriya, B., Maiyalagan, T., Vijayakumar, T.: Investigation on the electrochemical properties of hydrothermally synthesized pure and Nickel doped Zinc Sulfide microspheres for supercapacitor electrode applications. J. Mater. Sci. Mater. Electron. **31**, 19204–19212 (2020). https://doi.org/10.1007/s10854-020-04456-7
17. Wang, C., Hu, B., Xu, L.: Preparation and luminescence of ZnS:Mn nanoparticles on GaN substrates by hydrothermal method. Optik (Stuttg). **231** (2021). https://doi.org/10.1016/j.ijleo.2021.166329
18. Loghman-Estarki, M.R., Bastami, H., Davar, F.: Synthesis of one-dimensional MS (M = Zn, Cd, and Pb) nanostructure by MAA assisted hydrothermal method: A review. Polyhedron. **127**, 107–125 (2017). https://doi.org/10.1016/j.poly.2017.01.057
19. Bhushan, M., Jha, R., Bhardwaj, R.: Reduced band gap and diffusion controlled spherical n-type ZnS nanoparticles for absorption of UV-Vis region of solar spectrum. J. Phys. Chem. Solids. **135** (2019). https://doi.org/10.1016/j.jpcs.2019.05.018
20. Cholan, S., Shanmugam, N., Kannadasan, N., Sathishkumar, K., Deivam, K.: Effect of poly ethylene glycol (PEG) as surfactant on cerium doped ZnS nanoparticles. J. Mater. Res. Technol. **3**, 222–227 (2014). https://doi.org/10.1016/j.jmrt.2014.04.001
21. Devi, B.L., Chaitra, U., Hathwara, S., Kompa, A.: Influence of starch capping effect on optical absorption and photoluminescence behaviour of ZnS nanoparticles. Inorg. Chem. Commun. **149** (2023). https://doi.org/10.1016/j.inoche.2022.110374
22. Hathwara, S., Devi, B.L., Ramananda, D.: Optical and dielectric properties of poly(vinyl pyrrolidone-co-vinyl acetate)-capped ZnS nanoparticles. J. Electron. Mater. **50**, 5007–5012 (2021). https://doi.org/10.1007/s11664-021-09026-z
23. Liu, J., Ma, J., Liu, Y., Song, Z., Sun, Y., Fang, J., Liu, Z.: Synthesis of ZnS nanoparticles via hydrothermal process assisted by microemulsion technique. J. Alloy. Compd. (2009). https://doi.org/10.1016/j.jallcom.2009.07.109
24. Wang, Y., Ma, Q., Jia, H., Kong, J.: Controllable synthesis, characterization of ZnS nanostructured spheres. J. Mater. Sci. Mater. Electron. **27**, 7167–7173 (2016). https://doi.org/10.1007/s10854-016-4680-5
25. Kumar, S., Jeon, H.C., Kang, T.W., Singh, R., Sharma, J.K., Choubey, R.K.: Structural and optical properties of silica capped ZnS:Mn quantum dots. J. Mater. Sci. Mater. Electron. **26**, 3939–3946 (2015). https://doi.org/10.1007/s10854-015-2928-0

26. Nishimura, H., Enomoto, K., Pu, Y.J., Kim, D.G.: Hydrothermal synthesis of water-soluble Mn- and Cu-doped CdSe quantum dots with multi-shell structures and their photoluminescence properties. RSC Adv. **12**, 6255–6264 (2022). https://doi.org/10.1039/d1ra08491g
27. Nishimura, H., Lin, Y., Hizume, M., Taniguchi, T., Shigekawa, N., Sobue, S., Kawai, S., Okuno, E., Kim, D.: Hydrothermal synthesis of ZnSe:Mn quantum dots and their optical properties. AIP Adv. **9**, 025223 (2019). https://doi.org/10.1063/1.5085814
28. Sahoo, S.P., Tiwari, A.K.: A comparative study on the synthesis of a cationic gemini surfactant and a single tail surfactant-assisted ZnS quantum dots and their application to the degradation of methyl orange dye under visible light. J. Mol. Liq. **417**, 126640 (2025). https://doi.org/10.1016/j.molliq.2024.126640
29. Holzwarth, U., Gibson, N.: The Scherrer equation versus the "Debye-Scherrer equation". Nat. Nanotechnol. **6**, 534–534 (2011). https://doi.org/10.1038/nnano.2011.145
30. Zhou, D.-J., Xie, X.-Y., Zhang, Y., Guo, D.-Y., Zhou, Y.-J., Xie, J.-F.: Facile synthesis of ZnS nanorods in PEG and their spectral performance. Mater. Res. Express. **3**, 105023 (2016). https://doi.org/10.1088/2053-1591/3/10/105023
31. Chen, S., Li, Y., Lü, R., Wang, P.: Preparation of functionalized hollow ZnS microspheres and their analytical application. Appl. Mech. Mater. **151**, 281–285 (2012). https://doi.org/10.4028/www.scientific.net/AMM.151.281
32. Mehta, S.K., Kumar, S., Chaudhary, S., Bhasin, K.K., Gradzielski, M.: Evolution of ZnS nanoparticles via facile CTAB aqueous micellar solution route: A study on controlling parameters. Nanoscale Res. Lett. **4**, 17–28 (2009). https://doi.org/10.1007/s11671-008-9196-3
33. Shahi, A.K., Pandey, B.K., Singh, B.P., Gupta, B.K., Singh, S., Gopal, R.: Photo physical studies of PVP arrested ZnS quantum dots. Electron. Mater. Lett. **13**, 160–167 (2017). https://doi.org/10.1007/s13391-017-6132-7
34. Kumar, R.A., Geetha, K., Prabukanthan, P.: Synthesis of Mn^{2+} ions doped ZnS nanoparticles using PVP, TOPO and CTAB as capping agent and their characterization. Int. J. Appl. Sci. Math. **2**(5), 193–200 (2015)
35. Goktas, A., Tumbul, A., Aslan, F.: A new approach to growth of chemically depositable different ZnS nanostructures. J. Sol-Gel Sci. Technol. **90**, 487–497 (2019). https://doi.org/10.1007/s10971-019-04990-9
36. Bochev, B., Yordanov, G.: Room temperature synthesis of thioglycolate-coated zinc sulfide (ZnS) nanoparticles in aqueous medium and their physicochemical characterization. Colloids Surf. A Physicochem. Eng. Asp. **441**, 84–90 (2014). https://doi.org/10.1016/j.colsurfa.2013.08.077
37. Swami, S.K., Khan, J.I., Dutta, V., Lee, J., Laquai, F., Chaturvedi, N.: Spray-deposited aluminum-doped zinc oxide as an efficient electron transport layer for inverted organic solar cells. ACS Appl. Energy Mater. **6**, 2906–2913 (2023). https://doi.org/10.1021/acsaem.2c03858
38. Hu, Y., Wei, Z., Wu, B., Shen, B., Dai, Q., Feng, P.: Photoluminescence of ZnS:Mn quantum dot by hydrothermal method. AIP Adv. **8** (2018). https://doi.org/10.1063/1.5010833

Exploring the Structural and Optical Properties of Ni-Doped MgO Thin Films Deposited by E-Beam Technique

Pavan Kumar Meena, Prashant Meena, Ved Prakash Meena, Dhara Singh Meena, Harish Mahla, and M. K. Jangid(✉)

Department of Physics, Vivekananda Global University, Jaipur 303012, India
mahesh.jangid@vgu.ac.in, maheshjangid2008@gmail.com

Abstract. This study explores the structural and optical characteristics of MgO/Ni thin films, along with the impact of annealing. The films, doped with Ni, were deposited on glass substrates using the e-beam evaporation technique. The annealing was done in vacuum at 300 °C for 1 h. The XRD analysis of both as-deposited and annealed thin films revealed diffraction peaks at $2 = 42.1°$ and 62.3°. These peaks correspond to the (200) and (220) planes of cubic MgO nanostructures, confirming the formation of a well-defined cubic phase. Moreover, annealing improved the crystallinity, indicating enhanced grain growth. UV-Vis absorption spectroscopy revealed a distinct absorption peak occurring between 300 and 350 nm, which is linked to electronic transitions that move valence to conduction band. The bandgap noted to increase from 4.15 eV in the as-deposited films to 4.34 eV post-annealing, reflecting improved crystallinity. Hall measurements highlighted the effect of annealing on charge carrier dynamics and electrical properties, with both as-deposited and annealed films exhibiting n-type conductivity characteristic of MgO-based materials and conductivity increases from 3.65 $\Omega^{-1}cm^{-1}$ to 3.87 $\Omega^{-1}cm^{-1}$ after annealing. Surface morphological images revealed the enhanced surface uniformity in as-deposited as well annealed samples, characterized by smoother and more homogeneous surfaces with fewer defects. The results indicate that annealing plays a crucial role in enhancing the structural and optical properties of Ni-doped MgO thin films. This improvement makes them promising materials for advanced optical, optoelectronic, and electronic applications.

Keywords: E-Beam · XRD · UV-Vis · Hall Effect · Optical Microscopy

1 Introduction

The development of thin films with distinct optical and structural properties has gained substantial attention in materials science. Magnesium Oxide (MgO) thin films, in particular, have emerged as a versatile material, suitable for various applications such as optoelectronics, spintronics, and energy storage devices [1]. MgO is well-recognized for its excellent electrical insulation, high thermal stability, and chemical durability, making it highly desirable for advanced technological applications. MgO's inherent cubic

S. Kumari et al. (Eds.): ICMTP 2024, SPPHY 311, pp. 328–338, 2025.
https://doi.org/10.1007/978-981-96-6795-6_37

rock salt crystal structure ensures chemical and thermal stability, allowing it to function efficiently across diverse fields, including electronics, catalysis, and environmental remediation [3–5]. Its high optical transparency, capable of transmitting nearly 80% of incident light, makes it a prime candidate for applications in optoelectronics, photovoltaic devices, sensor technology, and light-emitting diodes (LEDs) [6]. The material's wide bandgap, ranging from 4.14 eV to 7.3 eV, further expands its functionality as an insulator in electronic devices [7, 8]. Additionally, MgO's ferromagnetic behavior at room temperature, attributed to structural defects in its nanostructures, positions it as a promising material for spintronic devices. The optical and structural properties of MgO can be significantly tailored by incorporating transition metal ions. Doped metal oxides possess improved properties such as higher conductivity, better visible light transmittance, and wider bandgaps. These materials are classified as transparent conducting oxides (TCOs). Researchers have extensively studied different dopants to enhance the optical and electrical properties of MgO films [9, 10]. Doping is the process of intentionally introducing foreign elements into a material to modify its structural and electronic characteristics. The doping elements nickel and zinc can improve the electronic, magnetic, and optical properties of MgO. Additionally, they enhance its biocompatibility and electrical conductivity [11–13]. Doping of MgO with elements such as nickel (Ni), aluminum (Al), and zinc (Zn) alters its electronic, optical, and morphological properties, enabling improved performance in devices. Ni-doped MgO thin films have gained significant attention due to their adaptable structural, optical, and magnetic properties, positioning them as promising materials for spintronic and optoelectronic applications. Studies reveal that Ni doping introduces lattice distortions in MgO, which enhances crystallinity and modifies its electronic band structures [14]. Annealing has been shown to further improve crystallinity, reduce defect density, and enhance grain size, leading to optimized optical and magnetic properties in MgO/Ni films [15, 16]. This tunability of properties is vital for advancing its use in modern technologies. MgO based thin films are easily fabricated using physical vapor depositions, e-beam methods to enable the production of high-quality thin films with controlled morphologies and tailored properties [17]. In this study, MgO/Ni thin films were fabricated using the electron beam (e-beam) method. The thin films were characterized using XRD, UV-Vis, Hall Effect and Optical Microscopy measurements for their potential application in advanced technologies.

2 Experimental

MgO/Ni hin films were deposited onto the glass substrates through the e-beam evaporation technique. High-purity MgO (99.99%) and Ni (99.98%) powders were used as source materials. The powders were weighed in a stoichiometric ratio of 95% MgO and 5% Ni, and then milled for 5 h to achieve a uniform mixture. The milled mixture was pressed into pellets to serve as targets for e-beam evaporation. The nano-structured films of thickness 400 nm were prepared using e-beam system (Model BC-300 HHV). During deposition, the vacuum chamber was evacuated, achieving a base pressure of $\times 10^{-5}$ mbar. Oxygen gas was introduced into the chamber through a manual regulator, and the pressure was $\times 10^{-4}$ mbar. The prepared thin films were annealed in vacuum at 300 °C for 1 h to investigate the effect of annealing. X-ray diffraction (XRD) analysis was carried out on a Panalytical X'Pert Pro diffractometer at room temperature for phase identification, utilizing CuKα radiation ($\lambda = 0.15406$ nm). The optical characteristics were evaluated through UV-Vis absorption spectroscopy (Shimadzu UV-1800 UV-Vis Spectrophotometer) across the 200–800 nm wavelength range. Electrical transport properties, including conductivity (σ), resistivity (ρ), sheet resistance (RS), Hall mobility (μ), and Hall coefficient (R_H), were measured using HMS-5000 series Hall Effect measurement equipment. Surface topography of the thin films was analyzed using optical microscope in reflected mode and micrograph images were captured using an input objective lens with a magnification of 50× in microscopic technique.

3 Results and Discussion

3.1 X-Ray Diffraction

X-ray diffraction (XRD) is a commonly used method to investigate the structural properties of the samples. It provides insights into the crystallographic orientations, phase transitions, and overall structural quality of the materials. In this study, XRD was employed to analyze the structural transitions and growth orientations of as-deposited and annealed MgO/Ni thin films. Figure 1 shows the XRD patterns of as-deposited and annealed thin films.

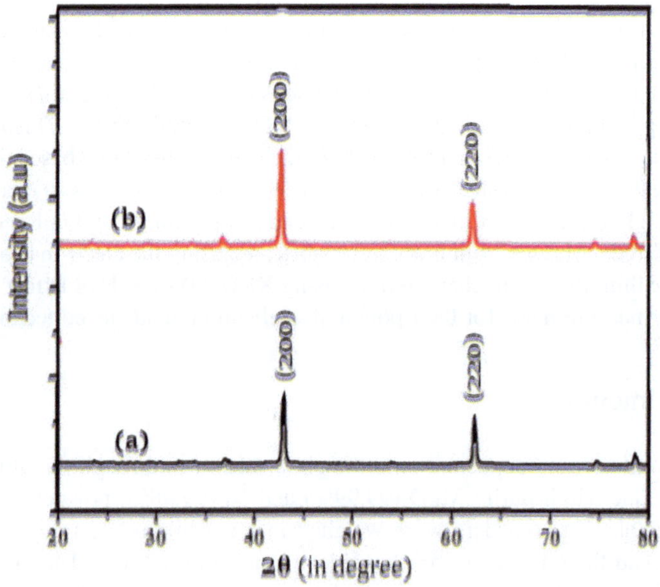

Fig. 1. XRD analysis of MgO/Ni thin films (a) as-deposited and (b) annealed at 300 °C.

It is shown in the figure that diffraction peaks were noted at angle 2 = 42.1° and 62.3° correspond to the (200) and (220) planes of cubic MgO nanostructures, confirming the formation of a well-defined cubic phase. These orientations are consistent with the typical peaks of MgO, representing that Ni addition were not change the primary crystal structure. The similarity in peak positions for both as-deposited and annealed samples suggests that the cubic structure is preserved post-annealing, with no significant phase transformation [18]. The observed structural features can be attributed to the incorporation of Ni impurities into the MgO matrix. Ni atoms are likely situated in the intergranular layers or at defect sites within the MgO lattice. Such interactions can enhance the structural stability and contribute to the observed diffraction patterns. Similar effects have been reported in the literature, where Ni additive concentrations in metal oxide systems were shown to occupy interstitial sites or intergranular regions, impacting the overall crystallographic properties [19–21]. Annealing typically improves crystallinity by facilitating atomic mobility and grain growth. However, in this study, the XRD analysis reveals that the basic cubic structure and orientation of the MgO/Ni thin films remain unchanged after annealing. This stability highlights the robustness of the MgO/Ni system against thermal effects, making it suitable for high-temperature applications. Furthermore, the presence of Ni impurities may contribute to enhanced mechanical and thermal properties by strengthening the lattice structure [12–24].

3.2 Optical Properties

The MgO/Ni thin films were examined through UV–Vis measurements to observe the effect of annealing on optical characteristics. The absorption spectra were recorded for all the fabricated thin films within the wavelength range of 300–800 nm. Figure 2 shows the absorption spectra of MgO/Ni thin films (a) as-deposited and (b) annealed at 300 °C. The results indicated a prominent absorption peak within the range of 300–350 nm. This prominent absorption band can be endorsed to the electronic transition of charges from valence to conduction band, as previously established in similar nanostructured systems [25]. Such transitions are characteristic of the optical behavior of MgO/Ni nanostructures and highlight their potential in optoelectronic applications. Several factors influence the absorption properties of thin films, including surface roughness, doping site distribution, and oxygen vacancies [26–28]. Variations in these factors can significantly affect the electronic structure. The annealing process plays a pivotal role in modifying these characteristics. After annealing, the sharpness of the absorption edges in MgO/Ni thin films was observed to diminish. This reduction in edge sharpness could be linked to thermal-induced structural changes, including increased crystallinity, reduced surface roughness, and improved interfacial uniformity between MgO and Ni layers. Additionally, annealing can alleviate defects such as oxygen vacancies, further influencing the optical absorption profile [29]. Previous studies have highlighted that annealing-induced modification in nanostructured thin films can also alter bandgap energy, shifting the absorption spectrum. For MgO-based thin films, these effects are often accompanied by changes in microstructural properties, such as grain size and film density, which are crucial for tailoring optical performance [30]. Furthermore, the observed changes in the absorption edge after annealing are consistent with reports on transition metal-doped oxide films, where thermal treatment enhanced film uniformity and reduced defect density [31].

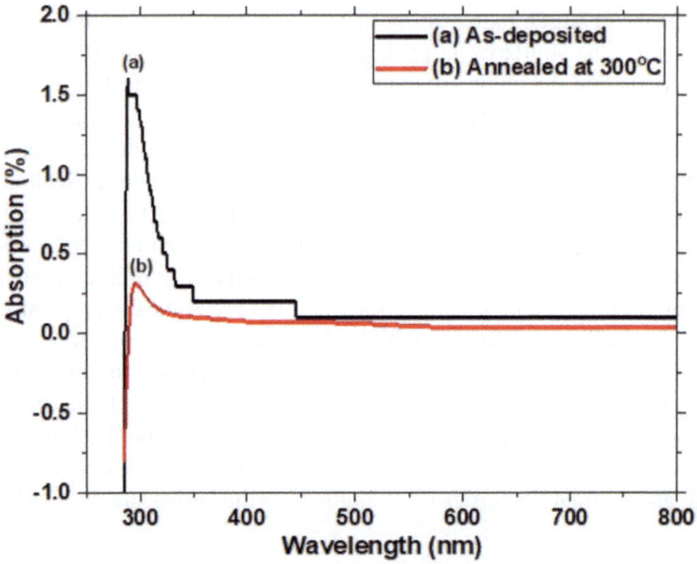

Fig. 2. Absorption spectra of MgO/Ni thin films (a) as-deposited and (b) annealed at 300 °C.

The optical bandgap of MgO/Ni thin films is a critical parameter influencing their suitability for optoelectronic applications. Bandgap energy is highly sensitive to structural, compositional, and morphological changes induced by annealing. The bandgap energy for the films was determined using optical absorption data and the Tauc relation [32], which is expressed as:

$$\alpha h\vartheta = A\left(h\vartheta - E_g\right)^m \tag{1}$$

where h is Planck's constant, v is the photon frequency, α is the absorption coefficient derived from the absorbance data, E_g is the bandgap energy, and m is the power factor of the transition mode. For direct transitions, m equals 2 was used to calculate the bandgap. Figure 3 shows the tauc plot of MgO/Ni thin films. For instance, the results films indicate a significant increase in the bandgap after annealing, for as-deposited thin film it was approximately 4.15 eV exhibited a shift to values near 4.34 eV after annealing.

This change is consistent with reduced defect density and improved structural order [33, 34]. When MgO/Ni thin films are subjected to annealing at 300 °C, notable modifications in their bandgap are observed due to alterations in crystal structure, defect density, and interface uniformity. Annealing at 300 °C often promotes grain growth, leading to increased crystallinity in the thin films. Improved crystallinity reduces the density of localized states in band structure, which in revolve minimizes the tail states in the bandgap region. Consequently, this can result in an increase in the optical bandgap energy, as the transitions become more defined and shift toward higher energy [35]. Another contributing factor is the reduction in oxygen vacancies during annealing. Oxygen vacancies introduce mid-gap states that reduce the effective bandgap energy by facilitating sub-bandgap absorption. Annealing at 300 °C provides sufficient thermal energy to heal these vacancies, thereby reducing defect-induced states and shifting the

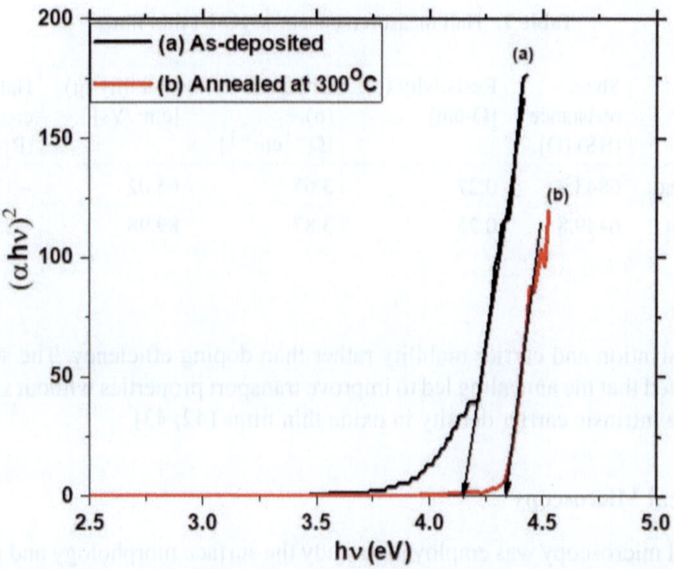

Fig. 3. Tauc's plot of MgO/Ni thin films (a) as-deposited and (b) annealed at 300 °C.

bandgap toward its intrinsic value [36, 37]. At 300 °C, diffusion processes are activated, improving the bonding at the interface and leading to better charge carrier mobility. This improved interface quality can influence the electronic structure, contributing to the observed variations in bandgap energy [38, 39]. Experimental studies on MgO/Ni thin films annealed at 300 °C have reported a modest increase in bandgap energy compared to as-deposited films.

3.3 Hall Effect Measurements

Hall Effect characteristics are pivotal to understand the electrical transport properties of thin films, particularly carrier concentration, mobility, and type (n-type or p-type). In the case of MgO/Ni thin films, Hall measurements provide critical insights into the influence of annealing on charge carrier dynamics and electrical behavior. Both as-deposited and annealed films exhibited n-type conductivity, which is characteristic of MgO-based materials. However, annealing significantly altered the electrical properties, highlighting the role of thermal treatment in optimizing film performance. For the as-deposited films, the Hall measurements revealed relatively low carrier mobility and high resistivity. These properties can be attributed to the presence of structural defects, grain boundaries, and impurities introduced during the deposition process. Upon annealing, a notable rise in carrier mobility and reduce in resistivity were observed (Table 1).

These changes can be linked to enhanced crystallinity, as evidenced by XRD analysis, and the reduction of defect density within the MgO lattice. Thermal annealing also aids in the redistribution of Ni impurities, which can influence carrier scattering mechanisms and contribute to the observed improvements [40, 41]. The carrier concentration remained relatively stable post-annealing, suggesting that the primary effect of annealing was on

Table 1. Hall measurements of MgO/Ni thin films.

Sample	Sheet resistance (RS) [Ω]	Resistivity (ρ) [Ω-cm]	Conductivity (σ) [$\Omega^{-1}cm^{-1}$]	Mobility (μ) [cm^2/Vs]	Hall coefficient (R_H) Cm^3/C]
As-deposited	6843.9	0.27	3.65	65.02	−17.8
Annealed at 300 °C	6449.8	0.25	3.87	89.98	−23.2

defect passivation and carrier mobility rather than doping efficiency. The studies have been reported that the annealing led to improve transport properties without significantly altering the intrinsic carrier density in oxide thin films [42, 43].

3.4 Optical Microscopy

The optical microscopy was employed to study the surface morphology and microstructural changes in MgO/Ni thin films before and after annealing. This technique offers a direct visualization of surface features, including grain size, roughness, and defect distribution, which play critical roles in determining the films' functional properties. Figure 4 shows the optical micrographs of MgO/Ni thin films. For as-deposited MgO/Ni films, optical microscopy revealed a relatively rough surface with small, irregularly shaped grains and noticeable defects such as voids and cracks. These features are indicative of incomplete crystal growth and strain accumulation during deposition. After annealing, the micrographs showed significant improvements in surface uniformity, with larger and more defined grains. This grain growth is a direct result of enhanced atomic diffusion during annealing, which promotes recrystallization and reduces surface energy [44]. The annealed films exhibited a smoother morphology with reduced defects, reflecting the healing effect of thermal treatment. These improvements correlate with the enhanced crystallinity observed in XRD analysis and contribute to better optical and electrical properties. Similar trends have been reported in other oxide thin films, where annealing led to smoother surfaces and fewer imperfections, enhancing material performance [45, 46].

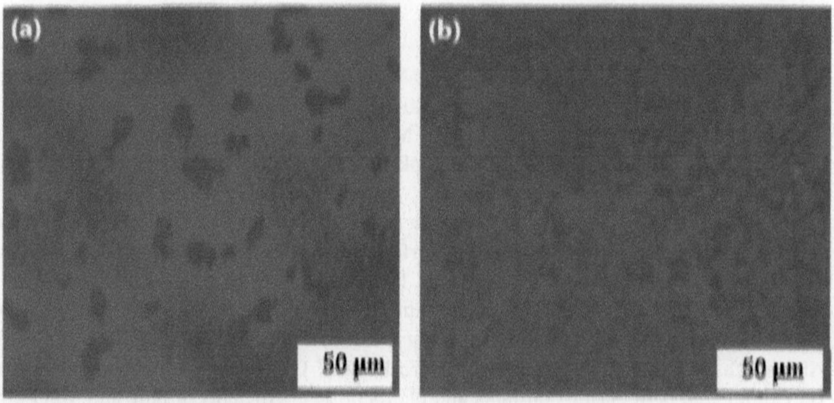

Fig. 4. Optical micrographs of MgO/Ni thin films (a) as-deposited and (b) annealed at 300 °C.

4 Conclusion

In conclusion the study of MgO/Ni thin films provides valuable insights into the impact of annealing temperature on their optical and structural characteristics. XRD analysis indicates that the observed orientations align with the typical peaks of MgO, confirming that the addition of Ni does not alter the primary crystal structure. The consistency in peak positions between as-deposited and annealed samples suggests that the cubic structure remains intact after annealing, with no significant phase transformation. The annealing at 300 °C promotes bandgap modifications attributed to improve crystallinity, a reduction in defects such as oxygen vacancies, and enhanced interfacial properties. The energy bandgap increases from 4.15 eV to 4.34 eV after annealing. This improvements are critical for optimizing the performance of MgO/Ni thin films in applications like UV detectors, transparent conductive oxides, and photocatalysts. Hall Effect measurements further highlight the impact of annealing on electrical transport properties. Both as-deposited and annealed films exhibited n-type conductivity. Increased carrier mobility (65.02 cm^2/Vs to 89.98 cm^2/Vs) and reduced resistivity (0.27 Ω-cm to 0.25 Ω-cm) achieved through annealing underscore the potential of these films in electronic applications, including thin-film transistors, sensors, and transparent conductive oxides. The research emphasizes the crucial role of annealing in improving the structural, optical, and electrical properties of MgO/Ni thin films. These enhancements make the material suitable for advanced applications.

Acknowledgements. The authors, with special acknowledgment to Pavan Kumar Meena, express their gratitude to CSIR for financial support provided under the JRF scheme. They also extend their appreciation to the Material Research Centre at MNIT Jaipur and the Department of Physics, University of Rajasthan, Jaipur, for providing access to experimental facilities during this research work.

References

1. Kim, J., Park, S., Lee, D., Kim, H., Cho, K.: MgO thin films for optoelectronic applications: a review. J. Alloys Compd. **741**, 120–133 (2018)
2. Lee, J., Kim, S., Park, Y., Jung, H., Choi, M.: Transition metal-doped MgO thin films: a review of their synthesis, properties, and applications. J. Mater. Sci. Mater. Electron. **31**(10), 9133–9146 (2020)
3. Luches, P., Benedetti, S., Liberati, M., Boscherini, F., Pronin, I.I., Valeri, S.: Absence of oxide formation at the Fe/MgO(001) interface. Surf. Sci. **583**, 191–198 (2005)
4. Jangid, M.K., Sharma, S.S., Ray, J., Yadav, D.K., Lal, C.: Structural, optical and electrical characterizations of Mg/Ti/Ni multilayer thin films deposited by DC magnetron sputtering for hydrogen storage. Int. J. Hydrogen Energy. **48**(96), 37921–37929 (2023)
5. Andrushchak, A., Mytsyk, B., Laba, H., Yurkevych, O., Solskii, I., Kityk, A., Sahraoui, B.: Complete sets of elastic constants and photoelastic coefficients of pure and MgO-doped lithium niobate crystals at room temperature. J. Appl. Phys. **106**, 73510 (2009)
6. Diachenko, O.V., Opanasuyk, A., Kurbatov, D.I., Cheong, H.: Investigation of optical properties of magnesium oxide films obtained by spray pyrolysis technique. In: IEEE 7th International Conference on Advanced Optoelectronics and Lasers (CAOL), pp. 31–33. IEEE (2016)
7. Araujo, C.M., Kapilashrami, M., Jun, X., Jayakumar, O.D., Nagar, S., Wu, Y., Århammar, C., Johansson, B., Belova, L., Ahuja, R.: Room temperature ferromagnetism in pristine MgO thin films. Appl. Phys. Lett. **96**, 232505 (2010)
8. Khamkongkaeo, A., Mothaneeyachart, N., Sriwattana, P., Boonchuduang, T., Phetrattanarangsi, T., Thongchai, C., Sakkomolsri, B., Pimsawat, A., Daengsakul, S., Phumying, S.: Ferromagnetism and diamagnetism behaviors of MgO synthesized via thermal decomposition method. J. Alloys Compd. **705**, 668–674 (2017)
9. Souza, H.J.D., Silva, E.D.D., Souza, N.D., Moger, S.N., Ashith, V.K.: Aluminium doped magnesium oxide thin films: a possible contender for transparent conducting oxides. Opt. Mater. **154**, 115651 (2024)
10. Raj, I.L.P., Valanarasu, S., Abdeltawab, A.A., Mohammady, S.Z., Ubaidullah, M., Shaikh, S.F., Shkir, M.: Improved UV photosensing properties of high crystalline nickel oxide thin films: role of yttrium doping. Optik. **248**, 168105 (2021)
11. Li, J., Hu, W., Hao, L., Gao, B., Wu, S., Zhang, J., Li, Y., Fan, H.: Electron-induced secondary electron emission of Zn-doped MgO/Au composite film. Mater. Lett. **229**, 360–363 (2018)
12. Taşer, A., Güldüren, M.E., Güney, H.: Tuning PL emission energy and bandgap with Ni dopant of MgO thin films. Ceram. Int. **47**, 15792–15800 (2021)
13. Mohan, A.C., Athira, A., Nair, B.P., et al.: Multifaceted properties of Ni and Zn codoped MgO nanoparticles. Sci. Rep. **14**, 32067 (2024)
14. Wang, Y., Zhang, L., Chen, H.: Structural and optical properties of Ni-doped MgO thin films prepared by sol-gel method. J. Appl. Phys. **128**(5), 054102 (2020)
15. Kim, J., Lee, T., Park, S.: Room-temperature ferromagnetism in Ni-doped MgO thin films: role of oxygen vacancies. Sci. Rep. **11**, 4567 (2021)
16. Singh, J.P., Gupta, L.K.: Annealing effects on MgO films grown using e-beam evaporation. Int. J. Math. Eng. Manag. Sci. **4**(3), 619–626 (2019)
17. Carta, G., El Habra, N., Crociani, L., Rossetto, G., Zanella, P., Zanella, A., Paolucci, G., Barreca, D., Tondello, E.: CVD of MgO thin films from bis(methylcyclopentadienyl) magnesium. Chem. Vap. Depos. **13**, 185–189 (2007)
18. Kulkarni, J., Ravishankar, R., Nagabhushana, H., Ananthraju, K.S., Basavraj, R.B., Renuka, L.: Green synthesis of Ni-doped magnesium oxide nanoparticles and its effect on photocatalysis. Indian J. Adv. Chem. Sci. **S1**, 64–67 (2016)

19. Dantas, S., Souza, A., Bohn, F., Lopes-Moriyama, A., Souza, C., Correa, M.: Magnetic properties of Ni-doped Mo_2C produced by fixed bed reactor. Mater. Lett. **2020**, 127916 (2020)
20. Jangid, M.K., Sharma, S.S., Ray, J., Jangid, S.: Effect of annealing and hydrogenation on optical and electrical properties of DC sputtered Mg-Ni bilayer thin films. Mater. Today Proc. **67**, 847–851 (2022)
21. Straumal, B., Mazilkin, A., Protasova, S., Myatiev, A., Straumal, P., Baretzky, B.: Increase of Co solubility with decreasing grain size in ZnO. Acta Mater. **56**, 6246–6256 (2008)
22. Kamarulzaman, N., Badar, N.: Pulsed laser deposition of MgO thin films. Adv. Mater. Res. **545**, 38–42 (2012)
23. Jangid, M.K., Jangid, S.K.: Structural, electrical, and optical properties of Mg-Ni thin films for hydrogen storage applications. Trends Sci. **19**(23), 2067 (2022)
24. Jangid, M.K., Nehra, S.P., Singh, M.: Hydrogenation effect on Mg/Mn multilayer thin films. AIP Conf. Proc. **1349**(1), 685–686 (2011)
25. Zhou, X., Liu, Y., Tang, Q., Wang, Z., Chen, H.: Optical properties of nanostructured metal oxides. J. Mater. Sci. **53**(12), 8450–8460 (2018)
26. Gupta, S., Mehra, R., Sharma, T., Patel, V.: Influence of defects on the optical behavior of thin films. Thin Solid Films. **704**, 137963 (2020)
27. Jangid, M.K., Nehra, S.P., Singh, M.: Preparation and characterization of Mg/Mn bilayer metal hydrides thin films. AIP Conf. Proc. **1393**(1), 311–312 (2011)
28. Jangid, M.K., Sharma, S., Meena, V.P., Arya, V.P., Sharma, S.S.: Impact of annealing and hydrogenation on the optical characteristics of ZnSe/Mn and ZnSe/Co DMS thin films prepared via thermal evaporation method. Evergreen. **11**(1), 178–185 (2024)
29. Chen, L., Huang, K., Lin, S., Zhao, P., Wang, J.: Annealing effects on MgO-based thin films. Appl. Surf. Sci. **467**, 566–573 (2019)
30. Smith, D.J., Johnson, K., Evans, T., Richards, M.: Impact of annealing on nanostructured oxide films. Mater. Lett. **204**, 152–155 (2017)
31. Park, J., Lee, S., Kim, H., Choi, K.: Role of annealing in modifying optical transitions of doped thin films. J. Appl. Phys. **119**(14), 145302 (2016)
32. Tauc, J.: Optical properties and electronic structure of amorphous semiconductors. Mater. Res. Bull. **3**, 37–46 (1968)
33. Gupta, S., Mehra, R., Sharma, T., Patel, V.: Bandgap tuning in annealed MgO thin films. Thin Solid Films. **704**, 137963 (2020)
34. Jangid, M.K.: Optical characteristics of Mg-based (Mg/Co & Mg/Mn) thin films for hydrogen storage applications. Mater. Today Proc. https://doi.org/10.1016/j.matpr.2023.08.083
35. Smith, A., Brown, P., White, C., Davis, R.: Thermal effects on the optical properties of MgO-based thin films. J. Appl. Phys. **123**(5), 055302 (2018)
36. Zhou, X., Liu, Y., Tang, Q., Wang, Z., Chen, H.: Defect healing in oxide thin films: role of annealing temperature. Mater. Res. Express. **6**(9), 095012 (2019)
37. Jangid, M.K., Singh, M.: Effect of hydrogenation on Raman spectra, structural, and optical properties of Mg/Al thin films. Int. J. Phys. Res. **2**(4), 15–23 (2012)
38. Wang, L., Zhang, H., Chen, J., Zhao, X., Li, Y.: Interfacial phenomena in annealed metal-oxide thin films. Appl. Surf. Sci. **419**, 703–709 (2017)
39. Jangid, M.K., Sharma, S.S., Ray, J.: Study on hydriding kinetics, structural properties, and electrical conductivity of DC magnetron sputtered Mg/Al thin films. Solid State Phenom. **338**, 83–90 (2022)
40. Gupta, S., Mehra, R., Sharma, T., Patel, V.: Electrical transport in MgO-based thin films: effects of annealing. Thin Solid Films. **704**, 137963 (2020)
41. Zhang, H., Li, M., Yang, W., Zhao, X.: Hall effect analysis of annealed metal oxide thin films. J. Phys. D: Appl. Phys. **52**(6), 065305 (2019)

42. Lee, J., Kim, S., Park, Y., Choi, H.: Impact of thermal treatments on the electrical properties of doped MgO thin films. Mater. Sci. Semicond. Process. **74**, 160–165 (2018)
43. Jangid, M.K., Nehra, S.P., Singh, M.: Optical, electrical, and Raman properties of annealed hydrogenated Mg/Co bilayer thin films. J. Nano-Electron. Phys. **3**(1), 460–468 (2011)
44. Wang, L., Zhang, H., Chen, J., Zhao, X., Li, Y.: Morphological evolution in annealed oxide thin films. Appl. Surf. Sci. **419**, 703–709 (2017)
45. Zhang, H., Liu, F., Xu, Q., Chen, K.: Surface analysis of annealed metal oxide thin films using optical microscopy. Mater. Charact. **162**, 110227 (2020)
46. Gupta, S., Mehra, R., Sharma, T., Patel, V.: Role of thermal annealing in improving oxide thin-film morphology. Thin Solid Films. **687**, 137231 (2019)

Effect of Annealing on Structural and Optical Characteristics of TiO_2/Zn Thin Film Prepared via D.C. Magnetron Sputtering Method

Prashant Meena, Pavan Kumar Meena, Ved Prakash Meena, Dhara Singh Meena, Harish Mahla, and M. K. Jangid(✉)

Department of Physics, Vivekananda Global University, Jaipur 303012, India
mahesh.jangid@vgu.ac.in, maheshjangid2008@gmail.com

Abstract. This study focuses on the preparation of TiO_2/Zn thin films using the d.c. magnetron sputtering technique. The impact of annealing on their structural and optical properties was investigated. The XRD analysis confirms the formation of a pure anatase TiO_2 phase. Notably, the anatase phase remains stable even after the annealing process. The XRD results also confirm the crystalline nature of the samples, with distinct peaks at $2\theta = 25.2°$, 37.19°, 47.57°, 55°, and 62.2° corresponding to (101), (004), (200), (211), and (220) planes, respectively. Fluorescence spectroscopy indicated that the annealing process significantly altered the optical characteristics of thin films. The pristine TiO_2/Zn thin films exhibited a broad emission peak in the visible region, centered around 420 nm. Annealing at 300 °C induced notable changes in the fluorescence spectrum, characterized by a reduction in the intensity of the broad emission peak. Further, annealed at 400 °C, the emission peak at 420 nm was significantly diminished, and a new peak emerged at around 480 nm. FTIR analysis provided valuable information on the structural changes and chemical composition of the films. The spectra showed that films annealed at 400 °C exhibited improved crystallization and the presence of surface hydroxyl groups, which are essential for enhanced photocatalytic and hydrophilic activity. AFM analysis demonstrated the effects of annealing on the morphology of films. The as-deposited films had a smooth surface with minimal roughness, but annealing increased roughness and promoted better structural development. In summary, annealing at 400 °C enhances the structural and optical characteristics of TiO_2/Zn thin films. These improvements make the films suitable for use in advanced applications. They hold significant potential in areas such as photocatalysis, optoelectronics, and energy storage.

Keywords: Thin films · Sputtering · XRD · Fluorescence spectroscopy · FTIR · AFM

1 Introduction

Titanium dioxide (TiO_2) has attracted considerable interest because of its versatility in technological applications. It is widely used in solar cells and optical filters. Additionally, it serves as an effective catalyst and is utilized in sensors for detecting humidity

S. Kumari et al. (Eds.): ICMTP 2024, SPPHY 311, pp. 339–350, 2025.
https://doi.org/10.1007/978-981-96-6795-6_38

and gases. The widespread use of titanium dioxide is due to its exceptional properties. It exhibits high durability, a significant dielectric constant, and a high refractive index. These characteristics make it an ideal material for numerous advanced applications [1–5]. To fully understand the potential of TiO_2, numerous studies have been conducted to investigate the effects of various factors on the physical and chemical characteristics of TiO_2 thin films using diverse analytical techniques [6, 7]. The performance of TiO_2 in optoelectronic devices is highly dependent on achieving the optimal film thickness, which is crucial for maximizing device efficiency. Furthermore, TiO_2 is chemically stable across different environmental conditions, non-toxic, and cost-effective, solidifying its status as a preferred material for a broad spectrum of applications. Recent research has particularly focused on nanoscale TiO_2 due to its enhanced light absorption capabilities, which arise from its high surface-to-volume ratio [8, 9]. This property not only increases photocatalytic activity but also boosts charge carrier generation efficiency. When photons with energies equal to or greater than the band gap of TiO_2 strike the crystal, electron-hole pairs (e^- and h^+) are generated. Effective trapping of these carriers during their journey to the surface is essential to minimize recombination and enhance photocatalytic performance. Doping TiO_2 with transition metal ions has emerged as an effective strategy to capture charge carriers and extend their lifespan, thereby improving photocatalytic activity [10]. By incorporating dopant atoms into the crystal lattice, researchers can modify the semiconductor's properties, making it more efficient for specific applications [11, 12]. Various metal and non-metal impurities, including Fe, Nb, Mn, Co, Sn, Cd, Zn, and Ni, have been introduced into the TiO_2 matrix to improve the functional properties of its thin films [13–15]. Among these dopants, zinc (Zn) stands out due to its abundance, low toxicity, and cost-effectiveness. The ionic radius of Zn^{2+} (0.74 Å) closely matches that of Ti^{4+} (0.68 Å), making Zn an effective substitutional dopant for TiO_2. The incorporation of Zn^{2+} ions into the TiO_2 structure often results in the formation of substitutional and interstitial defects, which enhance the material's photocatalytic and electronic properties [16]. Zn-doped TiO_2 is particularly appealing due to its low biological toxicity, making it safer for applications in various fields [17]. Zn-doped TiO_2 thin films exhibit several advantageous properties, including excellent transparency, high corrosion resistance, and enhanced semiconductor characteristics such as high electron mobility. Zinc doping alters the energy band positioning, improving the rate of photo-induced charge separation and, consequently, the photocatalytic efficiency [18]. Moreover, the close match between the ionic radii of Zn^{2+} (0.074 nm) and Ti^{4+} (0.061 nm) facilitates the substitution of Zn^{2+} ions into the TiO_2 lattice or their occupation of interstitial sites [19]. These attributes have made Zn-doped TiO_2 thin films a subject of extensive research and development in recent years. In this study, the structural and optical properties of TiO_2/Zn thin films were evaluated. Techniques such as fluorescence spectroscopy, atomic force microscopy (AFM), and Fourier-transform infrared (FTIR) spectroscopy were used for analysis. The findings reveal that annealing significantly enhances these properties. This improvement makes the films suitable for optoelectronic applications.

2 Experimental

The d.c. magnetron sputtering technique was used to deposit TiO_2/Zn thin films onto the glass substrates. Sputtering targets of TiO_2 (99.99%) and Zn (99.99%) of 10 mm dia and 2 mm thickness were used to prepare the thin films. High-purity argon gas was used for sputtering from the targets. An arrangement of diffusion pump and rotary pump was employed to achieve the required vacuum about 10^{-5} mbar. The total pressure for all depositions was 10^{-3} mbar, with a spacing of 140 mm maintained between the substrates and the target. The thin films of TiO_2/Zn (300 nm) were deposited onto the glass substrates with deposition rates of 0.14 Å/s for TiO_2 and 0.63 Å/s for Zn. To observe the annealing effect on structural and optical properties of the prepared thin films, subsequently, the prepared thin films underwent annealing in a programmable muffle furnace at 300 °C and 400 °C temperatures for an hour. The XRD analysis of the samples was obtained using a Panalytical X-ray diffractometer. The fluorescence characteristics were analyzed using a fluorescence spectrometer (LS 55, Perkin Elmer). Fourier-transform infrared (FTIR) spectroscopy measurements were taken using a Perkin Elmer spectrometer, with a wave-number range from 4000 to 400 cm^{-1}. Surface morphology and topography of the thin films were analyzed using Atomic Force Microscopy (AFM) with a Bruker Multimode Scanning Probe.

3 Results and Discussion

3.1 XRD Analysis of TiO_2/Zn Thin Films

Figure 1 shows the XRD analysis of TiO2/Zn thin films, in both pristine and annealed states (at 400 °C). The XRD spectra provide valuable insights into their crystal structure. The crystal phases of these films were studied using an X-ray diffractometer across the range of 10° to 90°. The results confirm the crystalline nature of the samples, with distinct peaks corresponding to the (101), (004), (200), (211), and (220) planes at 2θ angles of 25.2°, 37.19°, 47.57°, 55°, and 62.2°, respectively. These characteristic peaks confirm the anatase phase of TiO_2 in all samples. Notably, the XRD patterns of both pristine and annealed films are identical; indicating that annealing at 400 °C does not affect the anatase phase. The absence of additional peaks suggests the films are free from other phases or impurities, such as ZnO or zinc-based compounds. This indicates successful incorporation of Zn into the TiO_2 matrix while preserving the anatase structure. Furthermore, the consistent XRD patterns before and after annealing underscore the structural stability of the films, demonstrating the reliability of the d.c. magnetron sputtering technique in producing high-quality thin films. This stability is particularly important for applications in optoelectronics, where maintaining a stable and well-defined crystal structure is essential for optimal functionality [20]. The most prominent XRD peak for pristine and annealed TiO_2/Zn thin films were observed at $2\theta = 25.2°$. The results indicate that an increased in intensity and a reduced FWHM for annealed samples as compared to the pristine TiO_2/Zn thin films. This behavior indicates enhanced crystal growth and increased crystal size after annealing. The calculated values of crystallite size, dislocation density, and microstrain for TiO_2/Zn thin films are presented in Table 1.

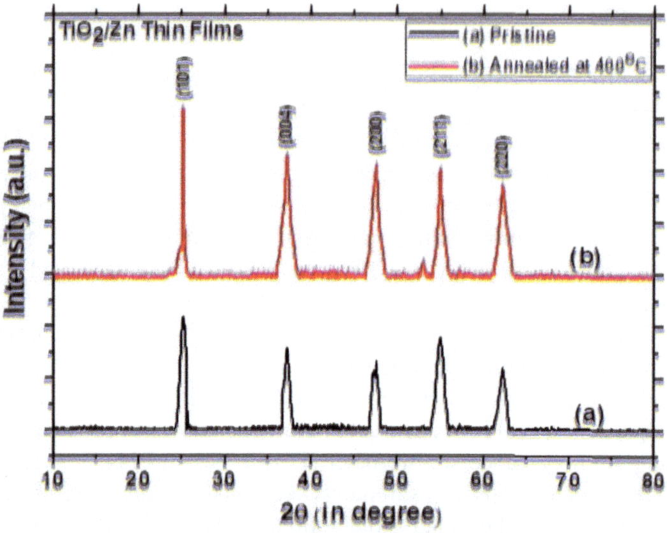

Fig. 1. XRD spectra of TiO_2/Zn thin films.

Table 1. FWHM, crystallite size, dislocation density and microstrain of TiO_2/Zn thin films.

TiO2/Zn	2θ (Deg.)	FWHM	Crystallite size (D) (nm)	Dislocation density (σ) (nm^2)	Micros train (ε) (μm)
Pristine	25.2	0.493	18.63	0.00315	0.0023
Annealed at 400 °C	25.2	0.294	27.79	0.00912	0.0035

3.2 Fluorescence Spectra of TiO_2/Zn Thin Films

Figure 2 illustrates the fluorescence spectra of TiO_2/Zn thin films, an essential tool for studying their optical properties and emission behavior. Fluorescence spectroscopy provides insights into the photoluminescence (PL) characteristics of these films under varying annealing conditions. This research examined the optical properties of TiO_2/Zn thin films under varying annealing conditions. Fluorescence spectra were recorded for pristine samples and those annealed at 300 °C and 400 °C. The impact of annealing temperature on their optical behavior was analyzed. The pristine TiO_2/Zn thin films showed a broad emission peak in the visible spectrum. This peak was centered at approximately 303 nm. It is commonly associated with the recombination of electron-hole pairs in TiO_2 and ZnO [21]. However, the emission intensity was relatively low, reflecting a high recombination rate and poor photo-carrier utilization, a common feature of untreated TiO_2-based materials [22]. Annealing at 300 °C induced notable changes in the fluorescence spectrum, characterized by a reduction in the intensity of the broad emission peak and obtained at 312 nm. This decrease suggests that thermal treatment at 300 °C facilitated crystallization, reducing surface defects and recombination centers responsible for non-radiative charge carrier recombination [23]. These findings align with prior studies on annealed TiO_2 films, where improved crystallinity and reduced trap states were observed following thermal treatments [24]. Further enhancements were evident in the fluorescence spectra of TiO_2/Zn thin films annealed at 400 °C. The emission peak was significantly shifted at 319 nm and intensity decreased, and also a new peak emerged at around 380 nm. This new shift is associated with enhanced excitonic emission resulting from improved crystallinity and the development of a more uniform metal-oxide network [25–27]. The reduction in recombination-related emissions peak indicate improved charge separation and transport in the material, key factors for photocatalytic applications [28]. Additionally, the 400 °C annealed TiO_2/Zn thin films exhibited stronger UV emission than those annealed at 300 °C, supporting the idea that higher annealing temperatures enhance the photonic response of the composite material [29]. The sharp emission peaks further suggest transitions of charge carriers from deep defect levels to the conduction band, evidencing improved crystallinity and reduced surface trap states [30, 31]. The energy bandgap were measured using fluorescence spectra of TiO_2/Zn thin films and found 4.10 eV for pristine, 3.97 eV and 3.88 eV for annealed at 300 °C and 400 °C respectively. The analysis indicates that the bandgap decreases with increasing annealing temperature for both the cases. This decrease is primarily endorsed to increase the grain size and the reduction of internal stress within the films due to the annealing process. Compared to the pristine and 300 °C annealed films, the 400 °C annealed TiO_2/Zn thin films demonstrated superior fluorescence performance, confirming the beneficial effects of higher annealing temperatures on their optical properties. These findings corroborate prior research, which indicates that thermal annealing effectively modifies the electronic structure and optical behavior of TiO_2/Zn thin films, enhancing their suitability for applications in optoelectronics and photocatalysis [32–34].

3.3 FTIR Analysis of TiO_2/Zn Thin Films

The FTIR spectra of pristine and annealed (300 °C and 400 °C) TiO_2/Zn thin films provide valuable insights into their chemical structure, functional groups, and potential applications. Figure 3 presents the FTIR spectrum of the pristine TiO_2/Zn thin film, revealing distinct vibrational peaks indicative of specific chemical groups. A prominent peak at 2956.38 cm^{-1} (85.95% T) corresponds to C–H stretching vibrations, often linked to organic residues or impurities. A sharp peak at 1730.84 cm^{-1} (72.16% T) represents C=O stretching, typically associated with carbonyl groups. Peaks at 2871.55 cm^{-1} (91.00% T) and 1376.04 cm^{-1} (92.82% T) are attributed to symmetric and asymmetric C–H bending vibrations, respectively. Additional peaks include one at 1456.14 cm^{-1} (88.54% T), corresponding to O–H bending vibrations, indicating the presence of hydroxyl groups. In the fingerprint region, a peak at 1242.31 cm^{-1} (84.67% T) suggests C–O stretching, while peaks at 1064.35 cm^{-1} (89.68% T) and 941.08 cm^{-1} (93.56% T) correspond to Ti–O–Ti or Ti–O–Zn stretching vibrations, confirming the formation of the composite matrix. Peaks at 821.70 cm^{-1} (96.2% T), 745.02 cm^{-1} (96.6% T), and 507.25 cm^{-1} (98.8% T) are linked to Zn–O and Ti–O bending vibrations, indicating the structural integrity of the TiO_2/Zn thin films. These results confirm

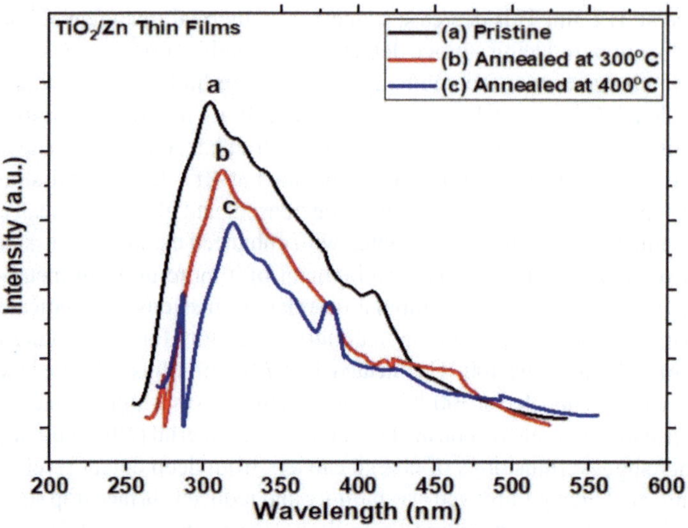

Fig. 2. Fluorescence spectra of TiO_2/Zn thin films.

the successful synthesis of the composite and its suitability for applications in opto-electronics and photocatalysis, consistent with earlier studies on similar materials [34, 36].

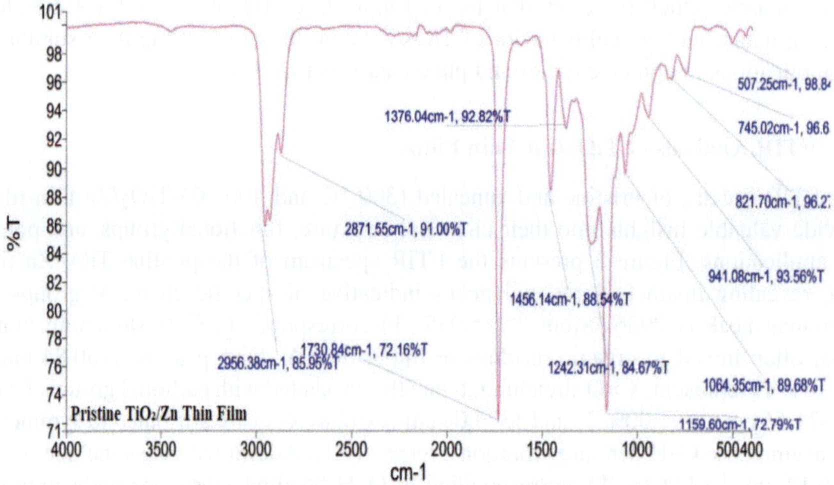

Fig. 3. FTIR spectrum of the pristine TiO_2/Zn thin film.

Figure 4 displays the FTIR spectrum of TiO_2/Zn thin films annealed at 300 °C. The spectrum indicates significant structural changes due to annealing. A broad band near 2921.49 cm^{-1} suggests C–H stretching vibrations, indicative of organic residues or surface-adsorbed species. The peak at 1581.69 cm^{-1} corresponds to O–H bending vibrations, often associated with adsorbed water molecules. A peak at 1392.21 cm^{-1} represents symmetric stretching vibrations of carbonate groups, likely formed through interaction with atmospheric CO_2 during preparation. Peaks at 873.22 cm^{-1}, 753.07 cm^{-1}, and 543.04 cm^{-1} correspond to Ti–O and Zn–O stretching vibrations, affirming the presence of metal-oxide bonds. A peak at 407.79 cm^{-1} indicates Ti–O lattice vibrations, in agreement with earlier observations in annealed thin films [37]. These results suggest that annealing at 300 °C promotes the crystallization of metal-oxide structures while retaining surface hydroxyl and carbonate groups, enhancing potential for photocatalytic and optical applications [38].

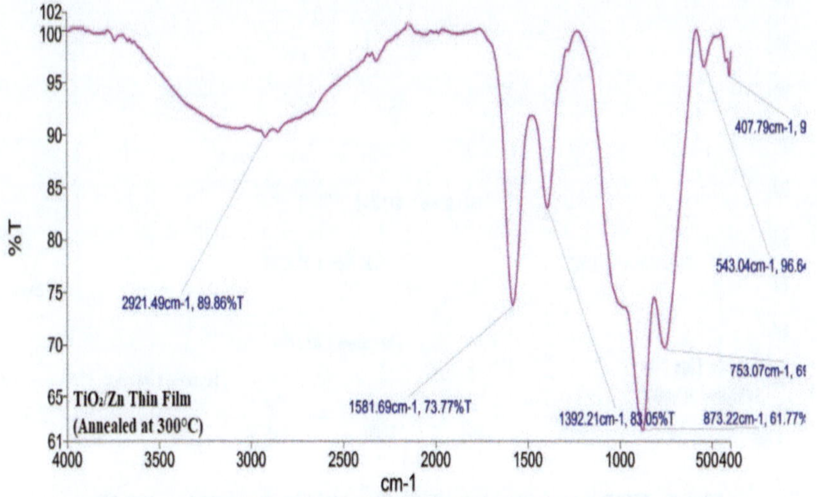

Fig. 4. FTIR spectrum of the TiO_2/Zn thin film annealed at 300 °C.

Figure 5 illustrates the FTIR spectrum of TiO_2/Zn thin films annealed at 400 °C, showing further improvements in structural definition. A broad peak at 3165.58 cm^{-1} (97.96% T) is attributed to O–H stretching vibrations, indicative of surface hydroxyl groups or adsorbed water. The peak at 1581.96 cm^{-1} (96.58% T) corresponds to bending vibrations of water molecules or Zn–O bonds, while the peak at 1406.76 cm^{-1} (97.42% T) suggests the presence of carbonate impurities or residual C=O stretching. Sharp peaks at 871.33 cm^{-1} (87.77% T) and 755.06 cm^{-1} (89.17% T) represent stretching and bending vibrations of Ti–O and Zn–O bonds, confirming the formation of metal-oxide linkages. The spectrum also shows a strong absorption at 466.25 cm^{-1}, attributed to Ti–O–Ti lattice vibrations, validating the formation of the TiO_2 framework. These findings indicate that annealing at 400 °C enhances the crystalline structure of TiO_2/Zn thin films, as observed in similar studies [39]. The presence of surface hydroxyl groups suggests improved photocatalytic and hydrophilic properties, making these films suitable for environmental remediation and surface engineering applications [40]. The comparative FTIR analysis of pristine, 300 °C, and 400 °C annealed TiO_2/Zn thin films reveals notable differences in structural characteristics. The 400 °C annealed sample demonstrates superior structural integrity, with well-defined Ti–O and Zn–O bonds and reduced organic impurities, as evidenced by sharper and more pronounced peaks. These features suggest enhanced photocatalytic and hydrophilic properties, consistent with prior research on high-temperature annealed composite films [41]. Conversely, the 300 °C annealed sample retains more organic and carbonate residues, while the pristine sample exhibits weaker structural integration. Thus, the 400 °C annealed TiO_2/Zn thin film offers optimal properties for advanced applications in photocatalysis, optoelectronics, and surface modification.

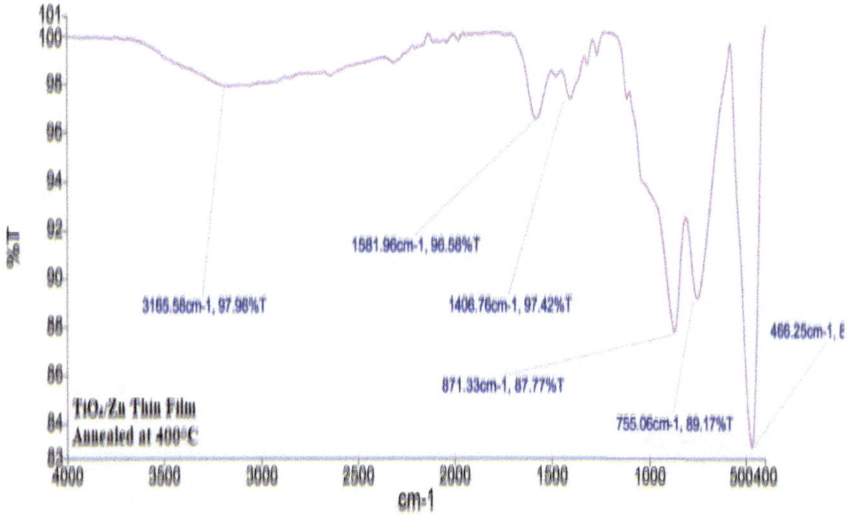

Fig. 5. FTIR spectrum of the TiO_2/Zn thin film annealed at 400 °C.

3.4 AFM Analysis of TiO_2/Zn Thin Films

Figure 6 illustrates the AFM images of TiO_2/Zn thin films (a) pristine, (b) annealed at 300 °C, and (c) at 400 °C. Atomic Force Microscopy (AFM) is a vital technique for examining surface morphology and topography of thin films with nanometer precision. In this study, the surface features of TiO_2/Zn thin films were assessed in their pristine state and after annealing at 300 °C and 400 °C. The 2D and 3D AFM images revealed significant changes in surface morphology and roughness with increasing annealing temperature, shedding light on the structural evolution of the films. The AFM images of the pristine TiO_2/Zn thin films displayed a relatively smooth surface with uniform distribution and minor granular structures. The 2D AFM scan depicted a homogeneous surface, while the 3D topography revealed minor variations, likely due to the presence of organic residues or impurities that contribute to the film's smoothness. These surface characteristics are typical of TiO_2/Zn composites and are slightly influenced by ambient conditions during deposition. Previous studies have also reported similar smooth surfaces with minimal roughness in TiO_2 thin films [42]. Annealing at 300 °C led to a notable increase in surface roughness, as evident in the AFM images. The 2D scans showed larger and more defined grain boundaries, while the 3D AFM images revealed enhanced surface texture, indicating crystallization and structural development within the composite material. This suggests the formation of metal-oxide bonds facilitated by thermal treatment, which improves the crystallinity of TiO_2 and Zn phases [34]. Additionally, the increase in grain size of 300 °C annealed films was larger, reflecting thermal activation of the TiO_2/Zn composite structure. These changes are advantageous for photocatalytic and optoelectronic applications. Similar observations have been reported in studies on TiO_2 films, where crystallization enhances surface roughness and morphology [35]. At 400 °C, the AFM images revealed a significantly rougher surface with well-defined grains and the highest surface roughness among the three conditions. The 2D AFM scans highlighted a highly crystalline surface with pronounced grain boundaries, indicative of a more ordered structure resulting from higher annealing temperatures. The 3D AFM topography showed extensive three-dimensional grain growth, characteristic of TiO_2 films annealed at elevated temperatures. This enhanced roughness is attributed to the increased crystallization of TiO_2 and ZnO phases at 400 °C, which promotes the formation of distinct nanoparticle aggregates [43]. The greater surface roughness observed at this temperature provides an increased surface area and active sites, making the material highly suitable for applications in photocatalysis and energy storage. These findings are consistent with previous studies on ZnO-TiO_2 composite thin films annealed at high temperatures [44].

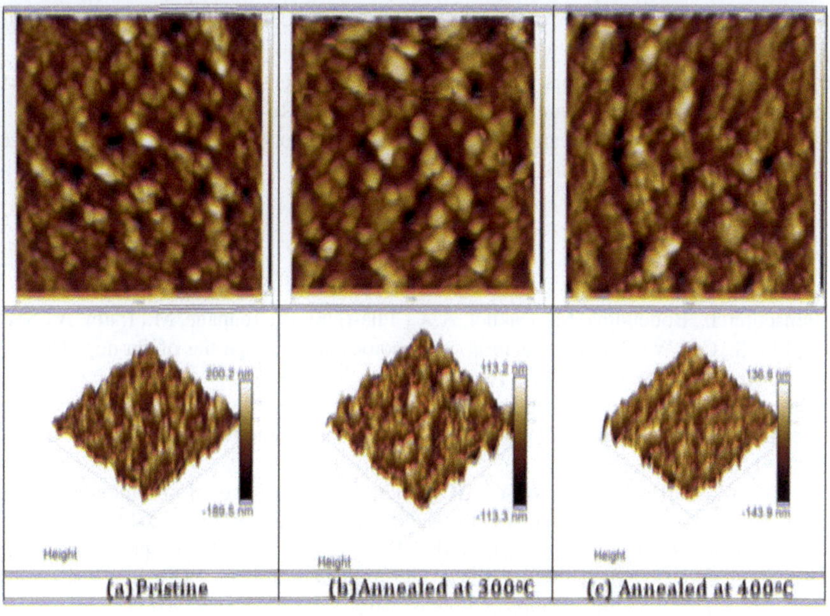

Fig. 6. AFM images of TiO_2/Zn thin films.

4 Conclusion

This work focused on the fabrication of TiO_2/Zn thin films using the d.c. sputtering technique. The influence of annealing temperature on their structural and optical properties was thoroughly examined. XRD analysis verified the formation of the anatase phase in TiO_2 thin films. Diffraction peaks corresponding to the (101), (111), (004), (200), (211), and (220) planes of the anatase lattice were observed. The intensity of these anatase peaks showed a slight increase after annealing, indicating improved crystallinity. Fluorescence spectroscopy revealed that annealing significantly enhances the optical properties of TiO_2/Zn thin films. The pristine TiO_2/Zn thin films exhibited a broad emission peak in the visible region, centered around 420 nm, typically endorsed to recombination of electron and hole pairs. Annealing at 300 °C induced notable changes in the fluorescence spectrum, characterized by a decrease in intensity of broad emission peaks. Further annealed at 400 °C, the emission peak at 420 nm was significantly declined, and a new peak also emerged at around 480 nm. This new shift is associated with enhanced excitonic emission resulting from improved crystallinity. FTIR analysis demonstrated a clear enhancement in crystalline structure and surface hydroxyl group density in films annealed at 400 °C, suggesting enhanced photocatalytic and hydrophilic properties. AFM analysis further supported these observations, showing that annealing significantly influences the surface morphology and roughness of the films. Pristine films displayed a smooth surface with minimal roughness, whereas annealing induced crystallization and increased surface roughness. This study emphasizes the importance of annealing temperature in modifying the structural and optical properties of TiO_2/Zn thin films. It identifies 400 °C as the ideal temperature for enhancing their performance. This makes the films suitable for a wide range of advanced optoelectronic applications.

Acknowledgements. The authors, with special acknowledgment to Prashant Meena, express their gratitude to UGC for financial support provided under the JRF scheme. They also extend their appreciation to the Material Research Centre at MNIT Jaipur for providing access to experimental facilities during this research work.

References

1. Bensouici, F., Bououdina, M., Dakhel, A.A., Tala-Ighil, R., Tounane, M., Iratni, A., Sourier, T., Liu, S., Cai, W.: Optical, structural, and photocatalysis properties of Cu-doped TiO_2 thin films. Appl. Surf. Sci. **395**, 110–116 (2017)
2. Dubey, R.S., Ganesan, V.: Reflectance modulation using SiO_2/TiO_2 multilayer structures prepared by sol-gel spin coating process for optical applications. Superlattice. Microst. **111**, 1099–1103 (2017)
3. Kim, W.-R., Park, H., Choi, W.-Y.: TiO_2 micro-flowers composed of nanotubes and their application to dye-sensitized solar cells. Nanoscale Res. Lett. **9**, 1–10 (2014)
4. Jangid, M.K., Sharma, S.S., Ray, J., Yadav, D.K., Lal, C.: Structural, optical and electrical characterizations of Mg/Ti/Ni multilayer thin films deposited by DC magnetron sputtering for hydrogen storage. Int. J. Hydrog. Energy. **48**(96), 37921–37929 (2023)
5. Jangid, M.K., Sharma, S.S., Ray, J., Jangid, S.: Effect of annealing and hydrogenation on optical and electrical properties of DC sputtered Mg-Ni bilayer thin films. Mater. Today Proc. **67**, 847–851 (2022)

6. Kitano, M., Funatsu, K., Matsuoka, M., Ueshima, M., Anpo, M.: Preparation of nitrogen-substituted TiO_2 thin-film photocatalysts by the radio frequency magnetron sputtering deposition method and their photocatalytic reactivity under visible light irradiation. J. Phys. Chem. B. **110**, 25266 (2006)
7. Jung, S.C.: Photocatalytic activities and specific surface area of TiO_2 films prepared by CVD and sol-gel method. Korean J. Chem. Eng. **25**, 364 (2008)
8. Anpo, M., Shima, T., Kodama, S., Kubokawa, Y.: Photocatalysis on TiO_2 surfaces. J. Phys. Chem. **91**, 4305–4310 (1987)
9. Jangid, M.K., Jangid, S.K.: Structural, electrical, and optical properties of Mg-Ni thin films for hydrogen storage applications. Trends Sci. **19**(23), 2067 (2022)
10. Zheng, S.K., Wang, T.M., Hao, W.C., Shen, R.: Structural properties of annealed TiO_2 films. Vacuum. **65**, 155–159 (2002)
11. Nithyadevi, D., Rajendrakumar, R.T.: Synthesis, characterization, and photocatalytic properties of TiO_2-SnO_2 composite nanoparticles. Adv. Mater. Res. **678**, 373–377 (2013)
12. Jangid, M.K., Nehra, S.P., Singh, M.: Preparation and characterization of Mg/Mn bilayer metal hydrides thin films. AIP Conf. Proc. **1393**(1), 311–312 (2011)
13. Lin, Y.T., Weng, C.H., Lin, Y.H., Shiesh, C.C., Chen, F.Y.: Effect of C content and calcination temperature on the photocatalytic activity of C-doped TiO_2 catalyst. Sep. Purif. Technol. **116**, 114–123 (2013)
14. Jangid, M.K., Sharma, S.S., Ray, J.: Study on hydriding kinetics, structural properties, and electrical conductivity of DC magnetron sputtered Mg/Al thin films. Solid State Phenom. **338**, 83–90 (2022)
15. Jangid, M.K., Sharma, S., Meena, V.P., Arya, V.P., Sharma, S.S.: Impact of annealing and hydrogenation on the optical characteristics of ZnSe/Mn and ZnSe/Co DMS thin films prepared via thermal evaporation method. Evergreen. **11**(1), 178–185 (2024)
16. Wang, B., Lu, X.Y., Xuan, J., Leung, M.K.H.: Facile synthesis and photocatalytic disinfection of boron self-doped TiO_2 nanosheets. Mater. Lett. **115**, 57–59 (2014)
17. Yeo, M.K., Kang, M.: The effect of nanoscale Zn-doped TiO_2 and pure TiO_2 particles on Hydra magnipapillata. Mol. Cell. Toxicol. **6**, 9–17 (2010)
18. Nguyen, T.B., Hwang, M.J., Ryu, K.S.: Effects of synthesis conditions on TiO_2 films. Bull. Korean Chem. Soc. **33**, 243–247 (2012)
19. Niaki, A.H.G., Bakhshayesh, A.M., Mohammadi, M.R.: Photocatalytic performance of TiO_2 films for solar applications. Sol. Energy. **103**, 210–222 (2014)
20. Paul, T.C., Podder, J.: Synthesis and characterization of Zn-incorporated TiO_2 thin films: impact of crystallite size on X-ray line broadening and bandgap tuning. Appl. Phys. A. **125**, 818 (2019)
21. Zhang, L., Sun, W., Liu, Y.: Photoluminescence properties of TiO_2/ZnO nanocomposite films. J. Lumin. **227**, 117457 (2020)
22. Wang, M., Chen, J., Zhang, L., Liu, X.: Enhancement of optical properties of TiO_2 thin films by annealing for photocatalytic applications. Appl. Surf. Sci. **441**, 146–152 (2018)
23. Zhou, Y., Sun, Q., Chen, W., Huang, Y.: Influence of annealing temperature on photoluminescence of TiO_2/ZnO nanocomposite films. Mater. Lett. **243**, 148–153 (2019)
24. Sharma, S., Gupta, N., Singh, R.: Effect of thermal annealing on the optical properties of TiO_2 thin films. J. Mater. Sci. Mater. Electron. **28**(6), 5063–5071 (2017)
25. Chen, Y., Wang, L., Zhang, H., Zhao, F.: Crystallization effects on the photoluminescence of TiO_2/ZnO thin films. J. Appl. Phys. **129**(3), 034301 (2021)
26. Swami, S.K., Khan, J.I., Dutta, V., Lee, J., Laquai, F., Chaturvedi, N.: Spray-deposited aluminum-doped zinc oxide as an efficient electron transport layer for inverted organic solar cells. ACS Appl. Energy Mater. **6**(5), 2906–2913 (2023)
27. Kumar, A., Swami, S.K., Dutta, V.: The role of electric field during spray deposition on fluorine doped tin oxide film. J. Alloys Compd. **588**, 546–550 (2014)

28. Liu, X., Zhou, M., Zhang, K., Wu, J.: Optical and structural properties of TiO_2/ZnO nanocomposite films annealed at different temperatures. Opt. Mater. Express. **9**(2), 736–745 (2019)
29. Yao, Y., Li, R., Chen, X., Zhang, M.: Improved photoluminescence properties of TiO_2/ZnO nanocomposite films for photocatalysis. Mater. Chem. Phys. **242**, 122483 (2020)
30. Huang, J., Zhao, L., Sun, J., Lin, W.: Fluorescence characteristics of TiO_2/ZnO nanocomposites synthesized by different annealing temperatures. J. Nanosci. Nanotechnol. **18**(11), 8201–8207 (2018)
31. Li, S., Zhang, W., Yang, N., Wu, X.: Effect of annealing on the optical properties and photocatalytic performance of TiO_2/ZnO thin films. Mater. Sci. Semicond. Process. **138**, 106267 (2022)
32. Jangid, M.K., Nehra, S.P., Singh, M.: Optical, electrical, and Raman properties of annealed hydrogenated Mg/Co bilayer thin films. J. Nano-Electron. Phys. **3**(1), 460–468 (2011)
33. Swami, S.K., Kumar, N., Radu, D.R., Cho, S.W., Lee, J.: Advanced materials for energy storage and conversion. ACS Appl. Energy Mater. **6**(16), 8599–8606 (2023)
34. Kumar, A., Singh, A., Kumar, D., Garg, A., Dutta, V., Swami, S.K., Chaturvedi, N., Kumar, M.: Spray deposited carbon nanotube embedded ZnO as an electron transport layer in inverted organic solar cells. Hybrid Adv. **4**, 100088 (2023)
35. Kumar, R., Pandey, R., Singh, K.: FTIR analysis of surface-functionalized ZnO/TiO_2 nanocomposites. J. Mater. Sci. Appl. **15**, 23–30 (2022)
36. Sharma, S., Gupta, N., Verma, P., Singh, R.: Structural and vibrational properties of TiO_2 nanostructures for photocatalytic applications. Spectrochim. Acta A Mol. Biomol. Spectrosc. **256**, 1193–1200 (2021)
37. Mukherjee, P., Das, A., Roy, S., Banerjee, K.: FTIR spectroscopic analysis of metal-oxide nanocomposites. J. Non-Cryst. Solids. **580**, 121389 (2022)
38. Tan, Y., Li, F., Wang, Z., Chen, H.: Annealing effects on the photocatalytic performance of TiO_2/ZnO thin films. Catal. Commun. **115**, 26–32 (2018)
39. Jiang, Y., Wang, L., Zhang, X., Liu, J.: Structural and photocatalytic properties of TiO_2/ZnO films annealed at high temperatures. Surf. Coat. Technol. **387**, 125556 (2020)
40. Zhang, X., Yang, W., Li, C., Wu, Q.: Surface engineering of TiO_2 films for photocatalytic applications. Surf. Sci. Rep. **74**(5), 193–211 (2019)
41. Singh, R., Patel, A., Sharma, S., Gupta, V.: Synthesis and characterization of TiO_2/ZnO thin films for environmental applications. Mater. Sci. Eng. B. **270**, 115160 (2021)
42. Swami, S.K., Barman, B., Kumar, A., Dutta, V.: Multilayer and thin transparent conducting oxide fabrication using RF magnetron sputtering on flexible substrates. Org. Mater. **5**(01), 59–65 (2023)
43. Li, X., Zhou, H., Zhang, M., Wu, S.: Surface hydroxyl groups on TiO_2 nanostructures: an FTIR study. Mater. Res. Bull. **149**, 121398 (2023)
44. Wang, T., Chen, J.: FTIR study of carbonate impurities in annealed TiO_2/ZnO composites. Appl. Surf. Sci. **549**, 149784 (2021)

Preparation and Characterization of TiO_2/SnO_2 Thin Films: Influence of MgO Doping on Structural and Optical Properties

Ved Prakash Meena[1], Dhara Singh Meena[1], Madhu Yadav[2], Chhagan Lal[2], S. S. Sharma[3](✉), and M. K. Jangid[1](✉)

[1] Department of Physics, Vivekananda Global University, Jaipur 303012, India
mahesh.jangid@vgu.ac.in
[2] Department of Physics, University of Rajasthan, Jaipur 302004, India
[3] Department of Physics (H & S), Government Mahila Engineering College, Ajmer 305002, India
shyam@gweca.ac.in

Abstract. This study explores the preparation and characterization of TiO_2/SnO_2, with a specific focus on influence of MgO doping on their structural and optical properties. TiO_2/SnO_2 and $TiO_2/SnO_2/MgO$ thin films were prepared by using the E-beam evaporation method. The X-ray diffraction (XRD) analysis of TiO_2/SnO_2 confirmed the crystalline nature of the prepared thin films, revealing the presence of anatase TiO_2 and rutile SnO_2 phases. The structural integrity was found to be influenced by the doping of MgO, as evidenced by variations in peak intensities. MgO doping can lead to development of new phases. The XRD analysis of $TiO_2/SnO_2/MgO$ reported the new phases, such as $MgTiO_3$, Mg_2TiO_4 and Mg_2SnO_4, which are formed due to the interactions between MgO, TiO_2 and SnO_2. The $TiO_2/SnO_2/MgO$ exhibit higher optical absorbance compared to TiO_2/SnO_2 samples. The impact of the MgO layer on the films' structural and optical characteristics is responsible for this variation in absorbance. The estimated band gaps were calculated using the tauc plots, revealing an increase in the bandgap from 4.10 eV to 4.32 eV after MgO doping. FTIR analysis provides valuable insights into structural and chemical properties of thin films. 2D and 3D AFM analysis highlights the significant role of MgO in tailoring surface morphology and topography of TiO_2/SnO_2 thin films and improving functional properties. These findings suggest that the prepared $TiO_2/SnO_2/MgO$ thin films may use in many applications such as optoelectronics, catalysis and environmental remediation.

Keywords: Thin film · E-Beam · XRD · UV-Vis · FTIR · AFM

1 Introduction

Titanium dioxide (TiO_2) is one of the most extensively used semiconductors because of its exceptional physiochemical characteristics, includes high availability, biocompatibility, chemical stability and corrosion resistance [1]. These attributes have made TiO_2

S. Kumari et al. (Eds.): ICMTP 2024, SPPHY 311, pp. 351–363, 2025.
https://doi.org/10.1007/978-981-96-6795-6_39

a preferred material for a variety of applications. Despite its wide range of uses, there is an ongoing search for novel optical applications that offer enhanced properties or overcome the limitations of TiO_2. One significant drawback of TiO_2 is its high electron-hole recombination rate, which reduces its efficiency. To address this, coupling TiO_2 with other semiconductors has been proposed as a strategy to improve its performance [2, 3]. Previous researches have reported that a combination of anatase and a tiny proportion of rutile phases can improve its efficiency compared to pure anatase, which is considered the most efficient TiO_2 phase [4, 5]. This phase coupling is successfully utilized in TiO_2 based optical devices. The certain cations, such as tin (Sn), facilitate the transformation of anatase into rutile. This occurs at much lower temperatures compared to undoped TiO_2 [6]. These cations act as promoters in developing coupled anatase/rutile systems. Such systems show improved efficiency for optoelectronic applications [7]. Tin (Sn), for instance, can be incorporated into the TiO_2 matrix to reduce its bandgap and increase visible light absorption [8]. As Sn^{4+} ions have a similar ionic radius and crystal structure to Ti^{4+}, they can substitute Ti^{4+} in the crystal lattice, altering the optical and electrical properties of TiO_2 and improving its optical behavior [9]. SnO_2, a material with high electron mobility and significant third-order nonlinear optical susceptibilities, is also a promising candidate for enhancing TiO_2 efficiency for optical devices [10]. SnO_2 exhibits unique characteristics, such as a large bandgap, oxygen vacancies, high carrier density, and these characteristics make it adaptable for various applications, including gas sensors, transparent electrodes for solar cells [11–13]. Moreover, SnO_2 is widely used in various industrial applications, such as heat-reflecting mirrors, varistors, electrochromic windows and catalysis [14, 15]. The preparation of fine SnO_2 thin films is crucial for advancing both basic research and industrial applications [16, 17]. Magnesium oxide (MgO) and titanium dioxide (TiO_2) are both widely recognized for their exceptional properties. These are making it suitable for use in high-temperature environments and applications requiring insulation and resistance to heat [18, 19]. TiO_2 stands out due to its excellent optical, electrical, and dielectric properties. It also exhibits remarkable catalytic performance. These attributes have made it valuable in advanced applications. Notable uses include optoelectronic devices and energy-related technologies like solar cells [20–25]. The incorporation of MgO into TiO_2 and SnO_2 systems has been shown to enhance the optoelectronic properties of these materials. MgO doping can direct to develop the new phases $MgTiO_3$ or Mg_2SnO_4, which may improve the absorption in visible range and reduce electron-hole re-combination rate. This doping not only improves the overall efficiency of TiO_2 and SnO_2based optoelectronic devices but also stabilizes the crystal structure and enhances the material's resistance to degradation under harsh environmental conditions. Therefore, the synergy between MgO, TiO_2 and SnO_2 creates a composite material suitable for environmental and energy applications. This research explores the fabrication of TiO_2/SnO_2 and TiO_2/SnO_2/MgO mixed metal oxide thin films and examines their structural and optical characteristics. The results demonstrate the potential for enhancing the performance of optoelectronic devices by utilizing these mixed metal oxide thin films, with MgO doping offering promising improvements in their overall functionality.

2 Experimental

The $TiO_2/SnO_2$and $TiO_2/SnO_2/MgO$ thin films of thickness 300 nm were produced onto the glass substrates using the e-beam technique. High-purity oxide materials, TiO_2 (99.99%), SnO_2 (99.98%), and MgO (99.98%), were used for the film preparation. To ensure uniform mixing, high-energy planetary ball milling was employed, with the oxides used in equal weight ratios. The milling process was conducted for 5 hours to achieve a homogenous mixture. After milling, the resulting powder was dried and compressed into pellets under a biaxial pressure of 10 tons. The resulting nano-structured thin films were deposited onto the substrates using the e-beam deposition method (Model BC-300 HHV). During deposition, the vacuum chamber was evacuated using rotary and diffusion pumps, achieving a base pressure in the order $\times 10^{-5}$ mbar. The substrates were put at room temperature and the space between the substrate and the target was maintained at 35 cm. X-ray diffraction (XRD) measurements were performed using a Panalytical X'Pert Pro diffractometer at room temperature for phase identification, utilizing CuKα radiation ($\lambda = 1.5406$ Å). The optical characteristics were estimated through UV-Vis absorption spectroscopy (Shimadzu UV-1800 UV-Vis Spectrophotometer) across the visible region. FTIR spectroscopy was conducted using a spectrometer from Perkin Elmer, with a wave-number range from 4000 to 400 cm^{-1}. Surface morphology and topography of the samples were obtained by Atomic Force Microscopy (AFM) with Bruker Multimode Scanning Probe.

3 Results and Discussion

3.1 XRD Analysis

Figure 1 presents the XRD patterns of TiO_2/SnO_2 thin films. For the TiO_2/SnO_2 samples, prominent peaks associated with TiO_2 were observed at $2\theta = 25.47°$, 38.19° and 78.97° corresponding to hkl (101), (004) and (224) planes of anatase-phase TiO_2 (JCPDS No. 21-1272), highlighting its dominant crystalline form within the film. The presence of these distinct peaks emphasizes the strong crystallinity of TiO_2 in the anatase phase, as reported in several studies on TiO_2 thin films [26, 27]. For SnO_2, peaks at $2\theta = 51.97°$, 58.05°, 66.24°, 71.51° and 83.83° are detected, corresponding to the (221), (310), (301), (220) and (302) planes of rutile-phase SnO_2. These are well matched with JCPDS No. 41-1445. These peaks confirm polycrystalline nature of SnO_2, which further verifies the incorporation of SnO_2 into the composite thin film, as demonstrated in previous research [28, 29]. Figure 2 presents the XRD patterns of $TiO_2/SnO_2/MgO$ thin films. Upon doping with MgO, additional peaks emerge in the XRD pattern at $2\theta = 34.04°$ and 43.01°, which corresponds to the (104) and (200) plane of the cubic MgO phase (JCPDS No. 65-0476). This peak confirms the successful incorporation of MgO into the thin film. Furthermore, new peaks appear at 2θ values of 26.78°, 54.95° and 62.10°, which are attributed to the (214), (223) and (440) planes of various magnesium-based phases, including $MgTiO_3$ (JCPDS No. 02-074), Mg_2TiO_4 (JCPDS No. 79-0830), and Mg_2SnO_4 (JCPDS No. 74-2423).

The emergence of these peaks signifies the formation of new phases such as $MgTiO_3$, Mg_2TiO_4 and Mg_2SnO_4, which are formed due to the interactions between MgO, TiO_2,

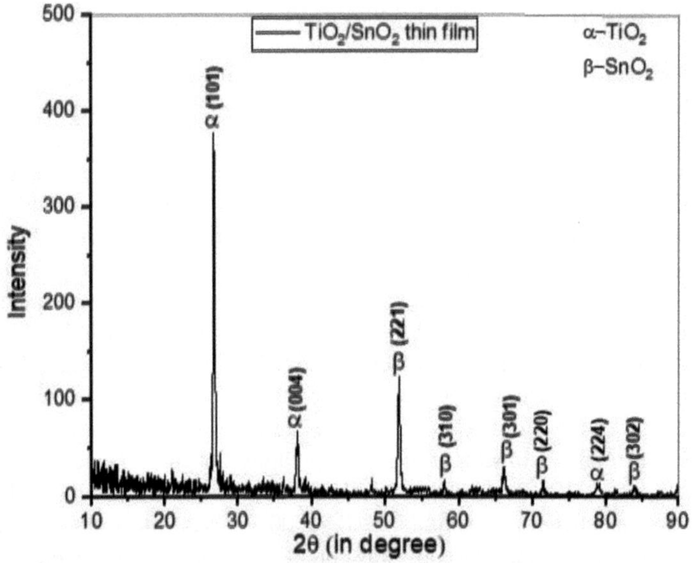

Fig. 1. XRD analysis of TiO_2/SnO_2 thin film.

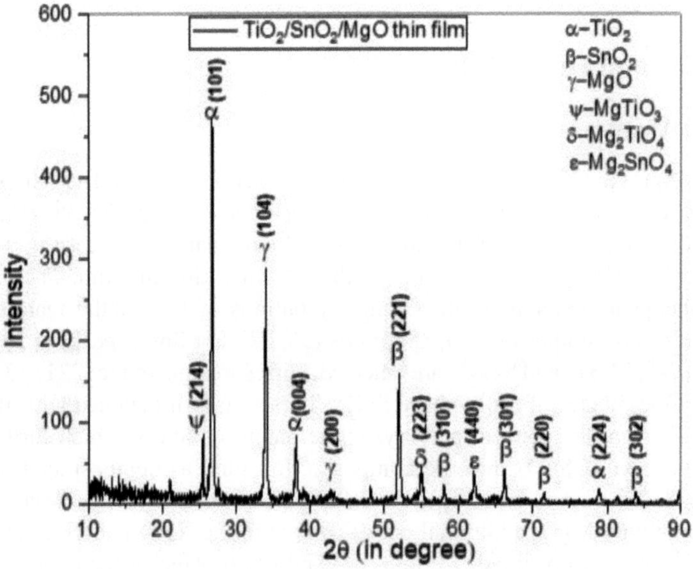

Fig. 2. XRD analysis of $TiO_2/SnO_2/MgO$ thin film.

and SnO_2 consistent with findings in similar studies [30, 31]. These results clearly indicate the presence of individual oxide components and the formation of new phases within the $TiO_2/SnO_2/MgO$ composite thin films. The XRD analysis provides crucial insights into the crystalline phases, confirming the structural evolution upon MgO incorporation, as demonstrated in other research on oxide composite films [32, 33].

3.2 Optical Properties

Figure 3 shows the absorption of TiO_2/SnO_2 and $TiO_2/SnO_2/MgO$ thin films. The data clearly shows that the optical absorbance of $TiO_2/SnO_2/MgO$ is high in respect to TiO_2/SnO_2thin film. This difference in absorbance can be attributed to the affect of MgO layer on structural and electronic characteristics of samples. With a wide bandgap of approximately 7.8 eV, MgO modifies the band formation and reduces defect state within thin films. The addition of MgO helps minimize interfacial defects and suppresses oxygen vacancies, which in turn reduces nonradiative recombination processes. As a result, the $TiO_2/SnO_2/MgO$ system exhibits enhanced optical transmittance and improved light-harvesting capability [34]. Additionally, the lattice-matching properties of MgO enhance the crystallinity of the underlying TiO_2/SnO_2 layers, reducing structural distortions and scattering losses [35]. The presence of MgO also optimizes the energy band alignment at the heterojunction, improving charge transport and light absorption [36].

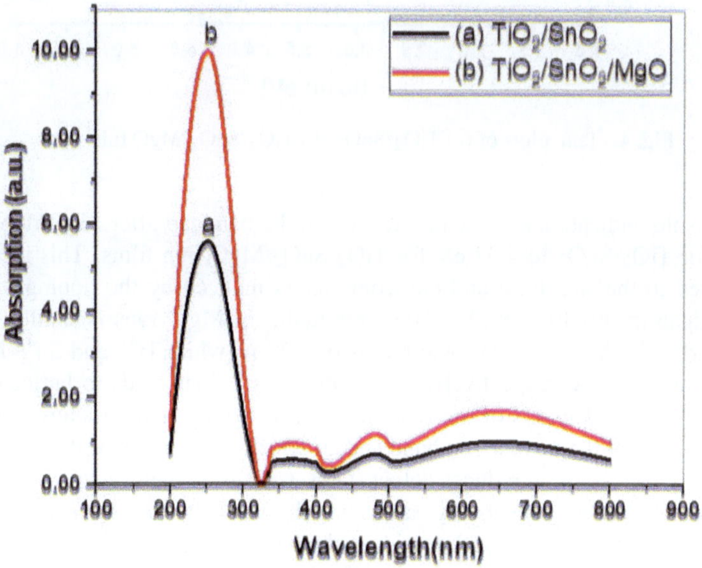

Fig. 3. Absorption of (a) TiO_2/SnO_2 (b) $TiO_2/SnO_2/MgO$ thin films.

Figure 4 shows the energy bandgap values for TiO_2/SnO_2 and $TiO_2/SnO_2/MgO$ thin films. The bandgap energy was calculated using the Tauc relation [37]. This relation is represented by the following equation:

$$\alpha h\nu = A\left(h\nu - E_g\right)^m \quad (1)$$

where h-Planck's constant, ν-photon frequency, α-absorption coefficient derived from absorbance data, E_g-the bandgap energy, and m is the power factor of the transition mode. For direct transitions, m equals 2 was used to calculate the bandgap.

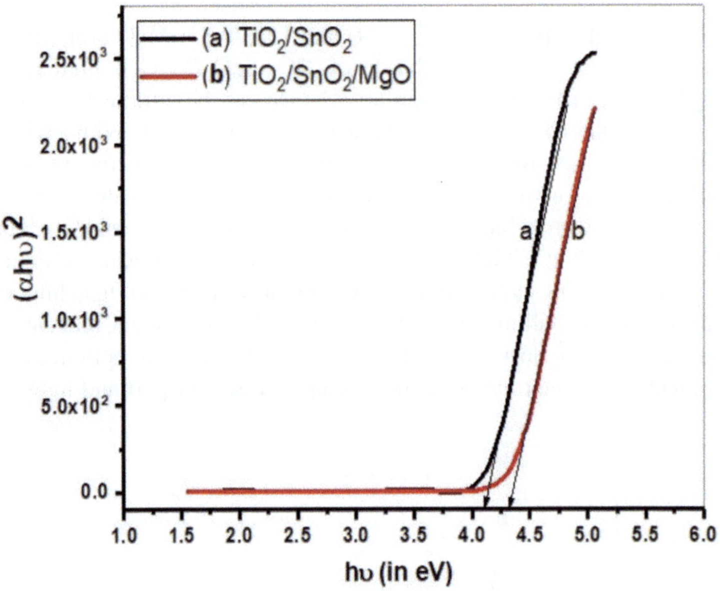

Fig. 4. Tauc plots of (a) TiO_2/SnO_2 (b) $TiO_2/SnO_2/MgO$ thin films.

The results indicate a significant increase in the bandgap after MgO doping, from 4.10 eV for TiO_2/SnO_2 to 4.32 eV for $TiO_2/SnO_2/MgO$ thin films. This increase can be endorsed to the structural and electronic alters induced by the doping of MgO, a wide-bandgap material (~7.8 eV). The ionic radius of Mg^{2+} ions is smaller than that of Ti^{4+} and Sn^{4+}. Mg^{2+} has an ionic radius of 0.72 Å, while Ti^{4+} and Sn^{4+} have radii of 0.74 Å and 0.83 Å, respectively. Substituting Mg^{2+} ions leads to lattice distortion and induces strain. This distortion modifies the electronic band structure, resulting in an up-ward shift of the conduction band and an effective widening of the bandgap [30]. Additionally, MgO doping reduces defect states, such as oxygen vacancies, which contribute to mid-gap energy levels. By passivating these defects, the MgO layer minimizes sub-bandgap absorption and further widens the bandgap [34, 38]. MgO also enhances the crystallinity of the films, reducing localized states at grain boundaries, which leads to an increase in the optical bandgap [36]. The Burstein-Moss effect, where increased carrier concentration due to MgO doping shifts the fermi level into the conduction band, may also contribute to the apparent increase in the bandgap [39]. A wider bandgap and controlled absorption properties make MgO doped TiO_2/SnO_2films promising for applications in UV photodetectors or transparent conducting oxides in optoelectronic devices.

3.3 FTIR Analysis

Fourier Transform Infrared Spectroscopy (FTIR) was used to study the fabricated thin films. It helped in analyzing their chemical bonding. Additionally, the structural properties were examined through this technique. The FTIR spectra for both samples reveal key peaks associated with the vibrational modes of functional groups and chemical bonds present in the films. Figure 5 represent the FTIR spectra of TiO_2/SnO_2 thin film. The peak at 2924.50 cm^{-1} shows to C–H stretching vibrations. The high transmittance of this peak suggests minimal organic residue [40]. The peak at 1632.41 cm^{-1} is associated with bending vibrations of hydroxyl (–OH) groups [41]. The strong absorption at 904.83 cm^{-1} is endorsed to Ti–O stretching vibrations within the TiO_2 lattice, confirming the existence of titanium dioxide in the sample [42]. The peak observed at 762.82 cm^{-1} corresponds to Sn–O stretching vibrations. This is a characteristic feature of the SnO_2 lattice, confirming SnO_2 as a major component of film [43]. The peaks at 536.05 cm^{-1} and 435.30 cm^{-1} are indicative of mixed vibrational modes involving Ti–O–Sn bonds, suggesting interactions and bonding between TiO_2 and SnO_2 in the composite structure [44]. These interactions are crucial for enhancing the functional properties, such as photocatalytic and optoelectronic performance.

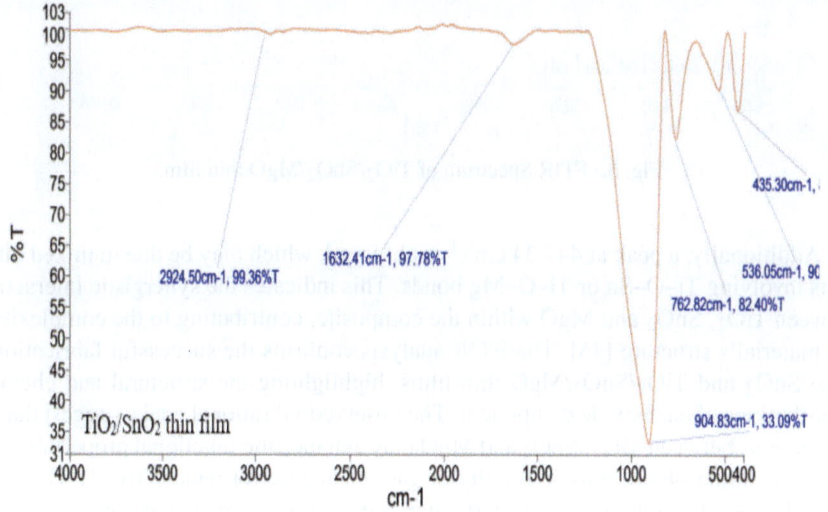

Fig. 5. FTIR Spectrum of TiO_2/SnO_2 thin film.

Figure 6 displays the FTIR spectra of $TiO_2/SnO_2/MgO$ thin film. The spectrum reveals a peak at 1629.94 cm^{-1}, which is endorsed to the bending vibrations. The high transmittance at this wavenumber suggests that the hydrophilic nature of the surface is not overly dominant, indicating minimal water absorption in the film [40, 41]. The intense peak observed at 910.70 cm^{-1} corresponds to Ti–O stretching vibrations within the TiO_2 lattice, confirms the existence of TiO_2 in thin film [42]. The lower transmittance at this wavenumber reflects strong absorption, further validating the incorporation of TiO_2 as a primary component of the material. At 764.07 cm^{-1}, the peak corresponds to Sn–O stretching vibrations, which confirms the incorporation of SnO_2 into the thin film. The moderate transmittance at this wavenumber indicates a significant presence of SnO_2 within the composite structure [43]. The peak at 548.04 cm^{-1} is associated with Mg–O stretching vibrations, confirming the successful doping of MgO into the thin film. The high transmittance at this peak suggests that MgO is incorporated in a controlled amount, potentially helping to tailor the properties of film [44].

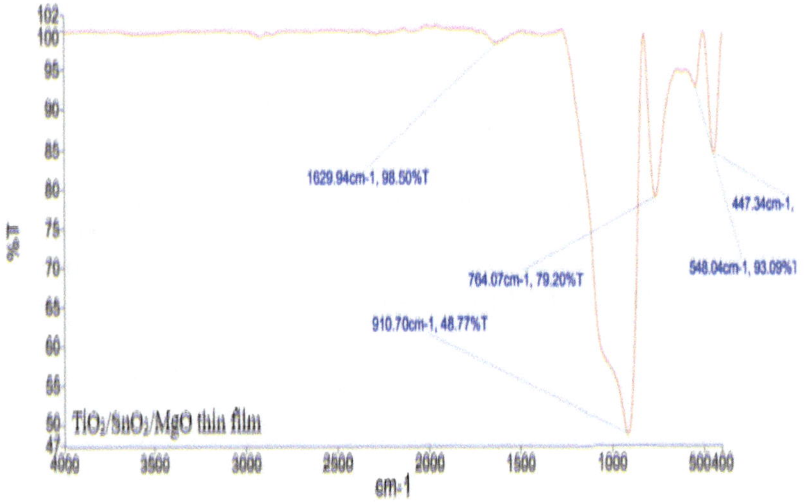

Fig. 6. FTIR Spectrum of $TiO_2/SnO_2/MgO$ thin film.

Additionally, a peak at 447.34 cm^{-1} is observed, which may be due to mixed vibrations involving Ti–O–Sn or Ti–O–Mg bonds. This indicates the synergistic interactions between TiO_2, SnO_2 and MgO within the composite, contributing to the complexity of the material's structure [45]. The FTIR analysis confirms the successful fabrication of TiO_2/SnO_2 and TiO_2/SnO_2/MgO thin films, highlighting the structural and chemical contributions of each oxide component. The observed vibrational peaks suggest that the interactions between TiO_2, SnO_2 and MgO may enhance the functional properties of the films, including photocatalytic activity or gas sensing performance. The Ti–O–Sn and Ti–O–Mg bonding interactions, in particular, suggest improved charge carrier dynamics, material stability and overall performance in relevant applications.

3.4 AFM Analysis

AFM is an essential tool for analyzing the surface morphology and topography of thin films, providing nanometer-scale resolution. Figure 7 shows the 2D and 3D AFM images of (a) TiO_2/SnO_2 and (b) TiO_2/SnO_2/MgO thin films. The AFM study of TiO_2/SnO_2 thin films demonstrates a granular and uniform surface. These films consist of densely packed nanostructures that are closely associated with their optical clarity and efficient optical properties [46–48].

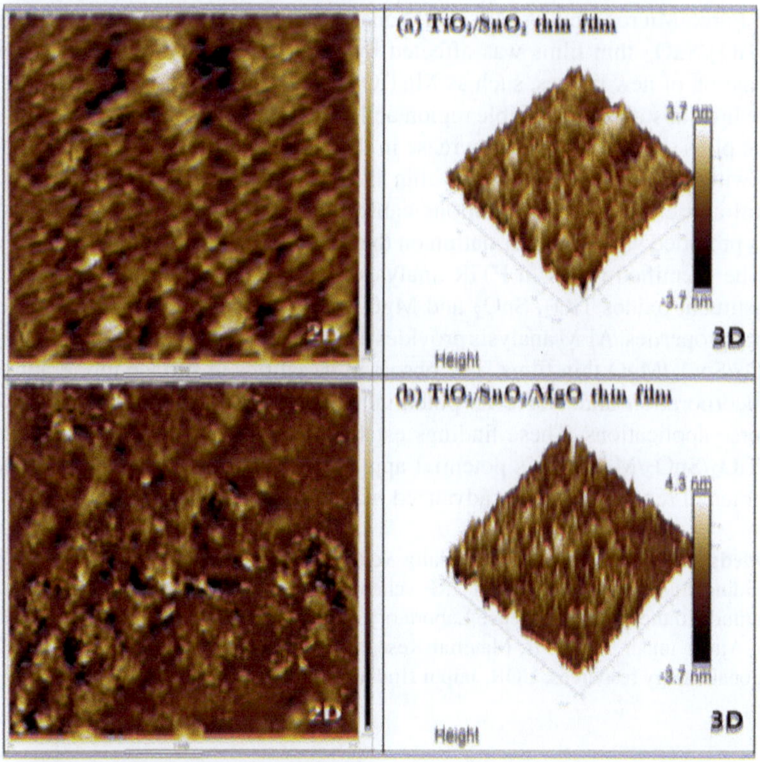

Fig. 7. AFM Images of (a) TiO_2/SnO_2 and (b) $TiO_2/SnO_2/MgO$ thin films.

The addition of MgO into the TiO_2/SnO_2 matrix notably modifies the surface properties of the films. The AFM image of $TiO_2/SnO_2/MgO$ thin film indicates that MgO reduces surface roughness by promoting grain boundary diffusion and enabling the formation of smaller and more uniform grains. This smoothing effect arises due to MgO's stabilizing role in the film structure, which enhances both mechanical and optical properties [49, 50]. Improved surface uniformity minimizes defects and enhances interfacial contact, making $TiO_2/SnO_2/MgO$ films particularly advantageous for optoelectronic applications [51, 52]. Three-dimensional AFM imaging further highlights the impact of MgO, showcasing compact nano-grains with spherical or irregular shapes. The integration of TiO_2, SnO_2 and MgO layers contributes to improved film stability and adhesion, especially under elevated temperatures. This combination is beneficial for applications such as solar cells, gas sensors and photocatalysts, where surface properties critically influence performance [53].

4 Conclusion

The TiO_2/SnO_2 and $TiO_2/SnO_2/MgO$ thin films were prepared using the e-beam deposition technique. Their structural and optical properties were analyzed using X-ray diffraction (XRD), UV-Vis spectroscopy, Fourier Transform Infrared (FTIR) spectroscopy, and

Atomic Force Microscopy (AFM). XRD analysis revealed that the structural integrity of the TiO_2/SnO_2 thin films was affected by the MgO doping, which contributed to the formation of new phases, such as $MgTiO_3$, Mg_2TiO_4 and Mg_2SnO_4. These phases enhance light absorption in visible region and help reduce electron-hole recombination. The tauc plots showed a notable increase in the bandgap from 4.10 eV to 4.32 eV after doping with MgO in the TiO_2/SnO_2 thin films. This change can be attributed to the structural and electronic modifications caused by the MgO incorporation. The FTIR analysis provided valuable information on the chemical structure of the synthesized thin films. The identified peaks in FTIR analysis confirm the presence and interaction of the constituent oxides TiO_2, SnO_2 and MgO, and validating the film's production and structural properties. AFM analysis provides comprehensive surface study of TiO_2/SnO_2 and TiO_2/SnO_2/MgO thin films. The observed variations in surface topography due to MgO incorporation underscore its potential to optimize the properties of these films for diverse applications. These findings establish a foundation for further exploration of the TiO_2/SnO_2/MgO film's potential applications in optoelectronics, catalysis, and environmental remediation or in advanced material applications.

Acknowledgements. The authors (Specially Ved Prakash Meena) are highly thankful to UGC for providing financial support under JRF scheme to carry out this work. They also express their gratitude to the Material Science Laboratory developed at Government Mahila Engineering College, Ajmer under TEQIP-II, Material Research Centre, MNIT Jaipur and Center for non-conventional energy resources, UOR, Jaipur (India) for providing experimental and measurement facilities.

References

1. Coronado, J.M., Fresno, F., Hernández-Alonso, M.D., Portela, R., Suárez, S., García Rodríguez, S., de la Peña O'Shea, V.A.: Design of advanced photocatalytic materials for energy and environmental applications. Springer (2013)
2. Khaki, M.R.D., Shafeeyan, M.S., Raman, A.A.A., Daud, W.M.A.W.: Application of doped photocatalysts for organic pollutant degradation - a review. J. Environ. Manag. **198**, 78–94 (2017)
3. Kumaravel, V., Mathew, S., Bartlett, J., Pillai, S.C.: Photocatalytic hydrogen production using metal doped TiO2: a review of recent advances. Appl. Catal. B Environ. **244**, 1021–1064 (2019)
4. Yuangpho, N., Le, S.T.T., Treerujiraphapong, T., Khanitchaidecha, W., Nakaruk, A.: Enhanced photocatalytic performance of TiO_2 particles via effect of anatase–rutile ratio. Phys. E Low. Dimens. Syst. Nanostruct. **67**, 18–22 (2015)
5. Hurum, D.C., Agrios, A.G., Gray, K.A., Rajh, T., Thurnauer, M.C.: Explaining the enhanced photocatalytic activity of Degussa P25 mixed-phase TiO2 using EPR. J. Phys. Chem. B. **107**, 4545–4549 (2003)
6. Hanaor, D.A.H., Sorrell, C.C.: Review of the anatase to rutile phase transformation. J. Mater. Sci. **46**, 855–874 (2011)
7. Maver, K., Arčon, I., Fanetti, M., Emin, S., Valant, M., Štangar, U.L.: Improved photocatalytic activity of anatase-rutile nanocomposites induced by low-temperature sol-gel Sn-modification of TiO2. Catal. Today. **361**, 124–129 (2021)

8. Chenaina, H., Messaadi, C., Jalali, J., Ezzaouia, H.: Study of structural, optical and electrical properties of SnO2 doped TiO2 thin films prepared by a facile Sol-Gel route. Inorg. Chem. Commun. **124**, 108401 (2021)
9. Liu, S., Li, L., Jiang, W., Liu, C., Ding, W., Chai, W.: Crystallinity and morphology-controlled synthesis of SnO_2 nanoparticles for higher gas sensitivity. Powder Technol. **245**, 168–173 (2013)
10. Wang, H., Song, L., Yu, L., Xia, X., Bao, Y., Lourenco, M., Homewood, K., Gao, Y.: Charge transfer between Ti^{4+}, Sn^{4+} and Pt in the tin doped TiO_2 photocatalyst for elevating the hydrogen production efficiency. Appl. Surf. Sci. **581**, 152202 (2022)
11. Jangid, M.K., Sharma, S., Meena, V.P., Arya, V.P., Sharma, S.S.: Impact of annealing and hydrogenation on the optical characteristics of ZnSe/Mn and ZnSe/Co DMS thin films prepared via thermal evaporation method. Evergreen. **11**(1), 178–185 (2024)
12. Zhang, H., He, Q., Zhu, X., Pan, D., Deng, X., Jiao, Z.: Surfactant-free solution phase synthesis of monodispersed SnO_2 hierarchical nanostructures and gas sensing properties. CrystEngComm. **14**(9), 3169 (2012)
13. Jangid, M.K., Nehra, S.P., Singh, M.: Optical, electrical, and Raman properties of annealed hydrogenated Mg/Co bilayer thin films. J. Nano-Electron. Phys. **3**(1), 460–468 (2011)
14. Gu, F., Wang, S.F., Lü, M.K., Zhou, G.J., Xu, D., Yuan, D.R.: Photoluminescence properties of SnO_2 nanoparticles synthesized by sol-gel method. J. Phys. Chem. B. **108**(24), 8119–8123 (2004)
15. Sain, S., Kar, A., Patra, A., Pradhan, S.K.: Structural interpretation of SnO_2 nanocrystals of different morphologies synthesized by microwave irradiation and hydrothermal methods. CrystEngComm. **16**(6), 1079–1090 (2014)
16. Wang, H.E., Xi, L.J., Ma, R.G., Lu, Z.G., Chung, C.Y., Bello, I., Zapien, J.A.: Microwave-assisted hydrothermal synthesis of porous SnO_2 nanotubes and their lithium ion storage properties. J. Solid State Chem. **190**, 104–110 (2012)
17. Chetri, P., Choudhury, A.: Investigation of optical properties of SnO_2 nanoparticles. Phys. E Low Dimens. Syst. Nanostruct. **47**, 257–263 (2013)
18. Filipovic, S., Obradovic, N., Kosanovic, D., Pavlovic, V., Djordjevic, A.: Sintering of the mechanically activated MgO-TiO_2 system. J. Ceram. Process. Res. **14**(1), 31–34 (2013)
19. Jangid, M.K., Sharma, S.S., Mathur, D., Sharma, Y.C.: Optical, electrical, and structural study of Mg/Ti bilayer thin film for hydrogen storage applications. Mater. Lett. X. **10**, 100076 (2021)
20. Sharma, S.S., Sharma, K., Sharma, V.: Nanographite-TiO_2 photoanode for dye sensitized solar cells. AIP Conf. Proc. **1728**, 020515 (2016)
21. Jangid, M.K., Sharma, S.S., Ray, J., Jangid, S.: Effect of annealing and hydrogenation on optical and electrical properties of DC sputtered Mg-Ni bilayer thin films. Mater. Today Proc. **67**, 847–851 (2022)
22. Photiphitak, C., Rakkwamsuk, P., Muthitamongkol, P., Thanachayanont, C.: Performance enhancement of dye-sensitized solar cells by MgO coating on TiO_2 electrodes. Int. Sch. Sci. Res. Innov. **6**(5), 463–467 (2012)
23. Sharma, S.S., Sharma, K., Singh, R., Srivastava, S., Rana, K.B., Singhal, R.: Natural pigments: origin and applications in dye sensitized solar cells. Mater. Today Proc. **42**, 1744–1748 (2021)
24. Sharma, S.S., Sharma, K., Sahu, J., Ray, J., Gupta, S.K., Dalela, S.: Role of rare-earth oxides, conjugated with TiO_2, in the enhancement of power conversion efficiency of dye sensitized solar cells. Environ. Sci. Pollut. Res. **30**, 98760–98772 (2023)
25. Sharma, K., Sharma, S.S., Ray, J., Chaure, N.B.: Investigation of Ru-complex free natural dye for the application of dye sensitized solar cells. J. Integr. Sci. Technol. **11**(3), 529–529 (2023)
26. Liu, X., et al.: Effect of annealing on the crystallization of TiO_2 films. J. Mater. Sci. **42**(2), 640–645 (2007)

27. Kumar, S., et al.: Optical and structural properties of TiO2 thin films. Thin Solid Films. **518**(10), 2795–2799 (2010)
28. Gharbi, M., et al.: Structural and optical properties of SnO2 thin films: Influence of deposition parameters. J. Mater. Sci. **44**(16), 4312–4319 (2009)
29. Miah, M.M., et al.: Structural and optical properties of SnO_2 thin films deposited by spray pyrolysis. J. Appl. Phys. **110**(9), 093510 (2011)
30. Jangid, M.K., Sharma, S.S., Ray, J., Yadav, D.K., Lal, C.: Structural, optical and electrical characterizations of Mg/Ti/Ni multilayer thin films deposited by DC magnetron sputtering for hydrogen storage. Int. J. Hydrog. Energy. **48**(96), 37921–37929 (2023)
31. Abo, G.S., Hong, Y.K., Park, J., Lee, J., Lee, W., Choi, B.C.: Spin dynamics in ferromagnetic thin films and nanostructures: a time-resolved magneto-optic Kerr effect study. J. Appl. Phys. **107**(9), 093901 (2010)
32. Benzarti, Z., Khalfallah, A.: Recent advances in the development of thin films. Coatings. **14**(1), 87 (2024)
33. Jangid, M.K., Nehra, S.P., Singh, M.: Hydrogenation effect on Mg/Mn multilayer thin films. AIP Conf. Proc. **1349**(1), 685–686 (2011)
34. Sakka, S.: Sol-gel science of oxides. J. Non-Cryst. Solids. **100**, 227–239 (1988)
35. Singh, R.D., Mehta, A.: FTIR study of hydroxyl absorption in metal oxides. Mater. Today Proc. **22**, 451–457 (2021)
36. Fujishima, A., Honda, K.: TiO2 as a photocatalyst. Nature. **238**, 37–38 (1972)
37. Tauc, J.: Optical properties and electronic structure of amorphous semiconductors. Mater. Res. Bull. **3**, 37–46 (1968)
38. Jangid, M.K., Sharma, S.S., Ray, J.: Study on hydriding kinetics, structural properties, and electrical conductivity of DC magnetron sputtered Mg/Al thin films. Solid State Phenom. **338**, 83–90 (2022)
39. Lu, J.G., et al.: SnO_2 thin films and gas sensing properties. Chem. Rev. **110**(10), 5133–5165 (2010)
40. Fleet, M.E.: Vibrational spectroscopy of MgO. Phys. Chem. Miner. **25**(7), 537–545 (1998)
41. Jain, P.K., et al.: Mixed oxide thin films and their FTIR studies. Adv. Mater. Lett. **12**, 15–20 (2019)
42. Li, X., Zhang, Y., Liu, W.: Surface morphology and optical properties of TiO_2-based thin films prepared by electron beam deposition. Appl. Surf. Sci. **514**, 146018 (2020)
43. Kumar, R., Singh, R., Yadav, P.: AFM and XRD analysis of TiO_2/SnO_2 thin films. J. Nanosci. Nanotechnol. **18**(5), 3127–3135 (2018)
44. Chen, Z., Huang, D., Wang, H.: The role of MgO doping in enhancing the properties of TiO_2-based thin films. Mater. Res. Bull. **141**, 111365 (2021)
45. Jha, R., Gupta, N., Sharma, V.: Nanostructured oxide thin films: characterization and applications. Surf. Coat. Technol. **368**, 120–130 (2019)
46. Patra, P., Mishra, S., Verma, K.: Advances in nanostructured oxide thin films: a review. Prog. Mater. Sci. **135**, 101073 (2023)
47. Jangid, M.K., Jangid, S.K.: Structural, electrical, and optical properties of Mg-Ni thin films for hydrogen storage applications. Trends Sci. **19**(23), 2067 (2022)
48. Chaturvedi, N., Swami, S.K., Dutta, V.: Spray deposition of poly(3-hexylthiophene) and [6,6]-phenyl-C_{61}-butyric acid methyl ester blend under electric field for improved interface and organic solar cell characteristics. Thin Solid Films. **598**, 82–87 (2016)
49. Naqvi, S.R., et al.: Structural and optical properties of TiO_2/SnO_2 composite thin films. J. Mater. Sci. **50**(3), 1450–1457 (2015)
50. Swami, S.K., et al.: Effects of electric field during deposition on spray deposited indium-doped zinc oxide films. Prog. Photovolt. Res. Appl. **24**(1), 74–82 (2016)
51. Kumar, A., Swami, S.K., Dutta, V.: The role of electric field during spray deposition on fluorine doped tin oxide film. J. Alloys Compd. **588**, 546–550 (2014)

52. Sertel, B.C., Sonmez, N.A., Kaya, M.D., Ozcelik, S.: Development of MgO:TiO_2 thin films for gas sensor applications. Ceram. Int. **45**(3), 2917–2921 (2019)
53. Hussein, F.M., Abdul Amir, Y.K., AL-Ani, R.R., AL-Mokaram, A.M.: Synthesis and photocatalytic activity of TiO_2-coupled SnO_2 nanoparticles prepared by sol–gel technique. Egypt. J. Chem. **65**(10), 551–559 (2022)

Ultrafast Room Temperature Chemi-Resistive Liquefied Petroleum Gas Sensor Based on TiO_2 Nanorods

Bhogendra Kumar Sah[1], Anita Patel[2], Vishwa Padia[2], Mitesh H. Patel[2], D. K. Dhruv[3], Abhay Dasadia[4], Shikha Varma[5], and Vanaraj Solanki[2](✉)

[1] St. Xavier's College, Kathmandu 44600, Nepal
[2] Dr. K C Patel R and D Centre, Charotar University of Science and Technology (CHARUSAT), Changa, Anand 388421, India
vanarajsolanki.rnd@charusat.ac.in
[3] Natubhai V. Patel College of Pure and Applied Sciences, The Charutar Vidya Mandal (CVM) University, Vallabh Vidyanagar, Anand 388120, India
dhananjay.dhruv@cvmu.edu.in
[4] A. D. Patel Institute of Technology, The Charutar Vidya Mandal (CVM) University, New Vallabh Vidyanagar 388121, Gujarat, India
[5] Institute of Physics, Sachivalaya Marg, Bhubaneswar 751005, Odisha, India

Abstract. Gas sensors play a vital role in making daily life easy and smooth. Researchers have exploited numerous sensor types for applications, but their measuring parameters are distinct. Amongst various sensors, gas sensors are unique in detecting minute gas leakage in industries or at home to avoid major accidents, which can even cause death. Liquefied Petroleum Gas (LPG) is widely used in household and industrial applications, and it is incredibly crucial to monitor it continuously. Here, chemi-resistive sensors, based on TiO_2 nanostructures, have been fabricated to monitor the LPG gas at room temperature. A simple and cost-effective chemical precipitation method was used to grow TiO_2 nanostructures. Grown titania nanostructures have rod-like morphology, with an average diameter of 650 ± 50 nm, and contain mixed rutile and anatase crystallite phases. Interestingly, the fabricated chemi-resistive sensor is highly sensitive towards LPG gas with a sensitivity of ~522% and a reversible and reproducible response with ultrafast response and recovery times. Overall, the results may help the researchers to fabricate TiO_2-based LPG sensors with predefined sensing parameters.

Keywords: Sensors · TiO_2 · Nanostructures · LPG

1 Introduction

Liquefied Petroleum Gas, a mixture of various hydrocarbons like propane, propylene, butylene, isobutene and n-butane, is widely used in comprehensive areas such as industries, agriculture, aerosol propellant, refrigerant, and in vehicles [1–2]. Furthermore, it is widely used as a cooking fuel in several countries. Since it is made up of hydrocarbons,

S. Kumari et al. (Eds.): ICMTP 2024, SPPHY 311, pp. 364–373, 2025.
https://doi.org/10.1007/978-981-96-6795-6_40

it leads to indoor air pollution upon leak, which may cause detrimental health problems in humans [3–5]. One can be exposed to LPG via eye and skin and by inhalation. Long-term exposure to low concentrations of LPG is pernicious and deleterious to the central nervous system and some other health complexes like nosebleeds, rhinitis, oral/nasal ulcerations, conjunctivitis, weight loss, etc. In contrast, short-term exposure to highly concentrated LPG may result in asphyxiation, seizure, coma, heart problems, and even death [3–5]. Therefore, monitoring or sensing LPG in all possible areas is essential for saving human life.

A sensor is a device that, fabricated in a specific way, is based on some particular principle to detect the physical condition of the chemical compounds. These sensors rely on distinct properties of the sensing elements, such as optical, acoustic, mechanical, and electrical. Chemi-resistive sensors proposed by different researchers, based on the electrical conductivity of metal oxides, are flourishing in the plethora of scientific research to enhance many technological inventions [4–7]. Various sensing materials like binary alloys, chalcogenides, metals and their oxides have been extensively explored by multiple means. Amongst different materials, metal oxides are superior for sensing applications regarding the sensor's sensitivity, response time, reversibility, selectivity and stability.

Titanium dioxide (TiO_2), a wide bandgap semiconductor, has captivated the great attention of researchers due to its excellent tunable electronic and optical properties [8, 9]. Flexibility in property modulation has opened a gateway for its utilization in many potential applications such as sensing catalysis, dye-sensitized-solar cells, supercapacitors, sunscreens, rechargeable batteries, etc. [8–11]. It is commonly known that the performance of the materials is strongly dependent on the shape and size of the grown materials in addition to its modified electronic and structural properties. The dependency of materials performance on phase compositions is not well explored but also play a crucial role in device activity [12]. TiO_2 exist in three different forms; rutile, anatase and brookite [13]. Amongst three different polymorphs, rutile and anatase phase of TiO_2 were explored widely by different research groups for gas sensing applications [14–18]. Rutile is the most stable form of TiO_2 and is proven to be useful for high temperature gas sensing whereas applications of the anatase phase, as a gas sensor, are related in the low temperature regime, due to its increased surface-to-volume ratio [12]. Therefore, the co-existence of these phases is expected to have better performance for gas sensing applications due to their synergic effect however is not well explored. These phases co-exist during samples prepared by different synthesis routes like sputtering [19] flame spray synthesis [20] and sol-gel method [21]. Titania nanostructures grown under mild reaction conditions make the synthesis method cost-effective. Therefore, we have utilized a simple chemical precipitation method to synthesize TiO_2 nanostructures at room temperature without using hazardous chemical compounds and acids.

Herein, we report the fabrication of a TiO_2-based chemi-resistive sensor for monitoring LPG gas at an ambient condition. The nanostructures of titanium oxide are grown via a simple chemical precipitation method. The nanorods are gown with an average diameter of 650 ± 50 nm and length 14 ± 2 μm. The fabricated TiO_2-based LPG sensor is highly stable and reproducible, with a high response of 522%. In addition, a fabricated sensor is also ultra-fast responsive, having response and recovery times of 11 and 11.3 s, respectively.

2 Experimental

2.1 Precursors

All reagents utilized in the study are of analytical grade and have been used without further purification.

2.2 Synthesis of TiO_2 Nanorods

Titanium oxide nanorods were synthesized through a simple and cost-effective chemical precipitation route. In a typical synthesis method, titanium tri-chloride ($TiCl_3$) was added dropwise to 50 ml deionized (DI) water under continuous stirring for 2 h, followed by an addition of 20 ml of polyethylene glycol (PEG) in the solution. Subsequently, a dropwise addition of NaOH, 5 M, under continuous stirring leads to a white precipitate in the solution. The formed precipitate was collected through filtration and washed a couple of times with DI water and ethanol. Finally, as-grown powder was dried at 100 °C for 5 h, followed by calcination at 650 °C for 2 h.

2.3 Characterization

The morphology of the grown material was analyzed with a Scanning Electron Microscope (SEM, JSM 6010 LA, Jeol Japan). A Bruker makes X-ray diffractometer (XRD) with Cu Kα source ($\lambda = 0.1542$ nm) was employed for structural characterization.

2.4 LPG Sensing

A home-made gas sensing system was used for the LPG sensing measurements. It consists of a test chamber, sensor and data acquisition system holder. The TiO_2 nanorod-based chemi-resistive sensor was built by glueing silver paste on both ends of the small chunk of TiO_2. Sensor current response was monitored under room temperature conditions, 50% relative humidity and 24 °C temperature through the Keithley 2611A source measure unit. A sample bias of 3 V is employed for all LPG sensing measurements. The sensitivity of the sensor was calculated from the current ratio measured in air to LPG gas exposing condition (Eq. (1)) [22],

$$\text{Sensitivity (S)} = \frac{I_{air}}{I_{gas}} \times 100\% \tag{1}$$

I_{air} and I_{gas} are currents from a sensor in the air and under LPG-exposed conditions, respectively.

3 Results and Discussions

Scanning electron microscopy image of TiO_2 nanostructures, grown via the chemical precipitation method, is shown in Fig. 1. SEM Image displays the formation of rod-like structures with an average diameter of 650 ± 50 nm and length of 14 ± 1 µm. In addition to nanorods, some sheet-like structures are observed in the SEM image. The formed nanorods are very important for many applications, such as gas sensing, catalysis, energy storage, etc., due to their enhanced surface area, which provides many active interaction sites.

Fig. 1. SEM image of TiO_2 nanostructure.

To analyze the grown TiO_2 structurally, the X-ray diffraction pattern was investigated and is presented in Fig. 2. The XRD plot depicts the formation of mixed phases, i.e. anatase and rutile, in the grown TiO_2 sample. The characteristic peaks appearing at positions 25.60^0, 36.01^0, 48.51^0, 52.69^0, 56.04^0 and 61.53^0 are respectively associated with the lattice planes (101), (112), (200), (105), (211) and (213) of anatase TiO_2. In addition, other peaks at 30.40^0, 32.15^0, 33.40^0, 44.04^0, and 58.03^0 correspond to (110), (101), (200), (211) and (220) lattice planes of rutile TiO_2, respectively [23, 24]. Similarly, mixed phases of rutile and anatase TiO_2 have also been observed by different research groups post-annealing of the grown TiO_2 at an elevated temperature, around 600 °C [23–25].

Sensors are crucial to achieve desirable scientific and technological advancement and make daily life easier and smoother. Considering the use of LPG in daily household and industrial applications and the associated risks, we have explored the applicability of the grown TiO_2 nanostructures for LPG sensing. Figure 3. shows the temporal current response curve (I-t) from TiO_2 nanorods obtained via switching between air and LPG

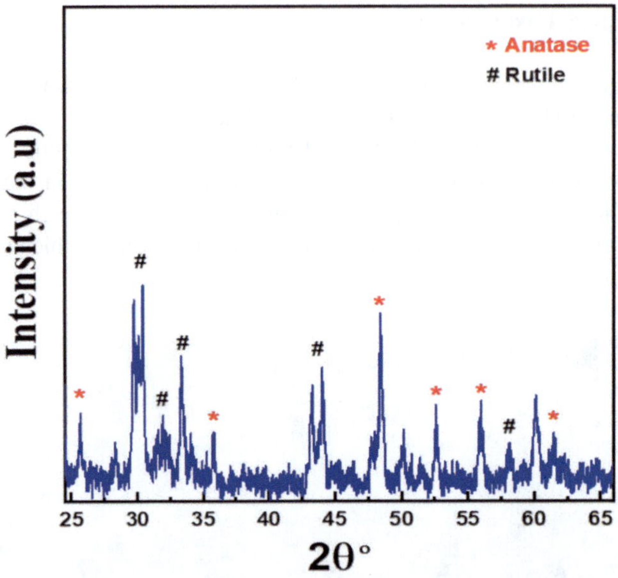

Fig. 2. X-ray diffractogram of TiO_2 Nanostructures.

environment at RT conditions. The response curve depicts the current decrease upon exposure to LPG gas and regaining its initial value under ambient conditions. This proves that the sensor response is reversible and reproducible. The sensitivity of the sensor was calculated using Eq. 1 and is found to be 522%, which is higher than many reported LPG sensors (Table 1). The observed high sensitivity from the fabricated LPG sensor might be due to the co-existence of the rutile and anatase phase, as observed in the XRD spectrum (Fig. 2). In this mixed phase, the band alignment difference between rutile and anatase phase, of TiO_2, play a crucial role in oxygen adsorption and electron transfer to the surface during gas sensing [12, 26–28]. The synergistic interaction between anatase and rutile phases facilitates improved charge separation and surface reactivity, leading to a higher response toward targeted gases. The decrease of sensor current, i.e. rise in resistance, upon exposure to LPG indicates the overall p-type behavior of TiO_2 nanostructures.

Generally, TiO_2 is known to exhibit n-type conducting behaviour; however, several researchers have also observed the existence of p-type conductivity in titania during growth under special conditions, doping, and transition from its n to p-type characteristics under specific environmental conditions [24, 25, 31, 32]. Gas sensing from any semiconducting materials can be explained by modulation of charge carrier density during the interaction of the analyte gas molecules with the sensor surface. At room temperature, the oxygen molecules from the environment interact with the sensor surface and get adsorbed via conversion to ionized oxygen species ($O-$ and O_2-) [33], which leads to the formation of charge accumulation (p-type) or charge depletion (n-type) region near the sensor surface [25, 33]. Upon exposure to LPG gas, the adsorbed oxygen species interact with the hydrocarbons as per Eq. 2 and release electrons in the

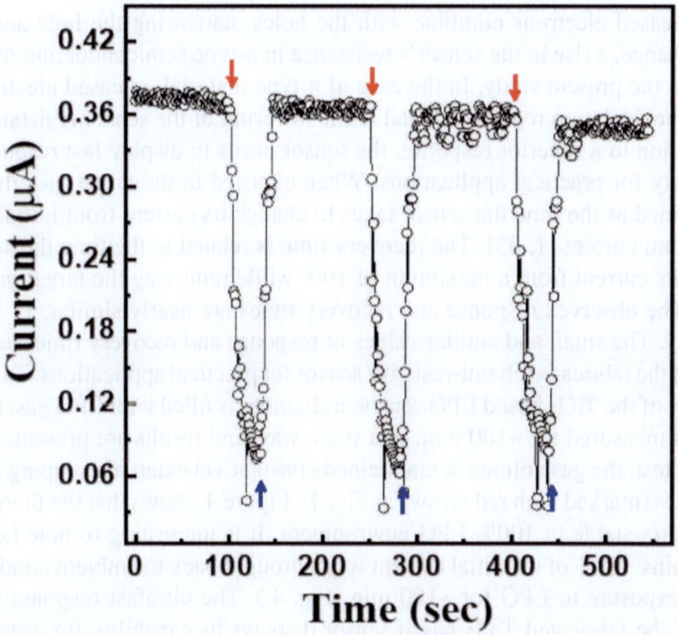

Fig. 3. Room temperature temporal response curve towards LPG sensing from TiO_2 based chemi-resistive sensor. Red arrows represent the insertion of LPG gas in the chamber whereas exhaust of the gas is marked with blue. arrows.

Table 1. Comparison of different metal oxides-based LPG sensors.

Sr. No.	Material	Operating temp.	Sensitivity	References
1	CdS- TiO_2 Thick film	300 K	35%	[14]
2	Pristine TiO_2 nanorods Pd decorated TiO_2 nanorods	598 K	27% 49%	[15]
3	Pristine TiO_2 thin film Irradiated chemically deposited TiO_2 thin film	673 K	7.45% 38.70%	[16]
4	TiO_2 nanowires	673 K	3.9%	[17]
5	Anatase TiO_2 film	698 K	35.80%	[18]
6	ZnO nanostructures	523 K	59%	[29]
7	p-polyaniline/n-ZnO Thin film	RT	81%	[30]
8	TiO_2 nanorods	RT	522%	This work

system [25].

$$C_nH_{2n+2} + O_2^- \rightarrow H_2O + C_nH_{2n} : O + e^- \qquad (2)$$

The released electrons combine with the holes, narrowing the hole accumulation layer and, hence, a rise in the sensor's resistance in p-type semiconducting materials, as observed in the present study. In the case of n-type material, released electrons, as per Eq. 2, fill the depletion region and lead to the lowering of the sensor resistance [25].

In addition to a superior response, the sensor needs to display fast responsivity and high stability for practical applications. When exposed to the target gas, the response time is defined as the time the sensor takes to change its current from initial to 90% of the maximum current [1, 33]. The recovery time is related to the time the sensor takes to change its current from a maximum of 10% while removing the target gas from the chamber. The observed response and recovery times are nearly similar, 11 and 11.3 s, respectively. The small and similar values of response and recovery times advocate the potential of the fabricated chemi-resistive sensor for practical applications. To investigate the stability of the TiO_2 based LPG sensor, a chamber is filled with LPG gas, and current response is measured for ~100 min, in a static way, and results are presented in Fig. 4. During the test, the gas volume is maintained constant via external pumping of the LPG several times (marked with red arrows in Fig. 4). Figure 4. shows that the fabricated LPG sensor is ultra-stable in 100% LPG environment. It is interesting to note here that the sensor regains ~95% of its initial current when brought back to ambient conditions after continues exposure to LPG for ~100 min (Fig. 4.). The ultrafast response and higher stability of the fabricated TiO_2 based sensor displays its capability for crucial sensing applications.

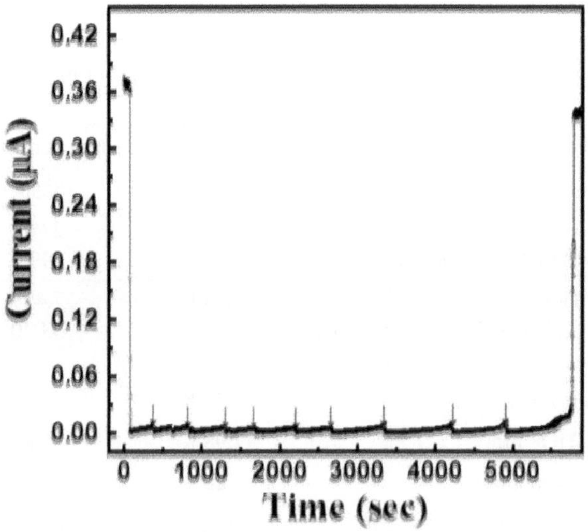

Fig. 4. Room temperature stability measurements from TiO_2 based LPG sensor

4 Conclusion

We have synthesized TiO_2 nanostructures via a simple and cost-effective chemical precipitation method. Grown TiO_2 nanostructures contain mixed crystallite phases of rutile and anatase, as depicted by the structural characterizations. A morphological study confirms the growth of TiO_2 in a rod-like shape. Interestingly, the fabricated TiO_2-based chemi-resistive sensor is highly sensitive towards LPG gas detection with a sensitivity of ~522%. The fabricated sensor shows reversible and reproducible response in addition to an ultrafast response and recovery times of 11 and 11.3 s, respectively. Overall, the results present the fabrication of a TiO_2-based LPG sensor with high selectivity, reproducibility and ultra-fast response/recovery times for room temperature operation.

Acknowledgments. The authors want to acknowledge Dr. K C Patel R & D Centre, Charotar University of Science and Technology, for the experimental facility. The authors also thank the Science and Engineering Research Board (SERB), Government of India, for the Core Research Grant (File Number: CRG/2023/001471).

References

1. Nemade, K.R., Barde, R.V., Waghuley, S.A.: Liquefied petroleum gas sensing by Al-doped TiO2 nanoparticles synthesized by chemical and solid-state diffusion routes. J. Taibah Univ. Sci. **10**(3), 345–351 (2016)
2. Aishwarya, K., Nirmala, R., Navamathavan, R., Navamathavan, R.: Recent advancements in liquefied petroleum gas sensors. **2**, 100091 (2021)
3. Yadav, B.C., Yadav, A., Shukla, T., Singh, S.: Experimental investigations on solid state conductivity of Cobaltzincate nanocomposite for liquefied petroleum gas sensing. Sens. Lett. **7**(6), 1119–1123 (2009)
4. Shukla, T.: Synthesis of tin oxide thick film and its investigation as a LPG sensor at room temperature. J. Sens. Technol. **02**(3), 102–108 (2012)
5. Kumar, V.A., Yadav, L.: Microstructural properties and sensing behaviour of SnO_2 gas sensor. Int. Res. J. Manag. Sci. Technol. IRJMST. **9**, 239–244 (2018)
6. Solanki, V., Banerjee, A., Nanda, K.K.: Conductometric room temperature ammonia sensor based on porous tin oxide. Sensors Actuators B Chem. **366**, 131942 (2022)
7. Solanki, V., Krupanidhi, S.B., Nanda, K.K.: Experimental evidence on RH-dependent crossover from an electronic to protonic conduction with an oscillatory behaviour. Appl. Phys. Lett. **110**(26), 263506 (2017)
8. Kanan, S.M., El-Kadri, O.M., Abu-Yousef, I.A., Kanan, M.C.: Semiconducting metal oxide based sensors for selective gas pollutant detection. Sensors. **9**(10), 8158–8196 (2009)
9. Solanki, V., Joshi, S.R., Mishra, I., Kanjilal, D., Varma, S.: Formation of anisotropic nanostructures on rutile TiO_2(110) surfaces and their photo-absorption properties. Metall. Mater. Trans. A Phys. Metall. Mater. Sci. **49**, 3117–3121 (2018)
10. Song, T., Paik, U.: TiO_2 as an active or supplemental material for lithium batteries. J. Mater. Chem. A. **4**(1), 14–31 (2015)
11. Zhang, D., Yu, S., Wang, X., Huang, J., Pan, W., Zhang, J., Meteku, B.E., Zeng, J.: UV illumination-enhanced ultrasensitive ammonia gas sensor based on (001)TiO_2/MXene heterostructure for food spoilage detection. J. Hazard. Mater. **423**, 127160 (2022)
12. Zakrzewska, K., Radecka, M.: TiO_2 based nanomaterials for gas sensing-influence of anatase and rutile contributions. Nanoscale Res. Lett. **12**, 89 (2017)

13. Mardare, D., Iftimie, N., Luca, D.: TiO_2 thin films as sensing gas materials. J. Non-Cryst. Solids. **354**(35–39), 4396–4400 (2008)
14. Sharma, A., Vishwakarma, A., Yadava, L.: Sensing and structural characteristics of TiO_2 thick film for LPG sensor. Mater. Today. **38**(5), 2123–2126 (2021)
15. Dattatray, S., Gujar, T., Chandrakant, D.: TiO_2 nanorods decorated with Pd nanoparticles for enhanced liquefied petroleum gas sensing performance. Anal. Chem. **89**(16), 8531–8537 (2017)
16. Dhawale, D., Salunkhe, R., Fulari, V., Rath, M., Shilpa, N., Sawant, D.: Lokhande liquefied petroleum gas (LPG) sensing performance of electron beam irradiated chemically deposited TiO_2 thin films. Sensors Actuators B. **141**(1), 58–64 (2009)
17. Dang, T., Dang Duc, V., Nguyen, D.: Synthesis and LPG-sensing properties of TiO_2 nanowires. J. Phys. **187**, 012086 (2009)
18. More, A., Gunjakar, J., Lokhande, C.: Liquefied petroleum gas (LPG) sensor properties of interconnected web-like structured sprayed TiO_2 films. Sensors Actuators B. **129**(2), 671–677 (2008)
19. Lrbl, P., Huppertz, M., Merge, D.: Nucleation and growth in TiO_2 films prepared by sputtering and evaporation. Thin Solid Films. **251**(1), 72–79 (1994)
20. Radecka, M., Rekas, M., Kusior, E., Zakrzewska, K., Heel, A., Michalow, K., Graule, T.: TiO_2 -based Nanopowders and thin films for Photocatalytical applications. J. Nanosci. Nanotechnol. **10**(2), 1032–1042 (2010)
21. Kusior, A., Klich-Kafel, J., Trenczek, A., Swierczek, K., Radecka, M., Zakrzewska, K.: TiO_2 –SnO_2 nanomaterials for gas sensing and photocatalysis. J. Eur. Ceram. Soc. **33**(12), 2285–2290 (2013)
22. Aleksanyana, M., Sayuntsa, A., Zakaryana, H., Aroutiouniana, V., Arakelyana, G.: Investigations of sensors for detection of hydrogen peroxide vapors under the influence of UV illumination. J. Contemp. Phys. **55**, 205–212 (2020)
23. Chaurasiya, N., Kumar, U., Sikarwar, S., Yadav, B.C., Yadawa, P.K.: Synthesis of TiO_2 nanorods using wet chemical method and their photovoltaic and humidity sensing applications. Sensors Int. **2**, 100095 (2021)
24. Li, Z., Li, Q., Zhang, A., Wen, W., Wang, L., Wang, Z., Wang, J., Lu, S., Li, X., Wang, Z.: Tuning surface conductivity and stability for high-performance Li- and Mn-rich cathode materials. New J. Chem. **43**(47), 18943–18950 (2019)
25. Chaurasiya, N., Singh, A., Kumar, K., Yadav, B.C., Yadawa, P.K., Singh, S.K., Dey, K.K.: A highly sensitive room temperature liquefied petroleum gas (LPG) sensor with fast response based on a titanium dioxide (TiO_2)-reduced graphene oxide (r-GO) composite. Sens. Diagn. **2**, 1215–1227 (2023)
26. Scanlon, D., Dunnill, C., Buckeridge, J., Shevlin, S., Logsdail, A., Woodley, S., Catlow, C., Powell, M., Palgrave, R., Parkin, I., Watson, G., Keal, T., Sherwood, P., Walsh, A.: Band alignment of rutile and anatase TiO_2. Nat. Mater. **12**, 798–801 (2013)
27. Mi, Y., Yuxiang, W.: Band alignment and controllable electron migration between rutile and anatase TiO_2. Nat. Mater. **12**, 798–801 (2013)
28. Xiagoan, L., Ramamoorthy, R., Prabir, K.: Study of the resistance behavior of anatase and rutile thick films towards carbon monoxide and oxygen at high temperatures and possibilities for sensing applications. Sensors Actuators B. **143**(1), 308–315 (2009)
29. Sivapunniyam, A., Wiromrat, N., Myo, M., Joydeep, D.: Dutta high-performance liquefied petroleum gas sensing based on nanostructures of zinc oxide and zinc stannate. Sensors Actuators B. **157**(1), 232–239 (2011)
30. Dhawale, D., Dubal, D., More, A., Gujar, T., Lokhand, C.: Room temperature liquefied petroleum gas (LPG) sensor. Sensors Actuators B. **147**(2), 488–494 (2010)
31. Lin, C.H., Liao, C.H., Chen, W.H., Chou, C.Y., Liu, C.Y.: Fabrication of p-type TiO_2 and transparent p-TiO_2/n-ITO p-n junctions. AIP Adv. **9**(4), 045229 (2019)

32. Wang, S., Pan, L., Song, J.J., Mi, W., Zou, J.J., Wang, L., Zhang, X.: Room-temperature ferromagnetism and remarkable photocatalytic performance. **137**(8), 2975–2983 (2015)
33. Solanki, V., Krupanidhi, S.B., Nanda, K.K.: Sequential elemental dealloying approach for the fabrication of porous metal oxides and chemiresistive sensors thereof for electronic listening. ACS Appl. Mater. Interfaces. **9**(47), 41428–41434 (2017)

Optimization of Growth Conditions for Single Crystal Diamond Substrates Using Microwave Plasma Enhanced Chemical Vapor Deposition

Pulkesh Prajapati[1], Abhay Dasadia[1](✉), Khyati Upadhyay[1], and Vanaraj Solanki[2]

[1] A. D. Patel Institute of Technology, The C.V.M University, New Vallabh Vidyanagar 388121, Gujarat, India
abhaydasadia@adit.ac.in

[2] Dr. K C Patel R and D Centre, Charotar University of Science and Technology (CHARUSAT), Changa, Anand 388421, India

Abstract. Microwave Plasma Enhanced Chemical Vapor Deposition (MPCVD) is a leading technique for producing high-quality diamond crystals under relatively low temperature and pressure conditions. This method was utilized to optimize various growth parameters, including substrate temperature (ranging from 800 to 950 °C), gas pressure (between 120 and 150 Torr), and the concentrations of H_2, CH_4, and N_2 during deposition. By simultaneously adjusting the growth parameters and analyzing them, the quality of the resulting sample was enhanced.

The substrate temperature was monitored with an infrared pyrometer, while gas flow rates were controlled through a mass flow controller. Raman Spectroscopy was performed to verify the diamond phase of the produced samples. Additionally, the optical absorption of the optimized samples was assessed using UV-VIS spectroscopy to determine their transparency over a wavelength range. The resulting diamond crystals exhibit exceptional material qualities, making them suitable for a variety of applications, including high-power electronics, radiation detectors, optical components for lasers, and windows for radio frequencies. New potential applications for high-quality single crystal diamonds may also include quantum sensing and erosion-resistant coatings in nuclear fusion reactors.

Keywords: Single Crystal Diamonds · Microwave Plasma Enhanced Chemical · Vapor Deposition (MPCVD) · Raman Spectroscopy · Technological Grade Diamonds

1 Introduction

Diamonds, in both natural and synthetic forms, have gained significant global interest. Over the past few decades, lab-grown diamonds have become increasingly popular across various industries due to their outstanding characteristics. The unique structural attributes of diamonds stem from their strong covalent bonds and rigid structure, making

S. Kumari et al. (Eds.): ICMTP 2024, SPPHY 311, pp. 374–379, 2025.
https://doi.org/10.1007/978-981-96-6795-6_41

them the hardest and stiffest material, as well as possessing the highest thermal conductivity at room temperature and one of the lowest rates of thermal expansion. However, their applications have been limited by their high cost, restricted size, and difficulties in controlling impurities [1]. Synthetic diamonds, also referred to as lab-grown or man-made diamonds, exhibit remarkable properties in mechanical, optical, electronic, and electrochemical domains, and are environmentally friendly as they eliminate the need for harmful mining practices [2]. Various methods exist for synthetic diamond production, including High-Pressure High-Temperature (HPHT) and Hot Filament Chemical Vapor Deposition (HFCVD), but Microwave Plasma Enhanced Chemical Vapor Deposition (MPCVD) stands out as the most promising technique for synthetic diamond creation. This method is advantageous for diamond production, requiring careful control of three key parameters: substrate temperature, gas pressure, and methane gas concentration during the synthesis process. We optimized these growth parameters for synthesizing single-crystal diamonds using the MPCVD method through a trial-and-error approach, resulting in high-quality diamonds with distinct properties verified by Raman spectroscopy.

2 Experimental

2.1 Growth of Single Crystal Diamond Substrate

Diamond seeds were utilized as a substrate, which underwent pre-cleaning using the Seki Diamond MPCVD system (Model SDS 6K) operating at 2.45 GHz and 3 kW microwave power for 20 min in hydrogen plasma at a concentration of 200 sccm. Growth was initiated in the deposition chamber with clean hydrogen gas (99.9998%) and ionized through the electric field generated by resonance between the microwave and the chamber. The adequate substrate temperature was achieved to the desired growth conditions using the trial-and-error method by adjusting the gas flow, gas pressure, and input power. Subsequently, methane with a purity of 99.9998% was introduced into the deposition chamber to initiate the growth of diamonds. The gas pressure in the deposition chamber during growth was maintained between 100–150 Torr, while the input power was adjusted within the range of 5–6 kW. The substrate temperature was controlled between 800 and 950 °C and it was measured using a two-color pyrometer (Williamson Make). After conducting several growth experiments using the trial-and-error method, the growth parameters for the four samples, D1, D2, D3, and D4, are outlined in Table 1.

With the used of mentioned parameters, we obtained the diamond substrates as shown D1, D2, D3, D4 in Fig. 1.

2.2 Raman Spectroscopy

Figure 2 shows the Raman spectra of grown samples, measured using the Horiba Micro-Raman Xplora Plus Model having resolution $\pm 1\ cm^{-1}$ illuminating at a 532 nm laser source for excitation. The presence of peaks near 1332 cm^{-1} is referred to as first-order Raman peaks with full width at half maximum (FWHM) ranging from 4.26 to 6.89 cm^{-1}.

Typically, the graphite structure exhibits wider peaks near 1332 and 1580 cm^{-1}, which have comparable intensities. The Raman spectra can vary considerably based on

Table 1. Growth parameters.

Sample	Substrate Temperature (°C)	Gas Pressure (Torr)	Microwave Power (kW)	Concentration H_2:CH_4: N_2 (sccm)	Growth period (h)	Dimensions (mm × mm × mm)
D1	800	120	5.5	450:30:1	98	8 × 8 × 1.8
D2	850	100	5.5	400:35:0.5	98	8 × 8 × 1.7
D3	900	130	5	450:40:1	98	8 × 8 × 1.8
D4	950	150	5.5	400:40:0.5	98	8 × 8 × 1.9

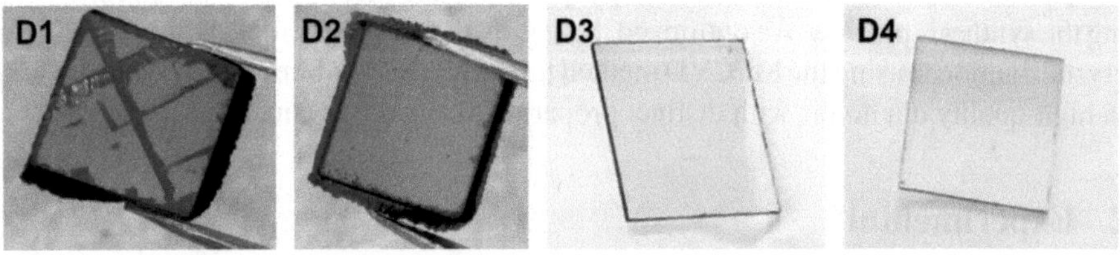

Fig. 1. Synthesized diamond seed substrates D1, D2, D3, D4

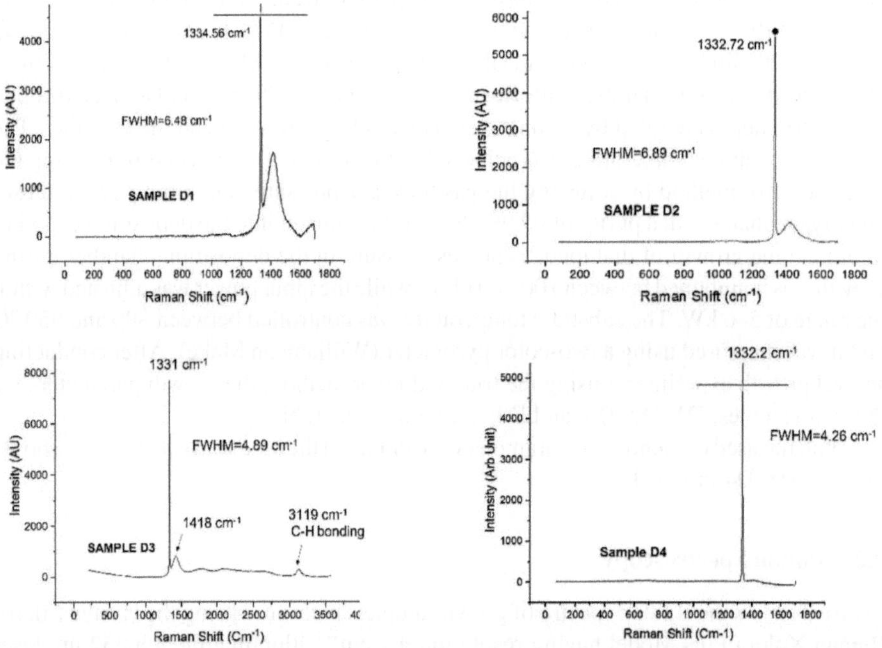

Fig. 2. Raman Spectra for samples D1, D2, D3 and D4; Excitation laser wavelength is 532 nm

the excitation wavelength of laser. The width of the Raman peaks indicates the amount of random stress present, while any shifts or splits in the lines reveal directional stress. The crystal could also include amorphous sp^2 bonded carbon (graphite) and sp^3 bonded carbon (diamond-like). The Raman peak at 1332 cm^{-1} corresponds to sp^3 bonded carbon [3].

2.3 UV-VIS Spectroscopy

The transmittance spectrum as shown in Fig. 3 of sample D4, measured using a LAMBDA 1050+ Perkin Elmer UV-Vis spectrometer, was analyzed due to its optimized quality.

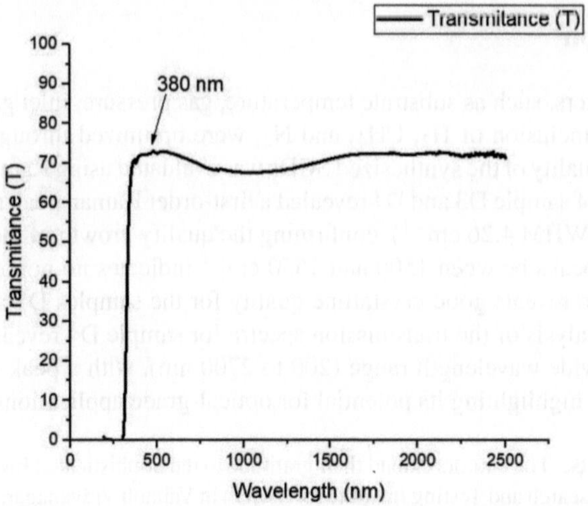

Fig. 3. Transmittance spectrum shows maximum transmission nearly 70% at 380 nm of D4

The absorption edge observed near 282 nm corresponds to the wide bandgap of diamond, resulting from the excitation of charge carriers during band-to-band transition [4].

3 Result and Discussion

As shown in Fig. 1, polycrystalline inclusions are encountered more during the growth cycle of samples D1 and D2 using MPCVD method. There was less growth of polycrystalline inclusions along edges during the entire process cycle in for samples D3 and D4. The dynamics of the vapor phase, including the flow rate and turbulence in the CVD chamber, can impact the uniformity and, consequently, the crystalline quality. Careful control of each parameter affects nucleation, growth rates, morphology, and impurity incorporation, all of which play critical roles in determining the final crystalline quality. Therefore, optimizing the growth parameters, a high-quality single-crystal diamond with dimensions of 8 mm × 8 mm × 1.9 mm (sample D4) was successfully grown. This was achieved over a process cycle lasting 98 h, with an excecated growth rate of 10–14 μm/h.

In the Raman spectra of samples D1 and D2, peaks appeared at 1334.56 cm^{-1} with FWHM 6.48 and 1332.72 cm^{-1} with FWHM 6.89 cm^{-1} respectively are responsible for the vibration of the sp^3 bonded carbon atoms. Additionally, peaks between 1400 to 1500 cm^{-1} are caused by to presence of low-order graphitic carbons in the D1 and D2 samples respectively [5]. The presence of graphitic carbons can be detected in the first-order region (1100–1800 cm^{-1}) and the second-order region (2200–3400 cm^{-1}) observable in sample D3. The absence of peaks in the region 1500–1550 cm^{-1} in samples D3 and D4 is generally considered strong evidence for the growth of the "pure" diamond phase without non-diamond components present [6]. The transmittance spectrum shown in Fig. 3 has an exceptionally high transmission over a very broad range of wavelengths from 200 to 2500 nm maximum of 70% at 380 nm for sample D4.

4 Conclusion

Growth parameters, such as substrate temperature, gas pressure, inlet gas ratios, plasma power, and the inclusion of H_2, CH_4, and N_2, were optimized through trial-and-error approach. The quality of the synthesized SCDs was evaluated using Raman spectroscopy. Raman spectra of sample D3 and D4 revealed a first-order Raman peak at approximately 1332.5 cm^{-1} (FWHM 4.26 cm^{-1}), confirming the quality growth of the samples, while the absence of peaks between 1500 and 1550 cm^{-1} indicates no non-diamond phases. This finding also reveals good crystalline quality for the samples D3 and D4. UV-VIS spectroscopy analysis of the transmission spectra for sample D4 reveals high transmittance across a wide wavelength range (200 to 2700 nm), with a peak transmittance of 72% at 380 nm, highlighting its potential for optical-grade applications.

Acknowledgments. The authors extend their gratitude to the Sophisticated Instrumentation Centre for Applied Research and Testing Institute (SICART) in Vallabh Vidyanagar, Gujarat, India, for granting access to instrumentation facilities for investigating the optical properties of developed SCDs. Additionally, they appreciate the support from ABD Diamonds Pvt. Ltd. in GIDC, Vatva, Ahmedabad, for facilitating the experimental work associated with SCD growth.

References

1. Yan, C., Vohra, Y., Mao, H., Hemley, R.: Geophysical Laboratory, Carnegie Institution of Washington, 5251 Broad Branch Road NW, Washington, DC 20015; and †Department of Physics, University of Alabama, Birmingham, AL 35294 (2002)
2. Naemura, K., et al.: Diamond and other possible ultradeep vidence discovered in the orogenic Spinel-Garnet Peridotite from the Moldanubian Zone of the Bohemian Massif, Czech Republic, Ultrahigh-Pressure Metamorphism, Elsevier, 77–111, 2011
3. Wagner, J., Wild, C., Koidl, P.: Resonance effects in Raman scattering from polycrystalline diamond films. Appl. Phys. Lett. **59**(7), 779–781 (1991)
4. Bennett, A.M., Wickham, B.J., Dhillon, H.K., et al.: Development of high purity, optical grade single crystal CVD diamond for intra-cavity cooling. Proceedings of SPIE 8959, Solid State Lasers XXIII: Technology and Devices (2014)
5. Ferrari, A.C., Robertson, J.: Interpretation of Raman spectra of disordered and amorphous carbon. Phys. Rev. B. **61**(20), 14095–14107 (2000)

6. Upadhyay K.: Enhancement in the quality of single crystal diamond seed substrate through various etching method. In: National Conference on Advances in Physical Science for Sustainable Development, August 27, 2022, Gandhinagar. Sector 15 Gandhinagar, 2023

Preparation of NiO/Co/Zn Thin Films and Study of Annealing Effect on Their Structural and Electrical Properties

Dhara Singh Meena[1], Arjun Kumawat[1], Ved Prakash Meena[1], Prashant Meena[1], Harish Mahla[1], Pavan Kumar Meena[1], Madhu Yadav[2], Chhagan Lal[2], and M. K. Jangid[1](✉)

[1] Department of Physics, Vivekananda Global University, Jaipur 303012, India
mahesh.jangid@vgu.ac.in, maheshjangid2008@gmail.com
[2] Department of Physics, University of Rajasthan, Jaipur 302004, India

Abstract. This research explores the fabrication of NiO-based (NiO/Co/Zn) thin films and examines the influence of annealing on their structural and electrical characteristics. The films were deposited on glass substrates using the E-Beam technique and subjected to vacuum annealing at 400 °C to analyze the annealing effects. X-ray diffraction (XRD) analysis confirmed the formation of the NiO phase, displaying peaks associated with cubic NiO. A decrease in crystallite size post-annealing indicates that the annealing process improved the crystallite structure. Scanning electron microscopy (SEM) revealed a uniform surface morphology with spherically shaped nanocrystalline grains. Energy-dispersive X-ray spectroscopy (EDS) confirmed the presence of NiO, Co, and Zn elements in the films. I-V characterization exhibited nonlinear behavior, indicating semiconducting properties. The annealed films demonstrated a notable rise in current, suggesting enhanced conductivity. These findings suggest that annealing plays a crucial role in improving both the structural and electrical properties of NiO/Co/Zn thin films, making them suitable for applications in resistive switching and sensing technologies.

Keywords: Thin Films · E-Beam · XRD · SEM-EDS · I-V

1 Introduction

Nickel oxide (NiO) thin films have received a lot of interest nowadays because of their promising properties, which include broad transparency, tunable conductivity through appropriate doping, chemical and thermal stability, and flexible fabrication techniques [1, 2]. One of the transparent conducting oxides (TCOs) that have been investigated the most is NiO. Gas sensors [3], photocathodes for dye-sensitized solar cells [4], electrochromic devices [5], memory devices [6], solar cells [7], and smart windows are just a few of its applications, etc. [8, 9]. At ambient temperature, NiO exhibits an insulating behavior in its pure stoichiometric state, with a high electrical resistivity. However, by doping with the right elements, nickel vacancies, and interstitial oxygen atoms can be introduced, greatly

S. Kumari et al. (Eds.): ICMTP 2024, SPPHY 311, pp. 380–387, 2025.
https://doi.org/10.1007/978-981-96-6795-6_42

lowering its resistivity and increasing conductivity [10, 11]. Researchers experiment with broad doping approaches utilizing different metal elements and manufacturing techniques to improve the electrical and optical properties of NiO in order to attain desired features for particular applications. NiO thin films with various dopants included metals like iron (Fe), copper (Cu), zinc (Zn), and cobalt (Co) have been created and reported [12–14]. Because of its close ionic radius parameter, zinc (Zn) is a potential transition metal element that can be employed as an appropriate additive for nickel oxide in addition to the dopants previously stated. Furthermore, there is currently a lack of comprehensive knowledge regarding the different physical characteristics of Zn-doped NiO film [15, 16]. NiO thin films have been created using various kinds of approaches, such as spray pyrolysis [17], chemical vapor deposition [18], thermal evaporation [19], sputtering [20, 21], electron beam evaporation [22, 23], and the sol-gel process [24]. Among these, the E-Beam method has drawn a lot of interest because of its many benefits, including ease of use, safety, affordability, low growing temperature, and excellent uniformity [25], is ideally suited for doping, and provides simple control over processing parameters. While similar studies have explored different deposition techniques and dopants, limited research has focused on the fabrication of Co, Zn-doped nickel oxide thin films using the E-beam method. In this work, the E-beam technique is employed to synthesize NiO/Co/Zn thin films, and the impact of annealing on their electrical, structural, and morphological characteristics is systematically analyzed.

2 Preparation Methods

The e-beam approach was used to create 400 nm-thick NiO/Co/Zn thin films on glass substrates. The film was prepared using high-purity materials, specifically NiO (99.99%), Co (99.98%), and Zn (99.98%). In order to guarantee consistent mixing, a planetary ball mill (Retsch PM 100 type) was used for integrating zinc oxide, cobalt, and nickel oxide at 250 rpm for 5 hours. The ratio of NiO (75%), Zn (15%), and Co (10%) is utilized to take the materials. The final powder was dried and compressed into pellets using hydraulic pressure, which produces regulated pressure, following grinding. The e-beam deposition technique (Model BC-300 HHV) was used to deposit the resultant nanostructured thin films onto the substrates. Diffusion and rotational pumps were used to empty the vacuum chamber during deposition, resulting in a base pressure of about 10^{-5} mbar. X-ray diffraction (XRD) analysis was performed at room temperature using a Panalytical X'Pert Pro diffractometer with CuKα radiation ($\lambda = 1.5406$ Å) to identify the crystalline phases. The surface morphology was examined using a Field Emission Scanning Electron Microscope (FESEM) equipped with Energy Dispersive Spectroscopy (EDS) (Nova Nano FE-SEM 450, FEI), providing ultra-high-resolution imaging and precise nanoscale characterization. Electrical properties were evaluated through I-V characterization using the MicroXact SPS-2200 Probe Station, which measures current as a function of applied voltage.

3 Results and Discussion

3.1 X-Ray Diffraction (XRD)

The crystalline structure of NiO/Co/Zn thin films was analyzed using X-ray diffraction (XRD) with CuKα radiation (λ = 0.1542 nm). The diffraction patterns displayed peaks at 2θ values of 31.8° and 34.6° corresponding to (100) and (002) planes of NiO cubic structure (JCPDS Card No. 47-1049) and at 2θ values of 37.2° and 43.2° corresponding to (111) and (200) planes of ZnO Hexagonal Wurtzite structure (JCPDS Card No. 36-1451). The results confirmed that both (a) as-deposited and (b) annealed NiO/Co/Zn thin films at 400 °C maintained a crystalline structure, as evidenced by sharp diffraction peaks (Fig. 1). After annealing additional peaks at 2θ values of 44.6°, and 51.9° corresponding to (222), and (101) planes were observed for NiO and ZnO, additionally, the diffraction intensities showed a slight increase following annealing at 400 °C (Fig. 1). Annealing at 400 °C may reduce lattice distortions, making peak shifts less pronounced.

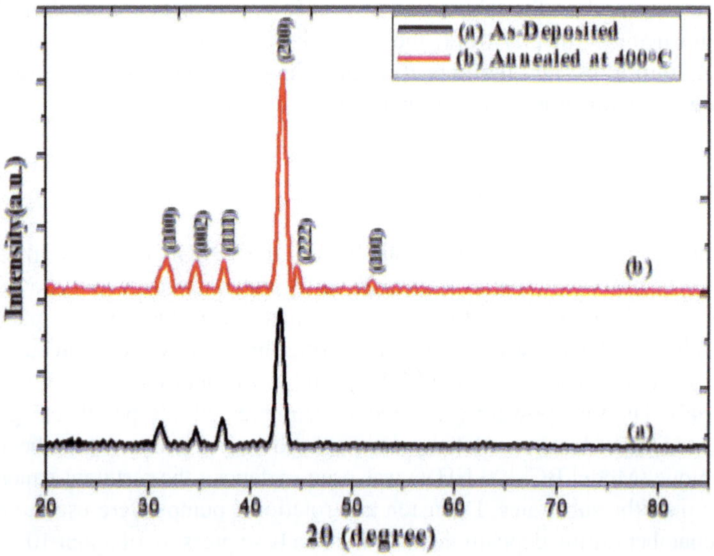

Fig. 1. XRD patterns of NiO/Co/Zn thin films (a) as-deposited and (b) annealed at 400 °C.

The increased intensity of the diffraction peaks is attributed to the grain size growth in NiO/Co/Zn thin films. To quantify this, the crystallite size was calculated using the Scherrer equation, which considers peak broadening in XRD patterns, providing insight into the structural refinement and crystalline quality of the films. The thin film's as-deposited crystallite size (D) is 12.45 nm; after annealing at 400 °C, it reduces to 9.46 nm, respectively. This reduction shows that the annealing procedure refined the crystallite structure. A denser film with improved structural integrity may result from annealing induction of stress relaxation and grain boundary reorientation, as seen by the smaller crystallite size at higher temperatures. With the annealing temperatures, the grain sizes decreased. The results of grain sizes are similar to the sizes obtained from SEM (Fig. 2). In addition, with the annealing temperatures, the (222), and (101) planes are also observed at high angles [26, 27].

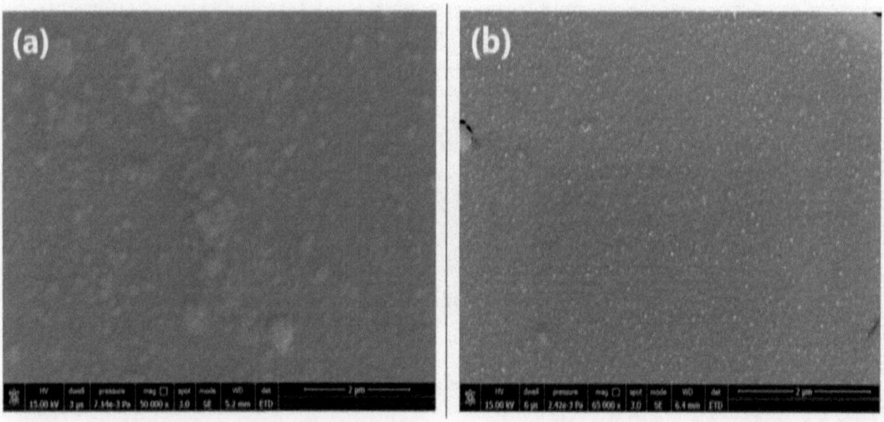

Fig. 2. FE-SEM images of NiO/Co/Zn thin films (a) as-deposited and (b) annealed at 400 °C.

3.2 FESEM and EDS Analysis

The rough surface and lack of grain development are seen in the FE-SEM picture of the as-deposited NiO/Co/Zn thin film (Fig. 2a). The surface morphology is showed tiny grain sizes when the annealing temperature was set to 400 °C (Fig. 2b). According to the micrographs, the film is made up of uniformly distributed nanocrystalline grains with randomly oriented morphologies on the substrate surface. At an annealing temperature of 400 °C, the size of the crystallites decreases. The crystallite size determined by XRD analysis and the crystallite size determined by SEM examination agree rather well. When the NiO/Co/Zn thin films were annealed at the temperature of 400 °C, the grain sizes slightly decreased as compared to the as-deposited NiO/Co/Zn thin films. Higher temperatures during annealing provide the substrate's surface atoms more energy, which allows them to shift to appropriate nucleation locations. Furthermore, the thin-film doped ions with low activation energy can easily move from trap sites to nucleation sites. As the annealing temperatures rose, the NiO/Co/Zn thin film thickness somewhat decreased [28, 29].

The compositional analysis of NiO/Co/Zn thin films (a) as-deposited and (b) annealed at 400 °C shown in EDS spectrum (Fig. 3) and Table 1. The EDS analysis confirms the presence of Ni, Co, and Zn components in the as-deposited as well annealed thin films. This attests to the nature of the produced NiO/Co/Zn thin films. It is crucial to remember that extra peaks that are ascribed to Si are seen in the glass substrate and may be the cause of the EDS spectra [30–32].

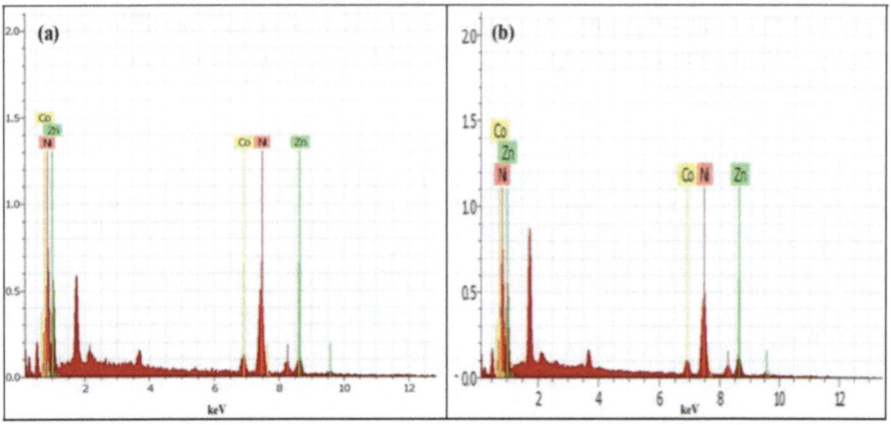

Fig. 3. EDS spectrum of NiO/Co/Zn thin films (a) as-deposited and (b) annealed at 400 °C.

Table 1. Elemental composition of NiO/Co/Zn thin films

Sample Type	As-deposited thin film			Annealed thin film at 400 °C		
Elements	Ni	Zn	Co	Ni	Zn	Co
Weight %	69.93	20.88	9.19	59.17	32.35	8.48
Atomic %	71.48	19.16	9.36	61.22	30.04	8.74

3.3 I-V Characteristics

I-V characteristics of NiO/Co/Zn thin films in Fig. 4 show the impact of annealing on electrical conductivity. In (a), the as-deposited film exhibits lower current values, indicating poor conductivity due to structural defects and grain boundary scattering. The nonlinear nature suggests semiconducting behavior, typical of NiO-based thin films. In (b), the annealed sample shows a significant increase in current, suggesting enhanced conductivity due to improved crystallinity, reduced defect density, and increased carrier mobility. Annealing likely promotes oxygen vacancy formation, which enhances charge transport [33, 34]. The nonlinear trend in both curves suggests conduction through space-charge-limited current (SCLC) or Poole-Frenkel mechanisms, common in metal oxides. Additionally, the asymmetry in the I-V curves suggests a Schottky barrier at the interfaces, which is modified by annealing, reducing contact resistance and improving charge injection [35, 36]. These results indicate that thermal treatment optimizes the electronic properties of NiO/Co/Zn thin films, making them suitable for resistive switching and sensor applications.

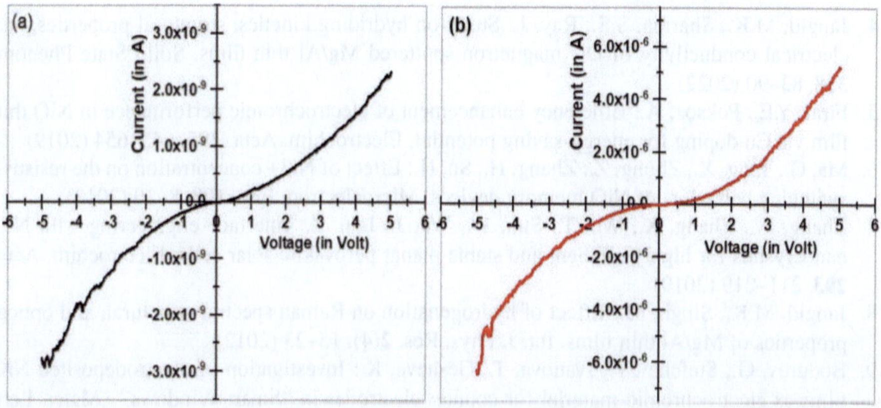

Fig. 4. I-V characteristics of NiO/Co/Zn thin films (a) as-deposited and (b) annealed at 400 °C.

4 Conclusion

NiO/Co/Zn thin films of 400 nm-thick were prepared successfully using the E-beam evaporation approach on glass substrates. NiO/Co/Zn thin film was vacuum annealed at 400 °C temperature. The XRD pattern exhibits a crystalline structure due to its sharp-intensity diffraction peaks at (200). The FE-SEM images show uniform agglomeration and distribution of the fine particles. The grain sizes and crystallite size decrease with the annealing of NiO/Co/Zn thin films at 400 °C temperature confirmed by FE-SEM images and XRD pattern. The EDS analysis also confirms the presence of Ni, Co, and Zn elements. The nonlinear nature suggests semiconducting behavior of NiO/Co/Zn thin films. The annealed sample shows a significant increase in current, suggesting enhanced conductivity. The findings confirm the promising nature of NiO/Co/Zn thin films. They exhibit excellent potential for application in advanced optoelectronic devices due to their improved crystalline quality, surface uniformity, and enhanced electrical properties.

Acknowledgements. The authors (specially Dhara Singh Meena) would like to thank CSIR for providing financial support under the JRF scheme for this work. Authors also would like to acknowledge Material Research Centre, MNIT Jaipur and Center for non-conventional energy resources, UOR, Jaipur (India) for providing experimental and measurement facilities.

References

1. Xu, J., Wang, M., Liu, Y., Li, J., Cui, H.: One-pot solvothermal synthesis of size-controlled NiO nanoparticles. Adv. Powder Technol. **30**(4), 861–868 (2019)
2. Zrikem, K., Song, G., Ait Aghzzaf, A., Amjoud, M., Mezzane, D., Rougier, A.: UV treatment for enhanced electrochromic properties of spin-coated NiO thin films. Superlattice. Microst. **127**, 35–42 (2019)
3. Zargouni, S., Derbali, L., Ouadhour, M., Rigon, M., Martucci, A., Ezzaouia, H.: Elaboration and characterization of PVP-assisted NiO thin films for enhanced sensitivity toward H2 and NO2 gases. Ceram. Int. **45**(4), 5779 (2019)

4. Jangid, M.K., Sharma, S.S., Ray, J.: Study on hydriding kinetics, structural properties, and electrical conductivity of DC magnetron sputtered Mg/Al thin films. Solid State Phenom. **338**, 83–90 (2022)
5. Firat, Y.E., Peksoz, A.: Efficiency enhancement of electrochromic performance in NiO thin film via Cu doping for energy-saving potential. Electrochim. Acta. **295**, 645–654 (2019)
6. Ma, G., Tang, X., Zhong, Z., Zhang, H., Su, H.: Effect of Ni3+ concentration on the resistive switching behaviors of NiO memory devices. Microelectron. Eng. **108**, 8–10 (2013)
7. Zhang, W., Zhang, X., Wu, T., Sun, W., Wu, J., Lan, Z.: Interface engineering with NiO nanocrystals for highly efficient and stable planar perovskite solar cells. Electrochim. Acta. **293**, 211–219 (2019)
8. Jangid, M.K., Singh, M.: Effect of hydrogenation on Raman spectra, structural, and optical properties of Mg/Al thin films. Int. J. Phys. Res. **2**(4), 15–23 (2012)
9. Bodurov, G., Stefchev, P., Ivanova, T., Gesheva, K.: Investigation of electrodeposited NiO films as electrochromic material for counter electrodes in "Smart Windows.". Mater. Lett. **117**, 270–272 (2014)
10. Jangid, M.K., Singh, M.: Hydrogenation and annealing effect on electrical properties of nanostructured Mg/Mn bilayer thin films. Int. J. Hydrog. Energy. **37**(4), 3786–3791 (2012)
11. Dirksen, J.A., Duval, K., Ring, T.A.: NiO thin-film formaldehyde gas sensor. Sensors Actuators B Chem. **80**(4), 106–115 (2001)
12. Wu, Z., Zou, Z., Huang, J., Gao, F.: Fe-doped NiO mesoporous nanosheets array for highly efficient overall water splitting. J. Catal. **358**, 243–252 (2018)
13. Jangid, M.K., Jangid, S.K.: Structural, electrical, and optical properties of Mg-Ni thin films for hydrogen storage applications. Trends Sci. **19**(23), 2067 (2022)
14. Zargouni, S., El Whibi, S., Tessarolo, E., Rigon, M., Martucci, A., Ezzaouia, H.: Structural properties and defect-related luminescence of Yb-doped NiO sol-gel thin films. Superlattice. Microst. **138**, 106361 (2020)
15. Dogan, D.D., Caglar, Y., Ilican, S., Caglar, M.: J. Alloys Compd. **509**(4), 2461–2465 (2011)
16. Noonuruk, R., Techitdheera, W., Pecharapa, W.: Thin Solid Films. **529**(4), 2769–2775 (2012)
17. Lou, X.C., Zhao, X.J., Xiong, Y.L., Sui, X.T.: The influence of annealing on electrochromic properties of Al–B–NiO thin films prepared by sol–gel. J. Sol. Gel Sci. Technol. **54**, 43–48 (2010)
18. Desai, J.D., Min, S.K., Jung, K.D., Joo, O.S.: Spray pyrolytic synthesis of large-area NiOx thin films from aqueous nickel acetate solutions. Appl. Surf. Sci. **253**(4), 1781–1786 (2006)
19. Jangid, M.K., Sharma, S.S., Ray, J., Jangid, S.: Effect of annealing and hydrogenation on optical and electrical properties of DC sputtered Mg-Ni bilayer thin films. Mater. Today: Proc. **67**, 847–851 (2022)
20. Park, J.W., Park, J.W., Kim, D.Y., Lee, J.K.: Reproducible resistive switching in nonstoichiometric nickel oxide films grown by rf reactive sputtering for resistive random access memory applications. J. Vac. Sci. Technol. A. **23**(4), 1309–1313 (2005)
21. Jangid, M.K., Nehra, S.P., Singh, M.: Hydrogenation effect on Mg/Mn multilayer thin films. AIP Conf. Proc. **1349**(1), 685–686 (2011)
22. Emam-Ismail, M., El-Hagary, M., El-Sherif, H.M., El-Naggar, A.M., El-Nahass, M.M.: Spectroscopic ellipsometry and morphological studies of nanocrystalline NiO and NiO/ITO thin films deposited by e-beams technique. Opt. Mater. **112**, 110763 (2021)
23. Jangid, M.K., Nehra, S.P., Singh, M.: Optical, electrical, and Raman properties of annealed hydrogenated Mg/Co bilayer thin films. J. Nano-Electron. Phys. **3**(1), 460–468 (2011)
24. Guendouz, H., Bouaine, A., Brihi, N.: Biphase effect on structural, optical, and electrical properties of Al-Sn codoped ZnO thin films deposited by sol-gel spin-coating technique. Optik. **158**(13), 42–1348 (2018)
25. Sahu, D.R., Wu, T.-J., Wang, S.-C., Huang, J.-L.: Electrochromic behavior of NiO films prepared by e-beam evaporation. J. Sci. Adv. Mater. Dev. (2017)

26. El-Nahass, M.M., Emam-Ismail, M., El-Hagary, M.: Structural, optical, and dispersion energy parameters of nickel oxide nanocrystalline thin films prepared by electron beam deposition technique. J. Alloys Compd. **646**, 937–945 (2015)
27. Madhavi, A., Harish, G.S., Reddy, P.S.: Effect of annealing temperature on optical and electrical properties of electron beam evaporated NiO thin films. IJSTE Int. J. Sci. Technol. Eng. **2**(11), 127 (2016)
28. Diao, C.-C., Huang, C.-Y., Yang, C.-F., Wu, C.-C.: Morphological, optical, and electrical properties of p-type nickel oxide thin films by nonvacuum deposition. Nano. **10**(4), 636 (2020)
29. Kumar, A., Sahay, P.P.: Lithium doping in spray-pyrolyzed NiO thin films: results on their microstructural, optical and electrochromic properties. Appl. Phys. A. **127**, 286 (2021)
30. Gavale, H.S., Wagh, M.S., Ahirrao, R.B., Gosavi, S.R.: Study of physical properties of nanocrystalline NiO thin films prepared by spray pyrolysis technique. J. Nanosci. Technol. **5**, 610–612 (2019)
31. Jangid, M.K., Sharma, S., Meena, V.P., Arya, V.P., Sharma, S.S.: Impact of annealing and hydrogenation on the optical characteristics of ZnSe/Mn and ZnSe/Co DMS thin films prepared via thermal evaporation method. Evergreen. **11**(1), 178–185 (2024)
32. Jangid, M.K., Sharma, S.S., Ray, J., Yadav, D.K., Lal, C.: Structural, optical and electrical characterizations of Mg/Ti/Ni multilayer thin films deposited by DC magnetron sputtering for hydrogen storage. Int. J. Hydrog. Energy. **48**(96), 37921–37929 (2023)
33. Jangid, M.K., Sharma, S.S., Mathur, D., Sharma, Y.C.: Optical, electrical, and structural study of Mg/Ti bilayer thin film for hydrogen storage applications. Mater. Lett.: X. **10**, 100076 (2021)
34. Lany, S.: Semiconducting transition metal oxides. Condens. Matter. **27**, 283203 (2015)
35. Jangid, M.K., Nehra, S.P., Singh, M.: Preparation and characterization of Mg/Mn bilayer metal hydrides thin films. AIP Conf. Proc. **1393**(1), 311–312 (2011)
36. Liu, J., Zhang, Z., Lin, J., Zhen, L., Jiang, Q., Shi, J., Deng, H.: Oxygen vacancy-modified fast charge transport channels at the interface of bismuth S-scheme heterojunctions promoting photocatalytic performance. Chem. Eng. J. **506**, 159887 (2025)

Study of Thermal Stability of Polyaniline Doped with Activated Charcoal

Manoj Kumar[1], Smriti Sharma[2(✉)], Rishi Pal[3], Sneh Lata Goyal[4], Ruchi[4], Vivek Gupta[4], Rajiv[5], and Supreet[1]

[1] Department of Physics, Amity School of Applied Sciences, Amity University, Gurugram 122413, Haryana, India

[2] Department of Physics, Government College for Women, Sector-14, Gurugram 122001, Haryana, India
Smritisharma09@gmail.com

[3] Department of Applied Science, Kalpana Chawla Government Polytechnic for Women, Ambala 134003, Haryana, India

[4] Department of Physics, Guru jambheshwar University of Science and Technology, Hisar 125001, Haryana, India

[5] Department of Physics, Hindu College, Sonepat 131001, Haryana, India

Abstract. Composites of Polyaniline and Activated Charcoal (PANI/AC) were synthesised through the oxidative polymerisation of aniline hydrochloride, utilising alter Wt.% (5, 10, 15, 20 wt. %) of Activated Charcoal particles in conjunction with the oxidant ammonium persulphate. The composites were subjected to thermogravimetric analysis (TGA) to assess their thermal stability. The thermal stability of the PANI/AC demonstrated an improvement over that of PANI, suggesting a significant interaction between the conducting polymer PANI and the activated charcoal particles in the composites. The Horowitz-Metzger method was employed to determine kinetic parameters such as free energy, entropy of activation, frequency factor and activation energy of change in composition. The values of all these parameters indicate a notable enhancement in the thermal stability of the composites.

Keywords: Polyaniline · TGA · Horowitz-Metzger method

1 Introduction

Among the well-researched polymeric materials for various applications [1–8], PANI is regarded as one of the most stable systems, with adjustable morphological, structural, and electrical properties [9]. By adding different types of loadings to the polymer matrix, these polymeric materials' characteristics can be further enhanced. Using a straight forward polymerization process, PANI can be formed in a variety of microstructural and nanostructural shapes, including core-shell, wires, rods, tubes, cables, fibers, and particles [10–15]. Though there are a number of reaction parameters, including surfactant types and concentrations [16–18], surfactant chain length [19], oxidant types and concentrations [20–22], protonic acid-based loadings [23–25], reaction temperature [21],

S. Kumari et al. (Eds.): ICMTP 2024, SPPHY 311, pp. 388–394, 2025.
https://doi.org/10.1007/978-981-96-6795-6_43

solvent [26], solution stirring [21], co-surfactants [20], and loading types like metal oxides, graphene, and carbon fibers, or carbon black [27, 28], etc. which affect the structural, morphological, electrical and other properties of nanostructured PANI.

The effects of various anionic, cationic, and non-ionic surfactants on PANI surface morphologies in the presence of different oxidants (FeCl3 and APS) were reported by Zhang et al. [29], while Omastava et al. [16] and Stejskal et al. [30] have examined the effects of these surfactants on PANI's structural, morphological, electrical, and thermal properties. Recent research has examined the impact of loading type and concentration on PANI's thermal characteristics.

2 Experimental

2.1 Synthesis of PANI/Activated Charcoal (AC) Nanocomposites

PANI(P0) was synthesised via in situ polymerisation, specifically by oxidising 0.2 M aniline hydrochloride (HCl) with 0.25 M ammonium persulfate (APS) in water medium as the oxidant. During the preparation of PANI/AC composites, Activated Charcoal (AC) (0.1 M) was added at weight percentages of 5, 10, 15, and 20 into the aniline hydrochloride solution while stirring vigorously. Following this, ammonium persulphate at a concentration of 0.25 M was incrementally introduced as an oxidising agent within a temperature range of 0–4 °C. The mixture was subsequently permitted to rest in the refrigerator for a duration of 24 h to promote polymerisation.

On the subsequent day, the PANI/AC nanocomposite precipitates were placed using filter paper and then treated with 1 M HCl and methanol until the excess became colourless. The precipitate was subjected to drying in air, followed by vacuum treatment at a temperature of 45 °C. Following this route, four peculiar PANI/AC nanocomposites i.e. P1,P2,P3,P4 were synthesised.

3 Theory

3.1 Thermal Properties Analysis through TGA

One important factor for the creation and use of devices in various environmental settings is the thermal stability of the produced samples. The loading states and process factors have a major impact on the polymeric materials' thermal stability [31–33]. Consequently, TG analysis has also been used to examine the produced materials' thermal stability. We can better comprehend the thermal characteristics of samples by using the TG analysis to compute additional kinetic parameters. Below is a discussion of these parameters:

3.1.1 Activation Energy (E_a)

The equation is used to examine the thermal activation energy (E_a) of the prepared samples is given below

$$\ln\left[\ln\left(\frac{w_0 - w_f}{w - w_f}\right)\right] = \frac{E_a \theta}{RT_s^2} \quad (1)$$

where w_o is initial, w is remaining weight (at temperature T) and w_f is final weight of sample, R is gas constant and E_a is activation energy and gas constant, respectively.

3.1.2 Frequency Factor (A)

The frequency factor values (A) may be estimated through the relation [34–36].

$$A = \frac{\beta E_a}{RT_s^2}\exp\left(\frac{E_a}{RT_s}\right) \tag{2}$$

where A is frequency factor and β is constant rate of heating, respectively.

3.1.3 Entropy of Activation (ΔS)

The entropy of activation values can be estimated through relation [34–37].

$$\Delta S = 2.303\ R\log\left(\frac{Ah}{kT_s}\right) \tag{3}$$

3.1.4 Free Energy of Change of Decomposition (ΔG)

The free energy of change of decomposition can be estimated through relations

$$\Delta G = E_a - T_s\Delta S \tag{4}$$

4 Results and Discussion

The degradation in weight % for the pristine PANI and its composite samples through TGA analysis in a nitrogen gas environment is shown in Fig. 1. According to TGA analysis of all manufactured samples, there are four phases involved in these samples' weight decrease, which are outlined below.:

1. The first phase of weight loss, known as initial dehydration, is seen at around 100 °C and may be caused by water molecules vaporizing from the prepared sample matrix [38].
2. At around 250 °C, a second stage of reduction in weight is seen, which may be the consequence of protonic acid particles being expelled [38].
3. As the temperature rises further, the third stage of weight loss is identified at around 520 °C, and the fourth stage is noted at about 700 °C. These two processes may have happened as a result of the prepared samples' polymer chains breaking. Around 500–700 °C temperature, the polymer samples begin to change from their solid state to a gaseous one [39].

Sample stability rises with increasing loading concentration in the PANI matrix, indicating a significant connection between PANI molecules and loading particles.

The value of entropy of activation **ΔS** rises with rise in loading concentration which implies that the rate of reaction increases. Further, negative value of ΔS implies that the structure of products is more ordered as compared with that of the reactants [40]. The values of ΔG reduces with the rise in wt.% concentration in PANI matrix for all loadings and this value for all the prepared samples is positive which signify that the

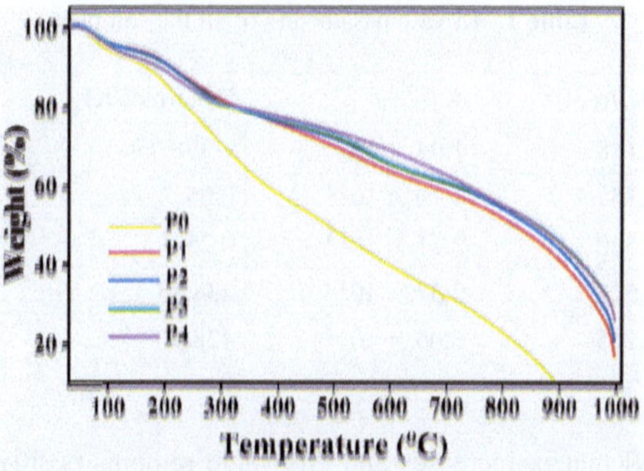

Fig. 1. Weight loss (%) as a function of temperature for samples P0-P4.

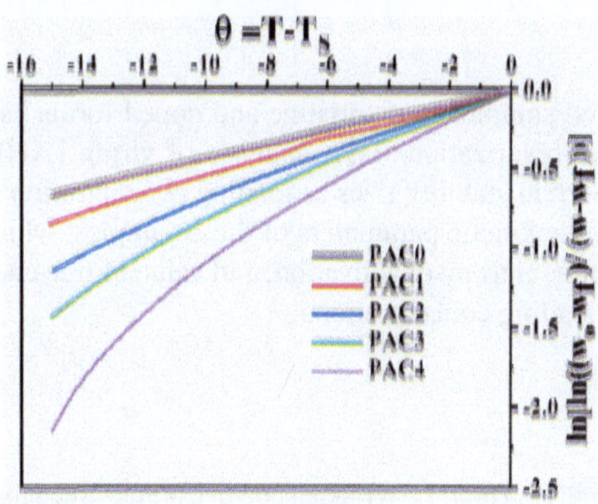

Fig. 2. Relationship between $\ln\left[\ln\left(\frac{w_0-w_f}{w-w_f}\right)\right]$ vs. θ for P0-P4 samples.

chemical reaction of degradation is non-spontaneous [40]. The slope of the line between $\ln\left[\ln\left(\frac{w_0-w_f}{w-w_f}\right)\right]$ and θ parameters, as shown in Fig. 2 and using Eq. (1), may be used to calculate the activation energy of the synthesised samples.

The Kinetic parameters of all the samples calculated by using the equations outlined in Table 1. Upon analysing the values presented in this table, it is evident that an increase in thermal activation energy corresponds with higher loading concentrations of AC across all types of loadings within the PANI matrix. This phenomenon may be attributed to the enhanced reorganisation of molecular arrangements and packing density, among other factors [41]. The frequency factor and thermal activation energy values increase with higher loading concentrations in all samples, likely due to an increased reaction rate and the scission of polymeric chains [41].

The activation entropy value increases with the increase in loading level of Activated Charcoal in PANI, indicating that the reaction rate is enhanced. Additionally, a negative value of ΔS indicates that the product structure exhibits greater order in comparison to that of the reactants [41]. The values of ΔG decrease as the various loading concentration

Table 1. Kinetic parameters of all the samples.

Sample	E_a(KJ/mol)	A (s^{-1})	ΔS(J/mol/K)	ΔG(KJ/mol)
P0	112,188	1.04×10^{10}	−178.77	213,041.8
P1	127,357	1.19×10^{12}	0.05	127,329.1
P2	161,530	2.71×10^{15}	0.25	161,394.8
P3	199,521	2.37×10^{19}	234.88	73,695.33
P4	246,245	6.05×10^{23}	428.96	14,179.07

of AC in the PANI matrix increases and this value remains positive for all prepared samples, indicating that the chemical reaction of degradation is non-spontaneous [41].

5 Conclusion

Various forms of PANI samples in its pristine and doped forms have been synthesized by chemical oxidative polarization. TGA analysis of virgin PANI and its composites shows that PANI's thermal stability rises as loading concentration does. The TGA data was used to compute the kinetic parameters of these samples, which include activation energy, frequency factor, entropy of activation, and value of free energy. The parameters increase with rise in loading concentrations.

References

1. Watanabe, M., Shirai, H., Hirai, T.: Wrinkled polypyrrole electrode for electroactive polymer actuators. J. Appl. Phys. **92**, 4631–4637 (2002)
2. Chang, J.B., Liu, V., Subramanian, V., Sivula, K., Uscombe, C., Murphy, A., Liu, J., Frechet, J.: Printable polythiophene gas sensor array for low-cost electronic noses. J. Appl. Phys. **100**, 014506-07 (2006)
3. Sivakkumar, S.R., Oh, J.S., Kim, D.W.: Polyaniline nanofibres as a cathode material for rechargeable lithium-polymer cells assembled with gel polymer electrolyte. J. Power Sources. **163**, 573–577 (2006)
4. Bundgaard, E., Krebs, F.C.: Low band gap polymers for organic photovoltaics. Solar Energy Mater. Sol. Cells. **91**, 954–985 (2007)
5. Ma, X., Zhang, X., Li, Y., Xu, H., Li, G., Wang, M., Chen, H.: Gas sensing behavior of nanostructured polypyrrole prepared by "carbon nanotubes seeding" approach. Nanopart. Res. **10**, 289–296 (2008)
6. Boix, P.P., Belmonte, G.G., Munecas, U., Neophytou, M., Waldauf, C., Pacios, R.: Determination of gap defect states in organic bulk heterojunction solar cells from capacitance measurements. Appl. Phys. Lett. **95**, 233302-03 (2009)
7. Reineke, S., Lindner, F., Schwartz, G., Gregor, S.N.S., Walzer, K., Lüssem, B., Karl, L.: White organic light-emitting diodes with fluorescent tube efficiency. Nature. **459**, 234–238 (2009)
8. Zhang, L., Meng, F., Chen, Y., Liu, J., Sun, Y., Luo, T., Li, M., Liu, J.: A novel ammonia sensor based on high density, small diameter polypyrrole nanowire arrays. Sensors Actuators B. **142**, 204–209 (2009)

9. Nalage, S.R., Navale, S.T., Patil, V.B.: Polypyrrole-NiO hybrid nanocomposite: structural, morphological, optical and electrical transport studies. Measurement. **46**, 3268–3275 (2013)
10. Pal, R., Goyal, S.L., Gupta, V., Rawal, I.: MnO_2-magnetic Core-Shell structured polyaniline dependent enhanced EMI shielding effectiveness: a study of VRH conduction. Chemistry Select. **4**, 9194–9210 (2019)
11. Xia, Y., Yang, P., Sun, Y., Wu, Y., Mayers, B., Gates, B., Yin, Y., Kim, F., Yan, H.: One-dimensional nanostructures: synthesis, characterization, and application. Adv. Mater. **15**, 353–389 (2003)
12. Rawal, I., Kaur, A.: Spectroscopic investigations of ammonia gas sensing mechanism in polypyrrole nanotubes/nanorods. J. Appl. Phys. **113**, 094504-1-11 (2013)
13. Kuemmeth, F., Ilani, S., Ralph, D.C., McEuen, P.L.: Coupling of spin and orbital motion of electrons in carbon nanotubes. Nature. **452**, 448–452 (2008)
14. Bai, H., Shi, G.: Gas sensors based on conducting polymers. Sensors. **7**, 267–307 (2007)
15. Ishpal, Kaur, A.: Spectroscopic and electrical sensing mechanism in oxidant-mediated polypyrrole nanofibers/nanoparticles for ammonia gas. J. Nanopart. Res. **15**, 1637-1-14 (2013)
16. Omastava, M., Trchova, M., Kovarova, J., Stejskal, J.: Synthesis and structural study of polypyrroles prepared in the presence of surfactant. Synth. Met. **138**, 447–455 (2003)
17. Kaur, A., Ishpal, Dhawan, S.K.: Tuning of EMI shielding properties of polypyrrole nanoparticles with surfactant concentration. Synth. Met. **162**, 1471–1477 (2012)
18. Rawal, I., Kaur, A.: Synthesis of mesoporous polypyrrole nanowires/nanoparticles for ammonia gas sensing application. Senors Actuat. A. **203**, 92–102 (2013)
19. Carswell, A.D.W., O'Rear, E.A., Grady, B.P.: Adsorbed surfactants as templates for the synthesis of morphologically controlled polyaniline and polypyrrole nanostructures on flat surfaces: from spheres to wires to flat films. J. Am. Chem. Soc. **125**, 14793–14800 (2003)
20. Dai, T., Yang, X., Lu, Y.: Controlled growth of polypyrrole nanotubule/wire in the presence of a cationic surfactant. Nanotechnology. **17**, 3028–3034 (2006)
21. Feng, J., Yan, W., Zhang, L.: Synthesis of polypyrrole micro/nanofibers via a self-assembly process. Microchim. Acta. **166**, 261–267 (2009)
22. Li, X., Guanglan, J., He, G., Zheng, W., Tan, Y., Xiao, W.: Preparation of porous PPy/TiO_2 composites: improved visible light photoactivity and the mechanism. Chem. Eng. J. **236**, 480–489 (2014)
23. Chitte, H.K., Bhat, N.V., Walunj, V.E., Shinde, G.N.: Synthesis of polypyrrole using ferric chloride (FeCl3) as oxidant together with some dopants for use in gas sensors. J. Sens. Technol. **1**, 47–56 (2011)
24. Chitte, H.K., Bhat, N.V., Gore, A.V., Shinde, G.N.: Synthesis of polypyrrole using ammonium peroxy disulfate (APS) as oxidant together with some dopants for use in gas sensors. Mater. Sci. Appl. **2**, 1491–1498 (2011)
25. Stamenov, P., Madathil, R., Coey, J.M.D.: Dynamic response of ammonia sensors constructed from polyaniline nanofibre films with varying morphology. Sensors Actuators B. **161**, 989–999 (2012)
26. Rai, A.R.U., Jun, A.P., Walaiporn, O., Sirisart, O.: Synthesis of highly conductive polypyrrole nanoparticles via microemulsion polymerization. J Met. Mater. Miner. **18**, 27–31 (2008)
27. Yang, J.E., Jang, I., Kim, M., Baeck, S.H., Hwang, S., Shim, S.E.: Electrochemically polymerized vine-like nanostructured polyaniline on activated carbon nanofibers for supercapacitor. Electrochim. Acta. **111**, 136–143 (2013)
28. Kwon, W.J., Suh, D.H., Chin, B.D.: Preparation of polypyrrole nanoparticles in mixed surfactants system. J. Appl. Polym. Sci. **110**, 1324–1329 (2008)
29. Zhang, X., Zhang, J., Song, W., Liu, Z.: Controllable synthesis of conducting polypyrrole nanostructure. J. Phys. Chem. B. **110**, 1158–1165 (2006)

30. Stejskal, J., Omastova, M., Fedorova, S., Prokes, J., Trchova, M.: Polyaniline and polypyrrole prepared in the presence of surfactants: a comparative conductivity study. Polymer. **44**, 1353–1358 (2003)
31. Hazarika, J., Kumar, A.: 160 MeV Ni^{12+} ion irradiation effects on the structural, optical and electrical properties of spherical polypyrrole nanoparticles. Nucl. Inst. Methods Phys. Res. B. **318**, 269–275 (2014)
32. Zhu, D., Guo, Q., Zheng, Z., Matsuo, M.: Thermal and dielectric investigations on the blends of κ-carrageenan and nano-structured polyaniline. Synth. Met. **187**, 165–171 (2014)
33. Horowitz, H.H., Metzger, G.: A new analysis of thermogravimetric traces. Anal. Chem. **35**, 1464–1468 (1963)
34. Gupta, R., Kumar, V., Goyal, P.K., Kumar, S., Goyal, S.L.: Thermal and electrical behaviour of 100 keV N+ and Ar+ ion implanted poly (methyl methacrylate) (PMMA) polymer. Adv. Appl. Sci. Res. **3**, 2766–2773 (2012)
35. Asha, G.S.L., Jain, D., Kishore, N.: Structural, thermal and electrical properties of polyaniline/CrO3 composites. J. Chem. Pharm. Res. **6**, 105–113 (2013)
36. Asha, G.S.L., Kumar, D., Kumar, S., Kishore, N.: Synthesis and characterization of polyaniline/TiO_2 composites. Indian J. Pure Appl. Phys. **52**, 341–347 (2014)
37. Ganesan, R., Gedanken, A.: Organic–inorganic hybrid materials based on polyaniline/TiO_2 nanocomposites for ascorbic acid fuel cell systems. Nanotechnology. **19**, 435709–435714 (2008)
38. Dumitrescu, I., Nicolae, C.A., Mocioiu, A.M., Gabor, R.A., Grigorescu, M., Mihailescu, M.: Synthesis and characterization of conductive polymers with enhanced solubility. UPB Sci. Bull. **71**, 63–72 (2009)
39. Kalsi, P.C., Mudher, K.D.S., Psey, A.K., Iyer, R.H.: Thermal studies on unirradiated and γ-irradiated polymer of allyl diglycol carbonate. Thermochim. Acta. **254**, 331–336 (1995)
40. Mallikarjun, K.G.: Thermal decomposition kinetics of Ni(II) chelates of substituted chalcones. E-J. Chem. **1**, 105–109 (2004)
41. Gupta, R., Kumar, V., Goyal, P.K., Kumar, S.: Thermogravimetric analysis of heat treated CR-39 polymer. J. Chem. Pharm. Res. **2**, 629–634 (2010)

Enhanced Dielectric Properties of SrTiO3-Ethyl Cellulose Nanocomposites

Bhavya Monga[1,2(✉)], Harish Kumar Meena[3], Karishma Jain[4], Ankur Jain[1,5], and B. Tripathi[2]

[1] Centre for Renewable Energy, Suresh Gyan Vihar University, Jaipur 302004, India
bhavayinsan@gmail.com
[2] Department of Physics, S. S. Jain Subodh P.G. (Autonomous) College, Jaipur 303007, India
[3] Department of Physics, University of Rajasthan, Jaipur 302004, India
[4] Department of Physics, School of Physical Sciences, Manipal University, Jaipur 303007, India
[5] Natural Science Centre for Basic Research and Development, Hiroshima University, Hiroshima 739-8530, Japan

Abstract. This study investigates dielectric properties of $SrTiO_3$/Ethyl Cellulose ($SrTiO_3$/EC) polymer nanocomposites. $SrTiO_3$ nanoparticles (1–5 wt%) were dispersed in EC using ultrasonication, and the resulting nanocomposites were characterized using XRD, FT-IR, SEM, and impedance spectroscopy. The incorporation of $SrTiO_3$ nanoparticles into the EC polymer matrix significantly improved the dielectric permittivity of the nanocomposites. This enhancement was attributed to the interfacial polarization between $SrTiO_3$ and EC, as well as the increased number of dipoles introduced by the nanoparticles. The dielectric response was found to be dependent on the $SrTiO_3$ loading content, with an optimal loading range identified for maximum dielectric performance. Additional factors influencing the dielectric behavior of $SrTiO_3$/EC nanocomposites included the size and distribution of $SrTiO_3$ nanoparticles, as well as the processing conditions used for nanocomposite preparation. This study aims to optimize the dielectric properties of $SrTiO_3$/EC nanocomposites for potential applications in energy storage devices, capacitors, and other electronic components.

Keywords: $SrTiO_3$ · Ethyl Cellulose · Dielectric response · Energy Storage

1 Introduction

Dielectric materials play an important role in modern electronics, enabling various applications such as capacitors, sensors, and energy storage devices. Polymer-based nanocomposites have emerged as promising candidates because of the lightweight nature, ease of processing, and tunable dielectric properties. In this study, we focus on $SrTiO_3$-Ethyl Cellulose ($SrTiO_3$-EC) nanocomposites, where $SrTiO_3$ nanoparticles are incorporated into an EC matrix. $SrTiO_3$, known for its high dielectric constant, is expected to enhance the dielectric properties of the polymer matrix. Polymer nanocomposites exhibit high potential for advanced electronic applications due to their tunable dielectric properties

S. Kumari et al. (Eds.): ICMTP 2024, SPPHY 311, pp. 395–404, 2025.
https://doi.org/10.1007/978-981-96-6795-6_44

[20]. Among various polymer-ceramic systems, $SrTiO_3$, a perovskite oxide with high dielectric constant, offers significant potential when incorporated into polymer matrices. This study focuses on investigating the dielectric properties of $SrTiO_3$/Ethyl Cellulose ($SrTiO_3$/EC) polymer nanocomposites.

$SrTiO_3$ nanoparticles were homogeneously dispersed within the EC matrix at varying concentrations (1–5 wt%) through ultrasonication. The structural, morphological, and dielectric properties of the resulting nanocomposites were systematically characterized using techniques such as X-ray diffraction (XRD), Fourier Transform Infrared spectroscopy (FT-IR), Scanning Electron Microscopy (SEM), and impedance spectroscopy.

This research aims to elucidate the influence of $SrTiO_3$ nanoparticle loading on the dielectric permittivity of the EC matrix and identify the factors that govern the dielectric response of these nanocomposites. The findings of this study are expected to provide valuable insights for optimizing the dielectric performance of $SrTiO_3$/EC nanocomposites and exploring their potential applications in energy storage devices, high-performance capacitors, and other electronic components.

2 Experimental

$SrTiO_3$ nanoparticles were dispersed in EC using ultrasonication at varying concentrations (1–5 wt%). To investigate the properties of the resulting nanocomposites, we employed XRD, FTIR, SEM, and EDX. Dielectric properties were measured using impedance spectroscopy over a wide frequency range.

Placed EC solution for ultrasonication and added the measured amount of STO powder to the EC solution. Ultrasonication is a process that uses high-frequency sound waves to break down particles and create a more uniform dispersion. Ultrasonication in benzene can effectively disperse nanoparticles and improve their interaction with the solvent and other components in the sample. The cavitation generated by ultrasound can break down agglomerates and create a more homogeneous mixture. Here, the EC solvent is placed in an ultrasonication bath to facilitate the dispersion of the STO nanoparticles. Once the EC solution is subjected to ultrasonication, the measured amount of STO powder is carefully added to the solution. After that, carefully transferred the STO-EC dispersion onto a clean substrate ensuring uniform spreading of the dispersion to achieve a consistent film thickness.

The substrate could be a glass slide, silicon wafer, or any other suitable material. This is important for achieving reproducible results in subsequent experiments. Allowed the thin film to dry at room temperature or under controlled conditions to remove residual solvent. Finally, the thin film is allowed to dry to remove any residual solvent. This can be done at room temperature or under controlled conditions, such as in an oven or a vacuum chamber. The drying process ensures that the STO-EC film is stable and ready for further analysis or application.

Overall, the process is a standard procedure for preparing thin films of nanoparticles dispersed in a polymer matrix. This technique is widely used in various fields, including electronics, optics, and materials science.

3 Results and Discussion

3.1 XRD

XRD analysis evidenced the presence of $SrTiO_3$ peaks in the nanocomposites, indicating the crystalline nature of the nanoparticles. The XRD graph presents the diffraction pattern of a material, typically used to identify its crystalline structure. In this case, the graph likely represents the XRD spectra for ST/EC composites, as indicated in the caption. The Fig. 1 presents two distinct XRD spectra, labeled (a) and (b).

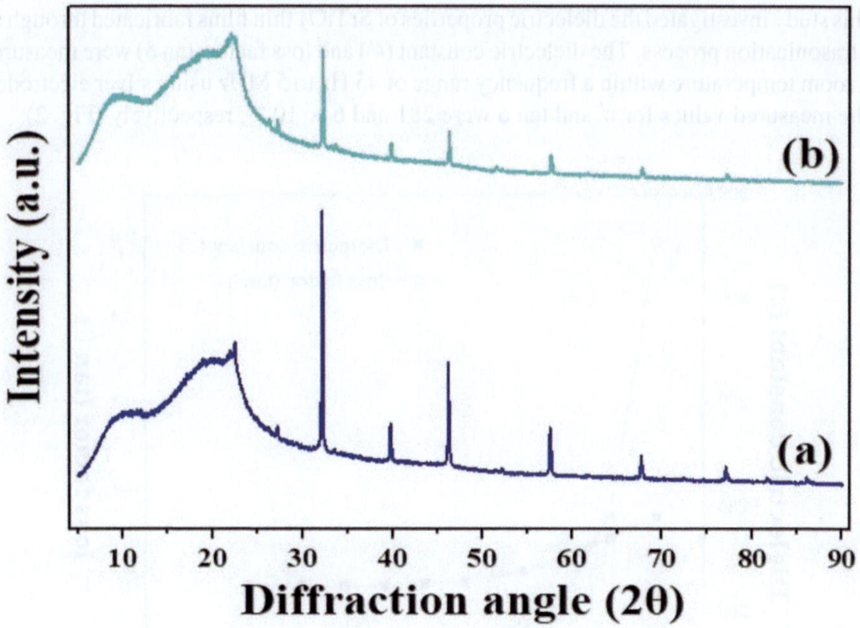

Fig. 1. XRD spectra for ST/EC composite

Both spectra exhibit numerous sharp peaks at specific diffraction angles (2θ). The prominent peaks are located around 2θ values of 22.8°, 32.4°, 39.9°, 46.5°, and 57.7°. These peaks are characteristic of crystalline materials, indicating the presence of well-defined crystal structures. The peak positions in both spectra are same with the known diffraction pattern of strontium titanate. This confirms the presence of SrTiO3 in the polymer matrix. The spectrum (a) likely represents the XRD pattern of the pristine ethyl cellulose polymer. The broader peaks and lower intensity compared to (b) suggest a less crystalline structure, which is expected for a polymer matrix. The (b) spectrum, with its sharper and more intense peaks, indicates a higher degree of crystallinity [13]. The presence of $SrTiO_3$ nanoparticles within the polymer matrix has likely induced this increased crystallinity. The nanoparticles can act as nucleation sites, promoting the formation of more ordered polymer chains. The broadening of some peaks in both spectra could be attributed to factors such as small crystallite size, strain within the material, or instrumental limitations [18]. The relative intensities of the peaks can provide information about the preferred orientation of the crystallites within the material. Any slight shifts in the peak positions compared to the standard $SrTiO_3$ pattern could indicate lattice strain or solid solutions may form between the $SrTiO_3$ and the polymer matrix.

Overall, the XRD spectra provide evidence for the successful incorporation of SrTiO3 nanoparticles into the ethyl cellulose polymer matrix and suggest that the presence of the nanoparticles has influenced the crystallinity of the polymer.

3.2 Impedance Spectroscopy

This study investigated the dielectric properties of $SrTiO_3$ thin films fabricated through an ultrasonication process. The dielectric constant (ε') and loss factor ($\tan \delta$) were measured at room temperature within a frequency range of 45 Hz to 5 MHz using silver electrodes. The measured values for ε' and $\tan \delta$ were 281 and 6×10^{-4}, respectively (Fig. 2).

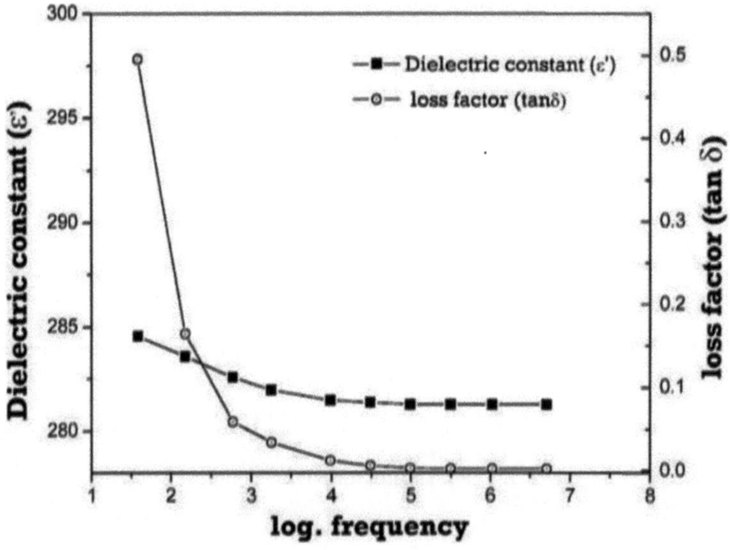

Fig. 2. Variation of dielectric constant (ε') and loss factor ($\tan \delta$) of $SrTiO_3$ with frequency

A review of existing literature reveals that the measured loss factor for this ultrasonically synthesized $SrTiO_3$ film is significantly lower than those reported for other $SrTiO_3$ materials. For instance, Krupka et al. [35] reported dielectric losses on the order of 10^{-4} for $SrTiO_3$ single crystals. Phillips [36] observed a loss factor of 6×10^{-2} at 100 K and 300 GHz. Studies on pulse-deposited [37] and RF-sputtered [38] $SrTiO_3$ thin films have shown loss tangents ranging from 0.001 to 0.01 at 1 kHz and 0.019 to 0.015 at 100 kHz, respectively. Tkach et al. [22] investigated the impact of non-stoichiometry on the dielectric properties of $SrTiO_3$ at low temperatures and reported loss factor values around 10^{-3} at temperatures below 100 K and a frequency of 10 kHz.

In summary, the measured loss factor of 6×10^{-4} for the $SrTiO_3$ film synthesized in this study using ultrasonication represents a remarkably low value compared to previously reported values for sintered $SrTiO_3$.

The dielectric constant increased with increasing $SrTiO_3$ loading up to an optimal concentration, followed by a slight decrease. This behavior is attributed to the interfacial polarization between $SrTiO_3$ and EC, as well as the increased dipole density introduced by the nanoparticles. The loss tangent remained low across the entire frequency range, indicating minimal energy dissipation and good dielectric performance.

3.3 (FT-IR)

The Fig. 3 presents FTIR spectra for three different ST/EC composite samples with varying ST (Strontium Titanate) content (5%, 3%, and 1%). The spectra exhibit distinct peaks and troughs, indicating the presence of various functional groups and chemical bonds within the composite materials.

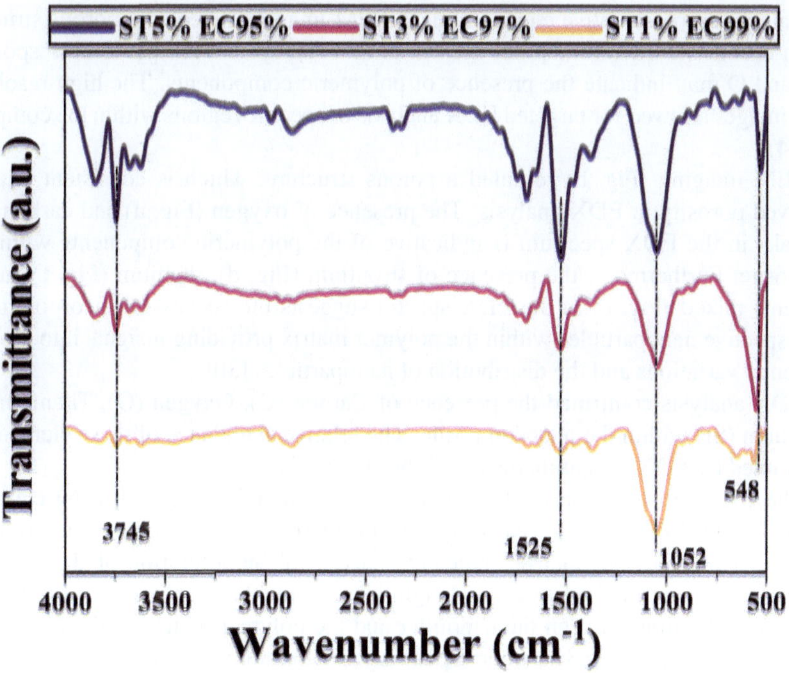

Fig. 3. FTIR spectra for ST/EC composites

The broad peak observed in all spectra around 3400 cm^{-1} is characteristic of the stretching vibration of the O-H bond. This peak is likely related with the presence of hydroxyl groups (O-H) within the polymer matrix or potentially from adsorbed moisture. Peaks in this region (2900–3000 cm^{-1}) are indicative of C-H stretching vibrations, suggesting the presence of aliphatic hydrocarbon chains within the polymer matrix.

The presence of a peak around 1700 cm^{-1} suggests the presence of carbonyl groups (C-O) in the polymer matrix. This could be due to the presence of carbonyl groups in the polymer backbone or from any carbonyl-containing additives or impurities. C-O Stretching Vibration (Peak around 1100-1200 cm^{-1}) The peak in this region is likely associated with the C-O stretching vibration of ether or ester groups within the polymer matrix [26]. The region below 1500 cm^{-1} is often referred to as the fingerprint region as it contains a complex set of peaks that are characteristic of the specific polymer structure and any additives present.

Here graph is representing O-H stretching and C=C bonding in the composite. Ethyl cellulose being a polymer, is likely held together by covalent bonds.

3.4 SEM and EDX

Elemental analysis of the polymer nanocomposite was performed using Energy Dispersive X-ray Spectroscopy (EDX) with Scanning Electron Microscopy (SEM). Prior to analysis, the nanocomposite sample was subjected to ultrasonication to ensure adequate dispersion and minimize particle aggregation [12]. This step is crucial for obtaining accurate and representative elemental data. SEM imaging revealed a porous structure, which is consistent with the pores present in EDX analysis. THE peaks corresponding to C and O may indicate the presence of polymeric components. The high-resolution SEM images allowed for targeted EDX analysis of specific regions within the composite (Fig. 4).

SEM imaging (Fig. b) revealed a porous structure, which is consistent with the observed porosity in EDX analysis. The presence of oxygen (Fig. a) and carbon (Fig. e) peaks in the EDX spectrum is indicative of the polymeric components within the composite. Furthermore, the presence of strontium (Fig. d), titanium (Fig. f), and all elements mixed (Fig. c) in the EDX spectra suggests the successful incorporation of the respective nanoparticles within the polymer matrix providing insights into the local elemental variations and the distribution of nanoparticles [30].

EDX analysis confirmed the presence of Carbon (C), Oxygen (O), Titanium (Ti), Strontium (Sr) within the nanocomposite. The relative abundance of these elements, as determined by EDX, is summarized in Table 1 (Fig. 5).

The spatial distribution of the elements was further investigated using elemental mapping. The EDX maps revealed a uniform distribution of titanium and strontium throughout the polymer matrix, indicating successful incorporation of the strontium titanate nanoparticles. The absence of significant elemental segregation suggests good interfacial adhesion between the nanofiller and the polymer matrix [10]. A prominent peak is observed around 0.3 keV, corresponding to the characteristic K-shell emission line of carbon. This is expected as carbon is a common element in organic materials. Another strong peak is visible around 0.5 keV, indicating the presence of oxygen. This could be due to the presence of oxygen-containing functional groups in the sample. A smaller peak is observed around 4.5 keV, suggesting the presence of titanium. This is likely due to the presence of titanium-containing compounds in the sample. A series of peaks are observed in the higher energy region (around 1.8 keV), corresponding to the L-shell emission lines of strontium. This indicates the presence of strontium in the sample. Based on the presence of these elements, the sample likely contains organic compounds (indicated

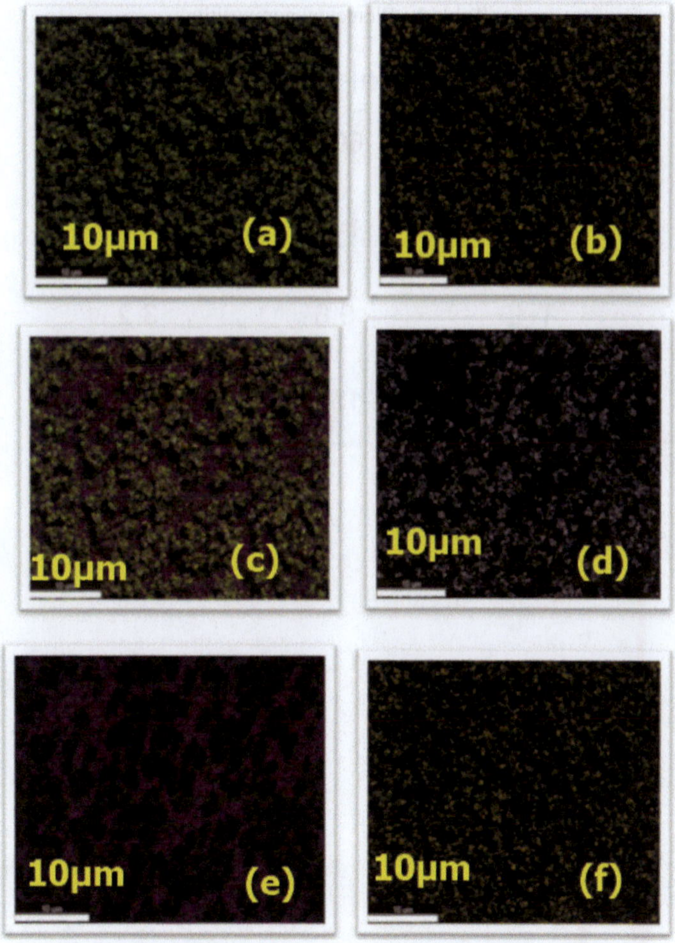

Fig. 4. SEM images for ST/EC composites

by the carbon and oxygen peaks) and inorganic compounds containing titanium and strontium.

The presence of Titanium (Ti) and Strontium (Sr) is attributed to the assimilation of strontium titanate nanoparticles into the polymer matrix. The accurate determination of the elemental composition and its spatial distribution is essential for understanding the chemical interactions between the polymer matrix and the nanofiller, which play a crucial role in influencing the electrochemical performance of the supercapacitor.

Table 1. Elemental composition

Element	Weight %	Atomic %
C K	60.7	79.9
O K	14.3	14.1
Ti L	9.6	3.2
Sr L	15.4	2.8

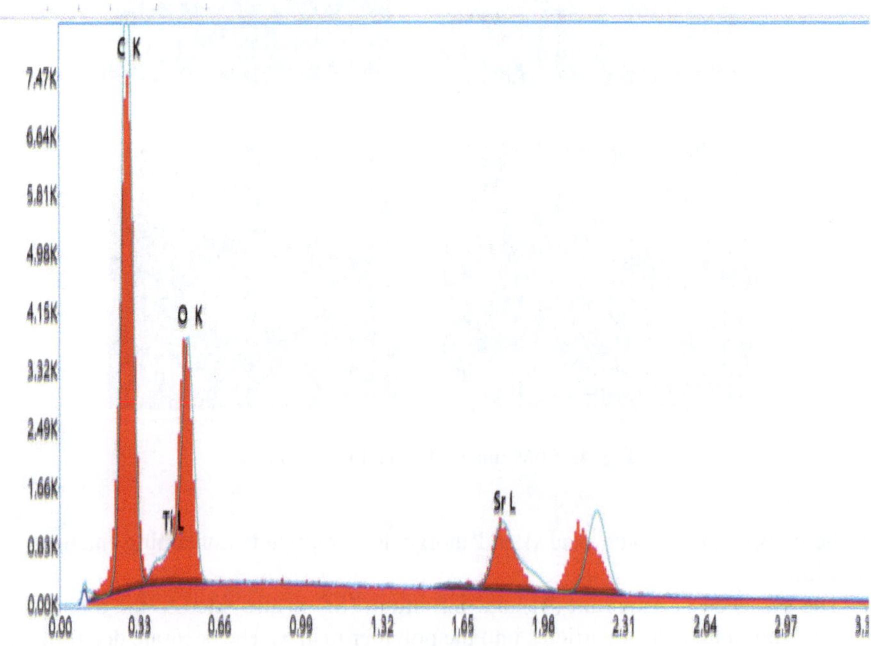

Fig. 5. EDX Mapping

4 Conclusion

This study successfully demonstrated the organization of $SrTiO_3$ nanoparticles into an ethyl cellulose polymer matrix. XRD analysis confirmed the crystalline nature of $SrTiO_3$ and its presence within the polymer, while SEM-EDX revealed the successful dispersion of nanoparticles and provided insights into the elemental composition and distribution. The dielectric properties of the nanocomposites exhibited promising results, with a low loss factor and a significant increase in dielectric constant attributed to the presence of $SrTiO_3$. FTIR analysis confirmed the presence of key functional groups in the polymer matrix. These findings suggest that the developed $SrTiO_3$/ethyl cellulose nanocomposites have the potential for applications in numerous fields, with emphasis on energy storage devices, dielectric materials, and sensor technologies. Further investigations into optimizing nanoparticle loading, exploring different polymer matrices, and evaluating the long-term stability of the nanocomposites are warranted to fully realize their potential.

References

1. Prateek, Bhunia, R., Siddiqui, S., Garg, A., Gupta, R.K.: Significantly enhanced energy density by tailoring the interface in hierarchically structured TiO_2–$BaTiO_3$–TiO_2 nanofillers in PVDF-based thin-film polymer nanocomposites. ACS Appl. Mater. Interfaces, **11**, 14329–14339. (2019)
2. Ming, Y., Zhou, Z., Hao, T., Nie, Y.: Polymer nanocomposites: role of modified filler content and interfacial interaction on crystallization. Eur. Polym. J. **162**, 110894 (2022)
3. Dutcher, J.R., Dalnoki-Veress, K., Nickel, B.G., Roth, C.B.: Instabilities in thin polymer films: from pattern formation to rupture. In: Macromolecular Symposia, vol. 159, pp. 143–150. WILEY-VCH.: Hoboken, NJ, USA, Weinheim (2000)
4. Peña, M.A., Fierro, J.L.: Chemical structures and performance of perovskite oxides, Chem. Inform., **101**, 1981–2017 (2001)
5. Nuraje, N., Su, K.: Perovskite ferroelectric nanomaterials, Nanoscale. **5**, 8752–8780 (2013)
6. Fletcher, N.H., Hilton, A.D., Ricketts, B.W.: Optimization of energy storage density in ceramic capacitor, J. Phys. D. Appl. Phys. **29**, 253–258 (1998)
7. Fujimoto, M., Kingery, W.D.: J. Am. Ceram. Soc. **68**, 169–173 (1985)
8. Chen, M.Y., Juuti, J., Hsi, C.S., et al.: Dielectric $BaTiO_3$–BBSZ glass ceramic composition with ultra-low sintering temperature, J. Eur. Ceram. Soc. **35**, 139–144 (2015)
9. Conway, B.E.: Electrochemical supercapacitors. Kluwer-Academic (1999)
10. Lokhande, C.D., Gujar, T.P., Shinde, V.R., Mane, R.S., Han, S.H.: Electrochemical supercapacitor application of pervoskite thin film. Electrochem. Commun. 9, 1805–1809 (2007)
11. Zeng, X., Song, H., Shen, Z., Moskovitsb, M.: J. Mater. **7**, 1198–1224 (2021)
12. Cao, Y., Tang, P., Han, Y., Qiu, W.: Chem. Eur. J., 26 (2020)
13. Tsuei, C.C., Kirtley, J.R.: Phys. Rev. Lett. **85**, 182–185 (2000)
14. Sun, C., Alonso, J.A., Bian, J.: Adv. Energy Mater., 11 (2021)
15. Lin, Y., Norman, C., Srivastava, D., Azough, F., Wang, L., Robbins, M., Simpson, K., Freer, R., Kinloch, I.A.: Thermoelectric power generation from lanthanum strontium titanium oxide at room temperature through the addition of graphene. ACS Appl. Mater. Interfaces **7**, 15898–15908 (2015)
16. Kan, D., Terashima, T., Kanda, R., Shimakawa, Y., Takano, M.: Nat. Mater. **4**, 816–819 (2005)
17. Srivastava, D., Norman, C., Azough, F., Schäfer, M. C., Guilmeau, E., Freer, R.: Improving the thermoelectric properties of $SrTiO_3$-based ceramics with metallic inclusions. J. Alloys Compd. **731**, 723–730 (2018)
18. Ohta, S. et al.: Large thermoelectric performance of heavily Nb-doped SrTiO3 epitaxial film at high temperature. Appl. Phys. Lett. **87**, 092108–092110 (2005)
19. Moreno, J., Dominguez, J.M., Montoya, A., Vicente, L., Viveros, T.: J. Mater. Chem. **5**, 509–512 (1995)
20. Thomas, J.K., Padmakumar, H., Solomon, S., George, C.N., Joy, K., Koshy, J.: Phys. Stat. Sol. (a). **204**, 3102–3107 (2007)

21. Zhang, D., Liu, W., Guo, R., Zhou, K., Luo, H.: High discharge energy density at low electric field using an aligned titanium dioxide/lead zirconate titanate nanowire array, Adv. Sci. **5** (2), 1700512 (2018)
22. Shen, Y., Lin, Y., Zhang, Q.M.: Polymer nanocomposites with high energy storage densities. MRS Bull., **40** (9) 753–759 (2015)
23. Hao, X.: A review on the dielectric materials for high energy-storage application. J. Adv. Dielectr. **03**, 1330001 (2013)
24. Peng, B., Xie, Z., Yue, Z., Li, L.: Improvement of the recoverable energy storage density and efficiency by utilizing the linear dielectric response in ferroelectric capacitors, Appl. Phys. Lett., **105**, 052904 (2014)
25. Burn, I., Smyth, D.M.: Energy storage in ceramic dielectrics, J. Mater. Sci., **7**(3), 339–343 (1972)
26. Love, G.R.: Energy storage in ceramic dielectrics, J. Am. Ceram. Soc., **73** (2), pp. 323–328 (1990)
27. Li, Z., Liu, F., Yang, G., Li, H., Dong, L., Xiong, C., Wang, Q.: Enhanced energy storage performance of ferroelectric polymer nanocomposites at relatively low electric fields induced by surface modified BaTiO 3 nanofibers, Compos. Sci. Technol. **164**, pp. 214–221 (2018)
28. Zhang, X., Jiang, J., Shen, Z., Dan, Z., Li, M., Lin, Y., Nan, C.W., Chen, L., Shen, Y.: Polymer nanocomposites with ultrahigh energy density and high discharge efficiency by modulating their nanostructures in three dimensions, Adv. Mater. **30** (16), p. e1707269 (2018)
29. Chen, G., Zhang, P., Pan, L., Qi, L., Yu, F., Gao, C.: Flexible nonvolatile resistive memory devices based on $SrTiO_3$nanosheets and polyvinylpyrrolidone composites. J. Mater. Chem. C. **5**, 9799–9805 (2017)
30. Zhang, W.F., Xing, Q., Huang, Y.B.: Mod. Phys. Lett. B. **14**, 709–716 (2000)
31. Hu, H.L., Zhang, F., Lim, S., Blanloeuil, P., Yao, Y., Guo, Y., Wang, C.H.: Surface functionalisation of carbon nanofiber and barium titanate by polydopamine to enhance the energy storage density of their nanocomposites. Compos. Part B Eng. **178**, 107459 (2019)
32. Li, W.J., Meng, Q.J., Zheng, Y.S., Zhang, Z.C., Xia, W.M., Xu, Z.: Electric energy storage properties of poly(vinylidene fluoride). Appl. Phys. Lett. **96**, 192905 (2010)
33. Gong, L., Chen, S.-H., Zhan, S.-P., Sun, X.-D., Yin, B.O., Liu, Z.-Y., Yang, M.-B., Yang, Y.I.: An enhancement on the dielectric performance of poly(vinylidene fluoride)-based composite with graphene oxide-$BaTiO_3$ hybrid. Nanocomposite Films. **5** (2), 61–66 (2019)
34. Hu, J., Liu, Y., Zhang, S., Tang, B.: Novel designed core–shell nanofibers constituted by single element-doped $BaTiO_3$ for high-energy–density polymer nanocomposite films. Chem. Eng. J. **428** (2022)
35. Dalwadi, S. et al.: The integration of biopolymer-based materials for energy storage applications: a review. Int. J. Mol. Sci.. **24**(4), 3975 (2023)
36. Wen, Q. et al.: Si-based polymer-derived ceramics for energy conversion and storage. J. Adv. Ceram. **11**, 197–246 (2022)
37. Behera et al. A review on polyvinylidene fluoride polymer based nanocomposites for energy storage applications. J. Energy Storage. **48**, 103788 (2022)
38. Li, Z., Shen, Z., Yang, X., Zhu, X., Zhou, Y., Dong, L., Xiong, C., Wang, Q.: Ultrahigh charge-discharge efficiency and enhanced energy density of the sandwiched polymer nanocomposites with poly(methyl methacrylate) layer. Compos. Sci. Technol. **202**, 108591 (2021)
39. Roy, P.K., Bera, J.: Mater. Res. Bull. **40**, 599–604 (2005)

Frequency Dispersion Dielectric and Electrical Properties of $MgFe_xCr_{2-x}O_4$ (x = 0.6, 0.8) Chromites

Yogita Sahu and Sadhana Agrawal(✉)

Department of Physics, National Institute of Technology Raipur, Raipur, Chhattisgarh, India
sagrawal.phy@nitrr.ac.in

Abstract. The properties of $MgFe_xCr_{2-x}O_4$ (x = 0.6, 0.8) chromite synthesized via solution combustion method were investigated. The XRD pattern confirmed phase formation of the compounds. The FTIR spectroscopy was used to obtain the transmittance spectra which determined the bands around 460.31 and 575.41 cm^{-1}, corresponding to the stretching and bending vibration of metal oxide bonds. The frequency dispersion dielectric parameters (ε′ and ε″) were examined using dielectric spectroscopy in various frequencies (100 Hz–100 kHz) and temperatures. The ε' and ε" values decreased with frequency owing to the ceasing effect of polarization. It can be explained according to Maxwell-Wagner's model on the basis of space charge polarization. The conductivity of $MgFe_xCr_{2-x}O_4$ chromites was increased due to charge carriers hopping between Cr^{2+} and Cr^{3+} ions, followed by Jonscher's power law. A single semicircle arc of cole – cole plot suggested non–Debye relaxation present in the chromites. The higher value of ε' was observed for x = 0.8 mol% of Fe^{3+} ions showing the utilization of $MgFe_xCr_{2-x}O_4$ chromites for various device applications such as electrode materials, capacitors, MOSFET and other energy storage devices, etc.

Keywords: Dielectric properties · AC conductivity · X-ray diffraction · FTIR spectroscopy

1 Introduction

The spinel-structured transition metal chromites have attracted continuous interest from researchers because of their chemical stability and durability. These materials have a potential application in many electronic and energy storage devices because of their low dielectric loss and high dielectric constant [1].

Chromites with the general formula ACr_2O_4, exhibit a cubic spinel structure with an Fd3m space group. In this structure, the A site allows for the substitution of various transition metals, alkali, and alkaline earth elements, providing structural flexibility to the chromites. Here, the oxygen (O^{2-}) ions form a fcc lattice, creating two types of interstitial sites: tetrahedral (A) and octahedral (B) sites, both surrounded by O^{2-} ions. In ACr_2O_4 chromites, A^{2+} ions generally occupy the A site, while Cr^{3+} ions occupy the B site due to their large CFSE [2, 3]. CFSE stands for crystal field stabilization energy, represents the

S. Kumari et al. (Eds.): ICMTP 2024, SPPHY 311, pp. 405–420, 2025.
https://doi.org/10.1007/978-981-96-6795-6_45

difference in energy among electrons of d orbitals. It is significant to understand various properties of chromites, such as their stability, reactivity, and dielectric behaviour. Owing to their interesting properties, chromites have various applications in electrical devices [4], multiferroic memories [5], solid oxide fuel cells [6], disk recording, and energy storage devices [7] and also proposed as photocatalyst materials for pollutant degradation and H_2 gas release [8].

Among spinel chromites, $MgFe_xCr_{2-x}O_4$ (x = 0.6, 0.8) chromites have attracted significant interest due to their high ε' [9]. The characteristics of charges and their distribution across various localized sites, significantly influence the dielectric polarization. The movement of electrons between Cr ions is important in this process [10]. In low-frequency regions, polarization is caused by electron exchange between Cr^{2+} and Cr^{3+} ions resulting in a higher dielectric constant. However, as frequency increases, the transfer of electrons between Cr ions is unable to follow the applied field, resulting in a decrease in polarization. Consequently, a lower dielectric constant is observed [11]. Using dielectric spectroscopy, various dielectric parameters (like ε', ε") and ac conductivity were studied at frequencies and temperatures.

In recent years, numerous methods such as solid-state methods [12], auto-combustion [13], hydrothermal [14], thermal decomposition [15], co-prepetition method [16], sol-gel, solution combustion [17] are available for the preparation of compounds. The solution combustion method is mostly utilized due to its low-temperature formation, cost effectiveness, and less time-consumption [18]. Therefore, this work reports the synthesis of $MgFe_xCr_{2-x}O_4$ chromites using the solution combustion method. The study also includes in depth analysis of the dielectric and electrical behaviour in the various frequencies and temperatures.

2 Material Synthesis

The stoichiometric amounts of raw materials i.e., Fe $(NO_3)_2$.$9H_2O$, Cr $(NO_3)_2$.$9H_2O$, Mg $(NO_3)_2$.$6H_2O$ (all having ≥99% purity) and citric acid, sourced from Sigma – Aldrich, were separately dissolved in 10 ml of distilled water. These individual nitrate solutions were combined and continuously stirred on a magnetic stirrer with a constant heat supply. Simultaneously, citric acid solution was added dropwise to the mixture in 11:molar ratio. As the temperature increased, the solution gradually evaporated and around 200 ° C ignition took place. The resulting dry, ignited powder was manually crushed for 20–25 min and then placed into a crucible. It was then calcined in a furnace at 1200 °C. Further, for the dielectric measurement, the calcined powder was mixed with 2–3 drops of PVA and converted into circular pellets (10 mm diameter) using a hot press machine at a pressure of 1500 kg/cm^2. These pellets were sintered at 1000 °C for 1 h. To ensure better ohmic contact, a thin layer of conductive paste was applied to both sides of pellets and dried at 200 °C.

3 Structural Analysis

3.1 XRD Analysis

The X-ray diffraction spectra (Fig. 1) of $MgFe_xCr_{2-x}O_4$ (x = 0.6, 0.8) chromites were obtained using a PANalytical X'Pert Powder Diffractometer (λ = 1.5406 Å). The pattern is well matched with the standard pattern of magnesium chromite with ICSD No. 98–007-5625, indicating the formation of (Fd-3m space group (227)) face-centered cubic crystal symmetry [19]. As seen in Fig. 1, all the diffracted peaks observed at 30.29, 35.59, 37.27, 43.35, 53.74, 57.15, 62.84, and 66.18 (different 2θ values) are indexed to the corresponding hkl values 220, 311, 222, 400, 422, 511, 440, and 531 respectively, with 311 prominent peaks. There are no extra peaks observed which confirms the single-phase spinel structure of $MgFe_xCr_{2-x}O_4$ (x = 0.6, 0.8) chromites [20]. The D (average crystallite size) of $MgFe_xCr_{2-x}O_4$ (x = 0.6, 0.8) was calculated using Debye Scherer's formula [21] (given in Eq. 1) and it was found to be 25.04 and 24.70 nm. The unit cell volume (V_{cell}) [9] and lattice parameters [22] of the prepared chromites were calculated using well-known equations (given in Eqs. 2 and 3). The values are shown in Table 1.

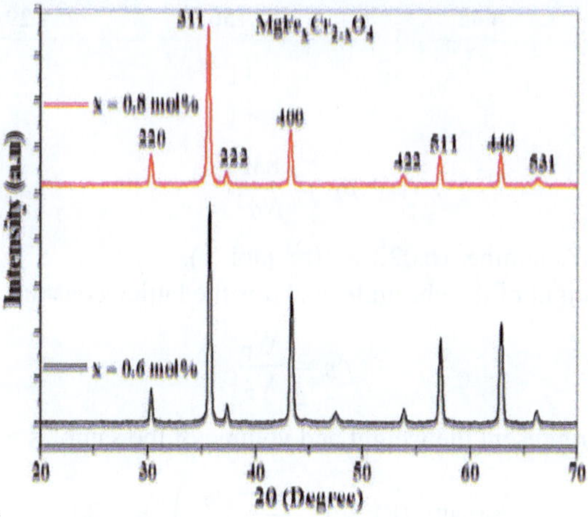

Fig. 1. XRD spectra of $MgFe_xCr_{2-x}O_4$ chromites.

Table 1. D, d spacing, V_{cell}, and lattice parameter (a) of $MgFe_xCr_{2-x}O_4$ chromites

x in mol %	a (in Å)	d spacing (in nm)	V_{cell} (in $Å^3$)	D (in nm)
x = 0.6	8.33717	0.2005	579.5033	25.04
x = 0.8	8.34953	0.2020	582.084	24.70

$$D = \frac{0.\ 9\lambda}{\beta\cos\theta} \tag{1}$$

Here β = full-width at half-maximum
λ = wave-length of X-ray, and
θ = diffraction angle

$$a = d\sqrt{h^2 + k^2 + l^2} \tag{2}$$

$$V_{cell} = a^3 \tag{3}$$

where hkl = Miller indices and.
d = inter-planer spacing.

The X-ray density (ρ_X), bulk density (ρ_B), and porosity (P) (Table 2) of the chromites were also calculated using the following relations [23, 24]:

Table 2. ρ_X, ρ_B (in g/cm^3), and P of $MgFe_xCr_{2\text{-}x}O_4$ chromites

x in mol %	ρ_X	ρ_B	P
x = 0.6	4.520	3.236	28.41%
x = 0.8	4.457	3.156	29.19%

$$\rho_X = \frac{8M}{Na^3} \tag{4}$$

here N = Avogadro's number (6.022×10^{23} mol^{-1}),
M = atomic weight of the chromite, and a is the lattice constant.

$$\rho_B = \frac{W_p}{V_p} \tag{5}$$

where W_p and V_p represent the weight and volume of the samples, respectively.

$$\text{Porosity (P)} = \left(\frac{\rho_X - \rho_B}{\rho_X}\right) \times 100 \tag{6}$$

The Williamson-Hall (WH) relation describes the relationship between the microstrain (ε) and average crystallite size (D) of the compound, as given by the equation [25]:

$$\beta\cos\theta = \frac{0.9\lambda}{D} + 4\varepsilon\sin\theta \tag{7}$$

The WH plot (Fig. 2(a, b)) is analyzed by plotting $\beta\cos\theta$ against $4\sin\theta$. The values of 'D' and 'ε' are determined using the intercept and slope of the linear fit of the WH plot [26] and values listed in Table 3. It is noticed that the lattice strain is positive for $MgFe_{0.6}Cr_{2\text{–}0.6}O_4$ chromite and negative for $MgFe_{0.8}Cr_{2\text{–}0.8}O_4$ chromite. A negative slope signifies compressive strain, whereas a positive slope indicates tensile strain within a crystal. Furthermore, Rietveld refinement pattern was applied to refine all the $MgFe_xCr_{2\text{-}x}O_4$ (x = 0.6, 0.8) chromites. The refinement patterns (Fig. 3(a, b)), confirm that all samples exhibit cubic crystal symmetry and the parameters are shown in Table 4.

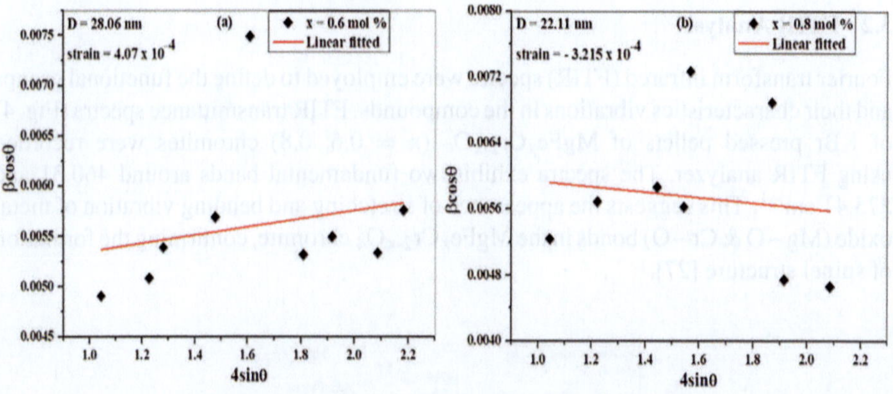

Fig. 2 **(a-b)**. WH plots of $MgFe_xCr_{2-x}O_4$ chromites.

Table 3. *D* and ε of $MgFe_xCr_{2-x}O_4$ chromites

x in mol %	D in nm	ε
x = 0.6	28.06	4.07×10^{-4}
x = 0.8	22.11	-3.215×10^{-4}

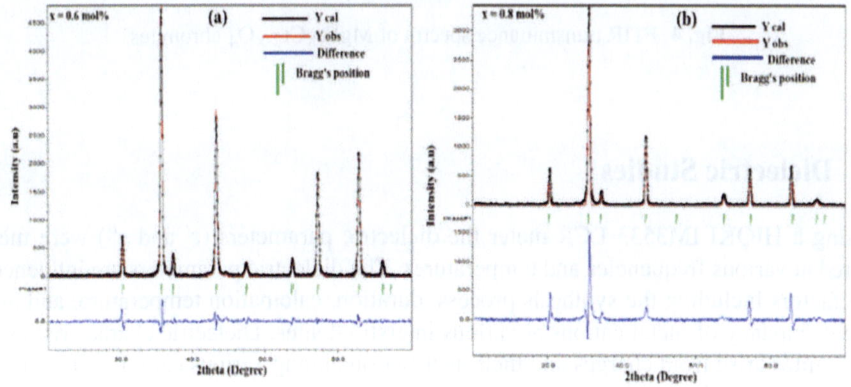

Fig. 3 **(a-b)**. Rietveld refinement of $MgFe_xCr_{2-x}O_4$ chromites.

Table 4. The parameters derived from the Rietveld analysis of the XRD pattern

x in mol %	a in Å	V_{cell} (Å^3)	χ^2
x = 0.6	8.3788	588.2430	2.80
x = 0.8	8.3269	577.3848	4.79

3.2 FTIR Analysis

Fourier transform infrared (FTIR) spectra were employed to define the functional groups and their characteristics vibrations in the compounds. FTIR transmittance spectra (Fig. 4) of KBr pressed pellets of $MgFe_xCr_{2-x}O_4$ (x = 0.6, 0.8) chromites were recorded using FTIR analyzer. The spectra exhibit two fundamental bands around 460.31 and 575.41 cm^{-1}. This suggests the appearance of stretching and bending vibration of metal oxide (Mg – O & Cr – O) bonds in the $MgFe_xCr_{2-x}O_4$ chromite, confirming the formation of spinel structure [27].

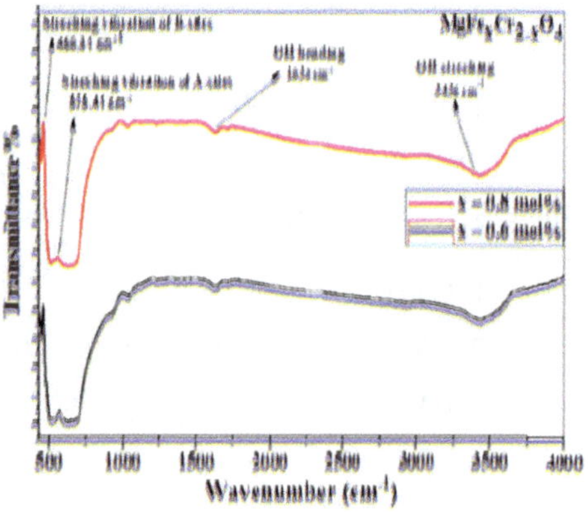

Fig. 4 FTIR transmittance spectra of $MgFe_xCr_{2-x}O_4$ chromites.

4 Dielectric Studies

Using a HIOKI IM3533 LCR meter the dielectric parameters (ε' and ε'') were measured at various frequencies and temperatures. The dielectric parameters are influenced by factors including the synthesis process, duration, calcination temperature, and also the occupancy of metal cations at various interstitial sites. Dielectric characteristics of chromites arise from charges and their distribution among various cationic sites. It can be explained using the Maxwell-Wagner model. According to this model, a dielectric system consists of two layers: a first conducting layer made of grains, surrounded by a second non-conducting layer of grain boundaries [28, 29].

4.1 Dielectric Constant (ε')

The ε' measures the material's efficiency to store electrostatic energy [30]. It can be calculated using relation (8)

$$\varepsilon' = \frac{Cd}{\varepsilon 0A} \tag{8}$$

Here, ε_0 = vacuum permittivity (8.854×10^{-12} F/m),
C = capacitance of pellets in Farad (F),
A and d are the area and thickness of pellets, respectively.

When $MgFe_xCr_{2-x}O_4$ chromites are subjected to an alternating electric field, dipoles, specifically (Cr^{2+} - Cr^{3+}) pairs, are induced within the material. This dipole formation arises due to the presence of Cr ions at both interstitial sites, which align under the influence of the external field. This alignment of dipoles contributes significantly to the materials' dielectric polarization and, consequently, to its dielectric constant. The frequency-dependent dielectric constant of $MgFe_xCr_{2-x}O_4$ (x = 0.6, 0.8) chromites are shown in Fig. 5(a, b). The ε' decreases with frequency and becomes saturated at higher frequency regions. The value of ε' is affected by various polarization like charge displacement (electronic and ionic), space charge, and orientation polarization [31]. The process of dielectric polarization is the same as that of the electrical conduction process, arising from voids, dislocations, grain and grain boundary defects, and oxygen vacancies, etc. [32, 33].

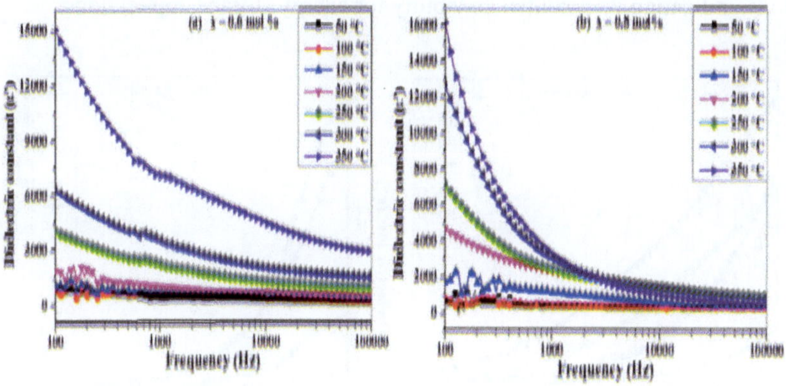

Fig. 5 (**a-b**). Frequency-dependent ε' for $MgCr_{2-x}O_4$: xFe (x = 0.6, 0.8) chromites at different temperatures.

In the lower frequency region, dipoles (Cr^{2+} - Cr^{3+}) respond quickly and accumulate at the grain boundary surfaces due to higher potential energy barriers. This accumulation of dipoles with higher relaxation time inhibits the migration of Cr^{2+} - Cr^{3+}, leading to significant space charge polarization and, consequently, a higher dielectric constant. With increasing frequency, Cr^{2+} - Cr^{3+} accumulated at the grain boundary surface are released due to shorter relaxation time. This results in a lower dielectric constant [34]. From Fig. 5(a, b), it is also noticed that for x = 0.8 mol% concentration of Fe^{3+} ions, value of ε' slightly increases compared to the x = 0.6 mol%. The higher substitution of Fe^{3+} ions introduces oxygen vacancies or other defects to maintain charge neutrality.

These vacancies or defects act as trapping sites for charges, leading to the formation of grain boundary defect barriers. This results in the creation of high degree of space charges that enhances the dielectric constant of the chromite.

4.2 Dielectric Loss (ε″)

The electrical energy dissipation of material is measured by dielectric loss (ε″) that may be caused by heating of material under the electric field [35]. It was calculated using Eq. 9:

$$\tan(\delta) = \frac{\varepsilon''}{\varepsilon'} \tag{9}$$

Figure 6(a, b) presents the frequency-dependent dielectric loss of $MgFe_xCr_{2-x}O_4$ (x = 0.6, 0.8) chromites at different temperatures. The behavior of ε″ shows a similar decaying trend to that of ε′. At lower frequencies, the ε″ is high and it decreases towards higher frequencies. In low-frequency range, the grain boundaries exhibit high resistance, requiring more energy for charge carriers exchange, resulting in higher dielectric loss. However, at higher frequencies, the low resistance grains become more effective, reducing the energy needed for charge carrier exchange and consequently lowering the dielectric loss [36, 37]. The dielectric loss decreased for x = 0.8 mol% concentration of Fe^{3+} ions, indicating its potential suitability for energy storage applications.

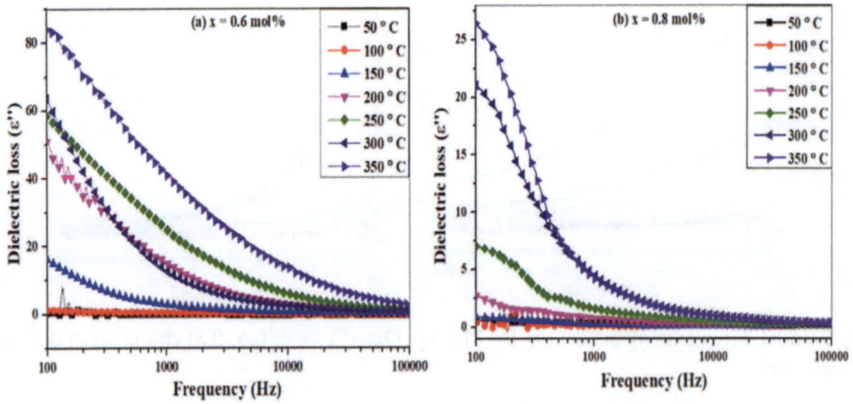

Fig. 6. **(a-b)**. Frequency dependence ε″ for $MgCr_{2-x}O_4$: xFe (x = 0.6, 0.8) chromites at different temperatures.

4.3 Temperature Dependence Dielectric Study

Figure 7(a, b) presents the effect of temperature on ε′ (dielectric constant) at different frequencies for $MgFe_xCr_{2-x}O_4$ (x = 0.6, 0.8) chromites. At low temperatures, insufficient thermal energy limits the charge carrier's mobility. However, with increasing temperature thermal energy becomes adequate to enhance the kinetic energy as well as mobility of the charge carriers. This facilitates charge exchange between Cr ions, thereby increasing the hopping rate. Consequently, polarization is enhanced, resulting in an increase in ε′ [38].

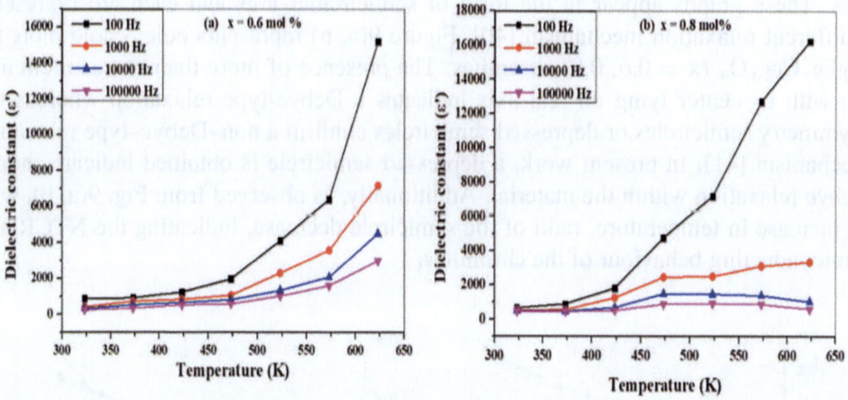

Fig. 7. **(a-b)**. Temperature-dependent ε′ for $MgCr_{2-x}O_4$: xFe (x = 0.6, 0.8) chromites at different frequencies.

Figure 8(a, b) presents the effect of temperature on ε″ (dielectric loss) at different frequencies for $MgFe_xCr_{2-x}O_4$ (x = 0.6, 0.8) chromites. With increasing temperature ε″ also increases, likely due to thermal excitation and enhanced mobility of charges [39].

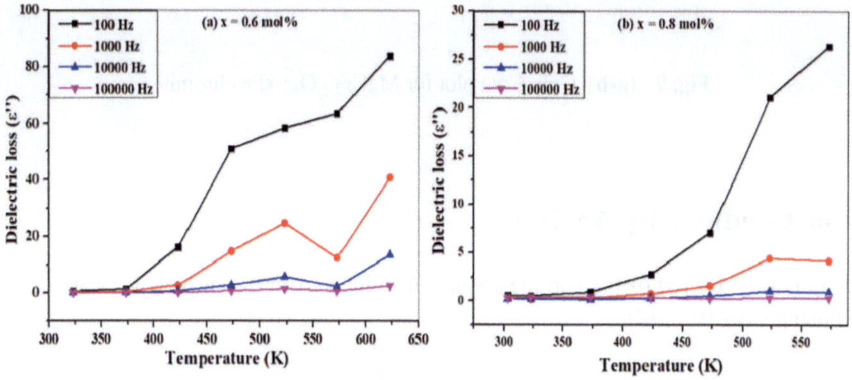

Fig. 8. **(a-b)**. Temperature-dependent ε″ for $MgCr_{2-x}O_4$: xFe (x = 0.6, 0.8) chromites at different frequencies.

5 Nyquist or Cole – Cole plot

The Nyquist plot is utilized to examine the grains (bulk), grain boundaries, and interfaces contribution to the overall conductivity of the compounds. It is plotted between the real part (Z') impedance in x-axis and the imaginary part (Z") impedance in the y-axis. These graphs appear in the form of semicircular arcs and each arc represents a different relaxation mechanism [40]. Figure 9(a, b) represents cole – cole plots for $MgFe_xCr_{2-x}O_4$ (x = 0.6, 0.8) chromites. The presence of more than one semicircular arc with its center lying on real axis indicates a Debye-type relaxation whereas the asymmetry semicircles or depressed semicircles confirm a non–Debye–type relaxation mechanism [41]. In present work, a depressed semicircle is obtained indicates non – Debye relaxation within the material. Additionally, as observed from Fig. 9(a, b), with an increase in temperature, radii of the semicircle decrease, indicating the NTCR and semiconducting behaviour of the chromites.

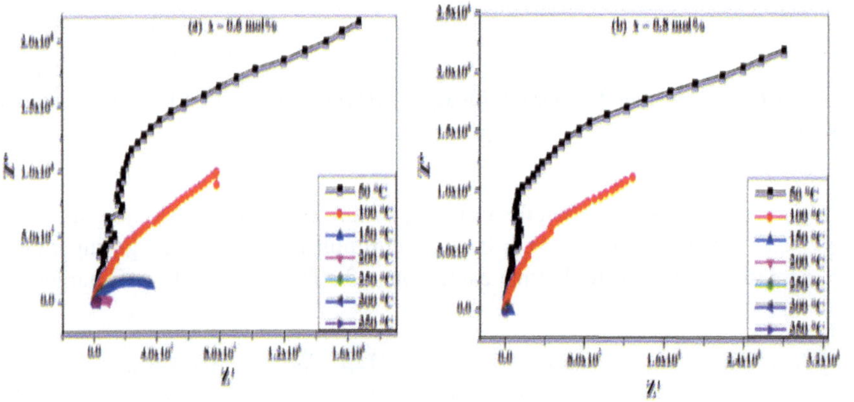

Fig. 9. **(a-b)**. Cole-Cole plot for $MgCr_{2-x}O_4$: xFe chromites.

6 ac Conductivity Analysis

The conductivity (σ) of the samples was determined from the impedance data (Z' and Z") using Eq. (10): [42],

$$\sigma = \frac{t}{A} \times \frac{Z'}{Z'^2 + Z''^2} \quad (10)$$

Here t = thickness of the pallets and.

A = area of the pallets.

The response of 'σ' with frequency and temperature for $MgFe_xCr_{2-x}O_4$ samples are represented in Fig. 10(a, b). It is noted that the σ increases with both frequency and temperature for all chromites. The frequency dispersion conductivity of $MgFe_xCr_{2-x}O_4$

is explained, based on Jonscher's power law. Based on power law, total σ is a sum of two parts i.e., the band conduction and hopping conduction [43]:

$$\sigma_{total} = \sigma_{dc} + \sigma_{ac} \tag{11}$$

Here, σ_{dc} = frequency-independent dc conductivity and.

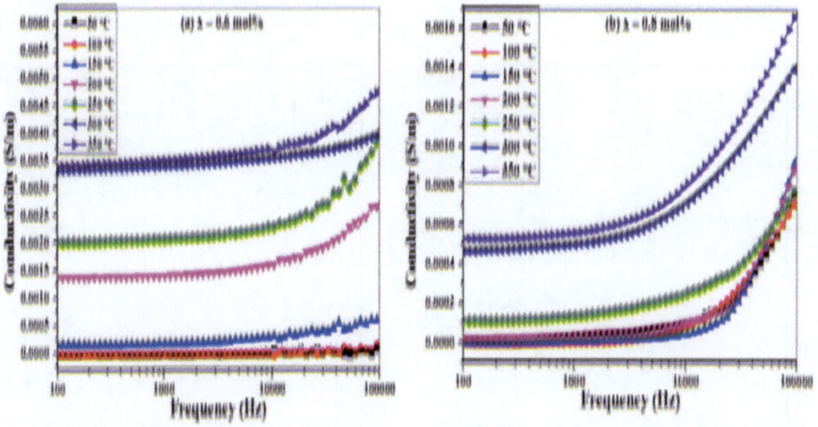

Fig. 10. **(a-b)**. ac conductivity plot for $MgCr_{2-x}O_4$: xFe (x = 0.6, 0.8) chromites at different frequency and temperatures.

σ_{ac} = temperature and frequency-dependent ac conductivity.

The frequency dispersion conductivity can be understood by the resistance of grains and grain boundaries within the material. In low-frequency range, the grain boundaries with high resistance are more prominent resulting in reduced inter-well hopping of charges. This leads to frequency frequency-independent conduction process (plateau region) and it is also known as dc conductivity. As frequency increases, the pumping force of the field increases, which enhances the electrons hopping between different localized states of chromium (Cr^{3+} and Cr^{2+}) ions [44]. Also, releases the free charge carriers from different trapping centers which correspond to ac conductivity i.e. the higher σ_{ac} is obtained in the high-frequency range owing to the released charge carriers generated by different localized states of chromium ions [45, 46]. From Fig. 10(a, b), it can be noticed that 'σ' decreases with increasing Fe^{3+} ions concentration. The substitution of Fe^{3+} ions leads to a reduction in the number of Cr^{3+} and Cr^{2+} ions, thereby disrupting the charge hopping mechanism between Cr^{3+} and Cr^{2+} ions. As a result, the conductivity decreases [47].

The dc conductivity of $MgFe_xCr_{2-x}O_4$ (x = 0.6, 0.8) chromites was determined by fitting Jonscher's power law to the frequency-dependent ac conductivity, follows the Arrhenius equation [16] given in Eq. (12)

$$\sigma_{dc} = \sigma(0)\ \exp\left(-\frac{E_a}{K_B T}\right) \tag{12}$$

where σ(0) = pre-exponential factor

K_B = Boltzmann constant
Ea = activation energy, and
T = Kelvin (K)

Figure 11(a, b) presents the correlation between σdc and temperature. Using the slope of the graph 'Ea' of $MgFe_xCr_{2-x}O_4$ (x = 0.6, 0.8) chromites are determined and found to be 0.49 eV and 0.33 eV. The E_a decreases with 0.8 mol% substitution of Fe^{3+} ions due to an increase in the concentration of charge carriers and their mobility.

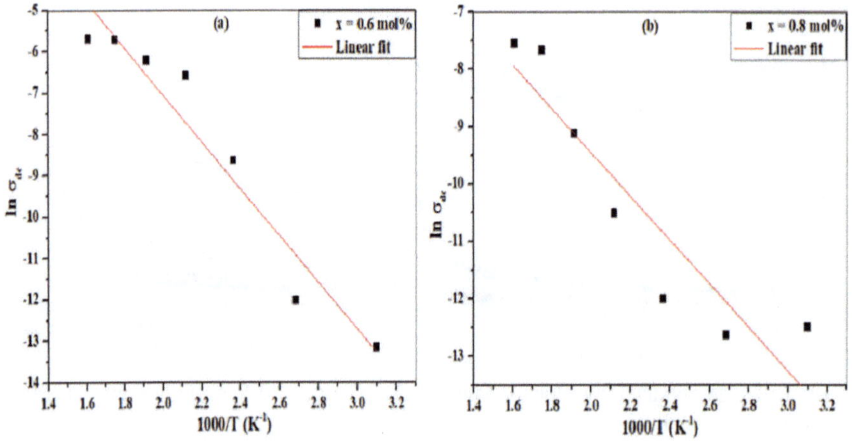

Fig. 11. **(a-b)**. Plot of activation energy for $MgCr_2O_4$ (x = 0.6 & 0.8 mol%) chromites.

7 Conclusion

Fe^{3+} substituted magnesium chromites $MgFe_xCr_{2-x}O_4$ were prepared via the solution combustion route. Structural analysis using XRD revealed the fcc spinel structure of chromites with average crystallite size of 25.04 and 24.70 nm. The FTIR transmittance spectra determined the bands around 460.31 cm^{-1} and 575.41 cm^{-1} confirms the functional groups and band formation in the compounds. The dielectric parameters showed a decreasing trend with frequency, attributed to the reduction in polarization as described by the Maxwell-Wagner model. According to Jonscher's power law, σac of chromites increased with both frequency and temperature, indicating semiconducting behaviour. The dielectric relaxation mechanism and NTCR behaviour of the chromites were investigated using complex impedance spectroscopy. With ε' of 15179.471 and 16227.784 for x = 0.6 and 0.8 mol %, respectively, these compounds are suitable for use in electric circuits as capacitors, insulating layers in MOSFETs, and components for electric power loss, etc.

References

1. Javed, M., Khan, A.A., Kazmi, J., Mohamed, M.A., Ahmed, M.S., Iqbal, Y.: Impedance spectroscopic study of charge transport and relaxation mechanism in MnCr2O4 ceramic chromite. J. Alloys Compd. **854** (2021). https://doi.org/10.1016/j.jallcom.2020.156996
2. Kassem, M.A., El-fadl, A.A., Nashaat, A.M.: Structure , optical and varying magnetic properties of insulating M Cr 2 O 4 (M ¼ co , Zn , mg and cd) nanospinels. J. Alloys Compd. **790**, 853–862 (2019). https://doi.org/10.1016/j.jallcom.2019.03.189
3. Iimura, K., Pedro, J., De Araujo, E., Brezesinski, T.: Structural, optical, and magnetic properties of highly ordered mesoporous MCr 2 O 4 and MCr 2 −. 4, (2012)
4. Honeybourne, C.L., Rasheed, R.K.: Nitrogen dioxide and volatile sulfide sensing properties of copper, zinc and nickel chromite. J. Mater. Chem. **6**, 277–283 (1996). https://doi.org/10.1039/jm9960600277
5. Javed, M., Khan, A.A., Kazmi, J., Mohamed, M.A., Khan, M.N., Hussain, M., Bilkees, R.: Dielectric relaxation and small polaron hopping transport in sol-gel-derived NiCr2O4 spinel chromite. Mater. Res. Bull. **138**, 111242 (2021). https://doi.org/10.1016/j.materresbull.2021.111242
6. Stefan, E., Irvine, J.T.S.: Synthesis and characterization of chromium spinels as potential electrode support materials for intermediate temperature solid oxide fuel cells. J. Mater. Sci. **46**, 7191–7197 (2011). https://doi.org/10.1007/s10853-010-4489-1
7. Gingasu, D., Culita, D.C., Calderon Moreno, J.M., Oprea, O., Marinescu, G., Preda, S., Kuncser, A.: Soft chemistry synthesis method of ZnAl2−xCrxO4 spinel: structural, morphological, optical and photocatalytic properties. Mater. Today Commun. **33**, 104656 (2022). https://doi.org/10.1016/j.mtcomm.2022.104656
8. Naz, S., Durrani, S.K., Mehmood, M., Nadeem, M.: Hydrothermal synthesis, structural and impedance studies of nanocrystalline zinc chromite spinel oxide material studies of nanocrystalline zinc chromite spinel oxide material. J. Saudi Chem. Soc. **20**, 585–593 (2016). https://doi.org/10.1016/j.jscs.2014.12.007
9. Saleem, M., Varshney, D.: Influence of transition metal Cr2+ doping on structural, electrical and optical properties of Mg-Zn aluminates. J. Alloys Compd. **708**, 397–403 (2017). https://doi.org/10.1016/j.jallcom.2017.03.016
10. Choudhary, P., Varshney, D.: Structural, vibrational and dielectric behavior of Co1-xMxCr2O4 (M = Zn, Mg, Cu and x = 0.0, 0.5) spinel chromites. J. Alloys Compd. **725**, 415–424 (2017). https://doi.org/10.1016/j.jallcom.2017.07.162
11. Sahu, Y., Agrawal, S.: Effect of Fe3+ ions substitution on electrical and dielectric properties of MgFexCr2-xO4 (x = 0, 0.2, and 0.4 mol%) chromites prepared by solution combustion method. Chem. Phys. Impact. **8**, 100655 (2024). https://doi.org/10.1016/j.chphi.2024.100655
12. Banaj, L., Agrawal, S.: Dielectric behavior of Zr4+ doped MgFe2O4 spinel ferrite synthesized by solid-state reaction method. Curr. Appl. Phys. **56**, 47–65 (2023). https://doi.org/10.1016/j.cap.2023.10.001
13. Jebanisha, B., Devi, V.N.M., Varghese, J., Aswathy, N.R., Vinod kumar, R.: Magnetic and dielectric characteristics of rGO modified BFO nanoparticles produced using sol-gel via auto-combustion method. Ceram. Int. **50**, 36955–36963 (2024). https://doi.org/10.1016/j.ceramint.2024.07.085
14. Min, B., Moon, S.M., Cho, N.H.: Structural and dielectric features of Nb- doped nanosized BaTiO3 powders prepared by hydro-thermal synthesis methods. Curr. Appl. Phys. **11**, S193–S196 (2011). https://doi.org/10.1016/j.cap.2011.01.035
15. Yao, J., Li, J., Mao, F.F., Xu, W., Jiang, J., Li, Y.: Enhanced zinc ions storage performance of Mn2O3/Mn3O4/NiMn2O4 composite cathode material synthesized by a thermal decomposition method. Mater. Lett. **370**, 136729 (2024). https://doi.org/10.1016/j.matlet.2024.136729

16. Gupta, R., Agrawal, S.: Investigation of structural, morphological and dielectric properties of yttrium magnesium silicate oxyapatite doped with Gd3+ ions. Emerg. Mater. (2024). https://doi.org/10.1007/s42247-024-00682-5
17. Gupta, R., Agrawal, S.: Judd–Ofelt analysis and photoluminescence properties of europium-activated Y8−xSr2(SiO4)6O2 oxyapatite phosphors. J. Mater. Sci. Mater. Electron. 33, 17199–17211 (2022). https://doi.org/10.1007/s10854-022-08597-9
18. Sahu, Y., Agrawal, S.: Structural features, dielectric behaviour of spinel type magnesium aluminate synthesized by the auto-combustion method. J. Phys. Conf. Ser. **2576** (2023). https://doi.org/10.1088/1742-6596/2576/1/012006
19. Garg, T., Saleem, M., Kaurav, N., Choudhary, P., Yadav, A.: The Co1-xZnxCr2O4[x = 0.0, 0.1] chromite system: a study of structural and frequency dependent room temperature dielectric properties. Mater. Today Proc. **89**, 4–7 (2023). https://doi.org/10.1016/j.matpr.2023.05.539
20. Yazdanbakhsh, M., Khosravi, I., Goharshadi, E.K., Youssefi, A.: Fabrication of nanospinel ZnCr2O4 using sol-gel method and its application on removal of azo dye from aqueous solution. J. Hazard. Mater. **184**, 684–689 (2010). https://doi.org/10.1016/j.jhazmat.2010.08.092
21. Rehani, B.R., Joshi, P.B., Lad, K.N., Pratap, A.: Crystallite size estimation of elemental and composite silver nano-powders using XRD principles. Indian J. Pure Appl. Phys. **44**, 157–161 (2006)
22. El-Said Bakeer, D.: Investigation on optical, dielectric, and magnetic properties of CoAl2-xFexO4 nanoparticles. J. Supercond. Nov. Magn. **33**, 1789–1801 (2020). https://doi.org/10.1007/s10948-020-05437-x
23. Lal, G., Punia, K., Dolia, S.N., Alvi, P.A., Choudhary, B.L., Kumar, S.: Structural, cation distribution, optical and magnetic properties of quaternary Co0.4+xZn0.6-xFe2O4 (x = 0.0, 0.1 and 0.2) and Li doped quinary Co0.4+xZn0.5-xLi0.1Fe2O4 (x = 0.0, 0.05 and 0.1) nanoferrites. J. Alloys Compd. **828** (2020). https://doi.org/10.1016/j.jallcom.2020.154388
24. Choudhary, B.L., Garima, Hasan, P.M.Z., Darwesh, R., Kumar, S., Dalela, S., Dolia, S.N., Alvi, P.A.: Low temperature field dependent magnetic study of the Zn0.5Co0.5Fe2O4 nanoparticles. J. Magn. Magn. Mater. **536**, 168102 (2021). https://doi.org/10.1016/j.jmmm.2021.168102
25. Choudhary, B.L., Kumar, U., Kumar, S., Chander, S., Kumar, S., Dalela, S., Dolia, S.N., Alvi, P.A.: Irreversible magnetic behavior with temperature variation of Ni0.5Co0.5Fe2O4 nanoparticles. J. Magn. Magn. Mater. **507**, 166861 (2020). https://doi.org/10.1016/j.jmmm.2020.166861
26. Anchal, S., Palsaniya, K.K., Priya, Kumari, N., Dolia, S.N., Alvi, P.A., Choudhary, B.L.: Tailoring quantum dots through citric acid modulation of CoFe2O4 ferrite. Mater. Chem. Phys. **313**, 128820 (2024). https://doi.org/10.1016/j.matchemphys.2023.128820
27. Hcini, F., Hcini, S., Almoneef, M.M., Dhaou, M.H., Alshammari, M.S., Mallah, A., Zemni, S., Lefi, N., Bouazizi, M.L.: Thermal, microstructural, optical, magnetic and magnetocaloric studies for Ni0.5Mn0.5Cr2O4 chromite spinel prepared using sol-gel method. J. Mol. Struct. **1243**, 130769 (2021). https://doi.org/10.1016/j.molstruc.2021.130769
28. Majeed, A., Khan, M.A., ur Raheem, F., Ahmad, I., Akhtar, M.N., Warsi, M.F.: Morphological, Raman, electrical and dielectric properties of rare earth doped X-type hexagonal ferrites. Phys. B Condens. Matter. **503**, 38–43 (2016). https://doi.org/10.1016/j.physb.2016.08.041
29. Datt, G., Abhyankar, A.C.: Dopant driven tunability of dielectric relaxation in MxCo(1-x)Fe2O4 (M: Zn2+, Mn2+, Ni2+) nano-ferrites. J. Appl. Phys. **122**, 11 (2017). https://doi.org/10.1063/1.4990275
30. Khan, Q., Majeed, A., Ahmad, N., Ahmad, I., Ahmad, R.: Structural features and dielectric behavior of Al substituted Cu0.7Ni0.3Fe2O4 ferrites. Mater. Chem. Phys. **273**, 125028 (2021). https://doi.org/10.1016/j.matchemphys.2021.125028

31. Khan, I., Khan, S., Khan, W.: Temperature-dependent dielectric and magnetic properties of Mn doped zinc oxide nanoparticles. Mater. Sci. Semicond. Process. **26**, 516–526 (2014). https://doi.org/10.1016/j.mssp.2014.05.028
32. Bhargava, R., Khan, S.: Effect of reduced graphene oxide (rGO) on structural, optical, and dielectric properties of Mg(OH)2/rGO nanocomposites. Adv. Powder Technol. **28**, 2812–2819 (2017). https://doi.org/10.1016/j.apt.2017.08.008
33. Mahapatro, J., Agrawal, S.: Optical, dielectric and electrical properties of Gd3+ ions doped barium hexaferrite ceramic compounds for microwave device applications. J. Alloys Compd. **907**, 164405 (2022). https://doi.org/10.1016/j.jallcom.2022.164405
34. Ansari, M.M.N., Khan, S.: Structural, electrical and optical properties of sol-gel synthesized cobalt substituted MnFe2O4 nanoparticles. Phys. B Condens. Matter. **520**, 21–27 (2017). https://doi.org/10.1016/j.physb.2017.06.020
35. Sharma, A., Srivastava, A.K., Raina, S.: Nano-structures & nano-objects microstructural, dielectric, electrical, electromagnetic, and magnetic composites for electronic devices application. Nano-Struct. Nano-Objects. **40**, 101331 (2024). https://doi.org/10.1016/j.nanoso.2024.101331
36. Paray, A., AlMasoud, N., Al-Ahmary, K.M., Alomar, T.S., Alrahili, M.R., Mohammed, J., El-Bahy, Z.M., Srivastava, A.K.: High frequency enhancement of dielectric constant, crystal structure analysis, and magnetic properties of Pr3+- Zn2+ substituted Co2Y-type barium hexagonal ferrites. Mater. Chem. Phys. **322**, 129505 (2024). https://doi.org/10.1016/j.matchemphys.2024.129505
37. Arifuzzaman, M., Hossen, M.B., Afroze, J.D., Abden, M.J.: Structural and electrical properties of cu substituted Ni–Cd nanoferrites for microwave applications. Phys. B Condens. Matter. **588** (2020). https://doi.org/10.1016/j.physb.2020.412170
38. Thakur, V.N., Kumar, A., Kaushik, A., Gupt, G.D., Dhaka, R.S.: High temperature dielectric and impedance spectroscopy study of LaCo0.7Nb0.3O3. Mater. Res. Bull. **158**, 112070 (2023). https://doi.org/10.1016/j.materresbull.2022.112070
39. Padmaraj, O., Venkateswarlu, M., Satyanarayana, N.: Structural, electrical and dielectric properties of spinel type MgAl2O4 nanocrystalline ceramic particles synthesized by the gel-combustion method. Ceram. Int. **41**, 3178–3185 (2015). https://doi.org/10.1016/j.ceramint.2014.10.169
40. Ludhiya, V., Hari Kumar, N., Ravinder, D., Edukondalu, A.: Structural, optical, dielectric and magnetic properties of Nd3+ ion substituted Ni-Mg-Cu spinel ferrites. Inorg. Chem. Commun. **150**, 110558 (2023). https://doi.org/10.1016/j.inoche.2023.110558
41. Sahu, Y., Agrawal, S.: Study of the effect of Sr2+ ions substitution on the dielectric behaviour of BaCr2O4 chromites synthesized by solution combustion method. J. Mater. Sci. Mater. Electron. **35** (2024). https://doi.org/10.1007/s10854-024-13135-w
42. Mahapatro, J., Agrawal, S.: Effect of Eu3+ ions on electrical and dielectric properties of barium hexaferrites prepared by solution combustion method. Ceram. Int. **47**, 20529–20543 (2021). https://doi.org/10.1016/j.ceramint.2021.04.062
43. Mahato, D.K.: Ac conductivity analysis of nanocrystallite MgFe2O4 ferrite. Mater. Today Proc. **5**, 9191–9195 (2018). https://doi.org/10.1016/j.matpr.2017.10.043
44. Javed, M., Khan, A.A., Akbar, N., Kazmi, J., Dar, A., Mohamed, M.A.: Low-temperature dielectric relaxation mechanism and correlated barrier hopping transport in neodymium perovskite chromite. Mater. Res. Bull. **165**, 112303 (2023). https://doi.org/10.1016/j.materresbull.2023.112303
45. Samkaria, R., Sharma, V.: Effect of Y3+ substitution on the structural, dielectric, and electrical properties of nanosized ZnAl2O4 ZnAl 2O4 spinel. Appl. Phys. A Mater. Sci. Process. **115**, 697–704 (2014). https://doi.org/10.1007/s00339-013-7855-4

46. Farea, A.M.M., Kumar, S., Batoo, K.M., Yousef, A., Lee, C.G.: Alimuddin: structure and electrical properties of Co0.5CdxFe2.5-xO4 ferrites. J. Alloys Compd. **464**, 361–369 (2008). https://doi.org/10.1016/j.jallcom.2007.09.126
47. Siddique, M.N., Ali, T., Ahmed, A., Tripathi, P.: Enhanced electrical and thermal properties of pure and Ni substituted ZnO nanoparticles. Nano-Struct. Nano-Objects. **16**, 156–166 (2018). https://doi.org/10.1016/j.nanoso.2018.06.001

Frequency and Temperature Dependent Dielectric Properties of Yttrium Sodium Silicate Apatite Doped with Terbium Ion Prepared by Auto-Combustion Method

Ritu Gupta and Sadhana Agrawal(✉)

Department of Physics, National Institute of Technology Raipur, Raipur 492010, Chhattisgarh, India

sagrawal.phy@nitrr.ac.in

Abstract. The yttrium sodium silicate apatite doped with terbium ions viz., $Y_{9\text{-}x}Na(SiO_4)_6O_2$: xTb ($0 \leq x \leq 0.2$ mol%) was synthesized through auto combustion method. The formation of apatite with a hexagonal structure belonging to space group p63/m was verified using PXRD analysis. The SEM micrographs show the formation of crystals, random in shape and irregular in distribution. The presence of chemical bonds between the Si–O elements of the compounds is confirmed from the analysis of transmission spectra obtained through FTIR spectroscopy. The frequency and temperature (10^2 Hz – 10^2 kHz) dependent dielectric parameters were obtained and analyzed. The dielectric constant (ε') and dielectric loss (ε'') showed decreasing trend with increasing values of external field frequencies due to a reduction in space charge polarization. Over a wide range of temperature and frequency, low values of ε'' are sustained. The alternating conductivity showed dispersion with increasing values of frequency. The impedance analysis confirmed NTCR properties of compounds. These compounds can be utilized in electrical devices as loss components, and capacitors, etc.

Keywords: Combustion · Transmission · Dielectric · Polarization · Loss component · Capacitors

1 Introduction

The general formula $X_{10}(SiO_4)_6U_2$ corresponds to apatite group, and when U (an anion site) is occupied by oxygen, then it forms one of the subtypes, i.e., oxyapatite (OA). In the formula, cations are accommodated in X site, offering a broad range of material selection from any of alkali/rare-earth/transition elements. The structure of this group of compounds is hexagonal – p63/m (space group), which is comprised of isolated and covalent silicate tetrahedrons [1]. In order to create a rigid network, isolated SiO_4 tetrahedra share oxygen with X polyhedra. Ionic conduction occurs in these compounds due to the presence of O(5) oxide ions, which occupy the middle of 1-D channels that run along c-axis. Seven and nine coordinated sites, designated X^I and X^{II} sites,

S. Kumari et al. (Eds.): ICMTP 2024, SPPHY 311, pp. 421–437, 2025.
https://doi.org/10.1007/978-981-96-6795-6_46

respectively, contain X cations. The doping in the cation site significantly alters the electrical characteristics of oxyapatite, which can withstand a considerable number of vacancies, structural defects, and interstitial atoms [2, 3]. The oxyapatite compounds have shown stability in terms of chemical and thermal behavior [4]. The advancements in the field of microwave communication systems have led to the production of novel materials with moderate values of dielectric constant (ε'), low values of dielectric loss (ε''), and negative values of temperature coefficient (τ_f) [5–7]. All these properties make the suitability of the compounds as microwave dielectric materials and as noise-reducing elements in electrical circuits during signal transmission [8, 9]. So, the researchers are striving to produce more such materials with more efficiency and reliability [10, 11]. Some of the oxyapatite compounds have shown these properties, i.e., slightly low ε' and minimum ε'' at elevated temperatures with negative τ_f [12]. Here in this report, the study of dielectric and impedance properties of pure $Y_{9-x}Na(SiO_4)_6O_2$ OA and two compositions of Tb^{3+} doped $Y_{9-x}Na(SiO_4)_6O_2$ OAs, is done because this composition was not explored before to the best of knowledge. The impedance analysis was carried out to see resistive and conductive properties of OA compounds.

The property of a compound to impede the flow of alternating electrical current passing through it, is measured by impedance. The overall effect of current-limiting elements of the circuit, such as capacitance (C) and resistance (R), is known as impedance, and it is measured in ohms. The impedance is a combined effect of resistance and capacitive reactance present in the circuit. The capacitive reactance depends on frequency, and therefore there is an association of a phase angle in the total impedance. The resistance and reactance are 90° out of phase with each other, and the complex addition of them gives the total value of impedance (Z). Complex impedance spectroscopy is used to study the electrochemical properties such as ionic conductivity and the overall resistance due to grain, grain boundary, bulk, and electrode effects present in oxyapatite silicates. The doping in the cationic site causes charge imbalance and hence creates oxygen vacancies, which greatly contributes to the conduction phenomena in these compounds [13, 14].

Herein, a wet-chemical route, i.e., citric acid (fuel) - assisted auto-combustion synthesis (ACS), was employed to synthesize oxyapatite compounds having nominal composition $Y_{9-x}Na(SiO_4)_6O_2$: xTb (x = 0, 0.1, and 0.2 mol%). With the help of nitrate-based precursors and fuel, ACS is a desirable technique that produces nano-sized functional materials in less time and without any complex procedures. The oxidizers and fuels go through quick, self-sustaining redox reactions when exposed to an external heat source, further, which on calcination produces powder samples [15].

2 Synthesis

The chemical precursors used for the synthesis were $Y(NO_3)_3.6H_2O$ (purity ≥99.8%), $NaNO_3$ (purity ≥99.0%), $Tb(NO_3)_3.5H_2O$ (purity ≥99.9%), and citric acid ($C_6H_8O_7$), supplied from Sigma Aldrich, and SiO_2 (purity ≥99.5%), supplied from HIMEDIA, were collected and measured the weight in a weighing machine as per stoichiometric proportions. All the measured chemicals were put together in a beaker and placed on the top of a hot plate. As temperature rose and reached around 200 °C, the mixture of the chemical precursors burned. The as-burnt foamy powder was then collected, lightly ground, and put in an alumina crucible. The powders were then calcined in a programmed furnace set at 1200 °C with a heating rate of 5 °C per minute, held for 4 hours. After natural cooling of the powder samples, they were consolidated using polyvinyl alcohol to form pellets via hand press at a 2-ton load for 1 min. The pellets were Ag polished on their adjacent sides to form ohmic contact.

3 Results and Discussion

3.1 Phase Analysis

XRD patterns (Fig. 1, of pure and doped $Y_{9-x}Na(SiO_4)_6O_2$: xTb (x = 0, 0.1, and 0.2 mol%) OAs are presented. The patterns were collected on a PANalytical x-ray diffractometer in a range - 20° ≤ 2θ ≤ 60°. The diffraction patterns are shown in Fig. 1, where it is evident that the peaks are provided Miller indices values based on the matched pattern with JCPDS No. 96-900-9979. This verified the hexagonal structure (space group p63/m) of $Y_{9-x}Na(SiO_4)_6O_2$: xTb OAs. Additionally, it is noted that a little shift in the most intense peaks towards the lower angle side occurred as the doping concentration increased, as shown in Fig. 2. This occurred because Tb^{3+}, which has a higher ionic radius (0.923 Å), gets incorporated into the Y^{3+} site, which has a lower ionic radius (0.90 Å) [16, 17].

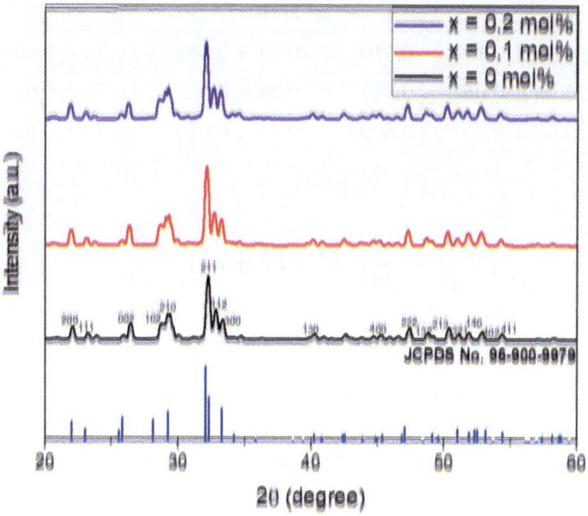

Fig. 1. XRD patterns of $Y_{9-x}Na(SiO_4)_6O_2$: xTb (x = 0–0.2 mol%) OAs

Using Eqs. (1), (2), and (3), crystallite size, lattice parameters, and volume of $Y_{9-x}Na(SiO_4)_6O_2$: xTb OAs were estimated which are listed in Table 1.

$$D = \frac{K\lambda}{\beta \cos\theta} \tag{1}$$

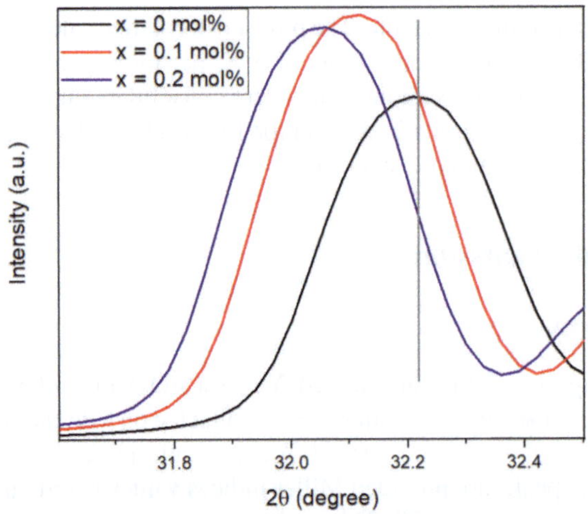

Fig. 2. Magnified image of '211' peak shift observed in XRD pattern

Table 1. Crystallographic parameters (crystallite size, strain, lattice parameters, and volume) of $Y_{9-x}Na(SiO_4)_6O_2$: xTb (x = 0–0.2 mol%) OAs

x in mol %	D (nm)	ε	Lattice Parameters (Å)		V (Å^3)
			a(=b)	c	
x = 0	22.16 ± 0.11	0.6101	9.29 ± 0.003	6.74 ± 0.006	503.74
x = 0.1	22.66 ± 0.05	0.5972	9.29 ± 0.003	6.76 ± 0.000	505.24
x = 0.2	22.67 ± 0.05	0.6094	9.33 ± 0.010	6.79 ± 0.010	511.86

$$\frac{1}{d^2} = \frac{4}{3a^2}\left(h^2 + hk + k^2\right) + \frac{l^2}{c^2} \tag{2}$$

$$V_{cell} = 0.8660\, a^2 c \tag{3}$$

Where λ = 0.154 nm, K = 0.9, β = FWHM of 211 peak, and θ = Bragg angle. A slight increase in the values of crystallite size, lattice parameters, and volume was found with an increase in Tb^{3+} concentration. Because of the lattice strain and distortion brought on by the substitution of larger dopant atoms (Tb^{3+} ions) in the host lattice (Y^{3+} ions), the crystallite size and lattice parameters typically increase. The increase in lattice parameters subsequently results in an increased value of volume.

Using Eq. (4), strain is calculated and listed in Table 1.

$$\varepsilon = \frac{\beta \cos\theta - 0.9\,\lambda/D}{2 \sin\theta} \tag{4}$$

As doping of larger ions increases, lattice is progressively stretched, which may alleviate the strain.

3.2 SEM and EDX Analysis

The microstructures of $Y_{9-x}Na(SiO_4)_6O_2$: xTb (x = 0 and 0.2 mol%) OAs have been obtained through scanning electron microscopy and presented in Fig. 3(a–d) at two different magnifications. The morphologies of undoped and Tb^{3+}-doped OAs are quite comparable; the particles are of uneven shape and well connected to one another, i.e., agglomerated, which may be due to high sintering temperature. The average particle size calculation was done by ImageJ software and was estimated to be 0.62 μm (620 nm) for x = 0 mol% and 0.82 μm (820 nm) for x = 0.2 mol%.

Fig. 3. **(a-d)**. SEM micrographs $Y_{9-x}Na(SiO_4)_6O_2$: xTb, x = 0 mol% - (a) and (b), x = 0.2 mol% - (c) and (d) OAs at two different magnifications

The compositional computation has been done through EDXS. The EDX spectra (Fig. 4a & b) confirm all initial precursors, i.e., Y, Na, Si, O, and Tb, to be present in undoped (x = 0 mol%) and doped (x = 0.2 mol%) OAs. The Table 2 shows atomic% and weight% of all elements along with some percentage of carbon, which may be incorporated during synthesis procedures.

3.3 FTIR Analysis

The FTIR spectra, as presented in Fig. 5, show the presence of bonds between the elements of OAs. The peaks in the spectra correspond to the oxyapatite formation [18]. Stretching vibrations of Y-O molecules are indicated by peak falling at 673 cm^{-1}. The Y-O stretching vibration showed slight shifting in the lower wavenumber side as a result of

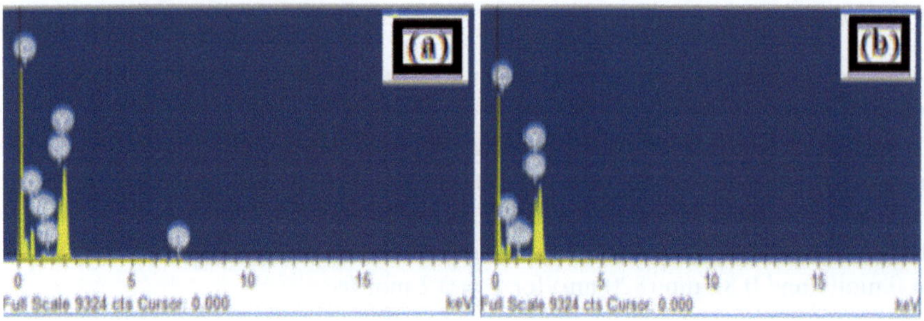

Fig. 4. (**a & b**). EDXS spectra of $Y_{9\text{-}x}Na(SiO_4)_6O_2$: xTb (a)x = 0 mol%, (b)x = 0.2 mol%, OAs

Table 2. Atomic (%) and Weight (%) of $Y_{9\text{-}x}Na(SiO_4)_6O_2$: xTb (a)x = 0 mol%, (b)x = 0.2 mol%, OAs

Element	X = 0 mol%		X = 0.2 mol%	
	Weight%	Atomic%	Weight%	Atomic%
C	4.42	9.89	6.16	14.14
O	37.92	63.77	34.98	60.29
Na	1.36	1.60	1.14	1.37
Si	11.76	11.26	9.76	9.58
Y	44.54	13.48	46.18	14.32
Tb	00.00	00.00	1.77	0.31
Total	100.00	100.00	100.00	100.00

the differences in atomic mass of Y^{3+} ions (88.91 amu) and Eu^{3+} ions (151.96 amu) [19]. Modes of stretching of Si-O are evident by the peak ranging from 851 cm^{-1} to 998 cm^{-1} [20]. The peak at 1631 cm^{-1} and 1850 cm^{-1} is associated with carbon impurity atoms' asymmetric stretching vibrations. The 3447 cm^{-1} peak is of OH stretching vibration of adsorbed water [21].

3.4 Dielectric Measurements

Using LCR meter of model HIOKI – IM3533, dielectric parameters were measured in air. The measurements were repeated as a step-wise function of frequency (from 10^2 Hz–10^2 KHz) and temperature (50 to 600 °C).

Complex Dielectric Constant

When a material is placed in an alternating electric field, ε(ω) shows a complex nature, as expressed in Eq. (5),

$$\varepsilon(\omega) = \varepsilon'(\omega) + i\varepsilon''(\omega) \tag{5}$$

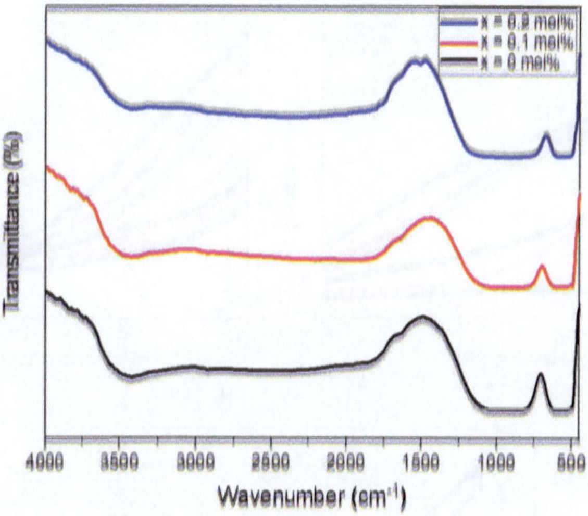

Fig. 5. FTIR spectra of $Y_{9\text{-}x}Na(SiO_4)_6O_2$: xTb (x = 0–0.2 mol%) OAs

Here, $\varepsilon'(\omega)$ stands for dielectric constant's real part, which describes the stored electrical energy, and $\varepsilon''(\omega)$ denotes dielectric constant's imaginary part, which describes about the power dissipated in material.

The value of real and imaginary part of dielectric constant were calculated using Eqs. (6) and (7) [22, 23],

$$\varepsilon'(\omega) = \frac{Cd}{A\varepsilon_0} \tag{6}$$

$$\varepsilon''(\omega) = \varepsilon'(\omega)\tan(\delta) \tag{7}$$

Here, ε_0 is vacuum permittivity whose value is 8.854×10^{-12} F/m, C denotes capacitance (Farad) of pellet, d and A: thickness and cross-sectional area of pellet, respectively, and tan (δ) denotes loss tangent.

Figure 6 represents frequency-dependent $\varepsilon'(\omega)$ for $Y_{9\text{-}x}Na(SiO_4)_6O_2$: xTb compounds. It is observed that $\varepsilon'(\omega)$ decreases with an increase in frequency, and beyond a certain frequency region (approximately up to 10^3 Hz), it reaches a saturated value. This frequency dispersion of the dielectric constant is explained by various types of polarization mechanisms including ionic, electronic, dipole, and space charge polarization. $\varepsilon'(\omega)$ is relatively high, in low frequency region, which may be due to the oxygen vacancies' presence, grain boundary defects, and space charges within the compounds. These factors increase the resistance of charge carriers and cause them to gather at grain boundary surface, and contribute significantly to space charge polarization. As a result, a high value of $\varepsilon'(\omega)$ is observed. As frequency increases, resistance of charges at grain boundary decreases, leading to the release of charges from the surface. Consequently, the space charge polarization decreases, which results in the reduction in dielectric constant [24, 25].

Additionally, the dielectric constant is higher at elevated temperatures compared to lower temperatures. In high-temperature region, bound charge carriers get enough

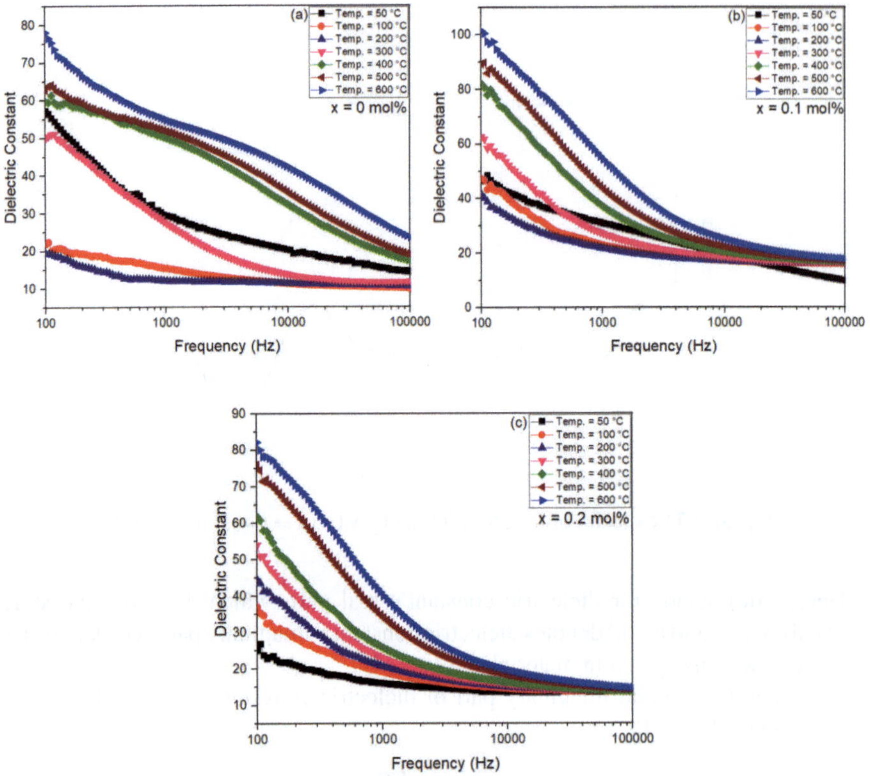

Fig. 6. (**a-c**) Variation of dielectric constant with frequency of $Y_{9-x}Na(SiO_4)_6O_2$: xTb (x = 0–0.2 mol%) OAs

thermal energy, which increases their mobility. This enhanced mobility contributes to the polarization, resulting in a higher dielectric constant [26].

$\varepsilon''(\omega)$ is related to the energy dissipation caused by the movement of charge carriers of the material under an external electric field. The presence of impurities and defects in the material also contributes to $\varepsilon''(\omega)$. Figure 7 represents the frequency-dependent dielectric loss for $Y_{9-x}Na(SiO_4)_6O_2$: xTb compounds. The values of $\varepsilon''(\omega)$ show a similar pattern to that of $\varepsilon'(\omega)$, $\varepsilon''(\omega)$ decreases towards higher frequency region. In low-frequency region, the charge carriers acquire more energy for the movement of charge carriers. Therefore, a higher dielectric loss is observed. On the other hand, as frequency increases, charge carriers require less energy due to low resistance of charge carriers at grain boundaries. Hence, a lower value of $\varepsilon''(\omega)$ is observed at high-frequency region [27]. Also, it is found that the dielectric loss values reduced around 10 times by doping Tb^{3+} ions in the host compound $Y_{9-x}Na(SiO_4)_6O_2$. The minimum loss was accounted by x = 0.2 mol% Tb^{3+} ions doped host compound. From Fig. 8, it is seen that value of temperature coefficient is found to be negative [28].

Cole – Cole Plots

Figure 9 represents the Cole–Cole plots, which illustrate relationship between real part

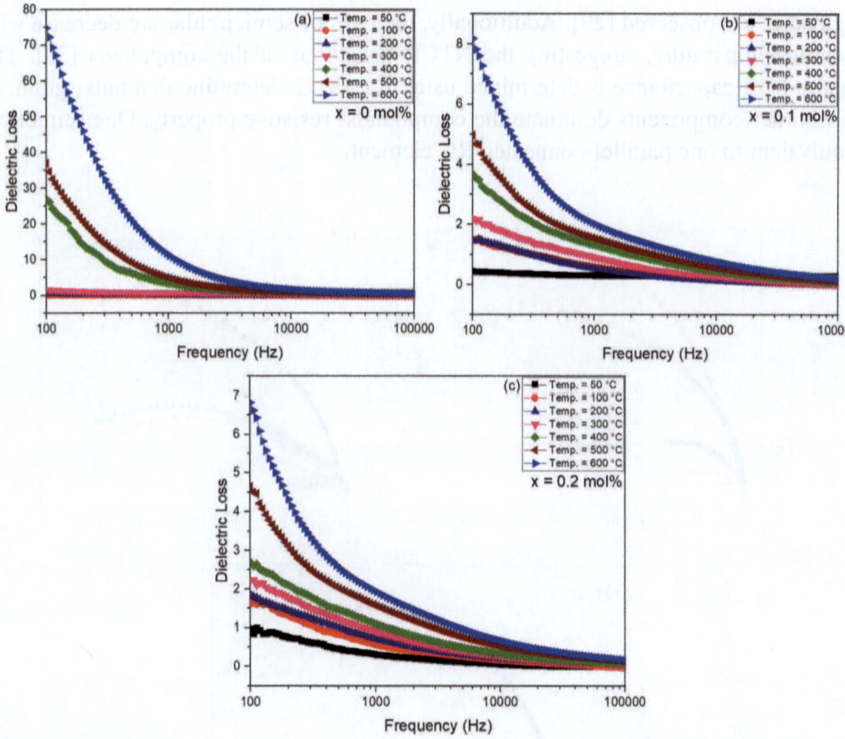

Fig. 7. **(a-c)** Variation of dielectric loss with frequency of $Y_{9\text{-}x}Na(SiO_4)_6O_2$: xTb (x = 0–0.2 mol%) OAs

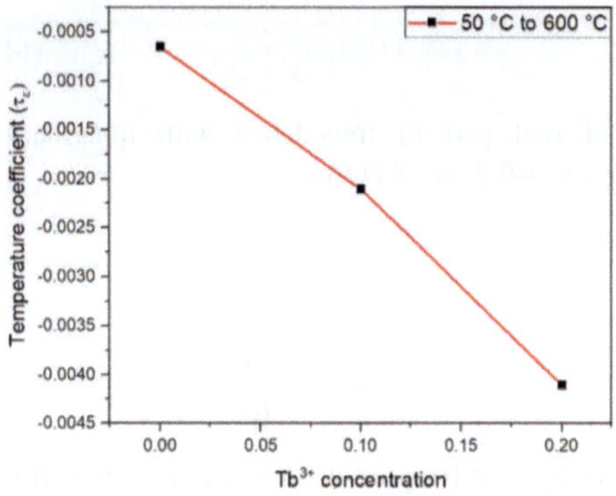

Fig. 8. Variation of temperature coefficient with doping (Tb^{3+}) concentration for 50—600 °C at 10^2 Hz frequency

(X-axis) and imaginary part (Y-axis) of complex impedance for $Y_{9\text{-}x}Na(SiO_4)_6O_2$: xTb compounds. The Cole–Cole plot appears as a semicircular arc, and the characteristics of these arcs, influenced by the frequency and temperature, provide valuable insights into the properties of the materials. In this study, asymmetric semicircular arcs at high frequencies are observed [29]. Additionally, the radii of semicircular arc decrease with elevating temperature, suggesting the NTCR behaviour of the compounds [30]. The magnitude of capacitance is determined using Eq. (8) to determine that bulk, grain, or grain border components dominate the compounds' resistive property. One semicircle is equivalent to one parallel-connected RC element.

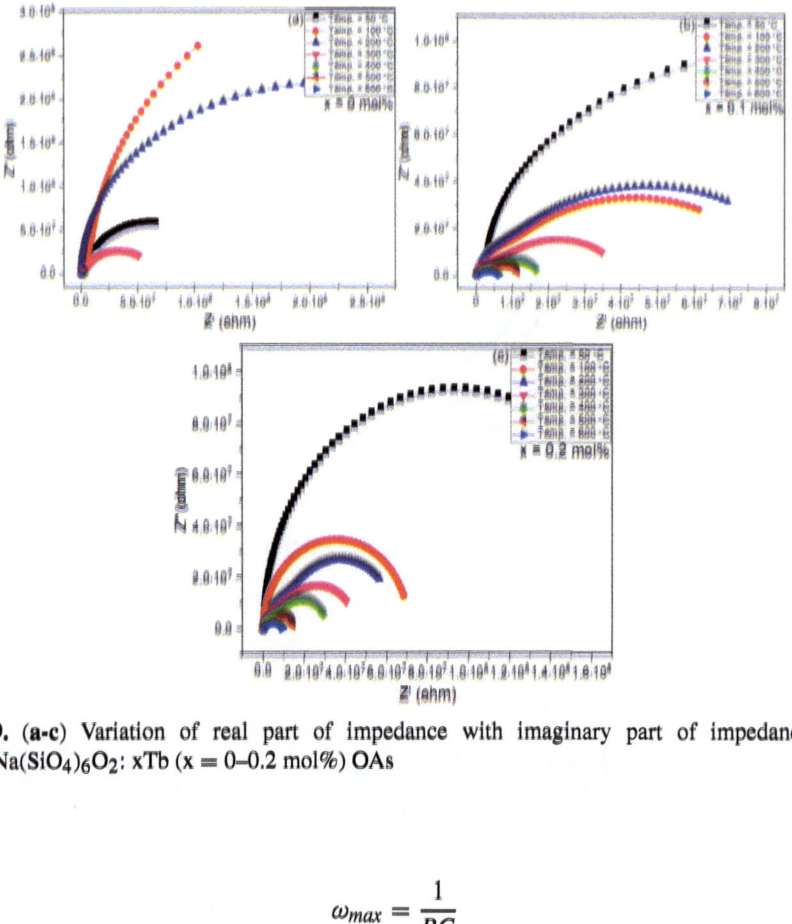

Fig. 9. (**a-c**) Variation of real part of impedance with imaginary part of impedance of $Y_{9-x}Na(SiO_4)_6O_2$: xTb (x = 0–0.2 mol%) OAs

$$\omega_{max} = \frac{1}{RC} \tag{8}$$

where ω_{max} denotes the highest frequency of the corresponding semicircle, C denotes the value of the capacitance, and R denotes the point of intercept on the real axis of the Nyquist plot. For the synthesized compounds, the calculated capacitance values (at 600 °C) are 2.1×10^{-11}, 1.05×10^{-11}, and 7.6×10^{-11}, respectively, for x = 0, 0.1, and 0.2 mol%. Since the order of capacitance value is of 10^{-11}, the resistance might be connected to the grain boundary contribution [31].

Conductivity Analysis

Figure 10 represents the frequency dispersion conductivity for $Y_{9-x}Na(SiO_4)_6O_2$: xTb

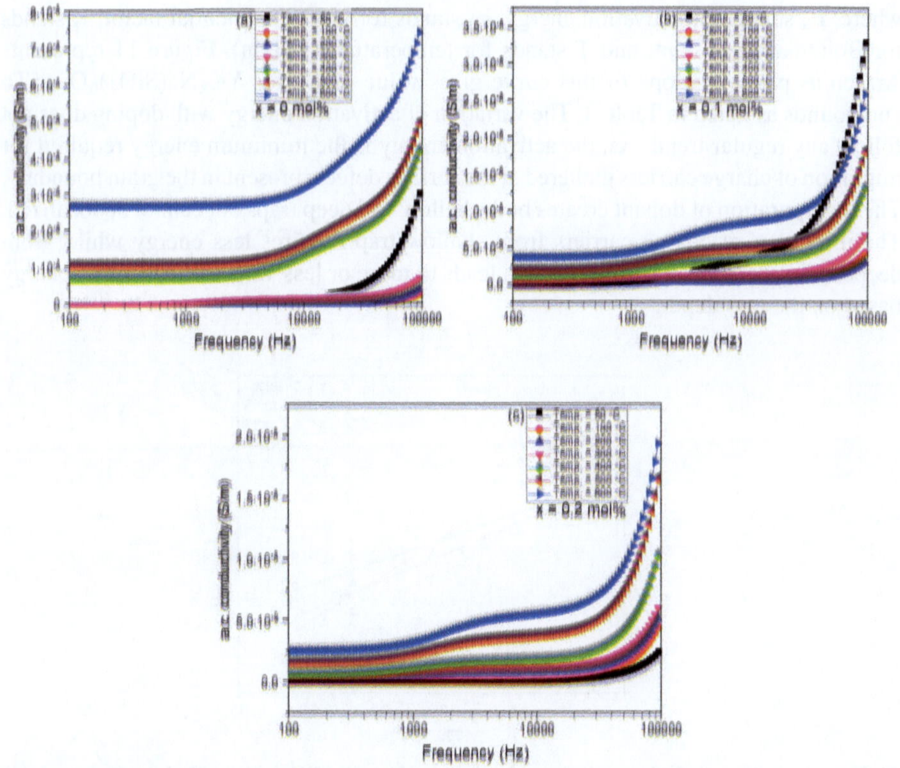

Fig. 10. **(a-c)** Variation of a. c. conductivity with frequency of $Y_{9-x}Na(SiO_4)_6O_2$: xTb (x = 0–0.2 mol%) OAs

compounds at various temperatures. Total conductivity is represented as sum of direct current (d. c.) and alternating current (a. c.) conductivity, as shown in Eq. (9),

$$\sigma_{total} = \sigma_{dc} + \sigma_{ac} \tag{9}$$

Here, σ_{dc} is the frequency independent d.c. conductivity and σ_{ac} is the frequency dependent a.c. conductivity. According to power law,

$$\sigma_{ac} = A\,\omega^{s} \tag{10}$$

Here, A is constant that depends on the material, ω represents angular frequency, and s represents frequency exponent taking values between 0 to 1. In low frequency regions (up to 10^3 Hz), conductivity increases linearly and after a particular frequency region it increases abruptly. In low-frequency regions (up to 10^3 Hz), the plateau curves are seen. This is represented as d.c. conductivity which depends on temperature, and expressed using Arrhenius Eq. (11) [32],

$$\sigma_{dc} = \sigma_0 \exp\left(\frac{-E_a}{kT}\right) \tag{11}$$

where, E_a stands for activation energy, σ_0 stands for pre—exponential factor, K stands for Boltzmann constant, and T stands for temperature (Kelvin). Figure 11 represents Arrhenius plot and slope of this curve gives value of E_a for $Y_{9-x}Na(SiO_4)_6O_2$: xTb compounds as listed in Table 3. The variation of activation energy with doping does not follow any regular trend. As, the activation energy is the minimum energy required for migration of charge carriers gathered at the surface defects present in the grain boundary. The incorporation of dopant creates both shallow and deep traps but cannot be localized. The migration of charge carriers from shallow trap requires less energy while, from deeper trap requires more energy. This leads to more or less values of activation energy based on the trap depth.

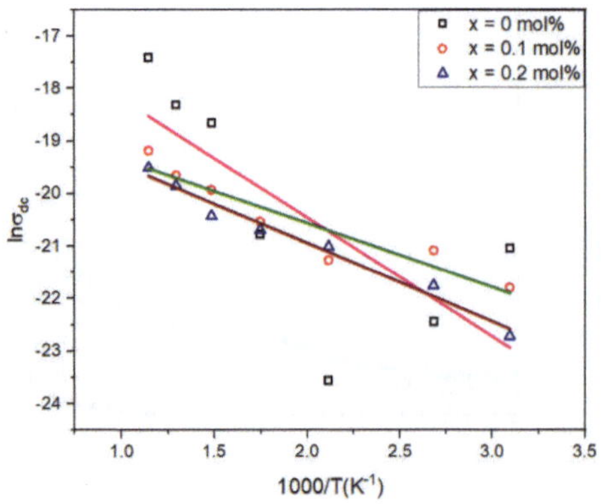

Fig. 11. Arrhenius plot of electrical conductivity of $Y_{9-x}Na(SiO_4)_6O_2$: xTb (x = 0–0.2 mol%) OAs

Table 3. Activation energy (E_a) of $Y_{9-x}Na(SiO_4)_6O_2$: xTb (x = 0–0.2 mol%) OAs

x in mol %	E_a (eV)
x = 0	0.194
x = 0.1	0.104
x = 0.2	0.128

On other hand, with increasing frequency, the strong frequency dispersion ac conductivity is observed. This frequency dispersion conductivity may be explained by two hopping mechanisms: successful hopping and unsuccessful hopping. In low-frequency region, excited ions hop to new neighbour sites, leading to a long-range movement of ions, and successful hopping gives rise to a d. c. conductivity (plateau regions). Whereas, in high-frequency region, ions are not able to stabilize at the new sites, and its jump back to its initial site leads to a short-range localized movement, and unsuccessful hopping gives rise to an a. c. conductivity (dispersive regions) [33].

Electric Modulus Analysis

The charge transport phenomena have been understood through the study of modulus spectroscopy. Figures 12 and 13 show the plot drawn between real and imaginary parts of modulus as a function of logarithmic angular frequency, respectively. The values of M' are seen to approach 0 at low frequencies, indicating that contribution arising from electrode polarization is negligible. At higher frequencies, however, M' portrays highest value. The dispersion between these frequencies may be occurring due to relaxation process of conductivity [34]. The M″ values increase as the frequency rises, reach a peak that indicates the relaxation frequency, and then decline. These peaks show the spreading of relaxation time as they move toward the higher-frequency side and widen as the temperature rises. Additionally, it also implies that the hopping mechanism occurs between several sites in low-frequency zone and within potential well in higher-frequency region [35]. Both M′ and M″ slightly lowers and does not get much affected with the increase in dopant concentration.

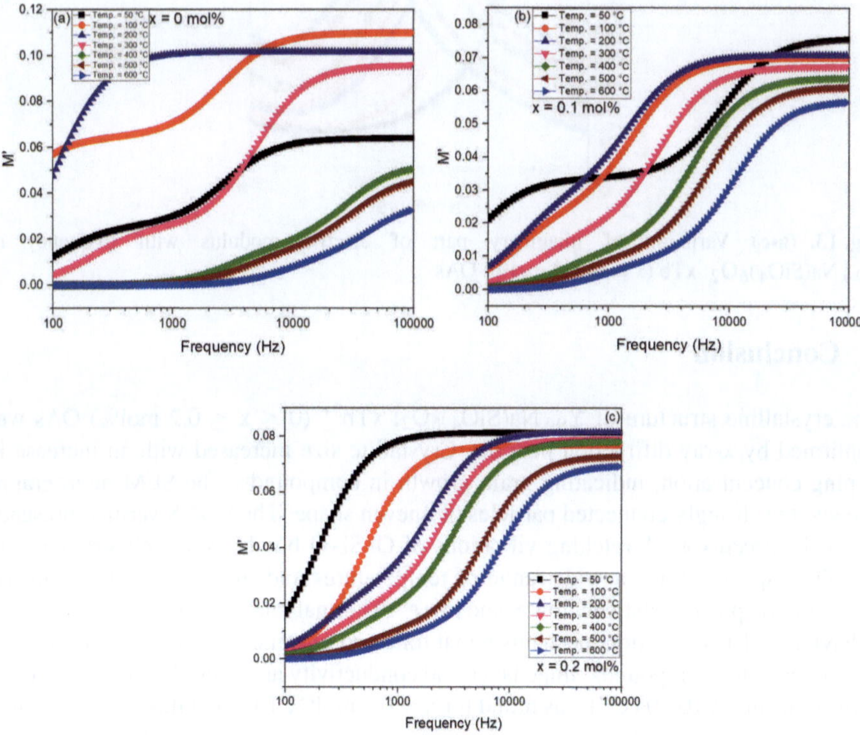

Fig. 12. **(a-c)** Variation of real part of electric modulus with frequency of $Y_{9\text{-}x}Na(SiO_4)_6O_2$: xTb (x = 0–0.2 mol%) OAs

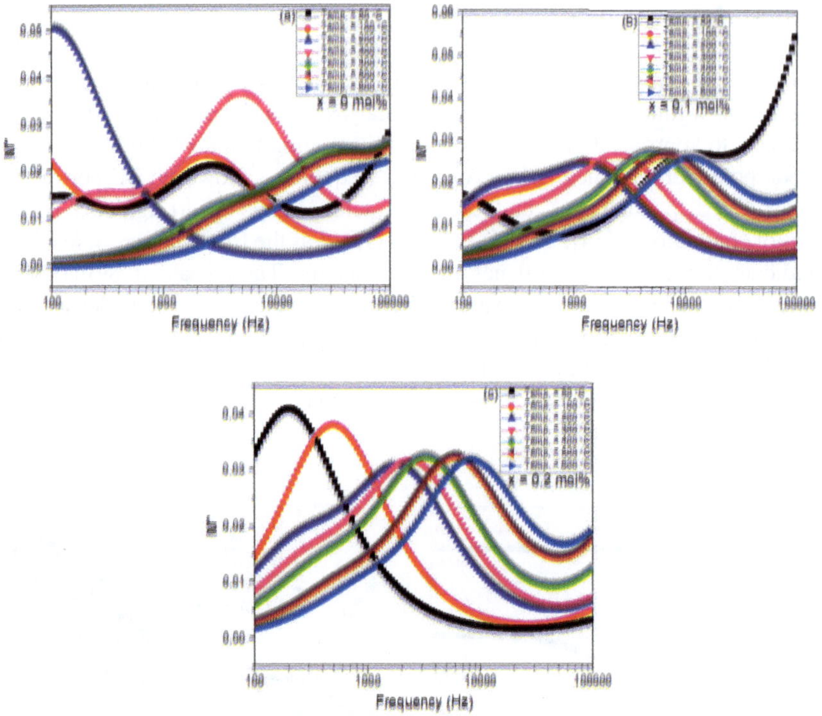

Fig. 13. (**a-c**) Variation of imaginary part of electric modulus with frequency of $Y_{9-x}Na(SiO_4)_6O_2$: xTb (x = 0–0.2 mol%) OAs

4 Conclusion

The crystalline structure of $Y_{9-x}Na(SiO_4)_6O_2$: xTb^{3+} ($0 \leq x \leq 0.2$ mol%) OAs was confirmed by x-ray diffraction patterns. Crystallite size increased with an increase in doping concentration, indicating grain growth in compounds. The SEM micrographs showed the strongly connected particles of uneven shape. The EDXS verified presence of all the precursors. Stretching vibrations of O–Si–O bonds were well structured in the FTIR spectra. Over a wide range of temperatures and frequencies, the dielectric properties displayed relatively low ε' and low ε''. Thermal relaxation phenomena, NTCR behavior, and a conduction mechanism that followed universal dielectric response were all shown by the compounds' impedance and conductivity analyses. The lowest value of activation energy (0.104 eV) was found for x = 0.1 mol%. The modulus spectra verified the charge carriers' short-range and long-range motions on high and low frequency sides, respectively. According to findings of dielectric tests, $Y_{9-x}Na(SiO_4)_6O_2$: xTb^3 OAs can be used in paraelectric capacitors and to shorten transmission delays for electrical signals.

References

1. Zhao, E., Liu, X., Liu, L., Huo, H., Xiong, Y.: Effect of La0.8Sr0.2Co0.2Fe0.8O3-δ morphology on the performance of composite cathodes. Prog. Nat. Sci. Mater. Int. **24**(1), 24–30 (2014). https://doi.org/10.1016/j.pnsc.2014.01.008
2. Abbassi, M., Ternane, R., Sobrados, I., Madani, A., Trabelsi-Ayadi, M., Sanz, J.: Ionic conductivity of apatite-type solid electrolyte ceramics Ca 2-xBaxLa4Bi4(SiO4) 6O2 (0≤x≤2). Ceram. Int. **39**(8), 9215–9221 (2013). https://doi.org/10.1016/j.ceramint.2013.05.026
3. Pasero, M., Kampf, A.R., Ferraris, C., Pekov, I.V., Rakovan, J., White, T.J.: Nomenclature of the apatite supergroup minerals. Eur. J. Mineral. **22**(2), 163–179 (2010). https://doi.org/10.1127/0935-1221/2010/0022-2022
4. Zhang, Z.W., Han, L., Zhao, Y., Wang, L.J., Yang, S.S., Chu, X.J.: Synthesis and photoluminescence properties of ALa9(SiO4)6O2: Eu3+ (A = Li, Na) red phosphor. Mater. Res. Bull. **94**, 147–153 (2017). https://doi.org/10.1016/j.materresbull.2017.05.060
5. Zheng, B., et al.: Ultra-wideband phosphor Mg2Gd8(SiO4)6O2:Ce3+,Mn2+: energy transfer and pressure-driven color tuning for potential applications in LEDs and pressure sensors. Chem. Eng. J. **427**(July 2021), 131897 (2022). https://doi.org/10.1016/j.cej.2021.131897
6. Randall, C.A. et al.: High temperature and high energy density dielectric materials. PPC2009 - 17th IEEE Int. Pulsed Power Conf., pp. 346–351, (2009), https://doi.org/10.1109/PPC.2009.5386292
7. Wang, D.H., Kurish, B.A., Treufeld, I., Zhu, L., Tan, L.S.: Synthesis and characterization of high nitrile content polyimides as dielectric films for electrical energy storage. J. Polym. Sci. Part A Polym. Chem. **53**(3), 422–436 (2015). https://doi.org/10.1002/pola.27445
8. Sebastian, M.T., Ubic, R., Jantunen, H.: Low-loss dielectric ceramic materials and their properties. Int. Mater. Rev. **60**(7), 392–412 (2015). https://doi.org/10.1179/1743280415Y.0000000007
9. Bi, J.X., Xing, C.F., Zhang, Y.H., Yang, C.H., Wu, H.T.: Correlation of crystal structure and microwave dielectric properties of Zn1−xNixZrNb2O8 (0 ≤ x ≤ 0.1) ceramics. J. Alloys Compd. **727**, 123–134 (2017). https://doi.org/10.1016/j.jallcom.2017.08.110
10. Wu, C.S., Lei, H., Huang, J.S., Yao, D.H., Cheng, Y.L., Huang, Z.L.: Preparation and characterization of Gd and Zn doped apatite lanthanum silicate electrolyte. IOP Conf. Ser. Mater. Sci. Eng. **678**(1), 2–6 (2019). https://doi.org/10.1088/1757-899X/678/1/012152
11. Niu, Y.J., Liu, M.T., Li, M.F., Bi, J.X., Wu, H.T.: Influence of Mg2+substitution on the crystal structure and microwave dielectric properties of ZnZrNb2O8ceramics. J. Alloys Compd. **705**, 399–404 (2017). https://doi.org/10.1016/j.jallcom.2017.02.196
12. Ulrich, R., Schaper, L., Nelms, D., Leftwich, M.: Comparison of paraelectric and ferroelectric materials for applications as dielectrics in thin film integrated capacitors. Int. J. Microcircuits Electron. Packag. **23**(2), 172–181 (2000)
13. Melkozerova, M.A., et al.: Intrinsic defects and their influence on optical properties of ALa9(GeO4)6O2 (A = Li, Na, K, Rb, Cs) oxyapatites prepared by spray pyrolysis. J. Alloys Compd. **839** (2020). https://doi.org/10.1016/j.jallcom.2020.155609
14. Xiang, J., Liu, Z.G., Ouyang, J.H., Yan, F.Y.: Synthesis, structure and electrical properties of rare-earth doped apatite-type lanthanum silicates. Electrochim. Acta. **65**, 251–256 (2012). https://doi.org/10.1016/j.electacta.2012.01.048
15. Reddy, P.M., Lakshmi, R., Dass, F.P., Sasikumar, S.: Synthesis, characterization and formulation of sodium calcium silicate bioceramic for drug delivery applications. Sci. Eng. Compos. Mater. **23**(4), 375–380 (2016). https://doi.org/10.1515/secm-2014-0057
16. Gupta, R., Agrawal, S.: Judd–Ofelt analysis and photoluminescence properties of europium-activated Y8−xSr2(SiO4)6O2 oxyapatite phosphors. J. Mater. Sci. Mater. Electron. **2** (2022). https://doi.org/10.1007/s10854-022-08597-9

17. Revathy, J.S., Priya, N.S.C., Sandhya, K., Rajendran, D.N.: Structural and optical studies of cerium doped gadolinium oxide phosphor. Bull. Mater. Sci. **44**(1), 1–8 (2021). https://doi.org/10.1007/s12034-020-02299-w
18. Perhaita, I., Muresan, L.E., Borodi, G., Popa, A., Nicoara, A., Tudoran, L.B.: Studies on morpho-structure and ionic conductivity of apatite-type lanthanum silicate doped with transitional metal cations. J. Electroceram. (2024). https://doi.org/10.1007/s10832-024-00374-w
19. Mahapatro, J., Meena, S.S., Agrawal, S.: Effect of Eu–Er substitution on structural, optical, dielectric, and electrical properties of Ba0.5Sr0.5EuxErxFe12–2xO19 hexaferrite. J. Rare Earths. (2024). https://doi.org/10.1016/j.jre.2024.01.010
20. Reddy, L.: The role of anions in rare - earth activated inorganic host materials for its luminescence characteristics, no. 0123456789. Springer US. (2024) doi: https://doi.org/10.1007/s10895-023-03561-0
21. Wang, X., He, X., Wang, X.: FTIR analysis of the functional group composition of coal tar residue extracts and extractive residues. Appl. Sci. **13**(8) (2023). https://doi.org/10.3390/app13085162
22. Chauhan, C.C., et al.: Structural, magnetic and dielectric properties of Co-Zr substituted M-type calcium hexagonal ferrite nanoparticles in the presence of α-Fe2O3 phase. Ceram. Int. **44**(15), 17812–17823 (2018). https://doi.org/10.1016/j.ceramint.2018.06.249
23. Torğut, G., Gürler, N.: Enhanced impedance, electrical conductivity, dielectric properties for colloidal starch-g-poly (methyl methacrylate) supported with semiconductor cadmium sulfide. Polym. Bull. **0123456789** (2024). https://doi.org/10.1007/s00289-023-05125-5
24. Chandekar, K.V., et al.: Significant and systematic impact of yttrium doping on physical properties of nickel oxide nanoparticles for optoelectronics applications. J. Mater. Res. Technol. **15**, 2584–2600 (2021). https://doi.org/10.1016/j.jmrt.2021.09.072
25. Afify, I.H., Farag, E.S.M., Abo-Ghazala, M.S., El-Zaidia, M.M.: AC conductivity and dielectric properties of InxTe20−xSe80 chalcogenides. Indian J. Phys. **98**(March), 967–973 (2023). https://doi.org/10.1007/s12648-023-02891-w
26. Gopalan, E.V., Joy, P.A., Al-Omari, I.A., Kumar, D.S., Yoshida, Y., Anantharaman, M.R.: On the structural, magnetic and electrical properties of sol-gel derived nanosized cobalt ferrite. J. Alloys Compd. **485**(1–2), 711–717 (2009). https://doi.org/10.1016/j.jallcom.2009.06.033
27. Ghosh, S.K., Singh, H., Mallick, K.: Low-dimensional lead chromate-based hybrid system for capacitance and polarization performances: a flexible device for pressure-induced voltage generator. Emerg. Mater. **7**(1), 235–245 (2023). https://doi.org/10.1007/s42247-023-00586-w
28. Bahel, S., Pubby, K., Narang, S.B.: X-band microwave reflection properties of samarium/bismuth-substituted barium lanthanum titanate ceramics. J. Electron. Mater. **46**(3), 1770–1776 (2017). https://doi.org/10.1007/s11664-016-5226-6
29. Thakur, S., Rai, R., Bdikin, I., Valente, M.A.: Impedance and modulus spectroscopy characterization of Tb modified Bi0.8A0.1Pb0.1Fe0.9Ti0.1O3 ceramics. Mater. Res. **19**(1), 1–8 (2016). https://doi.org/10.1590/1980-5373-MR-2015-0504
30. Mateyshina, Y., Slobodyuk, A., Kavun, V., Uvarov, N.: Conductivity and NMR study of composite solid electrolytes CsNO2-A (A = SiO2, Al2O3, MgO). Solid State Ionics. **324**(February), 196–201 (2018). https://doi.org/10.1016/j.ssi.2018.04.026
31. Irvine, B.J.T.S., Sinclair, D.C., West, A.R.: Electroceramics : characterization by impedance spectroscopy. **2**(3), 132–138 (1990)
32. Ojha, S., Ali, M.S., Roy, M., Bhattacharya, S.: Hopping frequency and conductivity relaxation of promising chalcogenides: AC conductivity and dielectric relaxation approaches. Mater. Res. Express. **8**(8) (2021). https://doi.org/10.1088/2053-1591/ac1d17
33. Almond, D.P., Duncan, G.K., West, A.R.: The determination of hopping rates and carrier concentrations in ionic conductors by a new analysis of ac conductivity. Solid State Ionics. **8**(2), 159–164 (1983). https://doi.org/10.1016/0167-2738(83)90079-6

34. Chérif, S.F., Chérif, A., Dridi, W., Zid, M.F.: Ac conductivity, electric modulus analysis, dielectric behavior and bond valence sum analysis of Na3Nb4As3O19 compound. Arab. J. Chem. (2020). https://doi.org/10.1016/j.arabjc.2020.04.003
35. Nobre, M.A.L., Lanfredi, S.: Phase transition in sodium lithium niobate polycrystal: an overview based on impedance spectroscopy. J. Phys. Chem. Solids. **62**(11), 1999–2006 (2001). https://doi.org/10.1016/S0022-3697(01)00042-7

Analytic Approaches to Study Thermodynamic Properties of Zirconium Carbides (ZrC)

Shraddha Chaudhari[1,2](✉), Amit Patel[3], Chirag Patel[1], and N. K. Bhatt[4]

[1] Faculty of Science and Humanities, Sankalchand Patel University, Visnagar, Gujarat, India
shraddhaphysics1@gmail.com
[2] M. N. College, Visnagar, Gujarat, India
[3] Government Science College, Gujarat University, Ahmedabad, Gujarat, India
[4] Department of Physics, Maharaja Krishnakumarsinhji Bhavnagar University, Bhavnagar, Gujarat, India

Abstract. Zirconium carbide (ZrC) have very high melting point and have high mechanical strength thus found wide application in nuclear reactors, machine industries. In present paper, UEOS (Universal Equation of State) in combined with the Mie-Grüneisen hypothesis is used to study temperature effect on isothermal EOS for Zirconium carbide. In addition, the variation of thermal expansion, bulk modulus and heat capacity with temperature have been discussed. Calculated results are compared with the available experimental and theoretical results.

Keywords: ZrC · Thermal properties · EOS · UEOS · Thermal expansion

1 Introduction

Transition Metal Carbides (TMCs) like ZrC, TiC have been found crucial attention in industrial applications [1–2] because of their significant physical and chemical properties. In this fact, most of the experimental as well as theoretical works were focused on this group four and five TMCs [3–5]. ZrC is exceptional among the TMCs due to its remarkable physical and chemical properties. At ambient condition, ZrC has been grow in B1 (rock salt) crystalline structure [6]. ZrC has several interesting properties like high melting point, exceptional hardness, chemically inertness and show metallic behavior in their electrical, optical and magnetic properties [7–10]. A complete ab initio calculation is restricted largely due to computational complexity of many-body system problem. We have been therefore used a simple analytical approach to incorporate thermal effect by combining the universal equation of state (UEOS) to the Mie-Grüneisen hypothesis. It has been demonstrated that from the knowledge of some cohesive properties at ambient condition various thermodynamic properties of ZrC can be obtained at an extreme environment, such as finite temperatures and pressures. Furthermore, we have been calculated the thermal pressure at different temperature as well as temperature variations of isothermal bulk modulus. Obtained results are compared with published ab initio calculation results as well experimental results.

S. Kumari et al. (Eds.): ICMTP 2024, SPPHY 311, pp. 438–444, 2025.
https://doi.org/10.1007/978-981-96-6795-6_47

2 Formulation

The isothermal equation of state at temperature T given by Eq. (1) [11],

$$p(V, T) = p_0(x) + \left(\frac{\gamma_{th}(V)}{V}\right) 3k_B T \tag{1}$$

Where p_0 (x) is the ambient condition pressure and is given by Vinet's UEOS scheme [11]. The thermal energy due to lattice vibration is assumed to be linear in temperature and taken as $-3k_BT$. Thermodynamic Grüneisen parameter (γ_o) at ambient condition is calculated from the knowledge of Poisson ratio [11]. While its volume dependence, which controls the thermal pressure is given in Ref. [11]. To include thermal effect, we assume the following conditions: (i) pressure derivative of bulk modulus does not change with temperature and pressure (ii) temperature dependence of bulk modulus is controlled by the Grüneisen parameter [11]. Using zero pressure condition, relative volume thermal expansion is defined by equation $V/V_o = (1+ \beta(T)\, T)$, Here, β (T) is the volume thermal-expansion coefficient. The constant volume specific heat C_v is calculated using Eq. (2) [12].

$$C_V(T) = \frac{V(T) B_T^{p=0} \beta(T)}{\gamma_{th}(T)} \tag{2}$$

3 Result and Discussion

In the present scheme, the input data as shown in Table 1 are given from the mentioned references to obtain the equation of state and introduce the thermal part as per discussed in formulation.

Table 1. The input data used in the equation of state in Eq. (1).

Material	Equilibrium Volume Å^3	Isothermal Bulk modulus (B_T) (GPa)	Pressure derivative of Bulk modulus (B'_T)	Poisson ratio, σ
ZrC	25.3639 [6]	218.0 [13]	3.5	0.280 [14]

Figure 1 shows isothermal equation of state (EOS) or P-V relation for zirconium carbide at different temperature and compared with the first principle calculated results due to Jun et al. [13]. The qualitative and quantitative agreement have been found for EOS at T = 300 K, 1600 K, 3200 K [13]. Figure 1 also explain that as pressure increases, the relative volume V/V_0 decreases for a particular temperature. As temperature increases the total pressure also increased. Similar nature of P-V graphs at higher temperature to corresponding T = 0 K EOS suggests that high-T equation of state are governed largely by ambient pressure only, and confirms the applicability of Vinet's UEOS to TMC family.

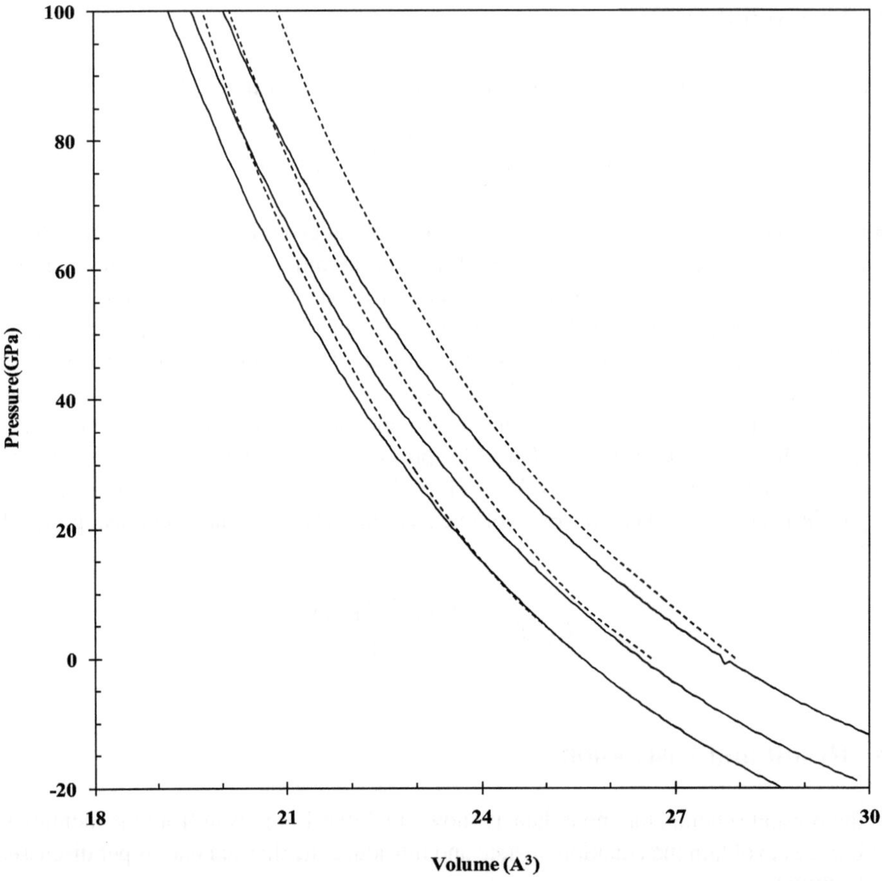

Fig. 1. Results show isothermal equation of state (solid lines) at T = 300 K, 1600 and 3200 K (from left to right) for ZrC and compared with the ab initio results (broken lines) due to Jun et al. [13] at corresponding temperature.

Effect of temperature shown in shifts of the P-V curve towards right indicating thermal expansion.

As a further test of the model, we have calculated temperature and pressure variations of isothermal bulk modulus. As shown in the Fig. 2, with temperature B_T decreases in qualitative agreement to the first principles [13], though, the rate of decrease is low.

Figures 3 shows the results of thermal expansion coefficient at different temperature for zirconium carbides. It depicts that calculated result is good agreement with other experimental and theoretical result. Which is in support to the present proposal scheme. Figure 3 also shows that calculated result lie between the experimental and first principles calculated result. Low temperature ambiguity is expected, where average over different vibrational eigen-modes to find thermodynamic gamma depends sensitively on quantum correlation effect. Like many high-temperature theories, here also this overlap effect is ignored and only classical treatment is adopted.

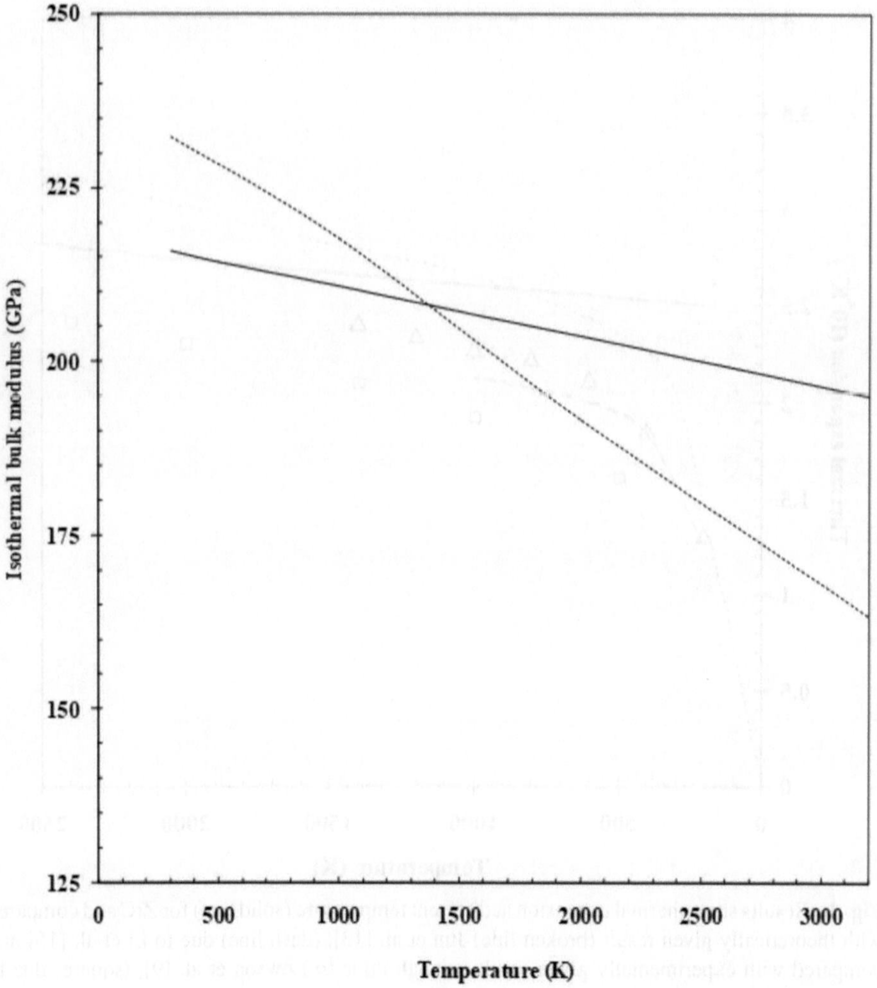

Fig. 2. Result shows isothermal thermal bulk modulus at different temperatures (solid line) for ZrC and compared with the ab initio calculated results (broken line) due to Jun et al. [13].

Experimentally most easily assessable thermodynamic physical property is the specific heat. Its evaluation is not only important as relates vibrational dynamics of the crystal but it also gives insight to anharmonicity. Presently calculated heat capacities at constant volume (C_V) is presented in Fig. 4 for ZrC and compared with the other results. Overall good agreement is observed for ZrC, low temperature inconsistency prevails, due to valid simplification only to high temperature regime. In our opinion, with temperatures atomic vibrations increase in amplitude leading to higher vibrational entropy, and corresponding specific heat increases too. Thus, anharmonicity at higher T (greater than Debye temperature), is observed as increase in C_V. Our results for C_V confirm that even with such a simple but realistic approach, anharmonic effect is taken care.

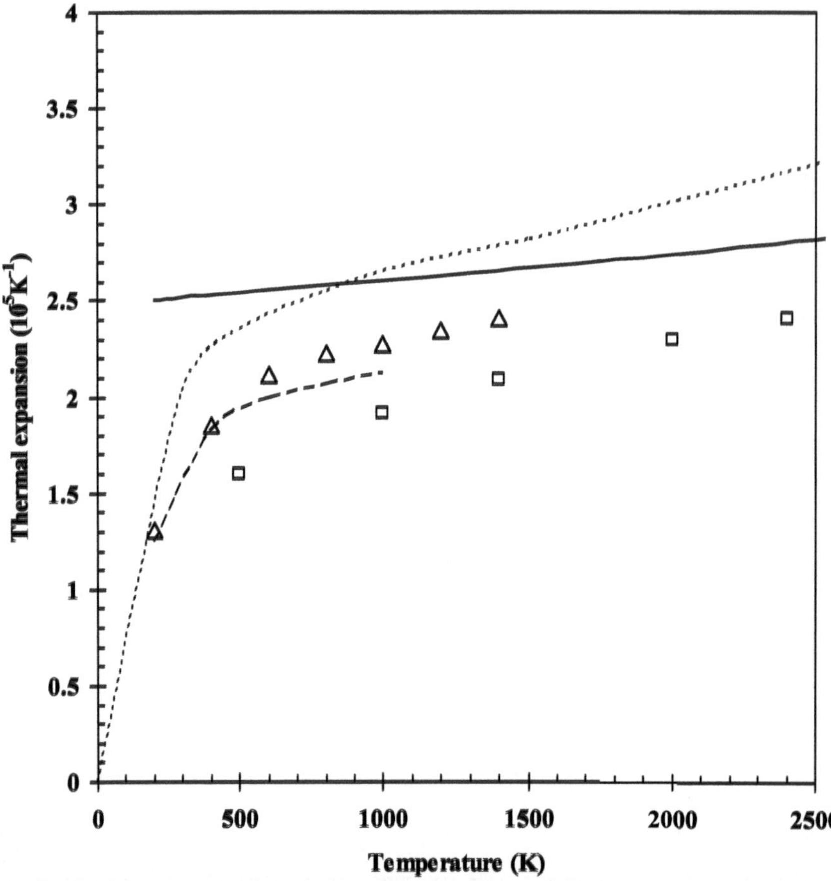

Fig. 3. Results show thermal expansion at different temperature (solid line) for ZrC and compared with theoretically given result (broken line) Jun et al. [13], (dash line) due to Li et al. [15] and compared with experimentally given result (triangle) due to Lowson et al. [9], (square) due to Singh et al. [16].

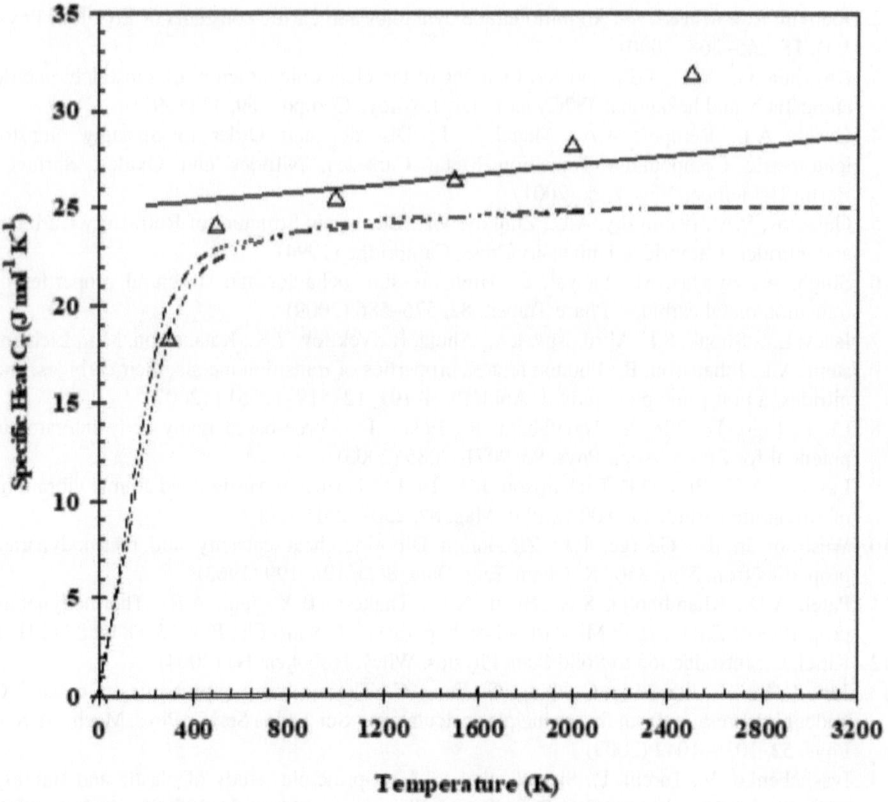

Fig. 4. Results specific heat at constant volume at different temperature (solid line) for ZrC and compared with theoretically given result due to (broken line with dash) Jun et al. [13], due to (broken line with double dash) Li et al. [15], and compared with experimental results [17] are given as open triangles.

4 Conclusion

In the summary, we can conclude that we have calculated some thermodynamic properties of transition metal carbide (ZrC) as prototype. Literature survey suggests that these carbides have potential applications in high temperature regime due to their hardness and thermal resistance, even though their thermodynamic investigation at high-T environment is limited. With minimal input data, we have plotted P-V curve at different temperature using UEOS formalism. Good accordance of our results with the available experimental and first principles data. Therefore, it justifies of the present coupling scheme.

References

1. Storms, E.K.: The Refractory Carbides. Academic, New York (1967)

2. Jochym, P., Parlinski, K.: Ab initio lattice dynamics and elastic constants of ZrC. Eur. Phys. J. B. **15**, 265–268 (2000)
3. Khyzhun, O.: XPS, XES, and XAS studies of the electronic structure of substoichiometric cubic TaCx and hexagonal Ta2Cy carbides. J. Alloys Compd. **259**, 47 (1997)
4. Gusev, A.I., Rempel, A.A., Magel, A.J.: Disorder and Order in Strongly Nonstoichiometric Compounds: Transition Metal Carbides, Nitrides and Oxides. Springer, Berlin/Heidelberg/New York (2001)
5. Gubanov, V.A., Ivanovsky, A.L., Zhukov, V.P.: Electronic Structure of Refractory Carbides and Nitrides. Cambridge University Press, Cambridge (1994)
6. Singh, A., Aynyas, M., Sanyal, S.: High pressure behavior and structural properties of transition metal carbides. Phase Transit. **82**, 576–586 (2009)
7. Isaev, E.I., Simak, S.I., Abrikosov, I.A., Ahuja, R., Vekilov, Y.K., Katsnelson, M.I., Lichtenstein, A.I., Johansson, B.: Phonon related properties of transition metals, their carbides, and nitrides: a first-principles study. J. Appl. Phys. **101**, 123519–123518 (2007)
8. Li, J., Liao, D., Yip, S., Najafabadi, R., Ecker, L.: Force-based many-body interatomic potential for ZrC. J. Appl. Phys. **93**, 9071–9085 (2003)
9. Lawson, A.C., Butt, D.P., Richardson, J.W., Li, J.: Thermal expansion and atomic vibrations of zirconium carbide to 1600 K. Phil. Mag. **87**, 2507–2519 (2007)
10. Westrum Jr., E., George, F.J.: Zirconium Diboride: heat capacity and thermodynamic properties from 5° to 350° K. Chem. Eng. Data. **8**(2), 193–199 (1963)
11. Patel, A.D., Khambholja, S.G., Bhatt, N.K., Thakore, B.Y., Jani, A.R.: Thermodynamic properties of ZnO with in Mie-Grüneisen hypothesis. J. Nano-Ele. Phys. **3**, 884–887 (2011)
12. Kittel, C.: Introduction to Solid State Physics. Wiley, Hoboken, NJ (2004)
13. Jun, Z., Bo, Z., Jian Ying, Q., Qing, G., Feng, C.: Thermodynamic properties of cubic ZrC under high pressure from first-principles calculations. Sci. China Ser. G: Phys. Mech. Astron. Phys. **52**, 1039–1042 (2009)
14. Ivashchenko, V., Turchi, P., Shevchenko, V.: First-principles study of elastic and stability properties of ZrC–ZrN and ZrC–TiC alloys. J. Phys. Cond. Matter. **21**, 395503–395509 (2009)
15. Li, J., Liao, D., Yip, S., Najafabadi, R., Ecker, L.: Force-based many-body interatomic potential for ZrC. J. Appl. Phys. **93**, 9072–9085 (2003)
16. Singh, M., Wiedemeier, H.: Estimation of thermal expansion behaviour of some refractory carbides and nitrides. J. Mater. Sci. **32**, 5749–5751 (1997)
17. Toth, L.E.: Transition Metal Carbides and Nitrides. Academic, New York (1971)

Finite Element Simulations of Thermal Properties of Multilayer Coatings

Lalit Kumar Sharma[1](✉), Neeraj Kumar Sharma[1], and Abhimanyu Singh Rana[2]

[1] Department of Mechanical Engineering, SoET, BML Munjal University, Gurugram 122413, Haryana, India

[2] Centre for Advanced Materials and Devices, SoET, BML Munjal University, Gurugram 122413, Haryana, India

lalit.sharma.18pd@bmu.edu.in

Abstract. In order to improve the hardness and toughness of cutting tools, coatings of metal nitrides are used. Titanium based hard coatings are used in many industrial applications. The high temperature application in many industrial areas makes the thermal insulation behavior of these coatings equally important. In the present work, we have investigated the thermal properties of titanium-based coatings (TiN and TiAlN) using Finite element (FE) simulations. FE models are prepared using alternative layers of TiN (Titanium Nitride) and TiAlN (Titanium Aluminium Nitride) thin film coatings on SS 301 steel substrate. These models are analyzed to predict the heat transfer through the thin film coatings. Various models using different numbers of layers on the substrate material are formulated. The effective thermal conductivity (ETC) of multilayer coatings has been predicted and the simulated results are compared with the reported experimental values. The ETC for 50 bilayer deposition of TiN and TiAlN has been reported as approximately to 3.85 W/m °C and the simulated results through FE simulations validate the same. The defects formed during deposition have been studied by incorporating the thermal contact conductance in FE models. These defects result in phonons scattering between layers and reduce the ETC values.

Keywords: Finite Element Method · Multilayer Coatings · Thermal Properties · Thin Film Coatings · FE Simulations

1 Introduction

Thermal characteristics play an important role in the service requirements of the material. The material becomes soft and prone to accelerated wear and tear at high-temperature service conditions [1–3]. Therefore, under thermo-mechanical loading conditions, it is crucial to have superior thermal properties in addition to mechanical characteristics. The coatings of thin film have yielded satisfactory results in thermal barrier applications [4–6]. The thermal barrier coatings require to impart the requisite hardness as well as the insulation against heat transport to the coated surface. Physical vapour deposition (PVD) is one of the popular methods of thin film coating. Coating parameters need to be optimized to get the desired characteristics.

S. Kumari et al. (Eds.): ICMTP 2024, SPPHY 311, pp. 445–452, 2025.
https://doi.org/10.1007/978-981-96-6795-6_48

Under thermo-mechanical working conditions, high hardness and less wear are desirable properties [7, 8]. TiN (Titanium Nitride) and TiAlN (Titanium Aluminium Nitride) thin film coatings are popular coatings to have such properties and are used in various applications such as machining. TiN coatings provide high hardness and TiAlN helps in achieving low thermal conductivity because of the presence of aluminum. Therefore, using alternate layers of TiN and TiAlN gives the advantage of superior mechanical and thermal response as compared to the substrate material. The heat conducting ability of a material depends on the availability of free electrons as well as the lattice vibrations. In thin film coatings the lattice vibrations play a significant role in heat transfer [9]. Heat flow is due to the phonons but phonons get scattered as a result of boundary resistance offered. Total resistance to heat transfer is achieved because of defects, dislocations, impurities, grain boundaries, and coating boundary interfaces. This effective resistance to heat flow which act as a thermal barrier is referred to as thermal contact conductance (TCC) [1]. Finite element (FE) analysis can provide important insight into thermal deformation of nano materials [10, 11]. FE simulations have been used to tailor the material microstructure of complex structured materials in order to yield the desired effective properties.

In the present work, we have analyzed the thermal transport behavior of TiN/TiAlN multilayer coatings on the SS301 substrate. The effects of phonon scattering, imperfect contact etc. have been studied by assigning a thermal contact resistance at the interfaces.

2 FEM Modeling

2.1 Material and Method

We have investigated the thermal properties of titanium-based coatings (TiN and TiAlN) using Finite element (FE) simulations. FE models are prepared using alternative layers of TiN (Titanium Nitride) and TiAlN (Titanium Aluminium Nitride) thin film coatings on steel substrate. These models are analyzed to predict the heat transfer through the thin film coatings. FE models with different numbers of layers: 5 layers, 10 layers, 20 layers, 50 layers deposited on the substrate material are generated using ANSYS software. A thermal contact conductance (TCC) value is assigned to incorporate the effects of thermal contact resistance because of various defects such as, missing atoms, voids, dislocations and phonon scattering at interfaces. The effective thermal conductivity (ETC) of multilayer coatings has been predicted and the simulated results are compared with the reported experimental values.

A three-dimensional model of substrate with coatings was prepared using ANSYS software. A substrate of 1000 nm × 500 nm × 100 nm dimensions was modelled and alternate layers of TiN/TiAlN was added on the substrate (Fig. 1). The material properties of TiN and TiAlN as reported in literature were used for FE analysis. A Dirichlet boundary conditions of specified temperatures was used and a temperature of 25 °C at the bottom of the substrate and 225 °C at the top coating layer are assigned. Steady state thermal analysis was conducted using ANSYS software. Nano-scale simulation models are used in the present study to take advantage of computation time and cost.

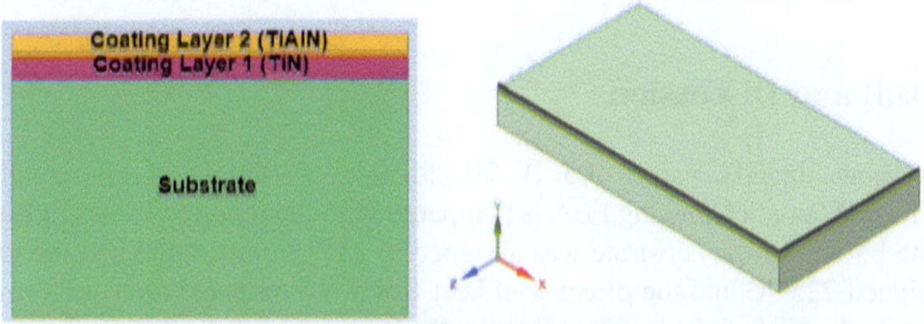

Fig. 1. Multilayer coating of TiN and TiAlN on substrate

A sufficiently fine mesh was used to ensure the mesh independent results (Fig. 2) Thermal conductivity value of TiN, TiAlN, and substrate is taken as 11.9, 4.25, and 16.3 W/m °C respectively for the FE study (Table 1).

Fig. 2. 50-layers model with finite element mesh

Table 1. Thermal properties considered for finite element analysis [1].

Material	Thermal Conductivity (W/m°C)
TiN	11.9
TiAlN	4.25
Substrate	16.3

The effective thermal conductivity was computed using Fourier's law as:

$$\dot{q} = k_{eff} \frac{\Delta T}{\Delta Y} \tag{1}$$

where $\dot{q}$ is heat flux, k_{eff} is effective thermal conductivity, ΔT is the change in temperature and ΔY is wall thickness.

3 Result and Discussion

Figure 6 shows the ETC values of 5, 10, 20 and 50 layers samples at different values of TCC. The ETC predicted using FEA is compared with the reported values in Literature [12]. The bottom of the substrate was assigned to 25 °C temperature and the top layer was assigned 225 °C and the directional heat flux was computed after performing the steady-state thermal analysis. The effective thermal conductivity was computed using Fourier's law.

It can be observed that as the ETC reduces as the number of layers increases, also, with each layer the ETC reduces as the TCC reduces. A lower value of TCC indicated higher thermal contact resistance at due to phonon scattering and other defects. Various defects and discontinuities result in the scattering of phonons and decreases the net heat flow. These defects result in the path deviation of phonons and reduce the effective thermal conductivity (ETC) values (Fig. 3).

It can be observed that the ETC of 5, 10, 20 and 50 layers FEA models best match with the reported value [12] of ETC at TCC values of 1.2e9, 5e8, 2.6e8 and 6e7 w/m^2°C respectively. The reduced values of TCC for the higher number of layers are due to the increased number of interfaces and defects (Figs. 3, 4, 5 and 6).

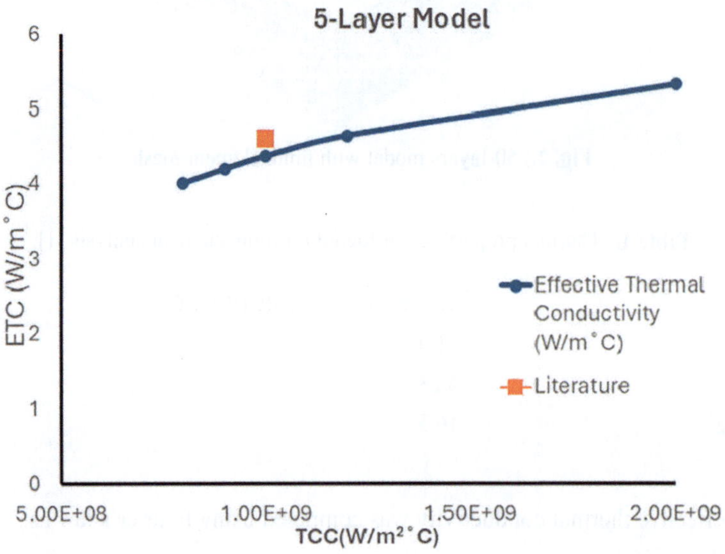

Fig. 3. ETC of 5 layers TiN & TiAlN thin-film model for different values of TCC

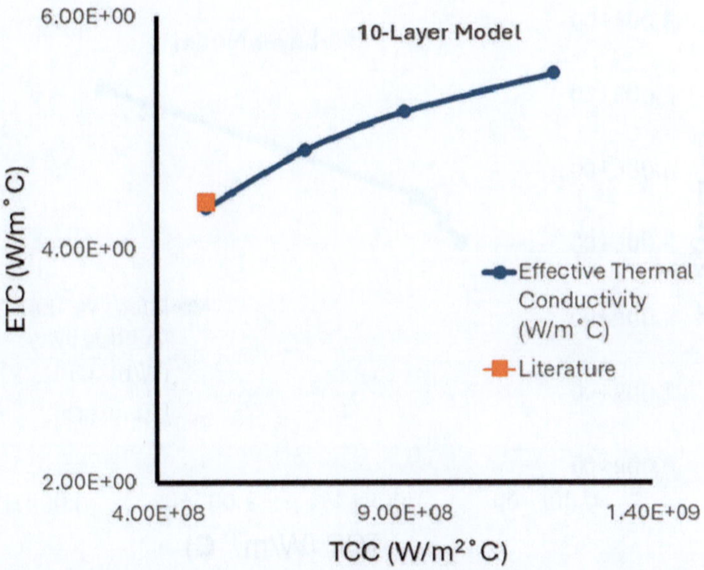

Fig. 4. ETC of 10 layers TiN & TiAlN thin-film model for different values of TCC

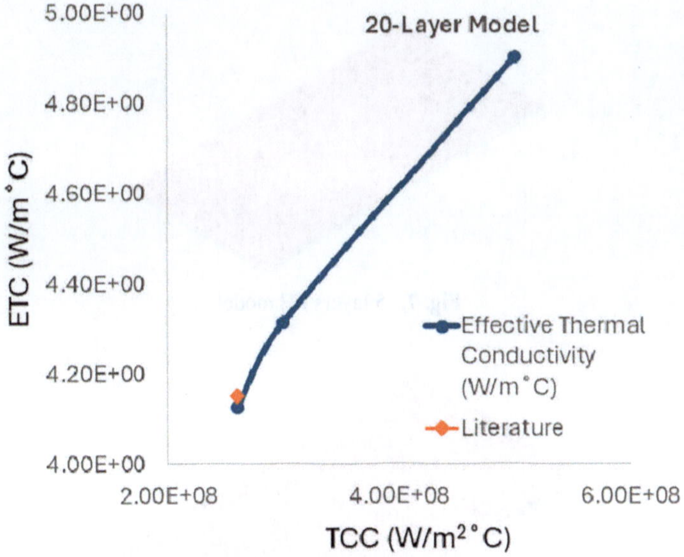

Fig. 5. ETC of 20 layers TiN & TiAlN thin-film model for different values of TCC

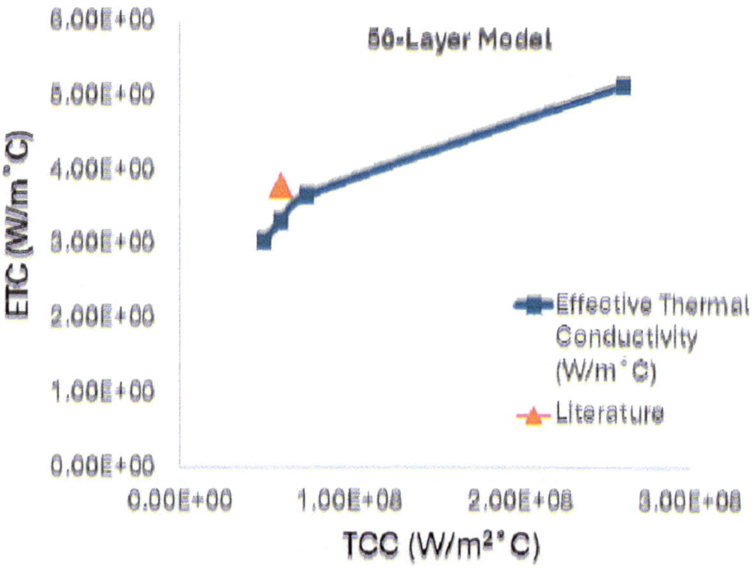

Fig. 6. ETC of 50 layers TiN &TiAlN thin-film model for different values of TCC

Fig. 7. 5 layers FE model

Fig. 8. 10 layers FE model

Fig. 9 20 layers FE model

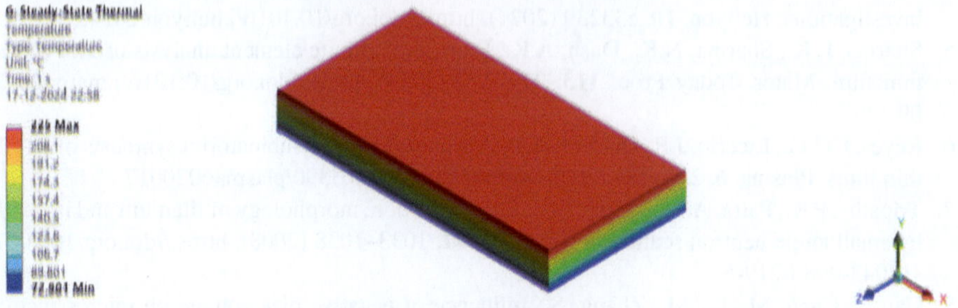

Fig. 10. 50 layers FE model

In model 1, a total of 5 layers of TiN and TiAlN are modelled. Results show the temperature of 180.41 °C at the exposed surface of substrate and an effective thermal conductivity of 6.80 W/m °C. Increasing the number of coating layers on substrate results in a greater number of interfaces which act as a barrier to the heat flow. Also, increased aluminum content results in phonon scattering and defects serve the same purpose. Therefore, substrate's exposed surface temperature for model 2, 3 and 4 are 151.99, 113.14, and 72.90 °C respectively(Figs. 7, 8, 9 and 10).

4 Conclusions

The effective thermal conductivity of multilayer coatings is determined by the thermal conductivity of each layer and the thermal boundary resistance between the layers. Differences in lattice constants of the TiN and TiAlN layers leads to the formation of defects and dislocations. Phonon scattering takes place because of the interface defects between the coating layers, resulting in a decreased value of thermal conductivity. It is reported that by increasing the coating layers on the substrate, the thermal conductivity valve of a multilayer film can be smaller as compared to their alloys. Effective thermal conductivity decreases with increased number of interfaces and dislocations of the multilayer coatings of TiN and TiAlN. It can be concluded from the study that ETC value of 50 multilayer coatings is closer to the reported value. Therefore, finite element analysis gives the advantage of simulations for various thermal conductivity finite element models and provides an approximately close solution to the problem.

References

1. Ding, X.-Z., Samani, M.K., Chen, G.: Thermal conductivity of PVD TiAlN films using pulsed photothermal reflectance technique. Appl. Phys. A. **101**, 573–577 (2010). https://doi.org/10.1007/s00339-010-5900-0
2. Bartosik, M., Holec, D., Apel, D., Klaus, M., Genzel, C., Keckes, J., Arndt, M., Polcik, P., Koller, C.M., Mayrhofer, P.H.: Thermal expansion of Ti-Al-N and Cr-Al-N coatings. Scr. Mater. **127**, 182–185 (2017). https://doi.org/10.1016/j.scriptamat.2016.09.022
3. Singh, J., Kumar, S., Mohapatra, S.K.: Tribological analysis of WC–10Co–4Cr and Ni–20Cr2O3 coating on stainless steel 304. Wear. **376–377**, 1105–1111 (2017). https://doi.org/10.1016/j.wear.2017.01.032
4. Sharma, N.K., Rana, A., Panwar, O.S., Rana, A.S.: Nanomechanical inhomogeneities in CVA-deposited titanium nitride thin films: Nanoindentation and finite element method investigations. Heliyon. **10**, e33239 (2024). https://doi.org/10.1016/j.heliyon.2024.e33239
5. Sharma, L.K., Sharma, N.K., Dash, A.K., Rana, A.S.: Finite element analysis of TiN coated thin film. Mater. Today Proc. **113**, 274–277 (2024). https://doi.org/10.1016/j.matpr.2023.09.149
6. Reyes, F.G.G., Licerio, J.P., Ontoria, A.B., Vasquez, M.R.: Combinatorial synthesis of AlTiN thin films. Plasma. **6**, 225–234 (2023). https://doi.org/10.3390/plasma6020017
7. Tripathy, P.K., Patra, A.K., Sastry, P.U.: Studies on pore morphology of titanium and its oxide by small angle neutron scattering. Pramana. **71**, 1033–1038 (2008). https://doi.org/10.1007/s12043-008-0219-5
8. Cai, F., Chen, M., Li, M., Zhang, S.: Influence of negative bias voltage on microstructure and property of Al-Ti-N films deposited by multi-arc ion plating. Ceram. Int. **43**, 3774–3783 (2017). https://doi.org/10.1016/j.ceramint.2016.12.019
9. Sharma, N.K.: Thermo-mechanical analysis of uniform, clustered and interpenetrating phase particulate composites using finite element method. J. Therm. Anal. Calorim. **148**, 5967–5984 (2023). https://doi.org/10.1007/s10973-023-12157-6
10. Sharma, N.K., Misra, R.K., Sharma, S.: Finite element modeling of effective thermomechanical properties of Al–B4C metal matrix composites. J. Mater. Sci. **52** (2017). https://doi.org/10.1007/s10853-016-0435-1
11. Sharma, N.K., Mishra, R.K., Sharma, S.: 3D micromechanical analysis of thermo-mechanical behavior of Al2O3/Al metal matrix composites. Comput. Mater. Sci. **115** (2016). https://doi.org/10.1016/j.commatsci.2015.12.051
12. Samani, M.K., Ding, X.Z., Khosravian, N., Amin-Ahmadi, B., Yi, Y., Chen, G., Neyts, E.C., Bogaerts, A., Tay, B.K.: Thermal conductivity of titanium nitride/titanium aluminum nitride multilayer coatings deposited by lateral rotating cathode arc. Thin Solid Films. **578**, 133–138 (2015). https://doi.org/10.1016/j.tsf.2015.02.032

Optimizing IoT Network Performance: Mobility-Aware Clustering and Energy-Efficient Multi-Hop Routing Using WSNs

G. Mariya Gnana Merlin(✉), R. Prema, K. Thangavel, A. Dhina Infant Maria, and K. Sathiyaseelan

Hindusthan College of Arts and Science, Coimbatore 641028, Tamilnadu, India
mariyagnanamerlin@gmail.com

Abstract. This paper introduces a Mobility-Aware Weighted Clustering Algorithm (MAWCA) designed to optimize the selection of cluster heads (CH) for mobile nodes in IoT networks by efficiently harvesting routing energy. MAWCA incorporates multiple factors such as node degree difference, discrepancies among neighbors, cumulative time, node mobility, and delay as weighting elements in the CH election process. To further enhance energy efficiency and ensure secure communication, the study integrates an Energy-Aware and Secure Multi-Hop Routing (ESMR) protocol, which employs a secret sharing mechanism. Additionally, the proposed approach performs a quantitative analysis of data connections to reduce routing disruption and improve overall network performance. The research aims to provide a lightweight yet effective solution for IoT-based wireless sensor networks (WSNs), facilitating efficient and secure multi-hop data routing while extending the network's operational lifespan. Through comprehensive simulations and evaluations, the proposed methods demonstrate significant improvements in energy efficiency, security, and network stability, making them highly suitable for resource-constrained IoT environments.

Keywords: Wireless Sensor Network (WSN) · Internet of Things (IoT) · Mobility-Aware Weighted Clustering Algorithm (MAWCA) · Energy-Aware and Secure Multi-Hop Routing (ESMR) protocol

1 Introduction

IoT devices are currently becoming more common and are being used in a variety of situations, including keeping track of a home's energy usage or the condition of an outdoor parking space. These networks frequently contain a large number of sensors that deliver incredibly brief information. Given these traits, their biggest issue arises from the battery life's extreme fluctuation and unpredictability when they are not hooked into a power source [1]. Energy efficiency techniques are used in wireless sensor networks to increase the system's energy savings and lifespan. In WSN, transmission is accomplished by the deployment of nodes rather than any physical medium for communication. The data is sent from the deployed nodes to the sink, or the destination node. Clustering occurs

S. Kumari et al. (Eds.): ICMTP 2024, SPPHY 311, pp. 453–467, 2025.
https://doi.org/10.1007/978-981-96-6795-6_49

when nodes are collected for improved communication. Clustering occurs when close nodes create a compatible cluster and transmit information to the cluster head, which aggregates the information from the various nodes and sends it to the sink. For the improvement of WSN, many clustering methods and approaches have been developed. The cluster head was the centre node in the network, and this node was restricted for network transmission. This single node consumes more energy than the other nodes, resulting in a dead node.

On the other side, because it takes the straight-line path, the routing is overloaded, using more energy and shortening the network's lifetime. Given the limited energy available in the powered sensor nodes, improving energy efficiency in IoT utilizing Wireless Sensor Networks (WSNs) is a fundamental design need. In order to maximize network lifespan, a lifetime-aware distributed clustering and routing technique is suggested in this study [2].

The suggested protocol, which is based on routing and clustering protocols, can successfully extend the network's lifetime. Through the use of a wireless sensor network, it is feasible to greatly boost the energy efficiency of IoT devices.

1.1 Clustering Protocols in Wireless Sensor Networks (WSNs)

Clustering protocols define the structure of hierarchical, non-overlapping sensor node clusters. Self-organizing sensor networks require a strong clustering approach. A good clustering strategy enables the formation of clusters with almost the same radius and cluster heads that are optimally positioned inside the clusters. Because every node in a clustered network is linked to a cluster head, route discovery among cluster heads is sufficient to build a workable network route. When compared to a flat, non-clustered network, clustering can simplify multi-hop route discovery and minimize the amount of transmissions in a large sensor network. Wireless sensor networks provide several issues [3]. The main problem is maximizing network stability and longevity. It is not possible to replace the batteries of hundreds or thousands of sensor nodes once they have been deployed.

Clustering is the process of grouping sensor nodes into a cluster in a sensor network. Every cluster has a leader known as the cluster head. A cluster head can be assigned or chosen by the cluster members. A cluster head takes data from cluster nodes and transfers it to a destination (the base station). Clustering methods are commonly employed by researchers to improve lifespan and scalability objectives. Many clustering techniques may be used to build a hierarchical structure that decreases the cost of communication with the base station.

1.2 Routing Protocols in Wireless Sensor Networks (WSNs)

The significance of wireless sensor networks (WSNs) is growing over time. Despite the widespread use of wireless sensor networks, only a few applications require rapid data transport with little to no interruption. Many apps prioritize throughput while paying little attention to latency [4]. Everything depends on the application's preferences for the preferred parameter. It is crucial to have knowledge of the routing protocol and network architecture that is appropriate for the usage. In the end, a WLAN and a wireless sensor network based on ZigBee are used to undertake a performance study of various routing protocols.

The routing protocol is a method for choosing an appropriate route for data to take from source to destination. Selecting the route, which depends on the network type, channel characteristics, and performance measures, is a challenging procedure. The base station that connects the sensor network to other networks (perhaps the internet) receives the data received by the sensor nodes in a wireless sensor network (WSN), which is then collected, processed, and appropriate action is taken. In multi-hop communication, the sensor nodes act as a conduit for other sensor nodes to get to the base station in addition to producing and delivering their own content. Routing is the process of determining an appropriate path from a source node to a destination node, and it is the network layer's main job.

2 Related Works

Recent advancements in energy-efficient clustering and routing protocols have significantly contributed to extending the lifespan of wireless sensor networks (WSNs). For instance, Yongcheng Wang (2022) introduced the Enhanced Multitier Energy-Efficient Clustering Protocol for IoT (EMEECP-IOT), tailored for heterogeneous WSNs linked to IoT systems. By evaluating parameters such as network stability, received packets, throughput, and residual energy, simulation results revealed a 35% improvement in network longevity and a 21% reduction in energy consumption compared to conventional protocols, making it particularly valuable for IoT-based applications.

R. Prema et al. proposed the Secured Power-Aware and Energy-Efficient Routing Protocol (SPAEERP) in 2013, which incorporates a novel cryptosystem alongside adaptive transmission power and routing strategies. Simulations using NS2 demonstrated SPAEERP's ability to enhance Quality of Service (QoS) while minimizing energy usage. The protocol's cryptosystem was validated for security using Cryptool.

Similarly, Muhammad Bilal et al. (2022) presented a distributed clustering and routing system that emphasizes battery health. By using a charge utility metric during cluster formation, their approach optimized energy usage and extended network lifetime significantly compared to existing methods. This study underscored the importance of battery-aware design in improving the efficiency and longevity of WSNs.

In 2022, Subha R. and Anandakumar H. proposed the Improved Emperor Penguin Optimization Algorithm-Based Clustering Protocol (IEPOACP). By mimicking emperor penguins' huddling behavior, this protocol identified optimal cluster heads, resulting in a 67% increase in network lifespan, improved throughput, reduced latency, and better packet delivery rates compared to traditional methods.

Salim El Khediri et al. (2021) extended the TEEN protocol by introducing WB-TEEN and WBM-TEEN, which incorporated multi-hop intra-cluster communication. These enhancements balanced node distribution and reduced energy consumption, achieving a 40% improvement in network lifespan compared to TEEN and LEACH protocols.

R. Prema et al. also proposed the Secured Power-Aware Routing Protocol (SPARP), which dynamically adjusts transmission power and incorporates a novel cryptosystem to enhance QoS while reducing energy costs. Simulations validated its ability to improve packet delivery ratios and reduce latency.

Lastly, Priyadarsini et al. introduced a source node-based routing method that eliminates repetitive routing information exchanges, optimizing network performance.

2.1 Literature Review

See (Table 1).

Table 1. Literature review

S. no.	Author	Title	Journal	Description
1	Chang Lei	An energy-aware cluster-based routing in the Internet of Things using particle swarm optimization algorithm and fuzzy clustering	Journal of Engineering and Applied Science, Springer, Article 135, 2024	Introduced PSO-based clustering, enhancing energy savings by 27%.
2	S. Regilan and L.K. Hema	Optimizing energy efficiency and routing in wireless sensor networks through genetic algorithm-based cluster head selection in a grid-based topology	Sage Journal, Volume 30, Issue 4, 2024	Employed GA-based CH selection, achieving a 22% reduction in power consumption.
3	Arun Kumar, Nishant Gaur, Aziz Nanthaamornphong	Wireless optimization for sensor networks using IoT-based clustering and routing algorithms	NLM-NCBI, doi: 10.7717/peerj-cs.2132, 2024	Developed IoT-based clustering and routing algorithms, improving network throughput.
4	Hoda Jalalinejad, Mahdi Rohani Hajiabadi, Ali Asghar Rahmani Hosseinabadi, Seyedsaeid Mirkamali	A Hybrid Multi-Hop Clustering and Energy-Aware Routing Protocol for Efficient Resource Management in Renewable Energy Harvesting Wireless Sensor Networks	IEEE, Vol 12, 2024	Proposed a hybrid multi-hop clustering and energy-aware routing protocol, improving resource efficiency.

3 Proposed Methodology

This study offers to extend the life of an IoT network. Using the mobility mobile nodes in an IoT network. Furthermore, we compare our suggested technique to an existing system known as the multi-objective fractional gravitational search algorithm in order to increase energy efficiency with multi-hop data security against malicious activities [5]. The NS2 simulator software provides the results of this research. Network similar

2 (NS2) provides significant functionality for modeling diverse protocols over wired and wireless networks. It offers a highly flexible framework for wired and wireless simulations that supports various network components, protocols, traffic, and routing types. This simulator generates output in two formats: NAM animation tool and X-Graph.

3.1 Clustering

One of the complex challenges with wireless sensor networks is extending the lifetime of the network. Many methods have been suggested to extend the sensor nodes' battery life. In this research, we emphasize the use of routing to increase battery life and lengthen network lifespan. The mobility-aware weighted clustering technique and the energy-aware and secure multi-hop routing protocol serve as the foundations for our protocol [6]. We propose a new protocol that increases network lifespan compared to these two basic protocols by taking the core operating principles of these two routing strategies.

3.2 Routing

Routing is a critical aspect of WSNs, particularly for prolonging the lifespan of sensor nodes. Numerous techniques have been proposed to optimize energy consumption and enhance network longevity [7]. This study focuses on routing strategies that integrate mobility-aware clustering and energy-efficient secure multi-hop routing to achieve significant improvements in network lifespan.

3.3 Proposed Protocol

In order to find the optimum cluster head (CH) for mobile nodes in an IoT network by harvesting routing energy, the mobility-aware weighted clustering algorithm (MAWCA) is proposed in this paper. The MAWCA analyzes the node's degree difference, the sum of discrepancies between neighbors, cumulative time, node mobility, and delay in selecting a cluster head as a weighting element during the CH election process [8]. Furthermore, we integrate an energy-aware and secure multi-hop routing (ESMR) protocol with a secret sharing mechanism to improve energy efficiency performance. Finally, the proposed work assesses the quantitative analysis of data connections in order to reduce routing disruption. The research proposed here proposes a lightweight solution for IoT-based limited wireless sensor networks with data routing in a multi-hop way (WSNs).

3.4 Steps Involved in Comparing Mawca & Mofgsa

3.4.1 Step 1 - Initial Node Deployment and Visualization

Figure 1 illustrates the initial deployment of nodes within the simulation environment. Users are represented as green circles, with the source and destination nodes marked as blue hexagons. Cluster heads (CH1, CH2, CH3) are identified and displayed for efficient data communication. This visualization enables a clear understanding of node distribution and CH positioning.

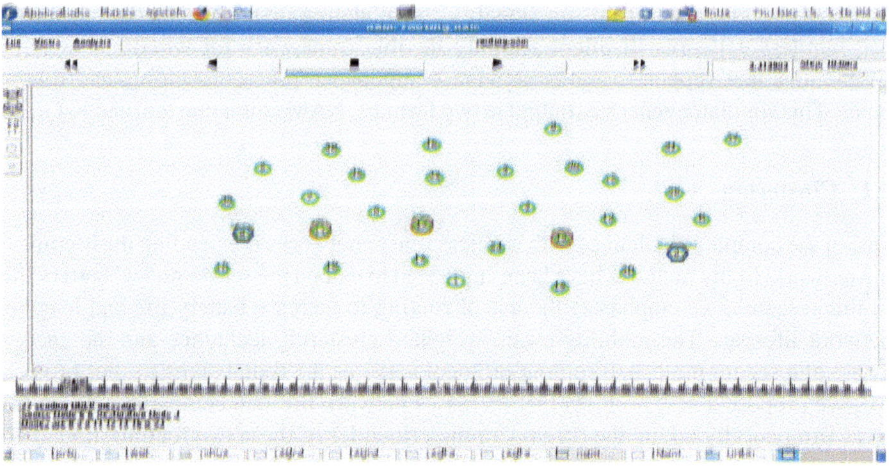

Fig. 1. Routing (NAM Animation tool)

3.4.2 Step 2 - Formation of Clusters

In Fig. 2, the clustering process begins with the selection of the source node, represented as a violet circle. This node initiates the clustering process by broadcasting messages to nearby nodes. Clusters form around CHs, aiming to minimize data loss before transmission. Small dots near the source node signify the initial phases of data exchange, indicating active communication links.

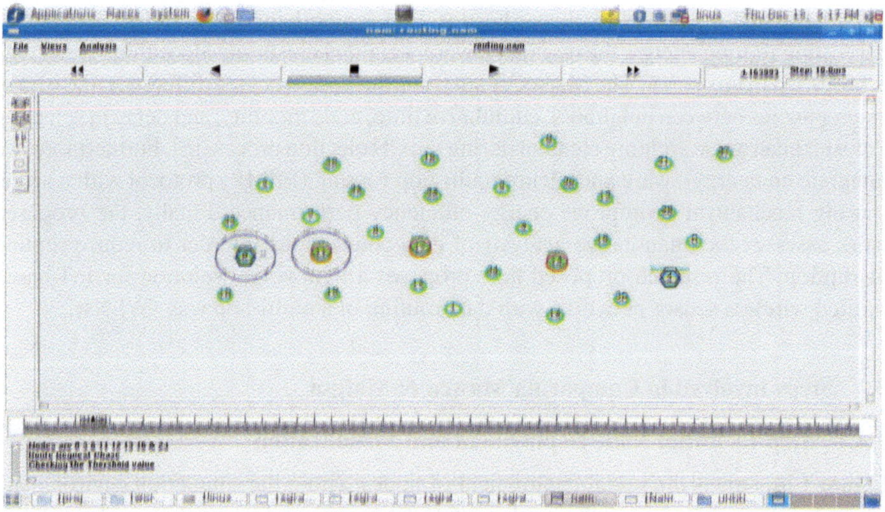

Fig. 2. Formation of cluster

3.4.3 Step 3 - Path Optimization for Routing

The routing protocol identifies the most efficient paths for data transmission. As depicted in Fig. 3, paths between nodes are established to ensure minimal energy consumption. For example, the route includes nodes 0, 3, 8, 11, 12, 16, and 24, forming an optimal chain for packet delivery. Distance between nodes is minimized to improve overall efficiency.

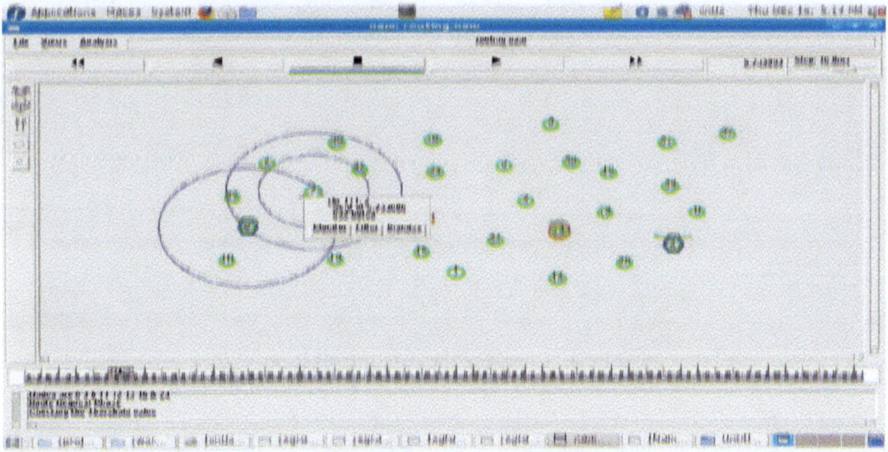

Fig. 3. Routing protocol

3.4.4 Step 4 - Monitoring and Data Segmentation

Cluster heads actively monitor data transfer within their clusters. As shown in Fig. 4, nodes moving out of range are detected, and CHs adjust their communication strategies accordingly. For instance, node 7 transmits data packets segmented into smaller units to maintain communication reliability. Key performance metrics, such as transmission duration (5.73 s) and bit rate (111), are recorded in a ".tr" trace file.

3.4.5 Step 5 - Complete Routing and Verification

The entire routing process is completed, as shown in Fig. 5. Nodes such as 7 and 25 are highlighted in the routing path leading to the destination node (node 4). This step includes sending route requests, verifying thresholds, and receiving route replies to ensure the secure and efficient transfer of data packets. This comprehensive routing approach enhances reliability and minimizes packet loss.

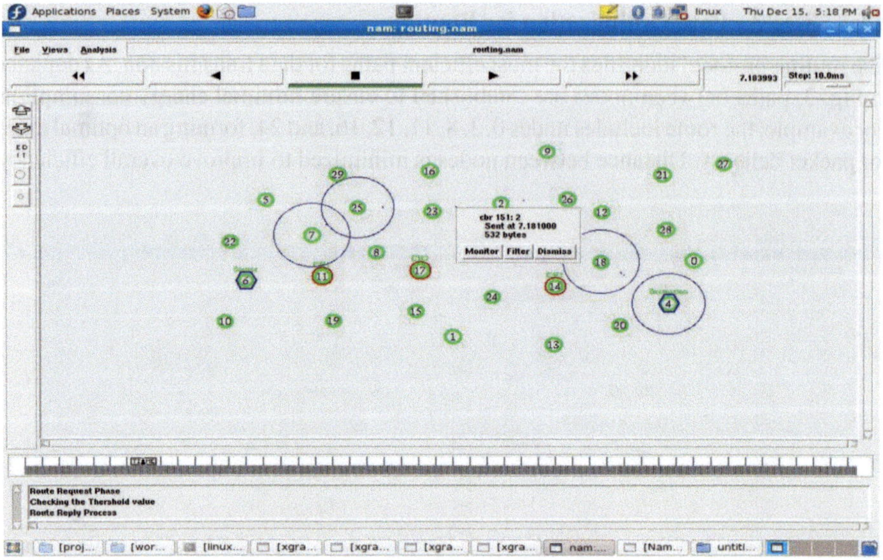

Fig. 4. Checking the data Information

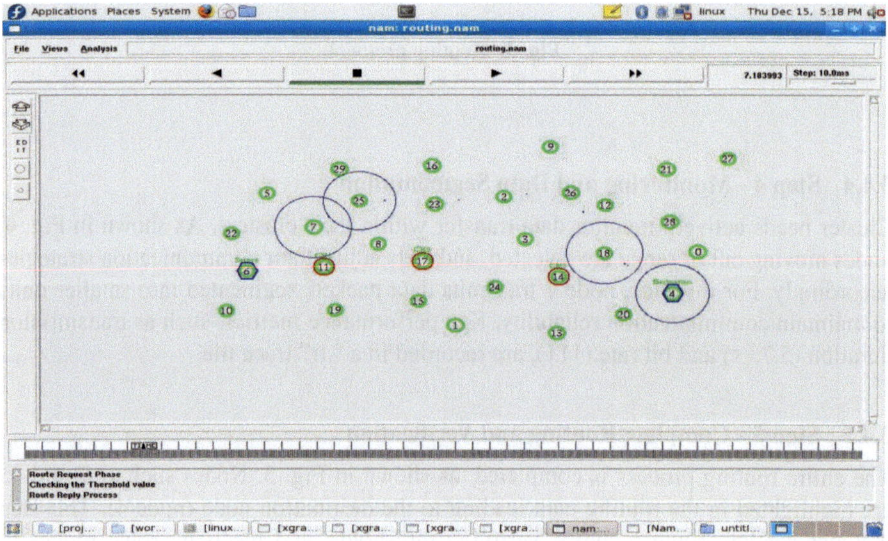

Fig. 5. Path of Routing

4 Software Implementation and Codings

4.1 NS-2 Software

Network Simulator (Version 2), often known as NS2, is simply an event- driven simulation programme that has been effective in understanding the dynamic dynamics of communication networks. NS2 can simulate both wired and wireless network operations and protocols (e.g., routing algorithms, TCP, and UDP). In general, NS2 offers users a manner of describing such network protocols and simulating their relevant behaviors. The basic architecture of NS2 [9].

NS2 offers users an executable command, "ns," which accepts one input argument, the name of a Tcl simulation scripting file. A simulation trace file is often produced and used to plot graphs and/or create animation (Fig. 6).

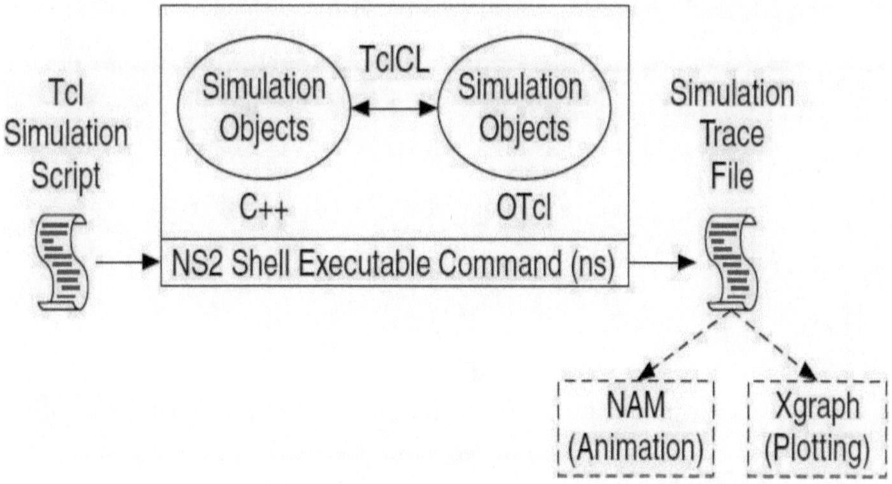

Fig. 6. Basic Architecture of NS2

Create a graph and/or an animation. NS2 consists of two essential languages: CCC and Object-Oriented Tool Command Language (OTcl). While the CCC describes the simulation fundamental mechanism (i.e., a backend), the OTcl configures the simulation by assembling and configuring objects and scheduling discrete events (i.e., a frontend). TclCL is used to connect the CCC and the OTCL. When mapped to a CCC object, variables in the OTcl domains are sometimes referred to as handles. A handle in the OTCL domain is simply a string (e.g., "_o10") that has no functionality. The functionality (for example, receiving a packet) is instead defined in the mapped CCC object (e.g., of class Connector) [10].

A handle in the OTcl domain serves as a frontend that communicates with users and other OTcl objects. To facilitate interaction, it may create its own methods and variables. It is worth noting that the OTCL domain's member procedures and variables are referred to as instance procedures (instprocs) and instance variables (instvars), respectively.

4.2 NAM Animation Tool

The Network Animator (NAM) is a Tcl/Tk-based tool used to visualize network simulations and analyze packet traces. It allows users to observe network topology, packet-level animations, and data flows. NAM was originally developed at LBL and later enhanced under the VINT project. By maintaining only essential information in memory and fetching additional data on-demand, NAM efficiently handles large-scale animation datasets. Trace files containing topology and packet trace information, typically generated by NS2, are used as input for NAM visualizations (Fig. 7).

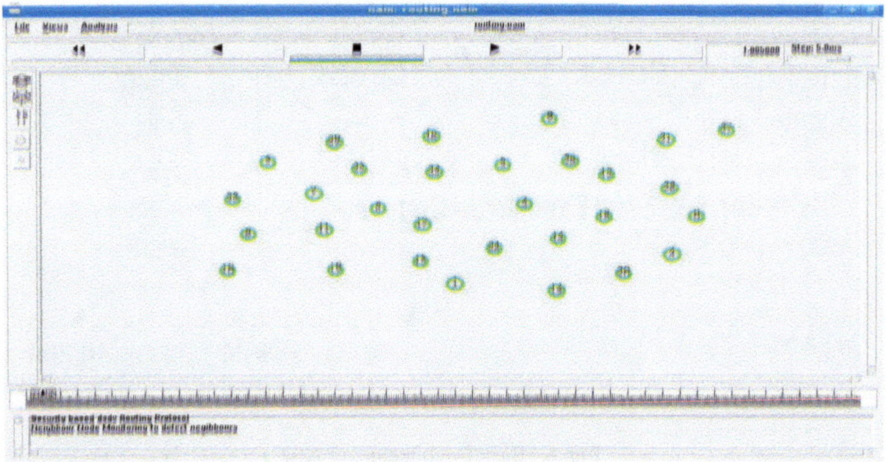

Fig. 7. NAM Animation Tool

4.3 X-Graph

X-Graph is a charting utility included in the NS2 package, enabling users to create graphical representations of simulation results [11, 12]. By generating output files in Tcl scripts, users can visualize performance metrics such as energy consumption, packet delivery rates, and network lifetime. The use of X-Graph in plotting simulation data and analyzing traffic patterns for network performance (Fig. 8).

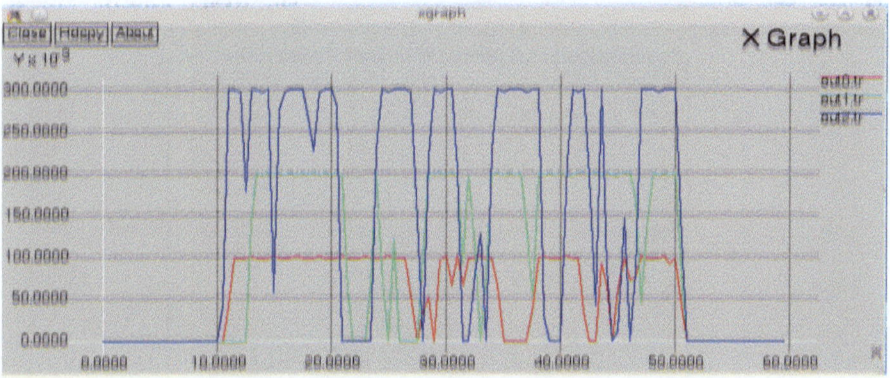

Fig. 8. Use X-Graph for Plotting Graphs in NS2

5 Result and Discussion

5.1 Comparison of MAWCA and MOFGSA Protocols

The mobility-aware weighted clustering algorithm (MAWCA) is developed in this study to discover the ideal cluster head (CH) for mobile nodes in an IoT network by harvesting routing energy. As a weighting element during the CH election process, MAWCA evaluates the node's degree difference, the total of differences between neighbors, cumulative time, node mobility, and delay in picking a CH. We have looked at and contrasted the planned MAWCA with the MOFGSA in this part. PDR, Network energy, packet loss rate, network life time based on living nodes, and other performance indicators are compared here for several simulation rounds.

In addition, we implemented an energy-aware and secure multi-hop routing (ESMR) protocol for limited WSNs, which is explained in this section. Based on the distance factor in the first component, network nodes are divided into inner and outer zones. Furthermore, the nodes inside each zone are sorted into multiple clusters using the k-nearest neighbors (k-NN) technique. To facilitate further data routing, all clusters are organized hierarchically. The fundamental goal of the second component is to balance the relationship between energy consumption and reliable data forwarding and to provide an energy-efficient, safe multi-hop routing strategy against malicious attacks.

5.2 Parameters Compared Graphically

5.2.1 Network Lifetime

The x axis in this graph stands for the number of rounds, and the y axis represents lifetime. The difference between the values of the existing and current systems, which are 13, 0000 and 19, 0000 respectively, clearly Fig. 9 demonstrates that the proposed method has higher lifetime efficiency than the existing one.

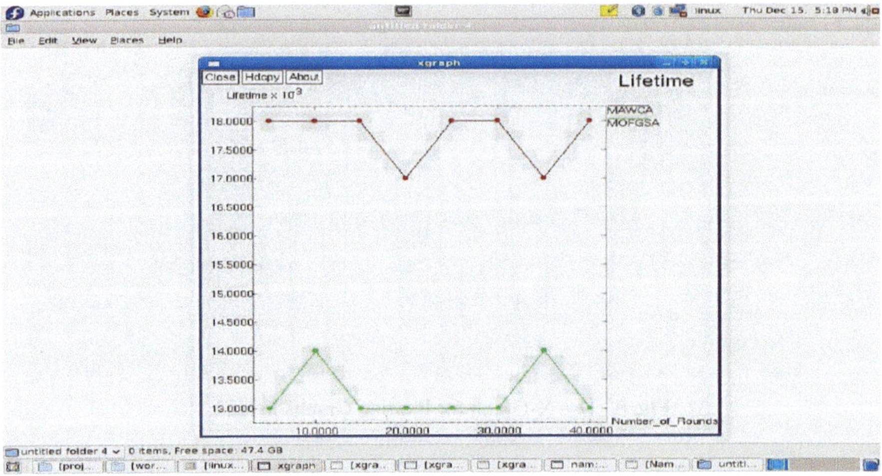

Fig. 9. Comparison of Network Lifetime

5.2.2 Energy Consumption

The energy utilized by the network is displayed in this graph. In this way, the x-axis represents the number of rounds, while the y-axis represents the energy acquired by the network. This Fig. 10 graphical representation makes it very evident that the proposed approach uses less energy than the current system.

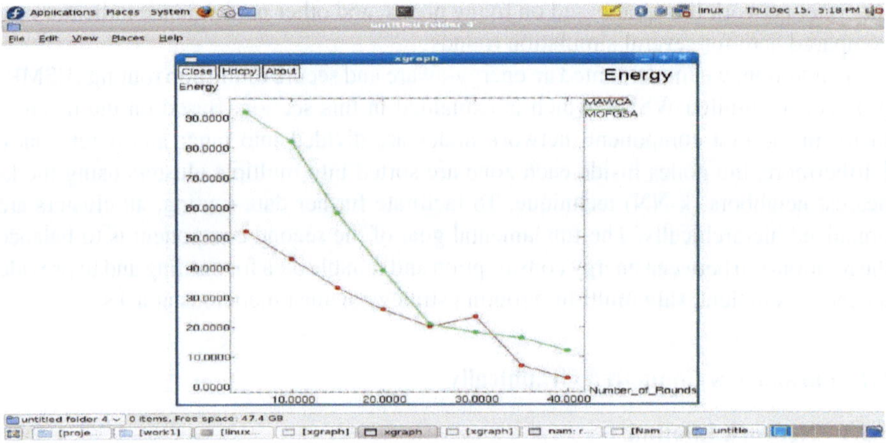

Fig. 10. Energy consumption

5.2.3 Packet Delivery Rate (PDR)

The proportion of data packets actually received at the receiver end to those that were initially sent by the sender is known as the PDR. So, another definition would be, where is the number of nodes really transmitted by the sender, and is the number of nodes actually received by the receiver. This Fig. 11 x-graph suggests that proposed work is more efficient than previously performed tasks.

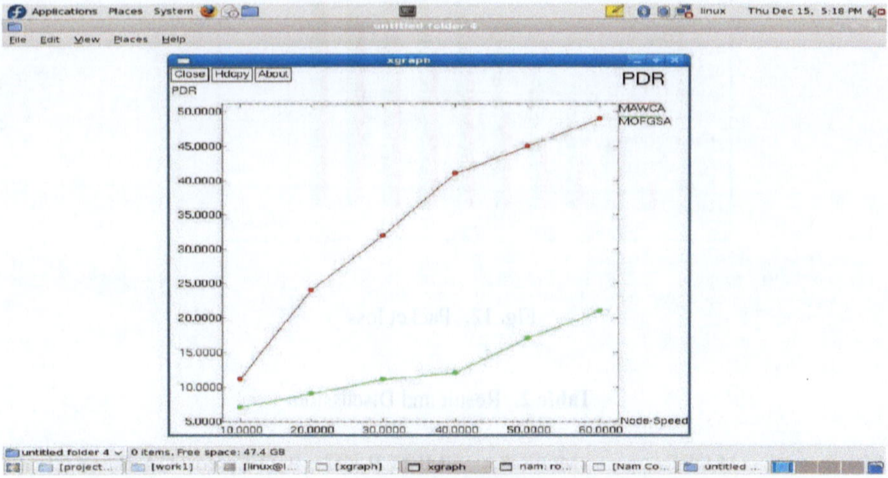

Fig. 11. PDR

5.2.4 Packet Loss

Data packets that are lost while being transferred across a network are referred to as "packet losses." Packet loss happens when data packets are dropped during data transmission due to network congestion, hardware problems, software defects, and several other reasons. The proposed system's packet loss is less than that of the existing system, as demonstrated in visual form as seen in Fig. 11 (Fig. 12).

The proposed MAWCA protocol is evaluated against recent clustering and routing algorithms, demonstrating superior performance in key metrics. MAWCA outperforms MOFGSA, EMEECP-IOT, and other clustering techniques by extending network lifetime and improving energy efficiency. The integration of a secret-sharing mechanism enhances security, addressing vulnerabilities in existing routing methods (Table 2).

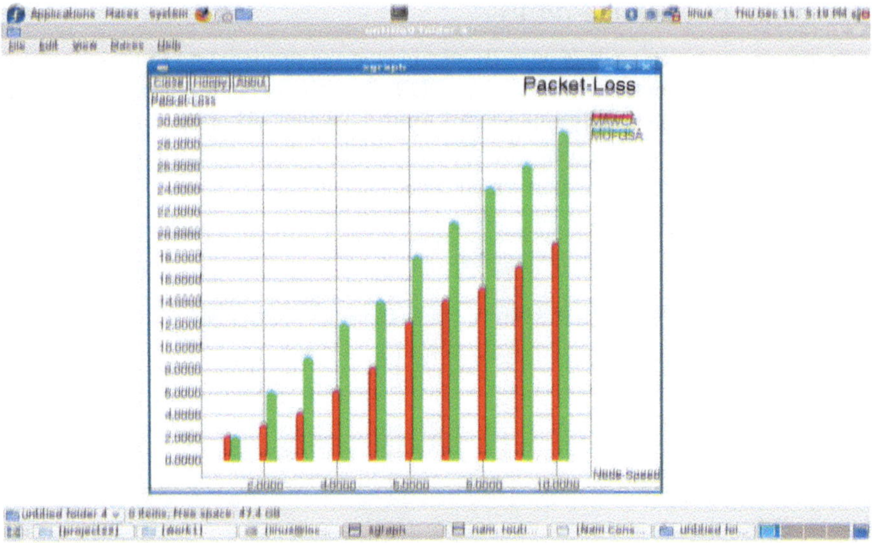

Fig. 12. Packet loss

Table 2. Result and Discussion

Metric	MAWCA + ESMR	MOFGSA	EMEECP-IOT	PSO-Based Routing	GA-Based CH Selection
Network Lifetime (rounds)	19,000	13,000	14,500	15,200	16,000
Energy Consumption	Lower	Moderate	Moderate	High	Low
Packet Delivery Ratio (PDR)	High	Moderate	Moderate	High	High
Packet Loss	Low	High	Moderate	Low	Low
Security Mechanism	Secret-Sharing	None	None	None	Encryption-Based

6 Conclusion and Future Work

This research has implemented to extend the life of the IoT network using wireless sensor networks. The mobility-aware weighted clustering algorithm with multi-hop routing protocol (MAWCA) makes routing energy to identify the optimal cluster head (CH) for mobile nodes in an IoT network to boost energy efficiency using multi-hop data. The proposed energy-aware and secure multi-hop routing protocol instructs forwarders to use the shortest path possible, which includes dependable and energy-efficient nodes, to transfer data. In addition, the proposed protocol analyzes quantitative data for a specific connection to determine whether it is congested, which reduces routing disruption and re-transmissions. In the future, upgrading the efficiency of WSN-based IoT environments will be studied to validate the benefits provided by the proposed methods and also security techniques were used to overcome the limitation.

References

1. Zagrouba, R., Kardi, A.: Comparative study of energy-efficient routing techniques in wireless sensor networks. Information. **12** (2021)
2. Al-Kaseem, B.R., Taha, Z.K., Abdulmajeed, S.W., Al-Raweshidy, H.S.: Optimized energy–efficient path planning strategy in WSN with multiple mobile sinks. IEEE Access. **9**, 82833–82847 (2021)
3. Mehta, D., Saxena, S.: MCH-EOR: multi-objective cluster head-based energy-aware optimized routing algorithm in wireless sensor networks. Sustain. Comput. Inf. Syst. **28**, 100406 (2020)
4. Darabkh, K.A., El-Yabroudi, M.Z., El-Mousa, A.H.: BPA-CRP: a balanced power-aware clustering and routing protocol for wireless sensor networks. Ad Hoc Netw. **82**, 155–171 (2019)
5. Reddy, V., Gayathri, P.: Integration of internet of things with wireless sensor network. Int. J. Electr. Comput. Eng. **9**(1), 439–444 (2019)
6. Prema, R., Rangarajan, R.: Power aware routing protocol (PARP) for wireless sensor networks. Wirel. Sens. Netw. **4**(5), 133–137 (2012)
7. Prema, R., Rangarajan, R.: Secured power aware routing protocol (SPARP) for wireless sensor networks. Int. J. Comput. Appl. **51**(17), 13–1851 (2012)
8. Krishna, K.P.R., Thirumuru, R.: Energy efficient and multi-hop routing or constrained wireless sensor networks. Sustain. Comput. Inf. Syst. Sci. Direct. **38** (2023)
9. Lei, C.: An energy-aware cluster-based routing in the internet of things using particle swarm optimization algorithm and fuzzy clustering. J. Eng. Appl. Sci.., springer, Article number. **135** (2024)
10. Regilan, S., Hema, L.K.: Optimizing energy efficiency and routing in wireless sensor networks through genetic algorithm-based cluster head selection in a grid-based topology. Sage J. **30**, 4 (2024)
11. Kumar, A., Gaur, N., Nanthaamornphong, A.: Wireless optimization for sensor networks using IoT-based clustering and routing algorithms. NLM-NCBI. (2024). https://doi.org/10.7717/peerj-cs.2132
12. Jalalinejad, H., Hajiabadi, M.R., Hosseinabadi, A.A.R., Mirkamali, S.: A hybrid multi-hop clustering and energy-aware routing protocol for efficient resource management in renewable energy harvesting wireless sensor networks. Institute of Electrical and Electronics Engineers (IEEE). **12** (2024)

Correction to: Temperature Dependent Structural and Optical Properties of $SnSe_2$ Nanosheets

Yogita Gendre, K. S. Ojha, and Chinmaya Mahapatra

Correction to:
Chapter 24 in: S. Kumari et al. (Eds.): *Proceedings of the 1st International Conference on Materials and Thermophysical Properties*, SPPHY 311, https://doi.org/10.1007/978-981-96-6795-6_24

The original version of the book, was inadvertently published with minor error. In chapter 24, the affiliation "National Institute of Technology Chandigarh" of author "Yogita Gendreis" has been changed to "National Institute of Technology Raipur Chhattisgarh". The erratum book has been updated with the changes.

The updated version of this chapter can be found at
https://doi.org/10.1007/978-981-96-6795-6_24

S. Kumari et al. (Eds.): ICMTP 2024, SPPHY 311, p. C1, 2025.
https://doi.org/10.1007/978-981-96-6795-6_50

Author Index

S. Kumari et al. (Eds.): ICMTP 2024, SPPHY 311, pp. 469–471, 2025.
https://doi.org/10.1007/978-981-96-6795-6

MIX
Papier aus verantwortungsvollen Quellen
Paper from responsible sources
FSC® C105338

If you have any concerns about our products,
you can contact us on
ProductSafety@springernature.com

In case Publisher is established outside the EU,
the EU authorized representative is:
Springer Nature Customer Service Center GmbH
Europaplatz 3, 69115 Heidelberg, Germany

Printed by Libri Plureos GmbH
in Hamburg, Germany